Excel

Get the Results You Want!

YEAR 12 BIOLOGY

Diane Alford & Jennifer Hill

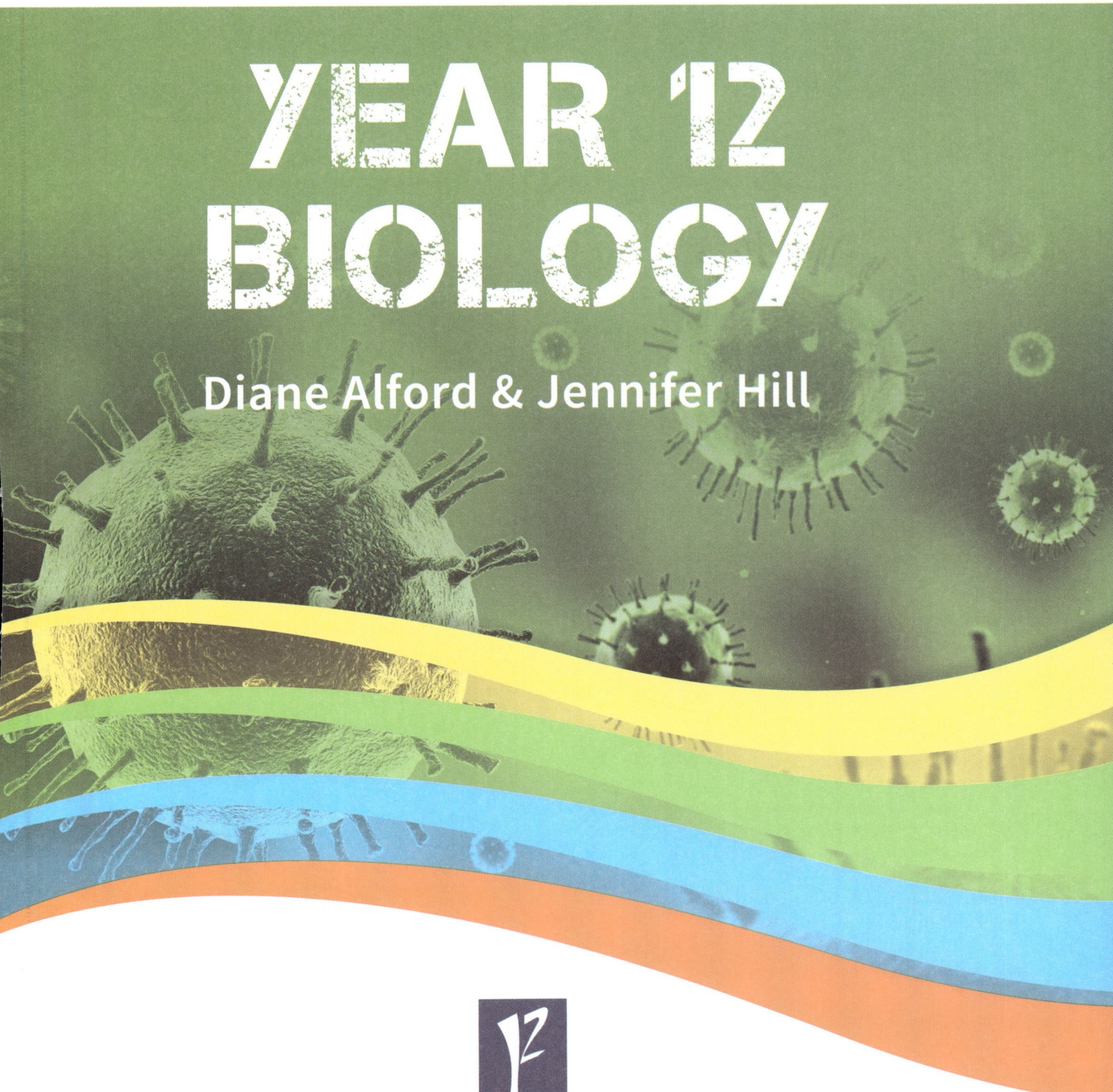

PASCAL PRESS

CONTENTS

MODULE 7: INFECTIOUS DISEASE

MODULE 8: NON-INFECTIOUS DISEASE

CHAPTER 1 REPRODUCTION

INQUIRY QUESTION:

How does reproduction ensure the continiuity of a species?

Life continues through the processes of reproduction and heredity. In this chapter you review reproduction in plants, animals, fungi, bacteria and protists.

1 The mechanisms of reproduction that ensure the continuity of a species

» Students explain the mechanisms of reproduction that ensure the continuity of a species, by analysing sexual and asexual methods of reproduction in a variety of organisms, including but not limited to animals: advantages of external and internal fertilisation; plants: asexual and sexual reproduction; fungi: budding, spores; bacteria: binary fission; protists: binary fission, budding.

Sexual and asexual methods of reproduction in a variety of organisms

- In Year 11 you learned about evolution through natural selection. A key process of natural selection is reproduction in which more 'fit' individuals produce more offspring with more beneficial characteristics.
- The biodiversity of life on Earth is almost matched by the great variety of ways that organisms reproduce and pass on their inherited characteristics. Continuity of a species occurs through asexual or sexual reproduction. Both types of reproduction require the replication of genetic material (DNA) and its transfer to the next generation. This transfer of genetic material also ensures the continuity of species.
- There are two main types of **reproduction**: **asexual** and **sexual**. The key differences are outlined in Table 1.1.

asexual reproduction: the production of offspring that are genetically identical to the parent

sexual reproduction: the production of genetically different offspring resulting from two parents, each producing gametes which fuse in a process called fertilisation

Table 1.1 Differences between sexual and asexual reproduction

Sexual reproduction	Asexual reproduction
an essential step is cell division by **meiosis**	processes include binary fission and/or **mitosis** cell division
requires two parents of different gender requires fertilisation, which results in offspring with a new, unique combination of genetic material, some from each parent fertilisation can occur internally or externally	results in offspring that are genetically identical to the parent
dominant form of reproduction in many multicellular plants and animals	dominant form of reproduction in many unicellular, colonial or simple multicellular organisms
animal examples include egg laying in insects, crustaceans, arachnids, amphibians, fish, birds, reptiles and monotremes; live birth in placentals; immature birth and pouch-raising in marsupials plant examples include production of flowers or cones	examples include spore production in fungi, and budding and binary fission in bacteria or protists plant examples include production of bulbs, tubers, rhizomes, stolons and corms

meiosis: a cellular reproductive process that occurs in eukaryotic cells; meiosis results in the formation of gametes that contain half the normal complement of genetic material

mitosis: a cellular reproductive process that occurs in eukaryotic cells, mitosis refers to the replication and division of the nucleus; mitosis results in the production of two identical daughter cells from the parent cell; cytokinesis, which follows mitosis, results in the formation of two cells with each containing a nucleus and cytoplasm

CHECKPOINT

Mitosis and meiosis, and the difference between eukaryotic and prokaryotic cells

➔ Before you begin your detailed review of reproduction it is important to check your understanding of the differences between both eukaryotic and prokaryotic cells and mitosis and meiosis.

➔ Mitosis and meiosis occur in eukaryotic cells because the DNA is carried on chromosomes. Both processes are outlined as a series of phases. Essentially there is the division of the nucleus (see Figures 1.1 and 1.2) followed by the separation of the cytoplasm contents (cytokinesis) to produce daughter cells.

➔ Prokaryotic cells (see Chapter 1 in *Excel Year 11 Biology*) carry their genetic material in different structures and arrangements. As a result, the process of cell division (duplication) is different in prokaryotic cells compared to that in eukaryotic cells. The overall result, however, is basically the same: offspring identical to their single parent. Later in this chapter you will review the differences in the processes of binary fission in bacteria (prokaryotes) with that in protists (eukaryotes).

➔ Many eukaryotic animals and some plants have somatic (normal body) cells containing chromosomes in pairs, with one member of each pair coming from each parent. Each pair of chromosomes is called a homologous pair. The number of homologous pairs is represented as n and is called the haploid number. The double set of chromosomes that occurs in somatic cells is represented as 2n and is called the diploid number. Humans have a diploid number of 46. The formation of gametes or germ cells by meiosis results in cells with half the diploid number of chromosomes; that is, they contain only one member of each homologous pair.

➔ Replication of the genetic material occurs as a first step in cellular reproduction in both eukaryotic and prokaryotic cells.

➔ You will review mitosis and meiosis in more detail in Chapter 2.

somatic cells: normal body cells excluding gametes (germ cells)
homologous pairs: chromosome pairs that each carry the same genes in the same location along their length and, in somatic cells, one member originates from the mother and the other from the father; homologous pairs look identical except for the sex chromosomes; humans have 46 chromosomes or 23 homologous pairs
haploid number: the number of homologous pairs, symbolised by n
diploid number: the number of chromosomes in the somatic cells of most organisms, symbolised by 2n

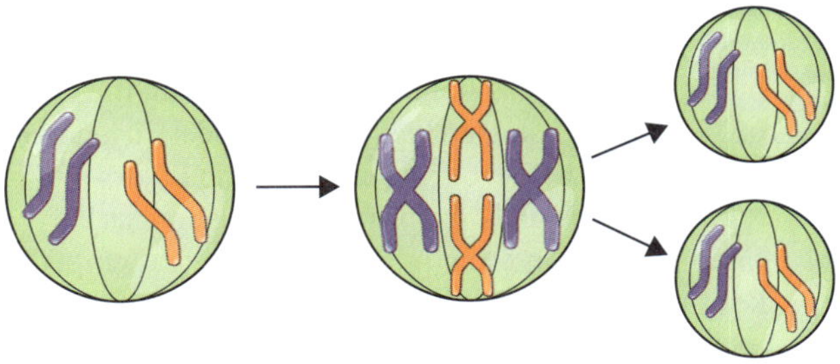

Figure 1.1 Summary of mitosis in a cell in which 2n = 4. It results in two daughter cells identical to the eukaryotic parent cell.

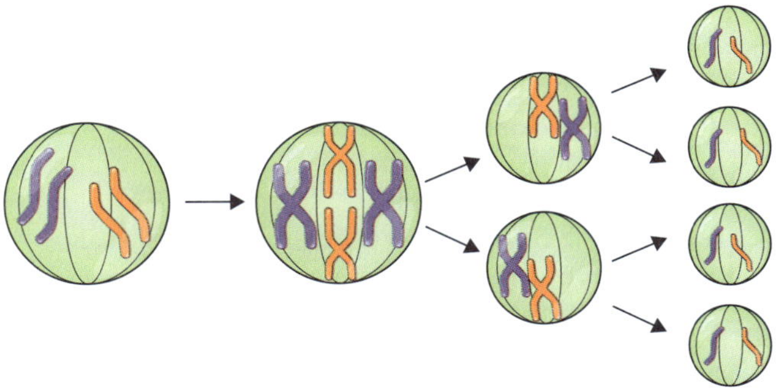

Figure 1.2 Summary of meiosis in a cell in which 2n = 4. It results in four unique daughter cells, each with n chromosomes (i.e. one member of each homologous chromosome pair).

➔ KEY QUESTIONS

1. **What are the two key processes in reproduction?**
2. **Outline the role of mitosis in reproduction.**
3. **Outline the role of meiosis in reproduction.**
4. **What is the difference between sexual and asexual reproduction?**

Answers ➲ p. 23

Asexual reproduction

➔ Asexual reproduction:
- occurs widely in plants, generally in the production of runners, rhizomes, tubers, bulbs, corms and spores
- results in clones of the original parent

spores: unicellular reproductive structures that can disperse, germinate and grow into a new organism; some are produced in adverse conditions and survive until more suitable conditions occur; spores are produced by algae, fungi, many plants (ferns and mosses), protists and some bacteria
clone: the identical offspring from a parent; clones occur naturally, mainly in plants, but human manipulation has resulted in cloned offspring of animals

- includes the regeneration of some simple animals such as starfish, sea anemones and flatworms; each fragment of the organism grows into a new organism; the Crown of Thorns starfish (see Figure 1.3) has destroyed parts of coral reefs by devouring corals; its ability to regenerate asexually from fragments as well as its rapid population growth has complicated attempts to keep numbers at manageable levels
- includes the development of an unfertilised ovum in a process called parthenogenesis in some species such as brine shrimp, water fleas, bees, ants and even in a few species of lizards, snakes, sharks and birds; parthenogenesis also occurs in plants: the rapid expansion of populations of sap-sucking aphids in summer is the result of female aphids giving birth to live female clones
- occurs as a result of a cell division called **binary fission** in unicellular organisms
- may involve **budding** as another form of asexual reproduction that results from an outgrowth or asymmetric cell division; budding occurs in some colonial organisms and unicellular fungi such as yeast.

➔ The advantages of asexual reproduction include:
- only one parent is needed
- the rapid population growth of identical members.

➔ The disadvantages of asexual reproduction include:
- the lack of biodiversity within population
- overpopulation, leading to possible negative ecological impacts.

binary fission: a mitotic-like cell division resulting in a parent cell becoming two identical offspring cells; occurs in bacteria and protists; some organelles with their own DNA, such as chloroplasts, also replicate through binary fission
budding: asexual reproduction resulting from an uneven cell division (in unicellular organisms) or outgrowth in which the parent is larger than the offspring

Figure 1.3 Crown of Thorns starfish
Source: image by Richard A Collato, [CC BY-SA 3.0 (http://creativecommons.org/licenses/by-sa/3.0)], via Wikimedia Commons

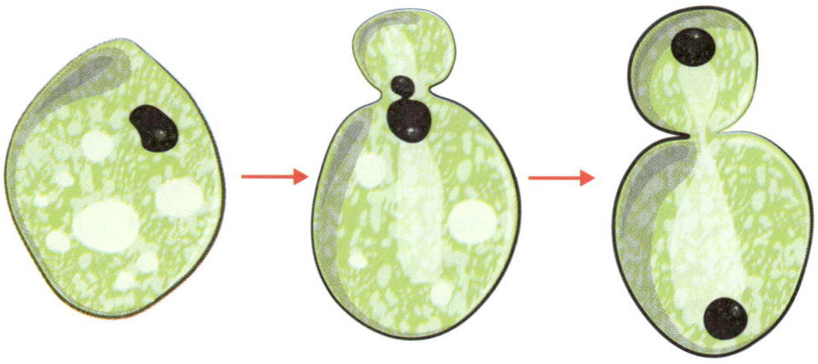

Figure 1.4 Yeast cell budding

Sexual reproduction

➔ Sexual reproduction is outlined in the points below.
- It occurs widely in animals and in seed-bearing plants; that is, angiosperms (flowering plants) and gymnosperms (cone-bearing plants). It also occurs in primitive plants such as ferns, mosses and liverworts. Fungi may reproduce sexually in times of adverse conditions.
- It results from the production of two different types of **gametes** (male and female). The male gamete is regarded as the smaller, more motile (mobile) gamete. The female gamete is usually larger, containing food reserves for the early development of the zygote into an embryo. The female gamete develops into the seed in plants or the egg in insects, fish, birds, reptiles, monotremes and amphibians. In placental and marsupial mammals it becomes the embryo which develops into a foetus in the pouch or uterus. Gametes contain the haploid number of chromosomes.
- It comes about due to male and female gametes uniting in a process called **fertilisation**. Fertilisation (or conception) results in a **zygote** which grows and develops through mitotic cell divisions and differentiation to produce an **embryo**. Fertilisation is the combination of two gametes or germ cells to produce a diploid zygote.
- It results from organisms of different gender. Population growth rates are often linked to the proportion of females, as female reproductive cycles become the limiting factor (one male may potentially fertilise many females). Animals are commonly either male or female gender, with each producing the appropriate gametes.

gametes: the sex cells produced in sexual reproduction that carry half the genetic material to the offspring; the egg or ovum is the female gamete and the spermatozoon or pollen is the male gamete
fertilisation: the union of the male and female gamete to result in the first cell of a new organism (zygote)
zygote: a fertilised egg or the first cell of an offspring resulting from fertilisation in sexual reproduction
embryo: an early stage in the development of an offspring resulting from sexual reproduction

- Some species such as earthworms and some snails contain both testes and ovaries, and are termed simultaneous hermaphrodites. Mangrove killifish are hermaphrodites that can self-fertilise internally. Earthworms are hermaphrodites that have mechanisms to ensure they cannot self-fertilise. Some organisms are sequential hermaphrodites that transition from one gender to the other. Banana slugs are hermaphrodites that prefer to mate. If, however, a partner is unavailable, banana slugs may self-fertilise. Most flowers are hermaphrodites as they contain both male and female organs. Some plants have separate male and female flowers or cones on the same plant (they are monoecious).
- It occurs in eukaryotes. As the genetic material is carried on chromosomes the gametes contain half of the normal number of chromosomes. This is called the haploid number. Normal (somatic) cells contain their chromosomes in pairs (one member of each homologous chromosome pair came from each parent). Somatic cells are diploid; that is, they contain chromosomes in pairs with twice the number of haploid chromosomes.

➔ The advantages of sexual reproduction include:
- the biodiversity of offspring; that is, each is different from the parent and usually each other
- the variation in offspring is the basis of evolution by natural selection
- the methods of sexual reproduction promote the dispersal of offspring in some species.

➔ The disadvantages of sexual reproduction include that:
- it generally requires two parents of different sex to produce viable gametes and fertilisation; cells and tissues differentiate to produce sex organs such as stamens (male) and pistil (female) in flowering plants
- it requires fertilisation—organs have evolved to ensure fertilisation via methods such as flowers using nectar to attract pollinators in flowering plants, and the development and dispersal of offspring such as seeds in fruit. In primitive plants, such as ferns, mosses and liverworts, sexual reproduction is restricted as external water is needed for fertilisation.

➔ KEY QUESTIONS

5 **Give five examples of asexual reproduction and the organisms in which it occurs.**

6 **Give five examples of sexual reproduction and the organisms in which it occurs.**

Answers ➲ p. 23

Advantages of external and internal fertilisation in animals

➔ In sexual reproduction fertilisation is either internal or external.

➔ **Internal fertilisation** occurs with the release of the motile male gamete within the female body, such as in the vagina of placental mammals, and the fertilisation of the female gamete. Internal fertilisation ensures that fewer male gametes are unproductive. Internal fertilisation is also associated with higher levels of development of the embryo internally or under parental care (see Table 1.2).

➔ In **external fertilisation**, ova and sperm are released into the external environment where fertilisation and subsequent development may occur (see Table 1.2).

internal fertilisation: the union of male and female gametes that occurs within the body of the female; in mammals this results from copulation (sexual intercourse) involving penetration of the female by the penis of the male and release of semen

external fertilisation: the union of male and female gametes after their release outside the body, often in an aquatic environment

Table 1.2 Advantages of internal and external fertilisation

Internal fertilisation	External fertilisation
gametes, zygote and embryo contained in a protected environment: controlled temperature, moisture, free from predators	gametes, zygote and embryo are usually in an aqueous environment for ready dispersal
relatively high success in development of offspring due to increased probability of fertilisation occurring and many instances of paternal care of offspring; less need for production of large numbers of gametes	development of offspring independent of parent; parent can put energy into the production of large numbers of gametes
well suited to a range of terrestrial and aquatic environments	well suited to moist or aquatic environments

Figure 1.5 External fertilisation resulting from Malabar gliding frogs spawning; they produce a foam nest above a small pool into which tadpoles drop and develop.

→ KEY QUESTIONS

7 What it the difference between internal and external fertilisation?

8 Describe the advantages of internal fertilisation.

9 Describe the advantages of external fertilisation.

10 Describe the disadvantages of internal fertilisation.

11 Describe the disadvantages of external fertilisation.

Answers ⊃ p. 23

Asexual and sexual reproduction in plants

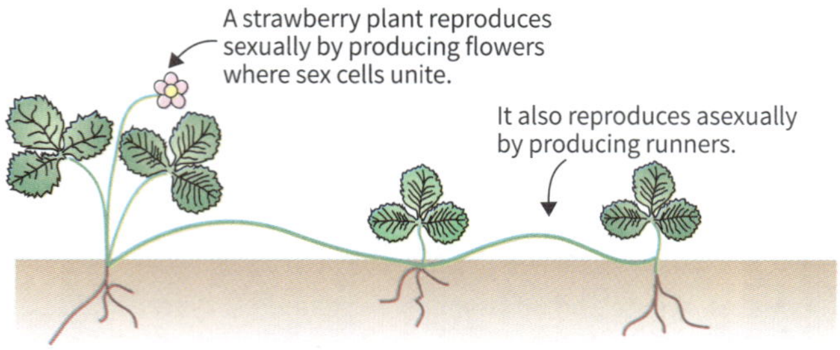

Figure 1.6 Strawberry plant showing reproduction by both asexual and sexual means

- Reproduction in terrestrial plants is both sexual and asexual.
- Most terrestrial plants are sessile so successful reproduction is a means of competition and a means of dispersal.
- Asexual reproduction in plants involves mitotic cell division in specific locations such as buds. At these points cellular differentiation can occur, resulting in the production of tissues and organs necessary for the formation of an independent organism. Plants with roots that produce adventitious buds (e.g. sweet potato and dahlias) often store food to ensure reserves for the growth of an independent plant. Underground stems may also have stored food, along with the adventitious buds, in corms (e.g. crocus), bulbs (e.g. garlic), rhizomes (e.g. ginger) and tubers (e.g. potatoes). Other examples of asexual reproduction in plants include aerial stems such as the runners or stolons in strawberry plants (see Figure 1.6) and many varieties of turf. Contact with the ground is often the stimulus for the production of adventitious buds which grow into roots and shoots to produce a new plant that may only become independent after the death of the parent plant. For example, begonia leaves produce adventitious buds that can grow into new plants when they make contact with the ground.
- Asexual reproduction is a form of cloning that humans have used extensively to help breed large numbers of plants with desirable characteristics. You will review the manipulation of plant reproduction for agriculture later in this chapter.
- The advanced and dominant terrestrial plants—angiosperms and gymnosperms—reproduce sexually by producing seeds that contain the embryo. There is a great variety of ways that the male gamete (pollen) is transferred to the female gamete (ovule) in a process called **pollination**.
- Angiosperms have ovules protected within the ovary.
- The term *gymnosperm* actually means 'naked seed' and, while it may not look like the seed is 'naked' as many gymnosperms are cone-bearing, the cone structure is actually made up of modified leaves or scales. Therefore the seed is not protected within an ovary.
- The seeds produced by angiosperms and gymnosperms usually contain food reserves for the **germinating** embryo, helping seeds survive adverse conditions. Seeds also have a huge variety of methods for dispersal (see Table 1.3).
- There are two main types of angiosperms: monocotyledons and dicotyledons. Monocotyledons and dicotyledons are distinguished from each other by their seed structure as well as important differences in their vascular arrangement, growth and floral structure. Monocotyledons include grasses, cereals, palms, orchids and lilies. Their flower parts are often in threes or multiples of three and many are wind pollinated. Dicotyledons include grevillea, callistemon, eucalyptus, proteas, hibiscus and citrus, stone and pome fruits. Their flower parts are in fours, fives or multiples of four or five.

pollination: the process by which the male gamete, pollen, is carried to female reproductive organs and hence to the ovum for fertilisation to occur

germination: the process of a seed growing into a new plant

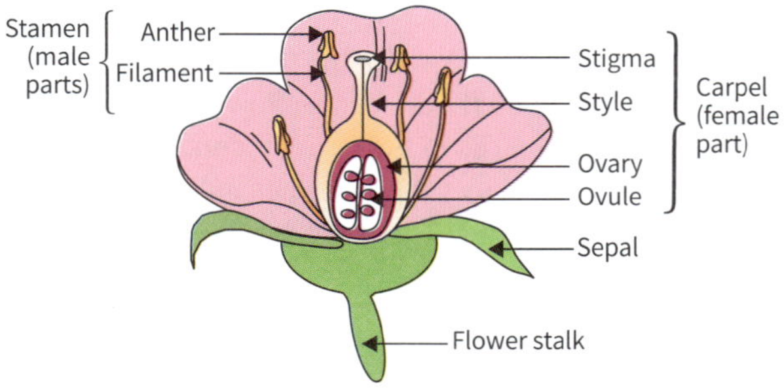

Figure 1.7 Generalised diagram of an angiosperm flower

Figure 1.8 Virginia pine (gymnosperm) showing female (left) and male (right) cones

Table 1.3 Summary of sexual reproduction in advanced terrestrial plants (angiosperms and gymnosperms)

Summary of sexual reproduction	Angiosperms	Gymnosperms
structures	flowers: both male (stamen) and female (carpel) organs may be contained in the same flower or there may be separate male and female flowers (rarely there are separate male and female plants such as *Casuarina*) some have brightly coloured and/or large petals; some produce large quantities of nectar or release strong odours to attract pollinators; some occur in spectacular clusters called inflorescences the fertilised ovary matures to form a structure called the fruit, which may or may not be the same as the fruit that we eat	cones made up of modified leaves (scales) produce male gametes (pollen) or female gametes (ovules) after fertilisation ovules develop into embryos in the seeds contained in female cones some species have plants that produce only male cones and others produce only female cones; male cones are generally smaller and more numerous than female cones some gymnosperms (e.g. gingko) have seeds that are not in a cone and have separate male and female plants; that is, are dioecious
pollination	insects, birds, small mammals and wind can transfer pollen some flowers are pollinated by a range of animals; others have one or several specific species that pollinate them some flowers can self-pollinate; others have male and female gametes that ripen at different times to avoid self-pollination some species self-pollinate only if cross-pollination has failed	generally wind pollinated and lack nectar and odour and have minimal petals stamens and stigma often have maximum exposure to wind for dispersal and entrapment of pollen
dispersal	wind, water or animals can disperse seeds/fruit some seeds pass through digestive systems and remain viable; others may depend on another species to initiate germination	many cones release winged seeds that are dispersed by the wind cycads (palm-like gymnosperms) and ginkgo produce fleshy seeds the ginkgo is called a living fossil and dates from before the age of dinosaurs; the ginkgo seed is foul smelling and it is speculated that carrion-feeding scavenging dinosaurs may have been responsible for their earlier widespread distribution
advantages	diversity is the key to their widespread success in terrestrial environments may compete strongly for pollinators and some have evolved precarious mutualistic relationships which are successful when both species survive	successful when growing in stands as pollination becomes relatively successful grow well in climates that may not support insects or other pollinating animals

➔ Many primitive terrestrial plants such as ferns, mosses and liverworts also reproduce sexually but rely on external water such as raindrops to transfer male gametes to the female organs. Their asexual reproduction occurs by spores and is largely responsible for plant dispersal. These plants have a life cycle that alternates between sexual and asexual reproduction stages. This process is called **alternation of generations**. The generation that produces the spores (asexual) is called the sporophyte and the generation that produces the gametes is called the gametophyte. Figure 1.9 shows the gametophyte of a tree fern.

➔ The tree fern plant we are familiar with is the sporophyte generation. If you look carefully at the underside of a tree fern frond you may see brown spots in a regular pattern. These contain the tiny spores. Figure 1.10 also shows another form of asexual reproduction in ferns. Being underground, rhizomes have the added advantage of being protected from the heat of bushfires.

alternation of generations: a life cycle that involves two different 'steps' (generations) and forms of the organism, one reproducing sexually and the other asexually; common in plants such as ferns

Figure 1.9 Prothallus, the tree fern sexual reproductive structure
Source: image by Velela, (public domain), via Wikimedia Commons

Figure 1.10 Fern frond showing spores; this is the asexual generation
Source: image by Anca Mosoiu, (own work, public domain), via Wikimedia Commons

Figure 1.11 Fern showing rhizome

➔ KEY QUESTIONS

12 **Describe examples of asexual reproduction in plants.**

13 **Describe examples of sexual reproduction in plants.**

Answers ➲ p. 23

Fungi: budding and spores

- ➔ Fungi include yeasts, mushrooms, moulds, mildew, rust and smut. Their roles include being decomposers and pathogens (disease-causing organisms). Some fungi live in mutual relationships within lichens or on the roots of plants. Some are unicellular (yeasts) but many consist of a body called a thallus that is made up of thread-like mycelia that in turn are made of branched tubular filaments called hyphae. Some filamentous fungi (mushrooms and toadstools) produce hyphae that make up visible fruiting bodies comprised of a cap and stem. The gills in the cap may release millions of tiny spores.
- ➔ Humans use fungi extensively in biotechnology (see Chapter 7 for more information).
- ➔ Asexual reproduction occurs through budding, spore production and **fragmentation**. Hyphae may break into segments that each grow into a new thallus in fragmentation. Fungal cells may produce a bulge that, after mitosis, detaches to grow into a new organism. See Figure1.4 for an image of a yeast cell budding. The production of a specialised reproductive sac called a **sporangium** can result in the release of large numbers of tiny spores, which is partly responsible for the successful widespread dispersal of many species of fungi. Some fungi produce spores as an extension of the hyphae.
- ➔ Adverse conditions may prompt initiation of sexual reproductive processes in fungi. Sexual reproduction may also result in the production of spores.

fragmentation: the process resulting in a fragment of an organism detaching and growing into an independent organism; occurs in fungi, plants, colonial filamentous cyanobacteria and sponges and some simple animals, including flatworms and sea stars

sporangium: the spore-producing structure in some species of fungi, ferns, mosses and liverworts

Figure 1.12 *Penicillium expansum* (blue mould spores on a rotting pear)
Source: image by HJ Larsen, Bugwood.org [CC BY 3.0 (http://creativecommons.org/licenses/by/3.0)], via Wikimedia Commons

14 Outline the role of asexual reproduction in fungi.

Answer ⊃ p. 23

Bacteria: binary fission

- Bacteria are prokaryotes so their duplication of DNA and cell division varies from the mitotic divisions in eukaryotic cells.
- Bacteria come in a wide range of types and roles, including microflora in the gut, decomposers and pathogens.
- Bacteria reproduce asexually in a process called binary fission. Occasionally they split into more than two cells in a process called multiple fission.
- Bacteria tend to have a short life cycle so they rapidly increase in population size. The *Escherichia coli* found in human intestines divide every 20 minutes. In seven hours, therefore, one bacterium could produce 2 097 152 bacteria.
- Some bacteria such as *Bacillus anthracis* (anthrax) produce endospores that are very resistant to heat, cold, dehydration, ultraviolet radiation and disinfectants. These adaptations are more a survival strategy than a method of reproduction.
- A bacterium may transfer its genetic material in the form of a plasmid or small circular piece of DNA to a recipient cell in a process called conjugation. This is believed to be a means of transferring antibiotic resistance (see Chapter 12).

→ KEY QUESTION

15 Distinguish binary fission from mitosis.

Answer ⊃ p. 23

Protists: binary fission and budding

- Protists are a diverse group of eukaryotic unicellular or colonial organisms. Examples are protozoa such as *Amoeba*, *Paramecium*, *Euglena*, algae, diatoms and slime moulds.
- Protists are relatively simple organisms and do not have differentiated tissues. Some protists cause diseases such as malaria and sleeping sickness so their life cycle and reproduction is associated with host organisms (see Chapter 9).
- Protists reproduce mainly through binary fission (see Figure 1.13) and budding. When budding occurs the offspring may stay attached to the parent, causing a colony to form. Toxoplasmosis is a disease that humans can get from cats that have eaten infected birds, rodents or other small mammals. The condition is only a serious threat because the protist that causes the disease can be passed on to the foetus in pregnant women. It lives inside the cells of its host and reproduces by budding in a number of ways. Internal budding results when it grows a pocket, produces buds into the pocket and they then break out through the cell membrane. The group of ciliate protists called suctoria reproduce by internal budding. They release swarmers; that is, cells that result from the internal budding. The budding may occur on the internal cell surface or in pouches which release the swarmers through an opening or inversion.
- As can be seen from Figure 1.13 the varied structures and arrangements in protists results in binary fission along different planes to ensure that the daughter cells not only contain a nucleus but also all other necessary structures for functioning.

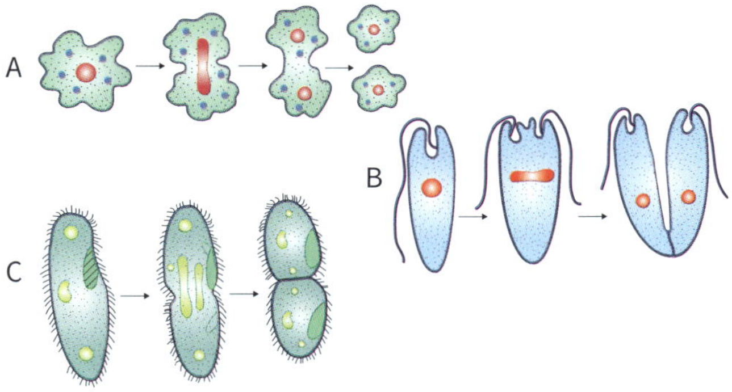

Figure 1.13 Binary fission in *Amoeba* (A), *Euglena* (B) and *Paramecium* (C)

Table 1.4 Binary fission in bacteria

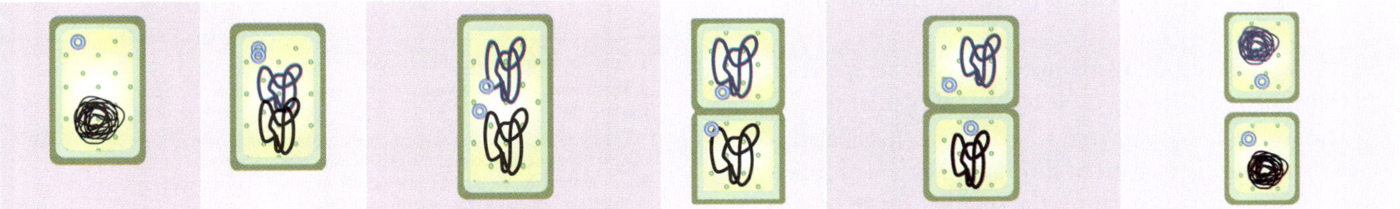

1 DNA tightly coiled in bacterium before fission	**2** DNA has replicated	**3** Bacterium increases in size and in the process the DNA is pulled to opposite poles	**4** New cell walls grow between the two poles	**5** New cell wall completes development, separating the two bacteria	**6** Two daughter cells each contain tightly coiled DNA, plasmids and ribosomes

Source: image by Ecoddington14 [CC BY-SA 3.0 (http://creativecommons.org/licenses/by-sa/3.0)], via Wikimedia Commons

➔ KEY QUESTION

16 **Compare binary fission with budding in protists.**

Answer ⊃ p. 23

2 Fertilisation, implantation and hormonal control of pregnancy and birth in mammals

» Students analyse the features of fertilisation, implantation and hormonal control of pregnancy and birth in mammals.

➔ Monotremes, marsupials and placentals are the three groups of mammals distinguished by their different reproductive processes and structures.

Fertilisation in mammals

➔ In mammals, copulation may result in internal fertilisation. Males release **semen** containing **spermatozoa** from the penis in a process called **ejaculation** into the cloaca/urogenital sinus of monotremes and marsupials, and into the vagina of placental mammals.

➔ The spermatozoa travel to the oviducts, which may contain an **oocyte** that has been released from a follicle in the female ovary. This journey involves changes to the spermatozoa so it is ready to enter the oocyte.

➔ Even with internal fertilisation the journey of the spermatozoa is uncertain. In the case of humans about 200 of the 500 million ejaculated spermatozoa reach the oocyte. This is despite the fact that the oocyte releases chemicals that help guide them.

➔ The testes produce the haploid spermatozoon through meiosis of the spermatocyte. The spermatozoon consists of:

- a head, which contains densely packed DNA and is capped by the acrosome, which releases enzymes to assist penetration of the oocyte
- a neck, which contains centrioles that support early divisions of the zygote
- mitochondria to supply energy
- a flagellated tail that helps the spermatozoon swim.

semen: a fluid containing spermatozoa and other secretions to transport, sustain and protect the male gametes

spermatozoa: male gametes; small motile (flagellated) cells carrying a haploid set of chromosomes and produced in the testes; spermatozoon is the singular form

ejaculation: the process of releasing semen from an erect penis

oocyte: an immature female germ cell that gives rise to the ovum or female gamete containing a haploid set of chromosomes and food reserves

➔ The ovary contains follicles in which meiosis occurs to produce the oocyte. This process occurs in two stages over an extended time. The first stage occurs in the female foetus. The second stage in humans occurs at intervals after puberty when the ovarian follicle ruptures to release the oocyte. The follicle becomes the corpus luteum, which plays a role in pregnancy (see Table 1.5). Meiosis is only fully completed after the entry of the spermatozoa. The oocyte is then called the ovum.

➔ The monotreme ovum is largest at about 4 mm in diameter. In marsupials it is 0.12 mm–0.28 mm and is smallest in placentals at 0.07–0.15 mm.

➔ The oocyte has two external layers: the inner jelly coat and the outer layer of cells that are remnants from the follicle. When the spermatozoon contacts the outer layer of the oocyte it releases enzymes that help it penetrate. As soon as the spermatozoon binds with receptors on the jelly coat the acrosome releases its contents, helping it penetrate the layer and change its nature. This ensures that only a spermatozoon from the same species can enter. The two plasma membranes fuse, allowing the spermatozoon to enter the cytoplasm of the oocyte. This causes the cortisol granules to release their contents to further ensure only spermatozoon enters the cell.

➔ The oocyte has suspended meiosis. The entry of the spermatozoon into the cytoplasm triggers the oocyte to complete its meiotic cell division. It is now called an ovum because it contains a pronucleus that is ready for fusion with the spermatozoon pronucleus. The membranes around the two nuclei dissipate and the two sets of haploid chromosomes join together. Once the pronuclei fuse, the diploid nucleus of the zygote has formed and fertilisation is complete.

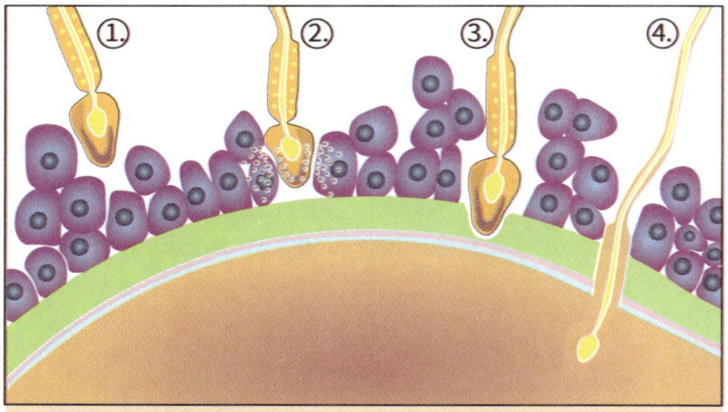

1. Spermatozoon approaches oocyte.
2. Spermatozoon releases enzymes to assist penetration of the outer layer of the oocyte.
3. The inner jelly coat changes structure in many mammals, preventing the entry of other spermatozoon and the cell membranes start to fuse together. It is believed there is some sort of molecular recognition so that only spermatozoon of the same species as the oocyte can enter.
4. The spermatozoon nucleus and tail enter the oocyte, the tail then disintegrates and the nucleus becomes the male pronucleus. The oocyte completes its meiosis cell division to become the ovum containing a pronucleus ready for the fusion. Once the pronuclei fuse the diploid nucleus of the zygote has formed and fertilisation is complete.

Figure 1.14 Approach and entry of the spermatozoon to the oocyte

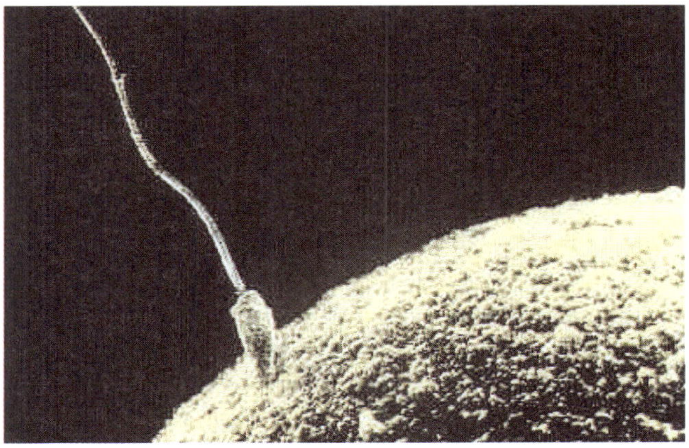

Figure 1.15 Scanning electron micrograph of spermatozoon about to penetrate the oocyte

Source: image from www.pdimages.com/web9.htm, (public domain), via Wikimedia Commons

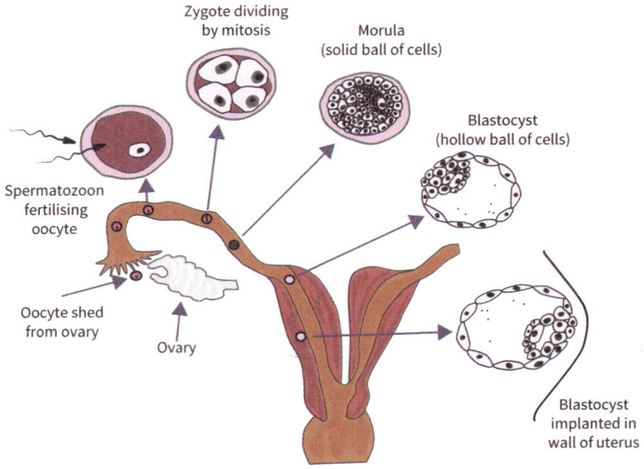

Figure 1.16 Ovulation, fertilisation and implantation

Source: adapted from image by Ttrue12, (own work), [CC BY-SA 3.0 (http://creativecommons.org/licenses/by-sa/3.0)], via Wikimedia Commons

➔ KEY QUESTIONS

17 **Describe the spermatozoon and its role in fertilisation.**

18 **Describe the oocyte and its role in fertilisation.**

Answers ➔ pp. 23–24

Implantation in mammals

- After fertilisation the zygote starts a process of mitotic cell divisions (cleavage) to become a small solid ball of cells called a morula. The morula continues travelling down the oviduct. In the **uterus** it divides and differentiates into the hollow fluid-filled ball of cells called the blastocyst. The blastocyst, as shown in Figure 1.16, embeds (implants) into the endometrial lining of the uterus. The jelly coat is shed prior to implantation.
- Implantation is a highly complex process because the zygote is a cell that is 'foreign' to the mother and as such should trigger an immune response (see Chapter 10). The process of accepting the foreign cells is often called immune tolerance during pregnancy and is still the subject of research.
- At this stage there is an inner cell mass that contains stem cells that will differentiate into the embryo, a cavity and an outer cell layer called the trophoblast.
- Implantation occurs in several ways, as outlined in the points below.
  - In humans and some rodents (e.g. guinea pigs) some of the trophoblast cells produce and release protein-dissolving enzymes that assist the erosion of the endometrium. This process assists the embedding of finger-like projections into the uterus wall. The enzymes also assist by digesting some cells in the endometrium that become the source of nutrition for the developing embryo.
  - In other mammals, such as rats, mice and hamsters, attachment into the endothelium of the uterus is via the surface that folds back on itself to form a pocket. This pocket surrounds the trophoblast.
  - In rabbits, dogs and some marsupials the blastocyst enlarges and binds directly with the surface of the epithelium.
- After the blastocyst has implanted, the stem cells start differentiating and the body plan of the new embryo starts to develop.
- After implantation in placental mammals the trophoblast develops into the placenta.
- In monotremes, implantation does not occur. The embryo develops in the egg which is nourished by secretions from the uterine wall of the mother for about 28 days before the shell is added and the egg is laid. The egg is incubated for about ten days before hatching. The egg shell is porous, allowing oxygen and nutrients to diffuse through. The egg yolk also assists with nutrient supply and waste removal.
- In marsupials **gestation** is only about one month. Early nutrition is from the yolk of the ovum and secretions of the uterus. Implantation only occurs towards the end of gestation and involves a loose connection through a yolk-sac type placenta.

uterus: the womb or organ in which the embryo and foetus develop in placental mammals for the period of gestation; in marsupials, the organ that holds and nurtures the embryo and foetus for about one month during the short gestation

gestation: the process or duration of the development of offspring in the uterus between conception and birth

- **Embryonic diapause** is the delayed implantation of the embryo, which remains in a suspended state at the blastocyst stage of development. In some species such as red kangaroos implantation occurs to allow reproduction in favourable environmental conditions. Some marine mammals (e.g. seals) may come ashore once a year to give birth and mate. Implantation is delayed to ensure the annual timing of birth or to delay implantation when environmental conditions are not favourable.

embryonic diapause: the delayed implantation and development of the embryo to ensure birth at the appropriate season or when conditions are favourable

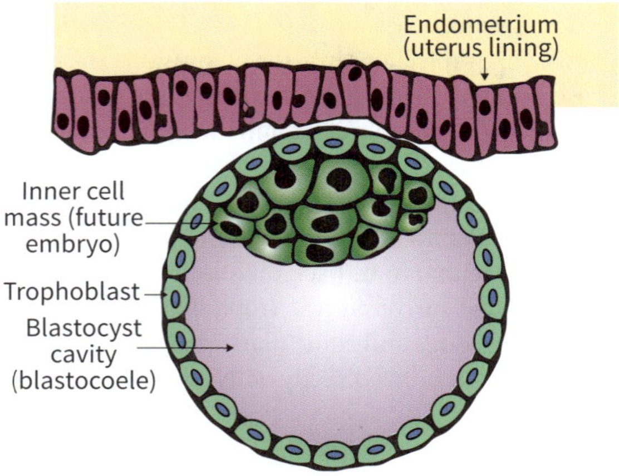

Figure 1.17 Implantation in mammals

Source: adapted from image by Seans Potato Business, Blastocyst.png, [GFDL 1.2, CC-BY-SA-3.0 (http://creativecommons.org/licenses/by-sa/3.0/)], via Wikimedia Commons

→ KEY QUESTION

19 **How is implantation different in placental, marsupial and monotreme mammals?**

Answer ⊃ p. 24

Hormonal control of pregnancy in mammals

- In mammals, **hormones** are released into the blood by **endocrine glands**. Hormones promote changes in cells and tissues such as increased growth, development of secondary sexual characteristics and control of the menstrual cycle in female primates, and the oestrous cycles in other groups of female mammals.
- Placental mammals, with their extended gestation and birth of relatively developed offspring, require complex hormonal coordination to ensure development of the embryo and foetus. In humans and other primates the blastocyst produces a hormone—(chorionic gonadotrophin (CG)—that prevents the corpus luteum from degenerating. The corpus luteum is the remnant of the ovarian follicle that released the oocyte and it produces hormones that maintain pregnancy.
- Table 1.5 summarises the current understanding of hormones that play a role in maintaining pregnancy, along with where they are produced and their effect. Some hormones promote continuation of pregnancy while others may terminate pregnancy or promote birth. In some cases the maintenance of pregnancy is a balancing act between two hormones with opposite impacts.

hormones: chemicals (steroids, proteins or oxylipins) released into body fluids, such as blood, that bind with protein receptors to activate specific cells or tissues to orchestrate long-term or occasionally short-term responses to stimuli and maintain body functioning

endocrine glands: glands such as the posterior pituitary gland that produce and secrete hormones into the blood supply

progesterone: a member of the steroid class of hormones called progestogens that are responsible for maintaining pregnancy; also called P4

Table 1.5 Hormones that control pregnancy in mammals

Hormone	Where produced	Role
progesterone	placenta and corpus luteum in humans, horses, sheep and cats the progesterone released from the placenta is sufficient to maintain pregnancy cattle, pigs, dogs and goats rely heavily on the progesterone released from the corpus luteum	progesterone is called the hormone of pregnancy because without it the endometrium would not be maintained or support the implantation of an embryo stops contraction of the smooth muscle of the uterus which is involved in birth; if occurring sufficiently before full gestation this may result in miscarriage prevents the pituitary secreting gonadotrophin luteinising hormone and follicle stimulating hormone to suppress ovulation during pregnancy helps induce immune tolerance, possibly mediated by a protein called Progesterone Induced Blocking Factor (PIBF) (see Chapter 10 for information about immune response)

Hormone	Where produced	Role
oestrogens, including oestrone (E1), oestradiol (E2), oestriol (E3) and oestetrol (E4); E3 and E4 are strongly associated with pregnancy	placenta, E4 from foetal liver	stimulates the development of mammary glands stimulates the development of uterine smooth muscle layers ready for birth; this counters the effect of the 'progesterone block' that suppresses uterine contractions late in pregnancy prevents the pituitary secreting gonadotrophin luteinising hormone and follicle stimulating hormone to suppress ovulation during pregnancy helps induce immune tolerance, possibly by expanding regulatory B and regulatory T or Tregs cells (see Chapter 10) oestriol (E3) is made from a precursor produced in the adrenal glands of the foetus; its levels late in pregnancy are considered an accurate indicator of the health of the foetus E3 promotes uterine growth and sensitivity to other pregnancy hormones and is believed to trigger labour when it becomes the dominate hormone E4 is believed to protect the foetus from maternal oestrogen
chorionic gonadotrophin (CG) e.g. human chorionic gonadotrophin (hGC); also called a luteotrophin	blastocyst (primates) placenta	prolongs the activity of the corpus luteum helps induce immune tolerance, possibly by expanding regulatory B and regulatory T or Tregs cells (see Chapter 10) induces production of relaxin and stimulates production of oestrogen and progesterone its presence in urine or blood indicates pregnancy has occurred
relaxin	corpus luteum of ovary and placenta	relaxes maternal muscle joints and ligaments to allow for the expanding foetus relaxes uterine muscles and prepares lining for implantation
corticotrophin-releasing hormone (CRH)	hypothalamus of brain and foetus and placenta during pregnancy	increases the production of the stress hormone cortisol (regulates metabolism, the immune response and helps body respond to stress) regulates duration of pregnancy and foetal maturation suppresses maternal immune response in early pregnancy improves blood flow between placenta and foetus in later pregnancy associated with 'late term cortisol surge' that is believed to support maternal bonding and care in humans and some animals
placental lactogen or chorionic somatomammotropin (identified in primates, ruminants and rodents)	placenta	possibly controls metabolism of foetus and mother, mobilising energy reserves for the foetus stimulates corpus luteum function promotes development of mammary glands before birth

oestrogens: a group of four steroidal hormones that bind with oestrogen receptors and have a major role in the development of secondary sexual characteristics and a minor role in pregnancy

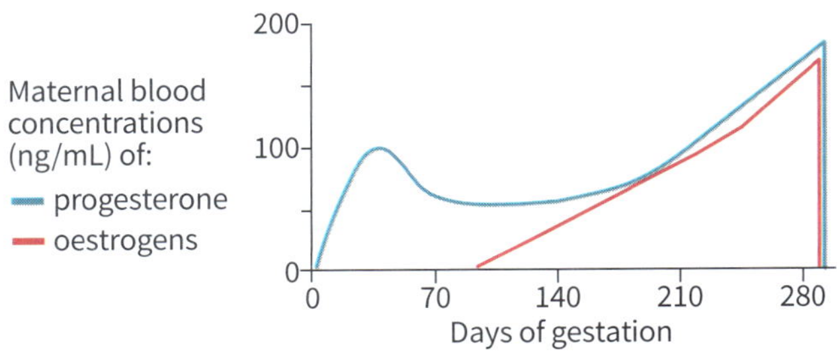

Figure 1.18 Changes in hormone levels in human females during pregnancy

→ KEY QUESTIONS

20 **Why is progesterone called the pregnancy hormone?**

21 **Describe the role of hormones in maternal immune tolerance.**

Answers ➲ p. 24

Hormonal control of birth in mammals

- In placental and marsupial mammals, **birth** is the process in which the foetus is delivered from the uterus to the outside world. It involves contractions of the uterine smooth muscles, widening or dilation of the cervix and the delivery of the foetus through the cervix and vagina (birth canal).
- Birth is orchestrated by a number of interacting hormones which may coordinate for the same role or balance each other in contradictory roles until one becomes dominant.
- In placental mammals, birth is generally followed by the expulsion of the placenta.

Table 1.6 Hormones that play a role in birth in mammals

Hormone	Where produced	Role
corticotrophin-releasing hormone (CRH) and cortisol (stress hormones)	foetus and uterus	believed to trigger a rise in oestriol (E3) and when this hormone becomes dominant, birth ensues; these stress hormones are the result of the increased pressure due to the growing foetus and their levels rise just before birth
oestriol (E3)—an oestrogen associated with birth	placenta	counters the role of progesterone on suppressing uterine contractions increases sensitivity of uterus to other hormones
oxytocin	hypothalamus and stored and released from posterior pituitary; also released from foetus	has a major role in birth as well as other functions such as ejection of milk while breastfeeding and maternal bonding initiates and strengthens rhythmic uterine contractions
prostaglandins	uterus	helps initiate labour and uterine contractions by reducing progesterone (which inhibits uterine contractions)
relaxin	ovaries	helps dilate (open) the cervix to allow passage of foetus; widens pubic bone and relaxes pelvic ligaments

birth: the process of delivery of the foetus from the uterus; also called parturition, delivery or labour

oxytocin: a protein-based hormone responsible for uterine contractions, release of milk and maternal and pair bonding

→ KEY QUESTION

22 **Why is oxytocin sometimes called the birth hormone?**

Answer ⊃ p. 24

3 The impact of scientific knowledge on the manipulation of plant and animal reproduction in agriculture

» Students evaluate the impact of scientific knowledge on the manipulation of plant and animal reproduction in agriculture.

- Agriculture must become **sustainable** for the long-term survival of humans and the environment. The ever-growing human population has increased demands on agricultural production. In your study of the impact of scientific knowledge on the manipulation of plant and animal production in agriculture you would have made judgements about the use of scientific knowledge and considered sustainability and ethical issues.

The impact of scientific knowledge on the manipulation of plant reproduction in agriculture

Asexual techniques

- Understanding of the variety of ways plants reproduce asexually has led to the wide use of cloning of plants through plant **propagation**, which allows the rapid development of plants with known qualities. This includes the propagation of plants from tubers, corms, bulbs, rhizomes and stolons. Some of these vegetative reproductive structures contain food reserves and these have been selected and propagated as sources of food. Examples include potatoes (tubers) and onions (bulbs).
- Plants often have the ability to regenerate after damage. Deliberately cutting a section of stem, leaf or root, and often treating it with plant growth hormones to stimulate the development of roots or mitosis and then planting these 'cuttings', results in plants identical to the parent. In the practice of layering, stems containing buds are pinned to the ground or growing medium to promote new plant growth.

sustainable: able to continue indefinitely, implying that resources are used at a rate that is renewable and wastes are produced at a rate that can be recycled or decomposed so there is no net deterioration of the environment and equitable social structures can exist

propagation: asexual and sexual techniques used to grow new plants; includes growing cuttings, grafting or seed germination

- Tissue culture is another method that uses knowledge of where cell division occurs in plants to produce new plants in a sterile growth medium.
- Knowledge of plant tissues and their roles has also been used to develop more sophisticated plant propagation practices. This includes grafting, where a segment of tissue including a bud and the cambium (dividing vascular tissue) or scion is inserted in a cut of another plant so that the cambium tissues align. The grafted plant, which may have been selected for properties of fruit quality or quantity, grows beyond the point of insertion. This technique can be used to grow disease-resistant plants. A disease-resistant root stock that is not very productive can have a more productive variety inserted so the resulting plant is both disease resistant and highly productive. Grafting can also be used in home gardens to produce a variety of fruits on the one plant. For example, a citrus tree may have lemon, lime and navel orange branches resulting from different grafts.
- All vegetative propagation techniques contribute to a lack of genetic diversity but produce large numbers of plants with predictable characteristics.

Sexual techniques

- Propagation of plants by the selection and germination of seeds (sexual methods of plant reproduction) results in plant biodiversity and produces large numbers of desirable agricultural plants. Cereal grains such as rice and wheat are staple foods in many cultures. Knowledge of the best conditions for seed storage, pre-treatment and germination has resulted in the proliferation of these plants.
- The ripened ovaries (fruits) and associated structures of many flowering plants provide many foods commonly called fruits (e.g. tomatoes, citrus) and vegetables (e.g. potatoes, zucchini). Other food plants that have complex or compound flower and fruit structures include strawberries, blackberries and mulberries.
- The **artificial selection** of plants with more desirable seeds and fruit has led to a greater diversity of plant foods. Seed banks now store seeds from a wide variety of original plant varieties so that their properties may be incorporated into current varieties as the need arises. Factors such as palatability, disease resistance, storage time, size and quantity of produce have been selected to increase agricultural production. For example, cabbage, broccoli, kale, cauliflower and Brussels sprouts are members of the same species that have different properties because of selective breeding.
- Understanding the structure and function of flowers has led to the ability to control pollination or to cross-pollinate or hybridise varieties of plants to benefit agriculture in a manner that has taken artificial selection one step further. Some of these techniques make use of the high natural incidence of **polyploidy** in plants. Two examples of this are outlined below.
  - Wheat is hexaploid; that is, it has six sets of chromosomes. Agriculture has used this understanding in its attempts to improve cereal crops by hybridising a wheat (*Triticum turgidum*) and rye (*Secalec ereale*) to produce triticale. The initial results combined the best characteristics of the parents but produced a sterile hybrid plant. Inducing polyploidy in wheat has resulted in a fertile hybrid.
  - Bananas are triploid (three sets of chromosomes) and as a result lack seeds in the fruit and therefore rely on vegetative propagation. Seedless citrus and watermelon are also triploid.
- Conventional selective breeding along with other changes in agricultural practices resulted in what has been termed the 'Green Revolution'. The Green Revolution occurred between the 1930s and late 1960s and stemmed from a number of initiatives in developing countries, including plant breeding. In Asia, for example, dwarf varieties of rice and wheat, selectively bred to be more responsive to fertilisers, resulted in a 32% increase in rice yields and a 51% increase in wheat yields. The impacts of the Green Revolution are still being felt. Table 1.7 on the following page summarises some of the issues and impacts that have arisen from these changes.

artificial selection: also called selective breeding, it involves selecting plants and animals with desirable characteristics to breed offspring with these traits; used in agriculture to increase efficiency, decrease costs and/or improve output (yields)

polyploidy: cells containing more sets than the diploid set of chromosomes

Figure 1.19 New varieties of grain, including wheat, were instrumental in the Green Revolution.

Source: image by User: H20, (own work, public domain), via Wikimedia Commons

Table 1.7 Economic, social and environmental issues arising from the Green Revolution

Economic issues	Social issues	Environmental issues
in India it transformed the country into a major rice exporter and more than halved the cost of rice	over one billion people are believed to have been saved from starvation	the ratio of crop production to energy input has decreased, with implications for increases in use of fossil fuels and pollution
Mexico became more self-sufficient in agricultural products	Mexican peasant farmers changed attitudes to become more open to the use of scientific ideas in agriculture	the widespread use of fertilisers, herbicides and pesticides, as well as mechanisation and irrigation, has impacted on biodiversity and caused water, soil and air pollution
soy bean production in Brazil and Argentina has dramatically improved as a result of soil treatment; Brazil is becoming a significant exporter of soy beans and, because they are used as stock feed, an exporter of poultry and beef	food security in some countries is at risk because governments are concentrating on export of crops at the expense of crops used to feed the local population	
the use of a new cultivar of rice (IR8), along with the use of fertilisers and pesticides, resulted in the Philippines being able to export rice for the first time in the 20th century	large-scale farming and ownership has meant many poor people have lost control of their land and lives	
corruption has been an impediment to adoption in Africa; Malawi has stood out with its improvements in agricultural production in the 21st century		

- Rice and maize are staple crops that have been the focus of hybridisation breeding programs. **Hybrids** often grow more vigorously and are much more productive (10–25% more) but new hybrid seed has to be supplied annually. Maize has separate male and female flowers so it naturally cross-pollinates. Removing the 'tassels' or stamens from male flowers is a simple strategy to produce hybrids. The use of cytoplasmic male sterility (CMS) on maize in the 1970s saved the need to 'detassel' but the resulting hybrid plants were susceptible to a disease (southern corn leaf blight).
- Rice has both male and female parts on the same flower and therefore self-fertilises (see Figure 1.20). Preventing self-pollination by physically removing the stamen is labour intensive and expensive but now the use of CMS and environment-sensitive genic male sterility (EGMS) has streamlined the process. The genes for CMS are cytoplasmic because they occur in the mitochondria and/or chloroplast. The flowers lack functional stamens, anthers or male gametes so while they cannot self-fertilise, they can be cross-pollinated using viable pollen from other plants. Hybrid seeds are produced and germinate into the productive but sterile hybrid plant.

hybrid: two genetically diverse parents are crossed and the first generation of offspring are the resultant hybrids; hybrids often demonstrate hybrid vigour via better growth and being more productive than either of the parental strains

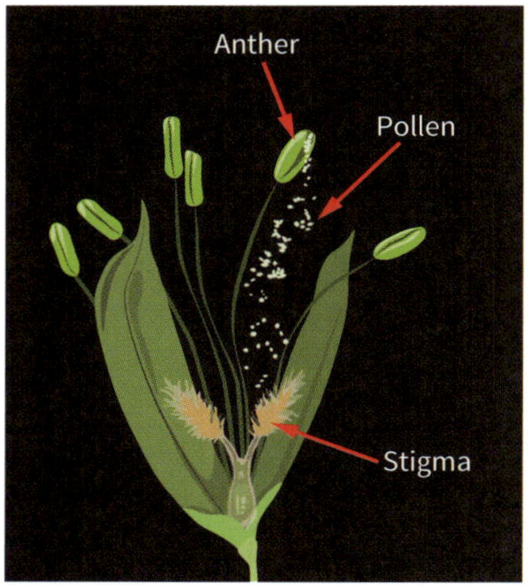

Figure 1.20 Rice self-pollinates.

- Conventional selective breeding used natural variation in populations. Since the 1930s there has been the deliberate attempt to produce variation by exposing plants to agents that bring about genetic change. These are called **mutagens**. The mutant plants are assessed in the hope of finding beneficial mutations. Various chemicals and ionising radiation were early mutagens but other more unusual ones have also been used. Other mutagens include the use of bacterial enzymes and even exposing plant varieties to space travel. One benefit that was reported was a higher rate of germination of the space-exposed wheat seeds.

mutagens: chemicals or energy forms such as radiation that promote changes in the genetic material (DNA) and hence mutations in offspring

- Insertion of genes from one species of organism into another is a relatively recent technique in plant breeding. Known as **gene technology** or genetic engineering, it results in **genetically modified organisms (GMO)**. One example of a genetically modified organism is the production of Bt cotton which contains a gene from the soil bacterium *Bacillus thuringiensis*. As in the bacterium, the gene in the Bt cotton produces a protein that is toxic to insect larvae. The benefit of this technology is the reduced need for pesticides. See Chapter 8 for information on gene technologies.
- In 2016 the University of Adelaide reported a potential new development in plant breeding: altering the expression of a gene. While studying durum wheat, Associate Professor Jason Able reported that small gene-regulating molecules found in plant cells (known as small RNA) are involved in stress adaptation. It is hoped these can be exploited to breed plants with favourable stress-tolerant traits.

gene technology: manipulation of DNA to change the genetic make-up of an organism
genetically modified organism (GMO): an organism that contains an artificially inserted gene or genes from another species of organism; a transgenic organism

→ KEY QUESTIONS

23 **Compare the use of asexual reproduction techniques with sexual techniques in plant propagation.**

24 **Use your scientific understanding to evaluate the social, economic and environmental impacts of the Green Revolution as a means of achieving sustainability.**

25 **Describe the potential impacts of application of scientific knowledge to the breeding of plants and animals.**

Answers ➲ p. 24

The impact of scientific knowledge on the manipulation of animal reproduction in agriculture

- Domestication of livestock has a long history but the need to rely on mainly sexual reproductive techniques has led to very different strategies for animal breeding compared with plant breeding. Ethics and animal welfare issues also become a concern in this field.
- Selective breeding of livestock has resulted in the development of strains of poultry, cattle, sheep, pigs, goats, horses, donkeys and dogs more suitable for specific agricultural purposes. Selective breeding was initiated by Robert Bakewell in Britain in the 18th century and contributed to the British Agricultural Revolution.
- Some disadvantages of selective breeding include that:
  - the selection for milk or meat traits has had negative effects on reproductive performance, including reduction in fertility in dairy cattle and reduced reproductive ability in meat poultry
  - purebred Belgian Blue cattle have a condition known as 'double muscling' due to a mutation in the muscle protein; this has been selected for because of the high production of lean meat, but the unintended consequence has been the need for caesarean delivery of a very high percentage of calves
  - inbreeding, with the resulting lack of genetic diversity, can increase susceptibility to disease, increased frequency of genetic diseases, or conditions such as hip dysplasia in large breeds of dogs.
- **Reproductive technologies** are based on the application of increased scientific knowledge to the manipulation of animal reproduction. They include **artificial insemination (AI)**, cloning by nuclear transfer, in-vitro embryo production, oestrus induction and synchronisation.
- Cloning by embryo splitting, which artificially creates an identical twin, has occurred for about 40 years. The resultant progeny then need to be assessed to determine whether they possess desirable characteristics for agriculture. Another method of cloning, **whole organism cloning**, was first successful in 1996 with the production of Dolly the sheep (see Figure 1.21). This cloning by somatic cell nuclear transfer (NT or SCNT) enabled the selection and replication of high-quality animals. Cloning can now make use of mature adult cells, such as skin cells, to obtain a nucleus. The nucleus is then transferred into an unfertilised oocyte and implanted into another female for development. The advantage with this method is that the nuclear material can be harvested after the beast is processed in an abattoir and its agricultural value assessed. To date, the major use of cloned animals has been in research.

reproductive technologies: technologies applied to improve reproduction
artificial insemination (AI): the collection, storage and transport of semen from male animals with desirable characteristics so that more female can be impregnated and larger numbers of offspring with the desirable characteristics produced
whole organism cloning: making genetically identical copies of whole organisms; an example of a reproductive technology

- Efficiency and effectiveness of selective breeding has been improved by the introduction of artificial insemination. This has the added benefit of circumventing quarantine restrictions on the movement of animals such as cattle and horses. Efforts have been made to determine the optimal time for insemination based on understanding of oestrus cycles in mammals in order to maximise fertilisation rates.

- This has been further developed by the use of artificial progesterone hormone implants followed by withdrawal to induce oestrus. When applied across a herd this is called oestrus synchronisation. This streamlines the use of artificial insemination as well as the timing of births.
- Excessive use of hormone treatments can induce multiple ovulation in cows, which is often followed with embryo flushing and transfer (MOET—multiple ovulation and embryo transfer). This technique expands the potential of desirable female animals to produce offspring but has the disadvantage of increased abnormalities.
- Recently alternatives to MOET have been developed, including the use of ultrasound devices through the vagina to detect and pick up the naturally produced ova (OPU), in-vitro maturation (IVM) followed by in-vitro fertilisation (IVF), in-vitro embryo culture (IVC) and embryo transfer. These reproductive technologies result in higher levels of complications in pregnancy, birth and the health of offspring compared with the natural reproductive processes.
- A technique called Juvenile In-Vitro Fertilisation and Embryo Transfer (JIVET) is used in dairy cattle and has been trialled in sheep. In this technique, the genetic potential of ewes is assessed. Oocytes are then extracted from those ewes with potential at between six–ten weeks old. This is followed by in-vitro fertilisation and embryo transfer. The technique results in reduced intervals between successive generations of sheep and more meat production.
- Aquaculture is becoming a more important source of seafood as natural stocks are overfished. Scientific knowledge of the breeding and life cycles of some fish is vital for planning for sustainable fishing. In aquaculture, reproductive technologies have been applied to brood stock as well as the harvested stock or even used to deliberately create sterile organisms. Exposing fertilised trout eggs to pressure has promoted the development of triploid fish. As with triploidy in plants, these organisms are infertile so there is a reduced risk of them interbreeding with and contaminating wild stocks when released for recreational fishing. Triploidy is also of benefit in aquaculture because it increases growth rates after the normal time of sexual maturity as energy is not expended in breeding.
- Single-sex populations of some fish (e.g. rainbow trout and tilapia) achieve higher growth rates in aquaculture. Aquaculture of tilapia has found that males grow faster than females because they are not expending energy in egg production. The early application of male hormones to tilapia fingerlings has resulted in male populations. Super-male tilapia have reportedly been bred so they only contain the sex chromosome Y (responsible for maleness) and so only produce males. This avoids the need for hormonal application to the fingerlings.
- Rainbow trout brood stock can be treated with hormones to obtain fast-growing populations of only females. Females that have been sex reversed into functional male rainbow trout produce sperm that when used to fertilise normal eggs, produce only female offspring.
- Human manipulation of fish reproduction was taken to a higher level in 2015 with the first case of a genetically modified fish species—AquAdvantage salmon—being approved for commercial production, sale and human consumption by the Food and Drug Administration in the United States. The inserted gene results in a doubling of the average growth rate and halving of the harvest time compared with unmodified salmon. AquAdvantage salmon are sterile because they are triploid.
- Scientific knowledge used for manipulation of animal reproduction also includes understanding of how stem cells in an embryo differentiate into the various cells and tissues that make up organs and organ systems. Workers in Atlanta have used adult stem cells, extracted from quality bulls, which are capable of differentiating into spermatozoa. The stem cells are inserted into a recipient bull that has been sterilised so it produces spermatozoa genetically identical to those produced by the donor bull. Calves can then be sired by natural mating or artificial insemination. This magnifies the breeding power of an elite male animal (as opposed to cloning, which attempts to magnify the qualities of an elite female animal). Issues of rejection are still to be overcome to improve success rates.
- A genetically engineered cloned sheep, Polly, followed shortly after Dolly. Polly was genetically modified to produce milk that contained the human clotting factor as a potential treatment for haemophiliacs. See Figure 1.21 on how Dolly was cloned.

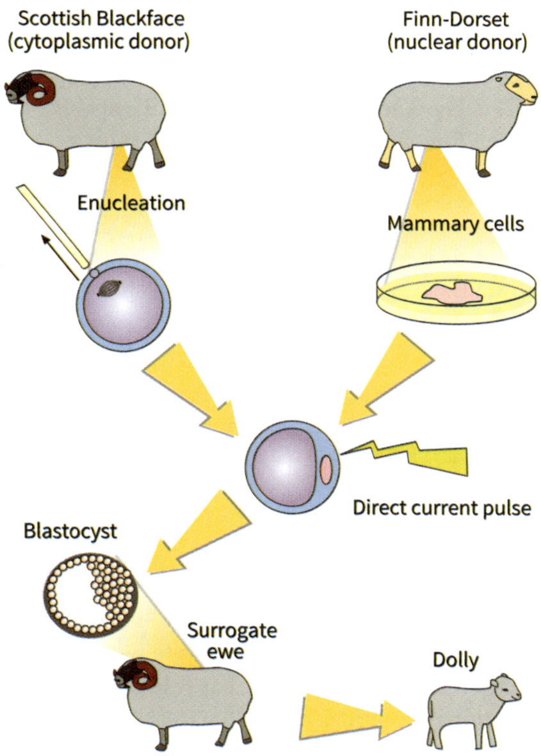

Figure 1.21 The process of creating Dolly, the first cloned animal, in 1996

Source: image by Squidonius, (own work, public domain), via Wikimedia Commons

→ KEY QUESTIONS

26 Outline the role of selective breeding in animal agriculture.

27 Define reproductive technologies.

28 Describe two methods of cloning animals.

29 Describe artificial insemination and technologies used to enhance its effectiveness in agriculture.

30 Outline the use of reproductive technologies in aquaculture.

Answers ➲ p. 24

CHAPTER SYLLABUS CHECKLIST

Are you able to answer these questions from the syllabus for this chapter? Tick each question as you go through the checklist if you are able to answer it. If you cannot answer a question, turn to the relevant page in the study guide to find the answer. For NESA key word meanings, go to www.educationstandards.nsw.edu.au and search 'key words'.

	FOR A COMPLETE UNDERSTANDING OF THIS TOPIC:	PAGE NO.	✓
1	Can I explain the mechanisms of reproduction that ensure the continuity of a species by analysing sexual and asexual methods of reproduction in a variety of organisms?	4–7	
2	Can I explain the mechanisms of reproduction that ensure the continuity of a species, including but not limited to animals: advantages of external and internal fertilisation?	7–8	
3	Can I explain the mechanisms of reproduction that ensure the continuity of a species, including but not limited to plants: asexual and sexual reproduction?	8–10	
4	Can I explain the mechanisms of reproduction that ensure the continuity of a species, including but not limited to fungi: budding, spores; bacteria: binary fission; protists: binary fission, budding?	10–12	
5	Can I analyse the features of fertilisation and implantation in mammals?	12–14	
6	Can I analyse the features of hormonal control of pregnancy and birth in mammals?	14–16	
7	Can I evaluate the impact of scientific knowledge on the manipulation of plant reproduction in agriculture?	16–19	
8	Can I evaluate the impact of scientific knowledge on the manipulation of animal reproduction in agriculture?	19–21	

A QUICK NOTE! All the **dark green panels** with information (like the one below) have the **syllabus inquiry questions** and all the **light green panels** (like the one below) have the **syllabus dot points** for each topic. Note the checklist questions in the table above are based on these.

INQUIRY QUESTION:

How does reproduction ensure the continuity of a species?

Life continues through the processes of reproduction and heredity. In this chapter you review reproduction in plants, animals, fungi, bacteria and protists.

1 The mechanisms of reproduction that ensure the continuity of a species

» Students explain the mechanisms of reproduction that ensure the continuity of a species, by analysing sexual and asexual methods of reproduction in a variety of organisms, including but not limited to animals: advantages of external and internal fertilisation;

Biology Stage 6 Syllabus

Reproduction

Inquiry question: How does reproduction ensure the continuity of a species?

- explain the mechanisms of reproduction that ensure the continuity of a species, by analysing sexual and asexual methods of reproduction in a variety of organisms, including but not limited to:
  - animals: advantages of external and internal fertilisation
  - plants: asexual and sexual reproduction
  - fungi: budding, spores
  - bacteria: binary fission (ACSBL075)
  - protists: binary fission, budding
- analyse the features of fertilisation, implantation and hormonal control of pregnancy and birth in mammals (ACSBL075)
- evaluate the impact of scientific knowledge on the manipulation of plant and animal reproduction in agriculture (ACSBL074)

HSC EXAM-TYPE QUESTIONS

Now for the real thing! The following questions are modelled on the types of questions you will face in the HSC Examination. Think about it: if you get extensive practice at answering these sorts of questions, you will be more confident in answering them in the actual HSC Examination. It makes sense, doesn't it?

Another advantage for your exam preparation is the format of the answers: they are deliberately structured to give you strategies on how to answer examination questions. This will help you aim for full marks!

- For each objective-response question you will have the correct answer and an explanation, and reasons why the other answers are incorrect.
- For each short-answer question you will have a detailed answer marked with ticks to indicate what part of the question gains which marks and also an examiner's plan (Examiner Maximiser/EM) to help you get full marks.

When you mark your work, highlight any questions you found difficult and earmark these areas for extra study.

Objective-response questions (1 mark each)

1 **What is a disadvantage of sexual reproduction in animals?**
- A It requires two parents of opposite sexes.
- B It produces diverse offspring.
- C Offspring are identical to the parent.
- D All adults can produce offspring.

2 **What is the main difference between asexual and sexual reproduction?**
- A Sexual reproduction results in identical offspring; asexual reproduction results in varied offspring.
- B Asexual reproduction results in identical offspring; sexual reproduction results in varied offspring.
- C Sexual reproduction only occurs in plants and asexual reproduction occurs in animals.
- D Asexual reproduction only occurs in plants and sexual reproduction occurs in animals.

3 **What is an advantage of internal fertilisation over external fertilisation?**
- A It is better suited to aquatic environments.
- B It is better suited to terrestrial environments.
- C It produces larger number of gametes and offspring.
- D It requires less care of young offspring.

4 **Which of the below is a correct statement about implantation in mammals?**
- A it occurs before fertilisation
- B it occurs in the oviduct
- C it occurs in all species of mammals
- D it occurs in the endometrium of the uterus

5 **Which of the below are examples of the application of understanding of reproduction to animal agriculture?**
- A crossbreeding, artificial insemination, in-vitro fertilisation
- B grafting, tissue culture, germination
- C mutation breeding, cytoplasmic male sterility, cross-pollination
- D induced polyploidy, cuttings, layering

Short-answer questions

6 **Outline the differences between reproduction of bacteria, protists and fungi.** (4 marks)

7 **Describe five key roles of hormones in mammals between conception and birth.** (5 marks)

8 **Outline the key structures in flowers and relate this to the process of crossbreeding in plant agriculture.** (4 marks)

9 **Explain how applications of scientific knowledge can make agricultural plant propagation and animal husbandry more sustainable.** (6 marks)

10 **Evaluate three ethical or animal welfare issues associated with recent developments in agricultural animal husbandry.** (6 marks)

ANSWERS

KEY QUESTIONS

Key questions ➲ p. 5

1 The two processes are the replication of the genetic material (DNA) and its transfer to offspring; that is, the next generation.

2 Mitosis is the process by which the genetic material of a cell replicates and the cell then divides to result in two genetically identical cells. Mitosis is a key process in asexual reproduction and in the growth of offspring after fertilisation in sexual reproduction.

3 Meiosis is the special cell division which results in cells containing half the amount of genetic material (i.e. number of chromosomes) in a unique combination so that upon fertilisation the full complement of genetic material is restored.

4 The next generation in asexual reproduction is genetically identical to the parent generation and only one parent produces the offspring. Sexual reproduction occurs after the formation of gametes (sex cells) by two parents of different gender and their fusion in the process of fertilisation. Offspring resulting from sexual reproduction are genetically unique.

Key questions ➲ p. 7

5 Asexual reproduction occurs as runners in plants (e.g. strawberry plants), tubers (e.g. potato plants), budding in yeast, spore production in fungi and binary fission in bacteria.

6 Sexual reproduction occurs in animals after mating by egg laying in frogs, fish and monotremes. It also occurs with the production and development of embryos in the uterus of placental mammals, and in the production and development of embryos in the pouch of marsupials. Sexual reproduction occurs in plants with pollination and the production of seeds in the ovary of flowers or in cones.

Key questions ➲ p. 8

7 Internal fertilisation is the union of two gametes inside the body. In external fertilisation the union occurs in the outside environment.

8 Internal fertilisation increases the chances of gametes meeting as the motile gamete is confined within a reproductive system. It also increases the chances of survival and development of the embryo/foetus as it develops within the reproductive system. Internal fertilisation is suited to terrestrial animals.

9 External fertilisation allows dispersal of offspring as it often occurs in aquatic environments and it is also less demanding of parental care.

10 Internal development of the embryo/foetus is more demanding on the female parent which needs to supply the requirements for development. Laying fertilised eggs can partly overcome this but the right conditions need to be provided for egg incubation and hatching. Internal fertilisation can limit the number of offspring produced, even though the chances of success may be higher.

11 Chances of fertilisation are reduced because of increased difficulties of gametes meeting. Offspring may need to be independent early as there may be little parental care. There is an increased risk of predation on the eggs/embryos/young.

Key questions ➲ p. 10

12 Asexual reproduction in plants results when a new plant grows from an existing plant. Stems may grow horizontally and, at points along them, buds develop into new plants. These stems may be called runners; for example, in strawberry plants or couch, buffalo or kikuyu lawns. Stems may be underground (rhizomes). Rhizomes in bracken fern produce new plants and are a way of the plant spreading. Other examples of asexual reproduction include corms, bulbs and tubers that may arise from underground stems or roots.

13 Angiosperms reproduce sexually through the production of flowers. The male gamete, pollen, must transfer to reach the female gamete, the ovum, through the stigma and style in the flower. Gymnosperms reproduce sexually through the production of seeds that are not enclosed in an ovary (hence the name 'naked seeds'). A common way of seed production in gymnosperms is in cones. Some cones release pollen that is transferred to the ovule, usually by wind.

Key question ➲ p. 11

14 Fungi reproduce commonly by asexual means and sometimes by sexual means. Asexual reproduction results from the processes of budding, spore production or fragmentation. Of these three methods, spore production is particularly important because it allows widespread dispersal of fungi as the spores are light and easily carried by the breeze.

15 Both binary fission and mitosis are processes that result in cell reproduction with offspring carrying the same genetic material as the parent cell. Mitosis occurs in eukaryotic cells. The replication of genetic material and its equal division into two nuclei, followed by cytokinesis, results in the production of two genetically identical cells. The formation and movement of the chromosomes that carry the genetic material relies on complex structures within eukaryotic cells such as the spindle. Bacteria are prokaryotic and do not have the same structures to carry genetic material and allow its replication and movement. The plasmids or circular chromosomes in bacteria still go through processes of replication and separation so that the result of binary fission is two identical cells, but the processes involved are different and replication and separation are simultaneous.

Key question ➲ p. 12

16 Both binary fission and budding are types of reproduction of single-celled organisms that involve replication of genetic material and its transfer to daughter cells. In binary fission the cytoplasm also divides between daughter cells, whereas in budding only a small amount of cytoplasm goes to the daughter cell, which will later grow to full size. In protists, the budding may develop internally in the parent cell and the release of the offspring results in the death of the parent.

Key questions ➲ p. 13

17 The spermatozoon is a haploid male gamete formed as a result of meiosis in the testes. It consists of a head that carries the DNA and some enzymes, a neck and midsection that contains centrioles and mitochondria, and a tail or flagella. The DNA is the male contribution to the genetic material in the zygote. The enzymes assist penetration of the oocyte. The centriole assists the early mitotic division of the zygote. The mitochondria supply energy for movement and the tail moves the spermatozoon so it can travel through the female reproductive tract to the oocyte.

18 The oocyte is a relatively large cell that has formed in a cell division that occurs in two distinct phases, often separated by many years. At fertilisation it completes the meiosis cell division. It is surrounded by two layers: one of cells and the inner zona jelly coat. It contains a large quantity of cytoplasm, a haploid nucleus and mitochondria. It is released from the ovary in ovulation and recommences and completes meiosis only after a spermatozoon has penetrated its cell membrane.

Key question ➲ p. 14

19 Monotremes lay eggs so the zygote develops within the egg and implantation does not occur. Marsupials give birth to an immature foetus, with development in the uterus taking up to about one month. The need for nutrition and the removal of wastes from the developing embryo/foetus is relatively small so implantation and attachment is only superficial. Placental mammals have an extensive gestation and so the young are nourished and wastes removed for an extensive period. The implantation into the uterus lining is rapidly followed by the development of the placenta to serve this purpose.

Key questions ➲ p. 15

20 Progesterone is a hormone that prepares the uterus for implantation and maintains it so that the developing embryo and foetus are not expelled.

21 Maternal immune tolerance is the failure of maternal cells to reject the embryo/foetus as it contains genetic material different from the mother. Progesterone is one hormone that may contribute to this by inducing a blocking protein. Other hormones that play a role, such as oestrogen and chorionic gonadotrophin, are believed to increase activity of regulatory T and B cells from the immune system. Chorionic gonadotrophin is a hormone that initiates the stress response and is also believed to prevent the mother's tissues from rejecting the foetus.

Key question ➲ p. 16

22 Oxytocin initiates and strengthens the contractions of the smooth muscle of the uterus, which is responsible for pushing the foetus along the birth canal. Many other hormones, such as oestriol, make the uterus more sensitive to oxytocin.

Key questions ➲ p. 19

23 Generally, asexual plant propagation is a natural method of cloning designed to produce large quantities of offspring that are genetically identical and therefore have predictable qualities. It may be limited to seasonal growth; for example, bulbs may only propagate at certain times of the year. Sexual techniques rely on an understanding of male and female cones or flower structure, methods of pollination, germination and assessment of offspring. Sexual methods produce diversity but if combined with artificial selection may eventually reduce overall genetic diversity. Sexual methods can be used to deliberately produce hybrids.

24 The Green Revolution greatly improved production, making agriculture more profitable. Because it involved the use of fertilisers, pesticides, irrigation, fossil fuels and mechanisation it was socially inequitable and had negative impacts on the environment.

25 Increased food production (reducing malnutrition and starvation) and increased food diversity are beneficial impacts of the application of scientific knowledge on plant and animal reproduction in agriculture. Potential negative impacts include disempowerment and reduction in social equity and damage to the environment because of more large-scale agriculture and the potential release of genetically modified organisms into the wild if controls are not managed.

Key questions ➲ p. 21

26 Selective breeding is the process of selecting breeding stock with desirable characteristics so that, over generations, the population changes to possess more of the sought-after characteristics. Because it uses sexual breeding methods there is always variation in offspring, making the improvement process gradual. The use of inbreeding (or line breeding) can result in the appearance of undesirable qualities, such as occurs in some dog breeds, or the loss of reproductive capacity in meat chickens.

27 These are technologies designed to assist reproduction by improving fertility, shortening generation intervals and producing more predictable offspring through practices such as cloning. Reproductive technologies are used to enhance the effectiveness of selective breeding.

28 Cloning can occur when an embryo is deliberately split to produce multiple embryos with identical genes. Unless this is performed on cultured embryos that have been genetically screened the offspring may or may not have the desirable attributes. Somatic cell nuclear transfer (SCNT or NT) is the transfer of the nuclear material of one female into an ovum that has had its nucleus removed and is then implanted into a surrogate mother. This theoretically produces offspring with predictable qualities but success is not always complete as the progeny often experience health issues.

29 Artificial insemination (AI) magnifies the reproductive potential of male livestock with desirable qualities by collecting, storing and transporting semen to impregnate females at the appropriate time. It can also be used to overcome quarantine issues that would restrict movement of animals. Attempts have been made to make AI more efficient through the application of hormones to female herds to induce ovulation synchronously. AI can also be used for in-vitro fertilisation along with embryo culture and transfer so that the beneficial qualities of female livestock can be more widely promulgated. Attempts have been made to sex segregate spermatozoa so more female progeny result in industries such as breeding dairy cattle.

30 Aquaculture relies on a supply of juveniles that result from the collection of milt (semen) and ova and their use for external fertilisation. Some brood stock fish have been administered sex hormones in their food, resulting in sex reversal. Sex reversed females (functional males) will only produce females when bred with normal female ova. Triploidy has been deliberately created in some species, either to make them sterile if they escape into wild populations (as in the case of genetically engineered salmon) or to produce stocks of one sex which have demonstrated faster growth rates and feeding efficiency than mixed populations.

HSC EXAM-TYPE QUESTIONS

Objective-response questions

1 **A**. This is a disadvantage, particularly in small or widely dispersed populations. **B** is wrong because that is an advantage. **C** is wrong because identical offspring applies to asexual reproduction. **D** is incorrect because males cannot produce offspring, whereas all adults can reproduce when asexual reproduction is possible.

2 **B**. This option correctly states the difference in genetic variation arising from the two methods. **A** is wrong as it reverses the correct statement. **C** and **D** are wrong because asexual and sexual reproduction occur in both plants and animals (even though asexual reproduction is more common in plants).

3 **B**. External fertilisation relies on water for the gametes to find each other for fertilisation to occur (hence **A** is wrong). **C** and **D** are incorrect because the statements apply to external fertilisation where larger numbers of gametes are produced and offspring are independent of parents.

4 **D**. Implantation occurs in placental mammals in the endometrium of the uterus. It does not occur in the oviduct (**B**) or in all mammals (**C**), making those options wrong. **A** is incorrect as implantation occurs after fertilisation.

5 **A**. Animal reproduction technologies include crossbreeding, IVF and AI, though crossbreeding also occurs in plants. **B**, **C** and **D** are wrong as they include examples of the application of scientific understanding to plant breeding.

Short-answer questions

6 **EM** You need to identify similarities in reproductive methods and outline the differences.

Both bacteria and protists can reproduce by binary fission (splitting in two). (✓) Bacteria lack a nucleus and have one circular chromosome but protists have a nucleus and many linear chromosomes, so while there are similarities in the steps of binary fission there are also many differences. Binary fission in protists is based around a mitotic cell division but, depending on the species, the cell itself may split vertically or horizontally (see Figure 1.13). The orientation of the splitting in protists ensures that both offspring have all necessary features such as a flagellum or oral groove. Bacteria cannot divide by mitosis. (✓)

Fungi such as yeast and some protists can reproduce by budding, in which the division is unequal. (✓) Fungi reproduce by spores, which has advantages for dispersal and resilience, whereas protists and bacteria do not. (✓)

7 **EM** For this question you need to combine your understanding of the role of hormones and their interactions throughout pregnancy and birth.

Progesterone maintains the uterus and its lining for initial nourishment and retention of the embryo. (✓) Oestrogens sensitise and prepare the uterine muscles for birth. (✓) Oxytocin causes the uterine muscles to contract during birth. (✓) Relaxin helps relax muscles, ligaments and tendons during pregnancy and birth to allow easier passage of the foetus. (✓) Chorionic gonadotrophin stimulates and enhances production and the effects of the other hormones through pregnancy and may play a role in preventing immune rejection of foetus. (✓)

8 **EM** This question requires you to outline the structure and function of a flower and then relate this to agricultural practices.

The stamen consists of the filament (stalk) and anther (which produces the male gamete called pollen). (✓) The carpel consists of the style (stalk) and stigma (sticky ending which accepts pollen), along with the female gamete, called the ovule-producing ovary. (✓)

Crossbreeding relies on pollen from one flower pollinating (and hence fertilising) an ovule from a different flower. This may occur naturally if the stamen and carpel mature at different times. (✓)

Crossbreeding in plant agriculture may require cytoplasmic male sterilisation so self-fertilisation is avoided. Alternatively, the anthers may be manually removed and pollen from another plant manually introduced. (✓)

9 **EM** You need to draw on your understanding of sustainability from Year 11 and apply this to new understandings on reproduction and applications in agriculture.

To be more sustainable production needs to continue or increase without the added use of resources or production of waste. (✓) Artificial selection in both plant and animal agriculture may result in more productive or disease-resistant and drought-tolerant plants and animals, which would contribute to sustainability. (✓) Asexual propagation of plants can be used to rapidly increase numbers of food crops that, if distributed fairly, could benefit developing countries and reduce starvation. (✓)

Seed banks are a form of insurance for maintaining sources of genetic diversity. In the future, this may help with breeding plants in a changing environment. The use of AI and stem cell technology allows breeding from elite male animals over a wider number of females. Semen can be transported and stored, allowing for wide distribution and potential future use. (✓) Understanding the factors influencing the success of external fertilisation in fish and crustaceans assists the success of aquaculture. Use of polyploidy, such as triploid sterile fish, may safeguard against the spreading of experimental animals and genes into wild populations. (✓) There is also the benefit of reducing short-term pressures on wild fish populations, allowing them to possibly re-establish. (✓)

10 **EM** You need to clearly identify three issues and discuss points associated with welfare issues or potential impacts. The answer below is a sample response but however you approach the question it is essential that you justify your argument.

Reduction of generation intervals by breeding animals at a younger age, more frequently or for longer with multiple pregnancies may have harmful impacts on female animals and their offspring. Procedures such as Juvenile In-Vitro Fertilisation and Embryo Transfer (JIVET) involve administration of hormones and procedures such as oocyte collection, which are invasive and may have detrimental impacts that are not yet fully understood. (✓) This aligns with the general increased concern for animal welfare, including the movement to free-range livestock and organic production. Care must be taken to fully research impacts on animals before adopting procedures such as JIVET. (✓)

In aquaculture, hormonal sex-reversal in brood stock fish has potential issues of hormones entering food webs and aquatic environments as well as possible issues for the health of the fish. (✓) The potential benefits of reducing starvation and providing access to farmers with dams or ponds in developing countries may, for some people, justify the potential harm to the fish. (✓)

Recent use of cloning in animal production has raised concerns about food quality and security as well as animal welfare. Generally, cloning has been applied to brood stock as opposed to animals used for human consumption. The greatest concern is the combination of cloning with genetic engineering, as this may result in new substances in food, leading to possible allergic reactions.(✓) The cost of some of these procedures is high and may lead to inequity and inaccessibility for those people in greatest need. Justification that the potential harm to animals is worth it because of the benefits to humans will continue to be an area of much debate and contention. (✓)

CHAPTER 2 CELL REPLICATION

INQUIRY QUESTION:

How important is it for genetic material to be replicated exactly?

Successful reproduction ensures the continuity of a species but just how does this occur at a cellular level? In Year 11 you learnt about cell structure and function, and how this related to organism structure and function. In this chapter you will review the cellular processes involved in ensuring genetic continuity and the processes involved in production of genetically diverse offspring. Together these allow the continuity of a species.

1 The processes involved in cell replication

›› Students **model** the processes involved in cell replication, including but not limited to mitosis and meiosis, DNA replication using the Watson and Crick DNA model, including nucleotide composition, pairing and bonding.

model: a conceptual or physical representation of something; models can simplify a concept to promote understanding or communication; a model can also be mathematical; models can be used to make predictions

Mitosis

- In Chapter 1 the difference between the processes and products of asexual and sexual reproduction was emphasised. The reason that there is such a difference is related to the differences that occur at the cellular level of each process.
- When modelling a process it is important to combine aspects of structure with function. The following descriptions of the processes of mitosis and meiosis are designed to help you review how you constructed models to represent the processes. Your model will include sequential explanations of the cellular or molecular structures involved.
- Mitosis is cell replication in which one cell becomes two identical cells. These daughter cells are somatic (body) cells that allow growth, repair, asexual reproduction and regeneration in multicellular organisms. They allow asexual reproduction in unicellular organisms.
- Mitosis and **cytokinesis** are the processes involved in cell replication in eukaryotes.
- Not all cells complete mitosis. Some cells, such as human red blood cells, do not have a nucleus so their renewal is dependent upon bone marrow cells dividing and differentiating into red blood cells. Alveoli cells in the human lung are replaced about every eight days. Muscle cells are multinucleate but special satellite cells divide and differentiate, fusing together in response to muscle injury. Even cells that can replicate spend most of their life performing roles other than mitosis.
- The stage in between cell divisions is called interphase or, misleadingly, the resting phase. Towards the end of interphase the cell prepares for mitosis by replicating the DNA and the centrosome (see Table 2.1).
- There are a number of structures both within the nucleus and in the cell that play an important role in cell division. These structures are outlined below, before the steps of mitosis are reviewed in detail.

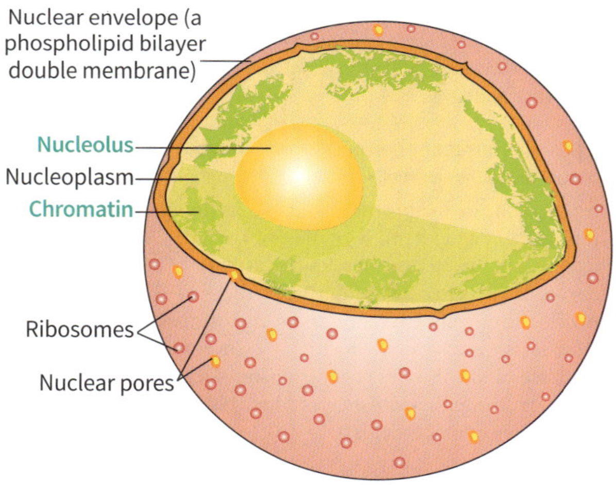

Figure 2.1 Structure of cell nucleus in eukaryotes
Source: adapted from image by Mariana Ruiz, numbered version by Felipe Fontoura, (public domain), via Wikimedia Commons

cytokinesis: the final stage of cell replication, involving the redistribution of cytoplasm between daughter cells
nucleolus: the dense structure within the nucleus associated with protein synthesis
chromatin: a complex of DNA bound up and protected by proteins

Table 2.1 Cellular structures involved in the nucleus or cell replication and their roles

Structure and description	Role
centriole: a cylinder of nine sets of triplets of microtubules held together by proteins; these are capable of replicating microtubules are hollow, straw-like tubes about 25 nm in diameter made up to two types of tubulin proteins that can rearrange, grow or shrink	involved in the organisation of the spindle in animal cells and are believed to make nuclear division more efficient not essential for mitosis because mitosis occurs in plants which lack centrioles centrioles are also associated with cilia and flagella that beat in order to help some animal cells move
microfilaments: smallest units (7 nm or 7 x 10^{-9}m diameter) in the cytoskeleton; made up of two actin proteins wound together; rapidly assemble and disassemble	forms tracks for the movement of the protein myosin so that a ring made of actin and myosin can help pinch the two daughter cells apart (cytokinesis) in the final stages of cell replication
centrosome: a pair of centrioles at right angles to each other that occur in animal cells and are surrounded by a matrix	microtubule organising centres that help the formation of spindle fibres that grow between each centrosome two centrosomes form before cell division the matrix is responsible for the formation of microtubules that make up the spindle fibres plants use microtubules in a different way and do not possess centrosomes
spindle: made up of microtubules; also called mitotic spindles and spindle apparatus	helps draw and push the sister chromatids apart in plant cells spindle fibres are capable of self-organisation
nucleolus: dense region of **RNA** and proteins within the nucleus	organises and synthesises ribosomes
nucleoplasm: aqueous jelly-like solution of salts and enzymes containing organic molecules in suspension	cushions and protects nuclear contents and provides a transport medium
nuclear envelope: phospholipid bilayer double membrane	provides shape and regulates the flow of materials through the nuclear pores
chromatin: the complex of DNA and histones (protein) that occur in different forms or levels of condensation during the cell cycle; when relatively uncondensed the DNA wraps around a histone in a structure often likened to beads on a string	during mitosis, chromatin condenses to form chromosomes that are visible through a light microscope; this protects the DNA as the chromosomes move around the cell
chromosomes: the dense form of chromatin that forms during cell division chromatids are two identical copies of a chromosome that are formed at the beginning of cell division and joined together through most of the cell division at a structure called a centromere chromosomes occur in pairs (the diploid or 2n number)	chromosomes carry the genetic material DNA so that it can be passed on intact to daughter cells chromatids are the copies of chromosomes that ensure full transmission of genetic material when a pair of chromatids separate they are called chromosomes
centromere: the structure of DNA and protein where the two chromatids are joined during mitosis and where the spindle attaches	centromeres are places of attachment for each chromatid and also for the spindle holds chromatids together and, when attached to the spindle, ensures separation of chromatid pairs, one to each daughter cell

centriole: a complex of microtubules, protein and a matrix that form part of the centrosome

centrosome: two centrioles combined at right angles and surrounded by a matrix that helps produce and organise the spindle fibres

spindles: lengths of microtubules that lengthen or contract to move the chromatids to different poles of the new cells

RNA: ribonucleic acid; a macromolecule related to DNA that plays an important part in protein manufacture in the cell

➔ Chromatin consists of DNA and histones (like 'beads on a thread') when not condensed tightly into a chromosome.

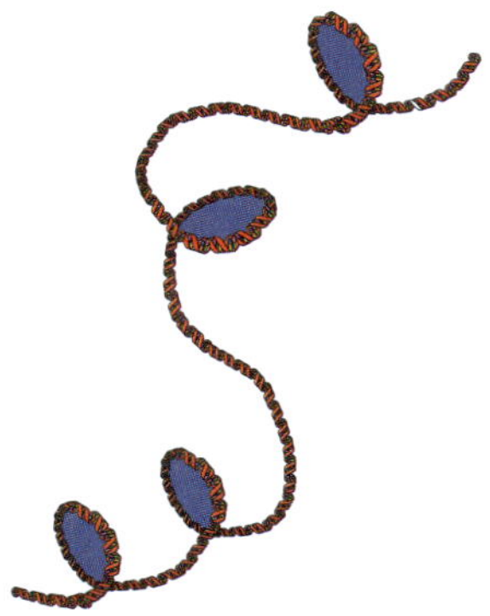

Figure 2.2 Make-up of chromatin in the least dense (uncondensed) form

Source: adapted from image by Richard Wheeler, [CC-BY-SA-3.0 (http://creativecommons.org/licenses/by-sa/3.0/)], via Wikimedia Commons

Table 2.2 Stages in mitosis and cytokinesis

Stages	Processes occurring
	prophase: the nucleolus disappears and the spindle fibres start to form as the centrosomes move apart; chromatin condenses into chromosomes that already consist of a pair of chromatids joined by a centromere
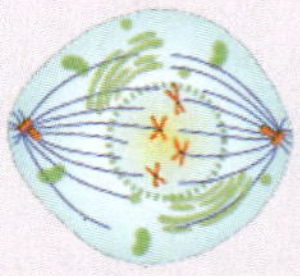	late prophase: the chromosomes are fully condensed and the spindle has begun to capture and organise the chromosomes; the nuclear envelope breaks down the spindle also radiates away from the centrosomes towards the cell membrane to form a structure called an aster
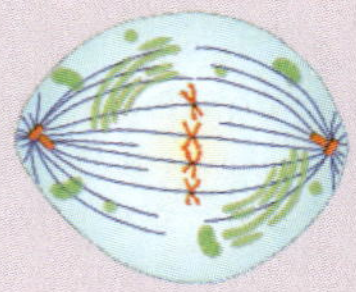	metaphase: chromosomes are all captured by the spindle and brought into alignment along the central axis of the cell (this is sometimes called the equator or metaphase plate though it is not an actual structure) each chromatid pair is attached at the centromere to different spindle fibres so they will move in opposite directions, ensuring that each daughter cell receives a full complement of chromosomes a checking process, called spindle checkpoint, occurs before division can continue
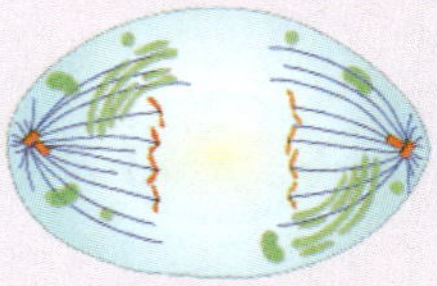	anaphase: the chromatids separate and move to opposite poles of the cell microtubules not attached to chromosomes extend to lengthen the cell, pushing the poles further apart
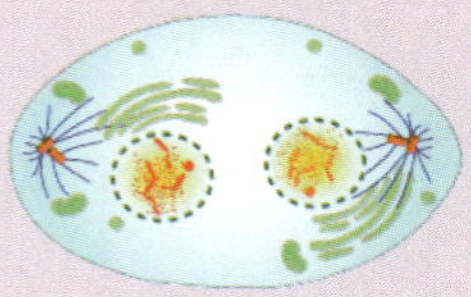	telophase: two new nuclear envelopes form, the chromosomes return to their more diffuse form, the nucleolus reappears and the spindle fibres disappear
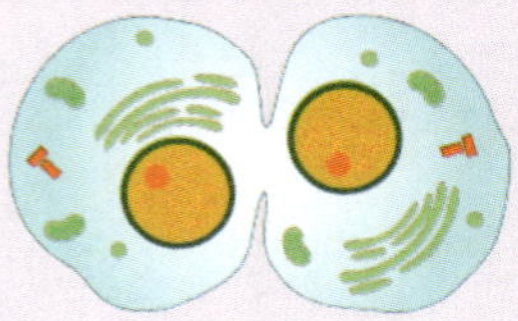	cytokinesis: the final division of cytoplasm occurs in animals this results from actin filaments working with myosin to pinch off the two cells plant cells grow a cell plate that separates the two daughter cells entirely aspects of cytokinesis may commence in late anaphase but complete after telophase

Source: images by LadyofHats (own work, public domain), via Wikimedia Commons

➔ KEY QUESTIONS

1 **Distinguish between mitosis and cytokinesis.**

2 **You are given a cork board, drawing pins, pipe cleaners (chenille stems or sticks), elastic bands, paper clips, drinking straws and other craft items and asked to model mitosis from late prophase to telophase and cytokinesis. Construct a table to suggest the three items you would select from this list (or your own selection) to represent various structures and the reasons for your choices.**

3 **Describe a stage of mitosis that is essential to ensure cell replication is exact.**

Answers ➲ p. 36

Meiosis

- ➔ Meiosis is cell division in which one diploid cell becomes four unique haploid cells. These cells have the potential to become gametes or germ cells in sexual reproduction.
- ➔ Strictly speaking, meiosis is a form of cell division that is not replication because the products of the division are not identical to the parent cell or each other.
- ➔ A key process during the interphase before both mitosis and meiosis is that the genetic material (DNA) is replicated. You will review DNA and its replication later in the chapter. While **DNA replication** is not foolproof and there is the occasional mistake, in the vast majority of cell divisions the DNA is carried intact to the daughter cells.
- ➔ Meiosis is regarded as having two main stages. The first is called the reduction division as it results in haploid cells; that is, cells with half the normal number of chromosomes. The second involves separation of the chromatids (see Table 2.3).

Table 2.3 Stages in meiosis and cytokinesis

Stages	Processes occurring
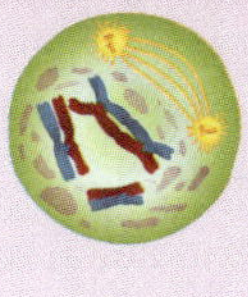	prophase: nuclear membrane breaks down; chromosomes condense as pairs of chromatids but also **homologous chromosomes** (i.e. chromosome pairs) come together adjacent chromatids from the pairs may engage in crossing over, in which exchange of chromosome fragments occurs at structures called chiasmata centrosomes form the spindle
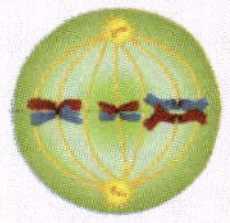	metaphase: homologous pairs of chromosomes move to the metaphase plate but each member of a pair attaches to opposite spindles; this lining up is random
	anaphase: random segregation of homologous pairs occurs to opposite poles
	telophase and cytokinesis: the nuclear membrane may reform around the two daughter haploid nuclei; chromosomes may diffuse into chromatin and the cytoplasm contents are split by cytokinesis the second part of a meiosis division is similar to mitosis but is occurring in haploid cells; there is no DNA replication in human females this stage of meiosis occurs in the female foetus and is held in suspension until after puberty, when at monthly intervals meiosis in one follicle resumes meiosis only completes after fertilisation (see Chapter 1)
	prophase II: nuclear membrane disappears and chromosomes condense centrosomes move apart and spindle forms and begins to capture chromosomes; sister chromatids are captured by spindle from opposite poles
	metaphase II: the haploid set of chromosomes moves to the metaphase plate as all chromosomes have been captured by the spindle
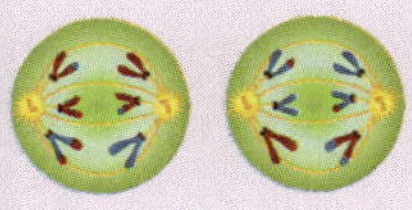	anaphase II: chromatids move apart to opposite poles
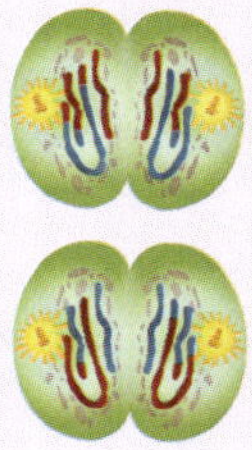	telophase II and cytokinesis: nuclear membrane reforms around the nuclei, the cytoplasm separates between cells

DNA replication: the process by which a DNA macromolecule makes an identical copy of itself

homologous chromosomes: a pair of chromosomes with the same genes, one originating from the mother and the other from the father in sexual reproduction

→ KEY QUESTIONS

4 Explain what is meant by the statement that meiosis is really two separate cell divisions.

5 What is a key difference to consider if constructing a physical three-dimensional model of meiosis compared with a physical three-dimensional model of mitosis?

Answers ➲ p. 36

Cell replication in prokaryotes

- Prokaryotic cells are much smaller than eukaryotic cells and their cell division (binary fission) is simpler.
- All essential genes are carried on the double stranded DNA in a circular chromosome that is often represented as an intertwined or tangled loop in the nucleoid, a central region of the cell that is part of the cytoplasm.
- The DNA in the circular chromosome is packaged by a binding protein (not histones as in eukaryotes).
- Prokaryotes also have double stranded DNA in a small circular plasmid. Plasmids can replicate autonomously. They are involved in some host-to-host transfer across different species of bacteria and carry genes that assist survival; for example, antibiotic resistance, virulence as a pathogen or the ability to kill other bacteria.
- Chapter 3 compares the forms of DNA in eukaryotes and prokaryotes.

Table 2.4 Cell division in prokaryotes

Stages	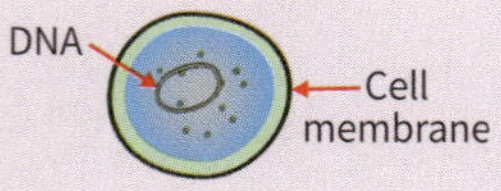	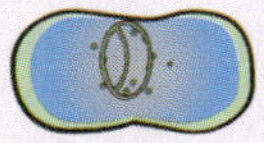 DNA duplicates	 Cell membrane indents	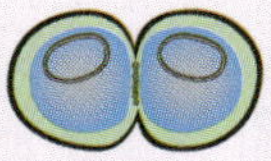 Septum forms	 Cell divides, two new cells form
Processes occurring	replication of the circular chromosome and its DNA starts from one point of origin and proceeds in opposite directions it proceeds around the circle until two new loops form; each loop contains one of the original strands of DNA and one new strand	the two circular chromosomes attach to different parts of the cell membrane	the cell elongates (lengthens), separating the two circular chromosomes	a cleavage furrow and septum begin to form across the divide between the two chromosomes	the cell pinches in two, resulting in two identical daughter cells formed in a process called binary fission

DNA replication using the Watson and Crick DNA model

The discovery of DNA

- While the discovery of the structure of DNA is attributed to Watson and Crick many scientists contributed to understanding its role and composition, and provided evidence for its structure. The following table summarises major findings in the discovery of the role and structure of DNA.

Table 2.5 Key events in the discovery of the structure and function of DNA

Researcher and date	Discovery
Gregor Mendel (mid-1800s)	discovered the genetic basis of inheritance in pea plants after his work was 'rediscovered' in the early 1900s it triggered the search for the cellular and molecular basis of inheritance
Theodor Boveri (1902) working with sea urchins and Walter Sutton (1903) working with grasshopper testes (independently)	that a full set of chromosomes was necessary for normal development and that Mendel's results can be explained if genes are carried on chromosomes
Thomas Morgan (1910)	discovered sex-linked inheritance such as white-eye colour in fruit flies; this can only be explained if genes are carried on chromosomes
Johann Friedrich Miescher (1869) and his pupil Richard Altmann (who named it nucleic acid in 1889)	Meischer discovered and isolated from nuclei of white blood cells a pure sample of DNA that he called nuclein; later postulated that the substance could be involved in heredity Altmann detected the acidic nature of DNA, changing the name to nucleic acid
Frederick Griffith (1928)	showed that a characteristic (e.g. virulence) could be passed on as a chemical from one bacterium to another (i.e. the bacterium was transformed)
Phoebus Levene (1929)	identified that the basic unit of a DNA macromolecule **polymer** was a **nucleotide** consisting of a sugar, phosphate and base identified the four **bases**: adenine (A), thymine (T), cytosine(C) and guanine (G)
George Beadle and Edward Tatum (1941)	showed experimentally that genes carried 'instructions' to build protein molecules
Oswald Avery (1944)	demonstrated that the bacterial transformation that Griffith discovered could be brought about if DNA was transferred; concluded that DNA is the molecule of inheritance

Researcher and date	Discovery
Edwin Chargaff (1940s)	found that there is always the same amount of adenine as thymine and cytosine as guanine; this is called Chargaff's rule
Rosalind Franklin and Maurice Wilkins (1953)	X-rayed crystallised forms of DNA to show a helical arrangement
James Watson and Francis Crick (1953)	built a model of DNA that showed the arrangement of nucleotides which explained ability of DNA to replicate

The Watson and Crick DNA model

➔ X-rays of DNA crystals completed by Rosalind Franklin and Maurice Wilkins provided the data that James Watson and Francis Crick needed to correct their model-building to demonstrate the structure of DNA as a **double helix** spiral of two sugar phosphate backbones running in opposite directions, with the internal bases held together in their matching pairs by hydrogen bonds. Watson and Crick constructed physical models and achieved their goal when the model accurately incorporated all data and was useful in predicting how DNA replication could occur.

polymer: a large molecule (macromolecule) made up of many similar smaller molecules (monomers) linked together in a chain-like manner
nucleotide: a monomer of DNA that consists of phosphate group joined to a deoxyribose sugar which is attached to a nitrogenous base of adenine, thymine, cytosine or guanine
base: also called a nucleobase or nitrogenous base, bases are the nitrogen-containing molecular units that make up a nucleic acid; the term *base* is historical (from acid-base reactions) and has little influence on the chemistry of DNA
double helix: a pair of parallel spirals intertwined around a common axis

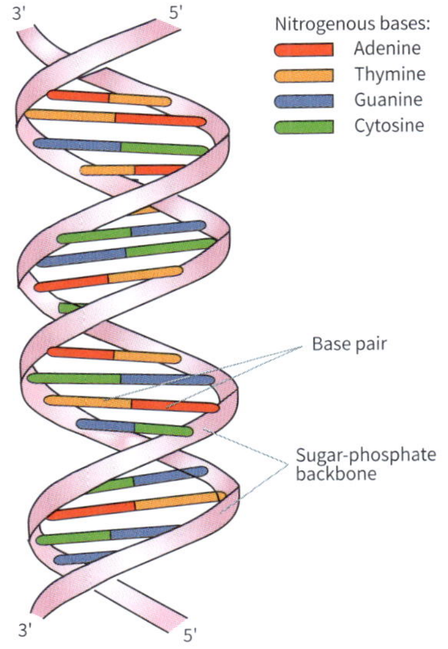

Figure 2.3 The double helical structure of DNA
Source: image by OpenStax College, via Wikimedia Commons

➔ KEY QUESTIONS

6 **Identify the components of a DNA nucleotide.**

7 **Outline the structure of the sugar phosphate backbone of DNA.**

8 **What is meant by the term *complementary strands of DNA*?**

9 **Give one example of how the discovery of the structure of DNA involved many branches of science.**

Answers ➲ p. 36

➔ The following descriptions and diagrams of the structure of DNA and its replication are designed to help you review how you constructed models to represent DNA and DNA replication. As with your models of mitosis and meiosis, your representation will include sequential explanations of the cellular or molecular structures involved. You will have to accurately convey relationships such as the paired relationship between adenine and thymine, and between cytosine and guanine bases. You may have used technology to create an animation, used drawings or physically constructed models. In some cases, modelling may involve simplification of complex processes so the use of keys, labels and/or an evaluation of the model can show a higher level of understanding or the ability of your model to help you make predictions.

Nucleotide composition

Nucleotides

➔ The nucleotide monomer in DNA consists of one deoxyribose sugar, one phosphate group and one nitrogenous base. There are four different bases so therefore there are four different nucleotides.

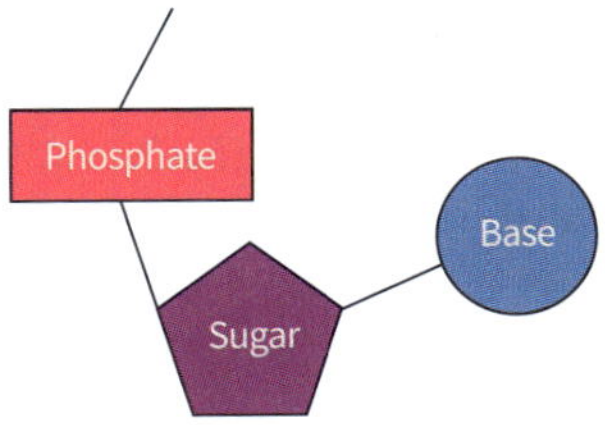

Figure 2.4 Simple representation of a nucleotide

➔ Deoxyribose sugar: sugars are carbohydrates consisting of carbon, hydrogen and oxygen atoms. The carbon atoms are the 'backbone' of the sugar with the five carbon atoms numbered as shown on Figure 2.5. Ribose and deoxyribose are sugars that form in a five-membered ring containing four carbon and one oxygen atom between carbon 1 and carbon 4. In deoxyribose sugar, carbon 2 does not have an oxygen atom attached as occurs in ribose sugar (deoxyribose sugar could be made from ribose sugar by removing one oxygen atom). Carbon 5 is not part of the ring and is joined to the phosphate group. Carbon 3 is also joined to a phosphate group and carbon 1 is joined to the nitrogen atom that is part of the nitrogenous base.

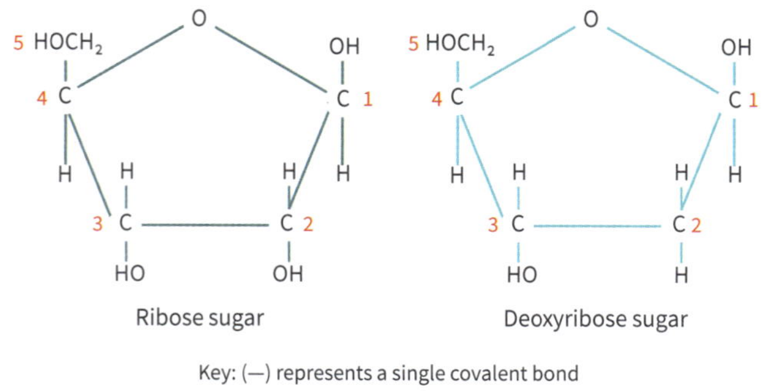

Figure 2.5 Deoxyribose sugar

Phosphate group

➔ The phosphate group is derived from phosphoric acid. Acids contain hydrogen ions (H^+) or protons which they release. Phosphoric acid is capable of donating three hydrogen ions. Two are released when phosphoric acid forms bonds with the sugar carbons. The phosphate group is the reason DNA is considered to be acidic.

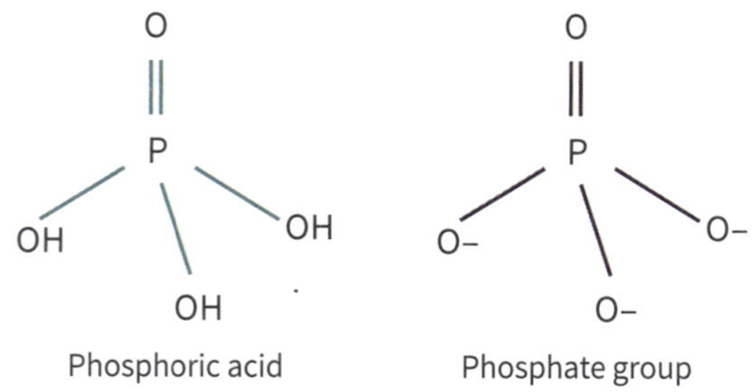

Figure 2.6 Phosphoric acid and a phosphate group

Sugar-phosphate backbone

➔ Alternating sugar and phosphate groups make up the sugar-phosphate backbone. Each sugar is attached to a base. Figure 2.7 shows the arrangement. One end of a strand has a phosphate group that is attached to carbon 5 of the deoxyribose sugar so it is called the 5' end. The other end of the sugar phosphate backbone has a hydroxyl group (i.e. an oxygen and a hydrogen) joined to the carbon 3 so it is called the 3' end.

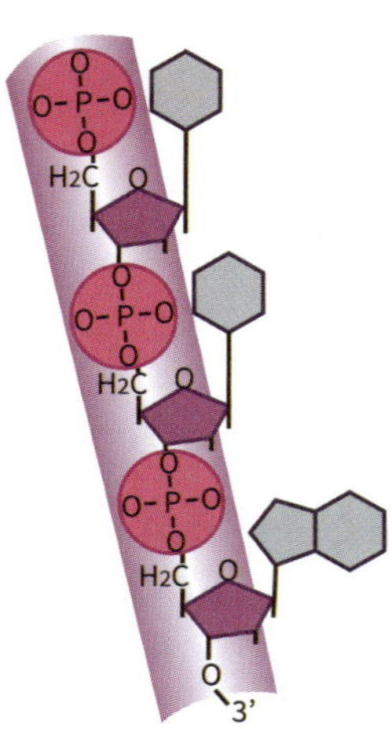

Figure 2.7 The sugar-phosphate backbone

Nitrogenous bases

➔ The nitrogenous bases that make up DNA exist in two forms: purines and pyrimidines, as outlined below.
- The pyrimidine bases contain four carbon and two nitrogen atoms in a hexagonal ring with other functional groups (see Figure 2.8).
- The purine bases contain two carbon-nitrogen ring bases totalling five carbon atoms and four nitrogen with functional groups attached (see Figure 2.8).

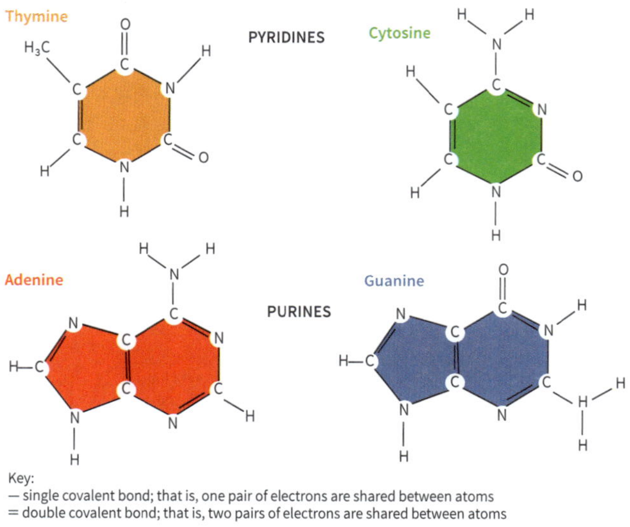

Figure 2.8 The nitrogenous bases

Pairing and bonding

➔ Table 2.5 identifies Chargaff's rule, which was a key discovery in solving the puzzle of the structure of DNA. He found that there was always the same number of adenine as thymine bases and the same number of cytosine as guanine bases. The explanation was that the bases occurred in pairs. The eventual explanation for how this occurred was that DNA consisted of two strands running in opposite directions, with the sugar-phosphate backbone on the outside and the bases internally arranged so that both A and T, and C and G, were matched.

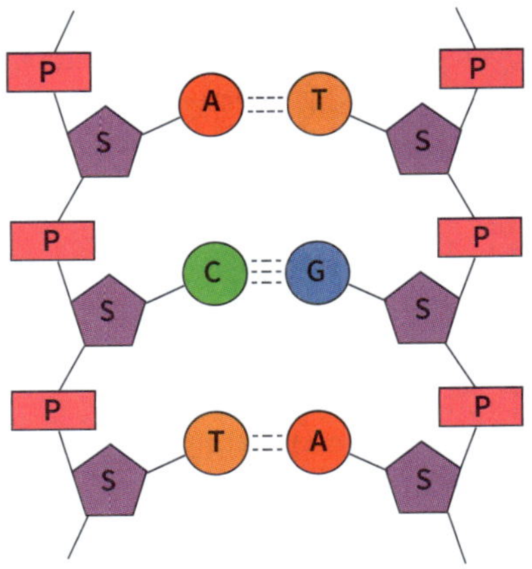

Key: dotted lines represent hydrogen bonds

Figure 2.9 The double-strand of DNA showing three matching base pairs

- The stability of the DNA macromolecule is due to the large number of hydrogen bonds that results from the pairing of A with T and C with G. These bonds being weaker than covalent bonds are the key to the processes of DNA replication and the role of DNA in protein synthesis (see Chapter 3). The actual shape of each base within the larger configuration of the macromolecule is such that only adenine and thymine align so that two hydrogen bonds between hydrogen and nitrogen can form and cytosine and guanine align so that three hydrogen bonds between oxygen and hydrogen and nitrogen and hydrogen can form.

hydrogen bonds: weak bonds that form between oxygen and hydrogen atoms and nitrogen and hydrogen atoms when they occur in adjacent molecules and have electrostatic differences

The processes involved in cell replication: DNA replication

- DNA replication occurs during the synthesis (S) phase of interphase; that is, well before the cell starts dividing either by mitosis or meiosis.
- The following steps summarise DNA replication (see Figures 2.10 and 2.11).

1 DNA double helix uncoils and the strands separate when an enzyme breaks the hydrogen bonds between the matching base pairs.

2 The strand terminating in a 5′ phosphate group is called the leading strand; that is, the strand that started uncoiling from its 3′ hydroxyl end. The opposite strand is called the lagging strand.

3 Units called **primers** bind with the DNA to start the copying. Primers bind with the DNA of the lagging strand at short intervals but only at the 3′ end of the leading strand.

4 Nucleotides slot into place in a continuous process along the leading strand (new strands can only build themselves from the 5′ carbon towards the 3′ end). Hydrogen bonding between base pairs helps to ensure faithful replication of the base sequence. Because the lagging strand is opposite, short segments of DNA slot into the segments between primers.

5 An enzyme removes all RNA primers and the vacant slots are filled with the appropriate nucleotides. The fragments associated with the lagging strand are fused together. The new strands are checked and any errors corrected by enzymes. The tip of DNA contains repeating segments of DNA called telomeres. A special enzyme catalyses the synthesis of the telomeres.

6 Two identical molecules of DNA have been formed. Each contains one strand from the DNA that acted as the template and one new strand. They each re-coil back into the double helix shape.

primer: a short segment of RNA (ribonucleic acid) that binds with a DNA single strand to start the replication process

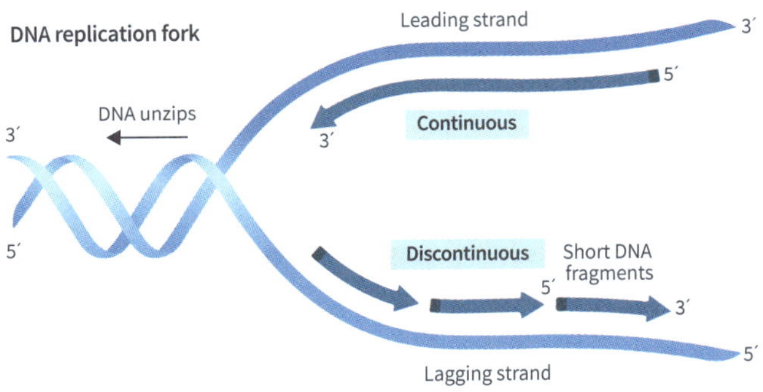

Figure 2.10 DNA replication overview

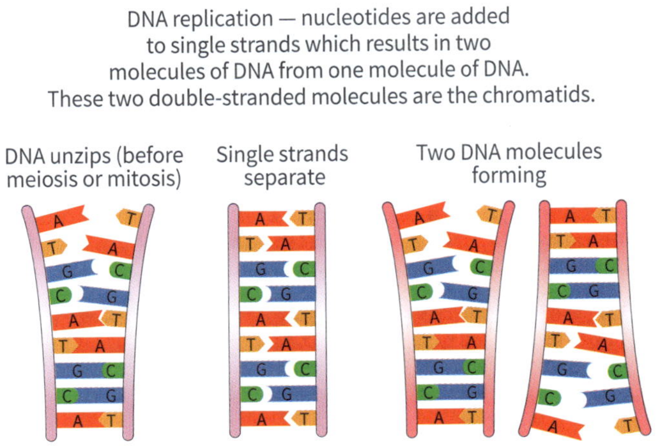

Figure 2.11 Summary of DNA replication showing the base pairs

→ KEY QUESTIONS

10 **Why is the hydrogen bonding between base pairs important for successful DNA replication?**

11 **Describe the role of enzymes in DNA replication.**

12 **Use your experience of modelling cell replication, DNA structure and its replication, and the work of Watson and Crick on DNA, to identify the benefits of modelling in biology.**

Answers ➲ p. 36

2 The effect of the cell replication processes on the continuity of species

» Students assess the effect of the cell replication processes on the continuity of species.

→ Continuity of a species depends on effective reproduction. Reproduction for most multicellular organisms depends on individuals reaching maturity and either sexually or asexually producing offspring. These processes depend on cell replication.

→ Continuity of a species in a changing environment also relies on diversity within the species. This natural variation is the basis of natural selection, which you learnt about in Year 11.This diversity depends on the cellular processes of meiosis and fertilisation.

→ During the life of a multicellular organism, cells replicate to provide growth along with repairing and regenerating cells to maintain function. These processes, and asexual reproduction in general, rely on the integrity of mitosis to ensure faithful copies of cells and their DNA. Key processes in mitosis that ensure continuity include:

- DNA replication, including the 'proofing' and repairing of mistakes
- chromosome duplication to form two chromatids, joined by a centromere
- chromatids faithfully separating into different daughter cells
- cytokinesis ensuring both daughter cells are viable by containing sufficient cytoplasm to function.

→ Sexual reproduction relies on the transmission of genes that are essential for normal functioning. Meiosis is the key cell division essential for production of gametes. Continuity of a species relies on the faithful replication of DNA prior to meiosis but also the continuation of processes that promote some variation in offspring. In meiosis these processes include:

- crossing over of paired chromatids, resulting in exchange of genes and DNA and new combinations of genes along a chromosome
- random and independent segregation of chromosomes
- halving of the number of chromosomes in the first reduction division to allow fertilisation of gametes.

→ Genetic diversity in sexual reproduction is also the result of fertilisation of gametes from different parents. You will review genetic variation in detail in Chapter 4.

→ KEY QUESTION

13 **Compare the roles of mitosis and meiosis in their contribution to the continuity of a species.**

Answer ➲ p. 36

CHAPTER SYLLABUS CHECKLIST

Are you able to answer these questions from the syllabus for this chapter? Tick each question as you go through the checklist if you are able to answer it. If you cannot answer a question, turn to the relevant page in the study guide to find the answer. For NESA key word meanings, go to www.educationstandards.nsw.edu.au and search 'key words'.

FOR A COMPLETE UNDERSTANDING OF THIS TOPIC:		PAGE NO.	✓
1	Can I model the processes involved in cell replication, including but not limited to mitosis and meiosis?	26–30	
2	Can I model the processes involved in cell replication, including but not limited to DNA replication using the Watson and Crick DNA model, including nucleotide composition, pairing and bonding?	30–34	
3	Can I assess the effect of the cell replication processes on the continuity of species?	34	

HSC EXAM-TYPE QUESTIONS

Objective-response questions

(1 mark each)

1 **What are the two types of cell replication?**

A mitosis and meiosis
B prokaryote and eukaryote
C mitosis and cytokinesis
D asexual and sexual

2 **What are the two key events in meiosis that generate varied daughter cells?**

A crossing over and random segregation
B cytokinesis and DNA replication
C centrosome duplication and spindle formation
D chromosome attachment to cell membrane and cell elongation

3 **Which of the below is an accurate statement about hydrogen bonds?**

A Hydrogen bonds form between hydrogen and oxygen and hydrogen and nitrogen.
B Hydrogen bonds only occur between hydrogen atoms.
C Hydrogen bonds are stronger than other types of bonds between atoms.
D Hydrogen bonds do not occur in organic macromolecules.

4 **The diagram below is a symbolic representation of a portion of DNA.**

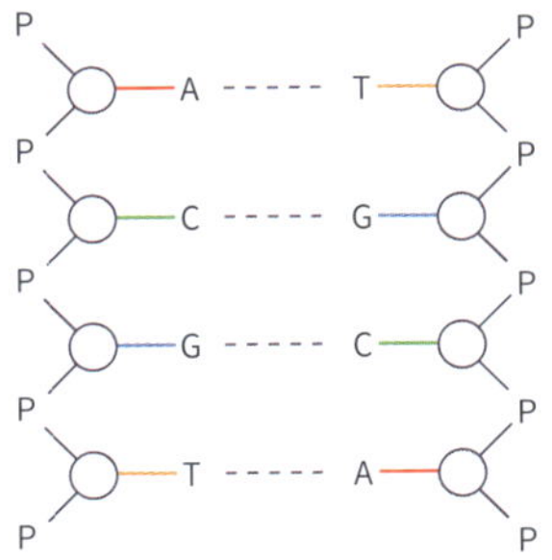

Figure 2.12 Portion of DNA

Which of the following statements accurately describes a key element of DNA structure?

A A always pairs with T.
B The P stands for a purine base.
C A, T, C and G represent nucleotides.
D The - - - - - line represents unbreakable bonds.

5 **Which statement best describes homologous chromosomes?**

A the diploid set of chromosomes
B the haploid set of chromosomes
C all the chromosomes derived from the maternal parent
D chromosome pairs that carry the same genes in the same location

Short-answer questions

6 **The diagram below shows a specific process in meiosis.**

Figure 2.13 Process in meiosis

Critically analyse this process as the source of variation in gametes. (4 marks)

7 **Outline the key steps in DNA replication.** (5 marks)

8 **Some people might argue that meiosis is not cell replication. Discuss.** (4 marks)

9 **Identify three similarities between mitosis and meiosis.** (3 marks)

10 **Using a wide variety of examples from other organisms, explain how cell replication contributes to continuity of a species.** (8 marks)

KEY QUESTIONS

Key questions ➲ p. 28

1 Mitosis is the replication and division of nuclear material into two nuclei; cytokinesis is the division of the remaining cytoplasm so that two complete daughter cells are the product. Cytokinesis follows mitosis.

2 **Table A2.1** Items useful for modelling

Item	Structure and reason
pipe cleaners	chromatid: long and thin; can be joined together easily at a 'centromere'
cut elastic bands	spindle fibres: can be joined to chromatids at centromeres and contracted to pull them apart
drawing pins	centrosomes: can be pinned to cork board to hold the elastic bands (spindle fibres) so they contract, pulling chromatids to opposite poles

3 Metaphase is the stage where chromatids line up along the metaphase plate so that each daughter chromatid goes to opposite poles of the cell and therefore the resulting nuclei have a full diploid set of chromosomes. As the chromosomes carry the DNA, each daughter cell must contain the exact full copy by having the full set of chromosomes. Alternatively you could argue that interphase is a critical stage (even though technically it is the stage in between cell divisions because in this stage the DNA is replicated and the chromatin starts to condense into chromatids).

Key questions ➲ p. 29

4 Meiosis consists of the first reduction division in which the chromosome number is halved; that is, changes from a diploid to two haploid cells. This is followed by a division of the chromatids in a similar manner to a mitosis cell division. In human females the two stages of division may occur up to about 50 years apart.

5 The key events that are unique to meiosis include the pairing of homologous chromosomes, crossing over and random segregation of homologous pairs. To show these in a model you need materials such as plasticine, which can be broken and re-joined so that chromatids can be seen to be a new combination. Homologous chromosome pairs can be represented as having the same length and point of attachment at the centromere. One member of each homologous pair has a maternal origin and the other paternal, so there needs to be a distinction. This could be achieved by either using different colours for paternal and maternal, or different materials such as plasticine and play dough. Random segregation of each homologous pair could be achieved by tossing a coin.

Key questions ➲ p. 31

6 A nucleotide consists of deoxyribose sugar joined to a phosphate group and, at the other side, joined to a nitrogenous base of adenine, thymine, cytosine or guanine.

7 The deoxyribose sugar is made up of five carbon atoms. It is joined to a phosphate group on its carbon 5 and to another phosphate group on its carbon 3 (carbon 1 is joined to the nitrogenous base). The sugar phosphate backbone is a zigzag of sugars and phosphate groups running in a sequence (usually considered from the phosphate on carbon 5 to the carbon 3 of the sugar).

8 The nitrogenous bases match faithfully in pairs across two strands of DNA running in opposite directions. The adenine always pairs with the thymine and the cytosine always pairs with the guanine. If the strands are separated the vacancies can be filled only by nucleotides that complement the one on the existing strand.

9 Physics was used to produce X-ray images of crystals of DNA. Crystallisation of DNA is a chemical procedure. The actual role of DNA is biochemistry.

Key questions ➲ p. 34

10 Hydrogen bonding only occurs between hydrogen atoms and highly electronegative atoms such as nitrogen and oxygen. The shape of the bases means that when cytosine comes near guanine, three hydrogen bonds will form and hold them together. When adenine and thymine come together the atoms align so that two hydrogen bonds form.

11 Enzymes are involved in unzipping the DNA strands, placing the RNA primers in position and forming bonds to link sugar and phosphates together to make the backbone and nucleotide segments in position on the lagging strand. Enzymes are involved in checking and correcting any pairing errors.

12 A model is a representation that describes, simplifies, clarifies or provides an explanation of the workings, structure or relationships within an object, system or idea. The use of physical models to show the stages involved in mitosis and meiosis helped to simplify, describe and clarify the steps, as well as highlighting the similarities and differences between the two types of cell division. Modelling the structure of DNA requires the use of simplified shapes that fit together to represent the pairing of A and T and G and C. This modelling was useful in explaining how in replication of DNA the two strands, once separated, could have complementary strands built so that replication was faithful. Watson and Crick used data provided by other workers (e.g. Franklin) as well as understanding the components of DNA to determine the structure. Their model was exploratory but explained all the facts about DNA and gave a mechanism for replication. Essentially, they used modelling as an investigative tool.

13 Mitosis is essential for asexual reproduction, which ensures rapid increase in populations of organisms in order to establish species in a relatively constant environment. Meiosis produces diverse offspring but requires more energy and the union of gametes. It allows variation so that species are more likely to continue surviving in a changing environment.

HSC EXAM-TYPE QUESTIONS

Objective-response questions

1 **A.** This is correct because cell replication occurs through mitosis or meiosis. **B** is wrong because they are different types of cells that may replicate differently because of different basic cellular structures. **C** is incorrect because they are the two parts of cell division for growth, repair and asexual reproduction. **D** is wrong because this is two types of reproduction, not cell replication.

2 **A.** This is correct as both crossing over and random segregation result in chromosomes carrying different genes in each daughter cell. **B** is wrong because the processes do not generate any variation. **C** is incorrect because they are processes that facilitate movement of chromosomes around the dividing cell. **D** is wrong because both processes refer to binary fission or cell division in prokaryotes.

3 **A.** Hydrogen bonds only occur between hydrogen and electronegative oxygen or nitrogen atoms. **B** is incorrect because covalent bonds hold hydrogen atoms together. **C** is incorrect because hydrogen bonds are weaker than covalent (and ionic bonds). **D** is wrong because hydrogen bonds are important in DNA and proteins.

4 **A.** The fact that A pairs with T and G pairs with C provides the codes that are the key to DNA being a genetic code. **B** is wrong because P stands for phosphate groups. **C** is wrong because A, T, C and G represent bases. A nucleotide is a base, a sugar and a phosphate unit together. **D** is wrong because the dotted line represents weak bonds that are broken when DNA replicates or is transcribed.

5 **D.** This option accurately defines homologous chromosomes. **A** is incorrect because the diploid is all homologous pairs. **B** is wrong because it contains only one member of each homologous pair. **C** is incorrect because all the chromosomes from the maternal parent are in effect a haploid set.

Short-answer questions

6 **EM** You need to identify the diagram as the process of crossing over and explain its significance in combination with other processes that contribute to diversity in meiosis.

The process illustrated in the diagram is crossing over. (✓) This occurs in meiosis and is one major source of variation as it results in chromosomes with new and unique combinations of genes. (✓) During meiosis there is also the random segregation of the members of homologous pairs of chromosomes. The resultant haploid cells contain unique combinations of chromosomes. The combination of one haploid cell with another (when fertilisation occurs) results in even greater variation. (✓) All three processes contribute to variation. (✓)

7 **EM** This is a recall question but it is not simple as DNA replication is a complex process. Use point form and identify the key processes before elaborating on what is occurring.

- DNA strands unzip: hydrogen bonds between matching base pairs are broken. (✓)
- Leading strand has RNA primer slot in at the end and commences filling with complementary nucleotides in a continuous manner. (✓)
- Lagging strand has numerous RNA primers slot in and fills in fragments of nucleotide sections from carbon 5 to carbon 3. (✓)
- Enzymes check for errors and make corrections if necessary. (✓)
- Two identical double helix strands. (✓)

8 **EM** This question challenges you to critically analyse what is meant by replication and identify the reasons why meiosis does not produce identical daughter cells.

If replication is defined as 'producing identical copies', meiosis cannot be classified as cell replication. One cell produces four daughter cells with half the number of chromosomes as the parent. (✓) In addition, crossing over results in new combinations of chromosome sections so this is another reason why it is not replication of the original cell. (✓)

Meiosis does produce gamete cells capable of fertilisation and development into a new organism and in doing so ensures the continuity of a species. (✓) The quantity and composition of the DNA may differ from the original cell but there is still a viable cell membrane, nucleus and cytoplasm, all of which allow for the continuity of life. (✓)

9 **EM** This question asks you to compare the processes in order to name the steps that are the same. Point form is an ideal way to answer.

- DNA replicates before both mitosis and meiosis. (✓)
- In mitosis and the second division of meiosis the chromatids separate to opposite poles to become chromosomes. (✓)
- Both processes rely on the duplication of centrosomes and formation of a spindle for attachment and separation of chromatids. (✓)

10 **EM** This question requires that you use the understanding of cell replication you gained in Chapter 1 as well as this chapter. One way is to explain how cell replication contributes to the continuity of a species and then link that to organisms such as bacteria, protists, fungi, plants and animals.

Asexual reproduction depends on mitosis, which replicates cells. This occurs in a wide variety of organisms; for example, binary fission in protists and binary fission and spore production in fungi. (✓)

Bacteria are prokaryotic organisms so their binary fission is different. While it still involves the duplication of DNA and its separation into separate daughter cells, it is more prone to changes, hence the rapid evolution of antibiotic resistance in some species. (✓)

Plant species that reproduce asexually by bulbs, corms, stolons, tubers and rhizomes also benefit in population expansion because of mitosis cell divisions. (✓)

Animals such as starfish that regenerate from fragments also benefit from mitosis cell divisions for maintaining population numbers. (✓) Animals such as aphids, which reproduce through parthenogenesis, proliferate because of mitosis cell divisions. (✓)

Sexual reproduction in flowering plants, gymnosperms, mammals, reptiles, fish, amphibians, birds, insects, crustaceans and arachnids increases or maintains populations and also produces varied offspring which respond through natural selection to environmental change. (✓) Without meiosis, gametes could not be produced so sexual reproduction would not occur. (✓) Meiosis allows the production of gametes with variation but also allows fertilisation so that genetic input from two parents increases the diversity, which also contributes to continuity of the species. (✓)

CHAPTER 3 DNA AND POLYPEPTIDE SYNTHESIS

INQUIRY QUESTION:

Why is polypeptide synthesis important?

In Chapter 2 you reviewed the importance of DNA and its replication, a process which enables it to carry genetic information from cell to cell and generation to generation. In this chapter you will review your understanding of how DNA functions as the genetic code through the synthesis of **polypeptides**.

1 Representations to model and compare the forms in which DNA exists in eukaryotes and prokaryotes

» Students construct appropriate representations to model and compare the forms in which DNA exists in eukaryotes and prokaryotes.

Modelling forms of DNA

- Modelling often involves simplification so that the key processes are understood and not overwhelmed by detail.
- Select the information that is most important to the different forms in which DNA exists in eukaryotes and prokaryotes You need to have constructed models to show:
  - no fundamental difference in the DNA but which shows that it occurs in different locations and arrangements and in different levels of combination with proteins in the two types of cells
  - the nuclear membrane that encloses the DNA in eukaryotic cells but not in prokaryotic cells.
- Your models need to have been produced in the form of labelled diagrams, physical models or an animation.
- In Year 11 you learnt about the differences between eukaryotes and prokaryotes. Table 3.1 summarises the differences, particularly in the arrangement and location of DNA.

Forms of DNA in eukaryotes and prokaryotes

Table 3.1 Key differences in the forms of DNA in prokaryotic and eukaryotic cells

Prokaryote	Eukaryote
most genetic material is contained as supercoiled DNA in a single circular chromosome found in a region of the cell called the nucleoid small amounts of associated protein help form some loops or bends, needed to create the supercoiling the level of supercoiling can change during the cell cycle extrachromosomal DNA occurs in small circular plasmids that can self-replicate	most DNA is found in the nucleus as chromatin, a macromolecule complex with protein and RNA the protein that packages the DNA is a histone the chromatin can change its form during the cell cycle DNA is more densely packed and found on linear chromosomes during mitosis. **extrachromosomal DNA** and ribosomes occur in some organelles (e.g. mitochondria and chloroplasts) mitochondrial DNA is an open or closed circle; chloroplast DNA is circular; these organelles can replicate and use their DNA to make at least some of their own polypeptides
DNA replication starts from a single origin and progresses in both directions (See Table 2.4) in some species DNA replication is continuous the DNA starts replicating from the origin while DNA replication at the other end of the circular loop is completing; this contributes to rapid cell divisions; e.g. *Escherichia coli* takes 40 minutes to replicate the DNA but only 20 minutes to divide	nuclear DNA replication occurs in interphase before mitosis proceeds human cells grown in culture will divide in 24 hours

polypeptide: a component of proteins, polypeptide is a polymer comprised of up to 20 different amino acid monomers; each polypeptide chain has a specific sequence of amino acids

extrachromosomal DNA: DNA that is not contained within the nucleus of eukaryotic cells or the long circular chromosome of prokaryotic cells

Prokaryote	Eukaryote
about 12% of DNA is non-coding (does not code for polypeptides)	over 98% of DNA is **non-coding**; non-coding sections may have a function in the regulation of cell processes
there are between 160 000 and 12.2 million base pairs in the cell DNA	humans have about 2.9 thousand million base pairs in DNA
the circular chromosome has one copy of each gene	homologous pairs of linear chromosomes contain two copies of each gene
DNA replication is fast: about 2000 base pairs per second	DNA replication is relatively slow: about 100 base pairs per second
no nucleus nor nuclear membrane but have nucleoid region where chromosomal DNA is located	DNA found in membrane-bound nucleus
no organelles; plasmids found in cytoplasm	organelles are membrane-bound; some contain DNA and can replicate themselves

non-coding DNA: sections of DNA that are not used directly for polypeptide synthesis; because their role was not understood, non-coding sections of DNA were once called 'junk' DNA

➔ An important similarity between prokaryotes and eukaryotes is that they both have ribosomes. DNA impacts through the manufacture of polypeptides by using ribosomes. However, the ribosomes of eukaryotes and prokaryotes are different. This difference is exploited by some antibiotics, which disable prokaryote ribosomes while leaving eukaryote ribosomes intact.

➔ KEY QUESTION

1 **Compare the structures where DNA is stored in prokaryotic and eukaryotic cells.**

Answer ➲ p. 50

2 Modelling the process of polypeptide synthesis

» Students model the process of polypeptide synthesis, including transcription and translation; assessing the importance of mRNA and tRNA in transcription and translation; analysing the function and importance of polypeptide synthesis; and assessing how genes and environment affect phenotypic expression.

Polypeptide synthesis

➔ In your study of polypeptide synthesis you will have constructed models that demonstrate a sequential explanation. You may have investigated and evaluated a number of existing models. You will also have distinguished between an illustrated lecture and an animated model. You can find examples of this on You Tube at www.youtube.com/watch?v=J3HVVi2k2No.

➔ In modelling polypeptide synthesis you will have shown interactions between different types of nucleic acids. You have already modelled DNA, including DNA replication, and modelled the structure of nucleotides and their arrangement, and compared the various forms DNA exists in within eukaryote and prokaryote cells.

➔ In Chapter 2 you reviewed types of cell division. The nucleolus is a region in the nucleus of eukaryotic cells which contains the nucleic acid RNA. RNA plays a large part in protein synthesis and will be part of your model.

Genes

➔ **Genes** carry the inherited information to give organisms their characteristics and enable them to function. In Year 11 you learnt about processes such as cellular differentiation that enables the development of tissues, organs and organ systems in multicellular organisms. These very complex processes occur because certain genes 'turn on' and become active while other genes are inactive. When a gene is 'turned on' and becomes active it is called **gene expression**.

➔ A major part of gene expression is polypeptide synthesis (see Figure 3.1).The actual processes in which a gene is turned on or off is called **gene regulation**. A growing field in biology is **epigenetics**, which investigates the mechanisms of gene regulation. Before reviewing the interaction between genes, environment and gene expression, we need to review the process of polypeptide synthesis. For more information see epialliance.org.au/what-is-epi/.

gene: the structural and functional unit of inheritance; a sequence of DNA carried on chromosomes in eukaryotes and processed to produce a polypeptide
gene expression: the total process between a gene and a functional gene product, including polypeptide synthesis and subsequent modification to result in a functioning protein
gene regulation: the process in which a gene is turned on or off; genetic research shows that genes may be activated or deactivated (called gene silencing) in many ways
epigenetics: the study of changes in inheritance patterns, such as changes in gene regulation, that are not a result of changes in genes or DNA

Polypeptides

- A polypeptide is a long chain of **amino acids**, which make up proteins. **Peptide bonds** join amino acids into the chain.
- Proteins are large, complex molecules containing one or more polypeptide chains. Proteins are critical to cell structure and function. You will review the structure and function of proteins later in this chapter (see pages 44–46).
- In studying this topic it is important to look at some of the groundbreaking experiments that led to understanding the process of polypeptide synthesis. Below is a summary of some of this work.
  - Early work on inheritance patterns was determined by Gregor Mendel (1822–84) in his work on crossing varieties of peas. Later work on inheritance patterns involved breeding fruit flies with different characteristics. You will review inheritance patterns in Chapter 4.
  - Up until 1941 the common belief was that genes controlled trivial factors such as bristle patterns in fruit flies or the height of pea plants. It was believed that some 'mysterious' processes in the cytoplasm determined the basic biochemistry of the cell.
  - In 1941 Beadle and Tatum experimentally demonstrated that genes coded for enzymes. At the time genes were believed to consist of protein or nucleoprotein. In 1944 Avery showed that genes are made of DNA.
  - Beadle's previous work had been on **metabolic pathways** of pigment synthesis in the eyes of fruit flies. Beadle and Tatum investigated strains of the haploid bread mould (*Neurospora crassa*) that had a short life cycle and could be cultured in a simple medium. They exposed the mould spores to X-rays and screened for defective genes. The irradiated mould could not grow on the simple growth medium so they added different supplements to achieve growth. They found that the defective genes had stopped the production of the amino acid arginine. They added various precursors of arginine and found that for each defective gene there was a non-functional enzyme. They concluded that a particular gene produced a specific enzyme.
  - Understanding of polypeptide synthesis has since been refined to the 'one gene, one polypeptide' hypothesis.

amino acid: a simple organic compound of carbon, hydrogen, oxygen and nitrogen; it is a monomer in polypeptides and contains an amine group (NH_2), a carboxyl group (COOH) and a side chain (R); the side chain may have other elements that influence the chemical properties

peptide bond: a covalent bond (electron-sharing bond) formed between the carbon in a carboxyl group of one amino acid with the nitrogen in the amine group of the adjacent amino acid

CHECKPOINT

Ribonucleic acid (RNA)

- Before reviewing transcription and translation and the role of mRNA and tRNA it is important to check your understanding of RNA (ribonucleic acid).
- As with DNA, RNA is a nucleic acid made up of nucleotides. The main differences are:
  - the sugar in RNA is ribose, not deoxyribose; ribose has an extra oxygen atom on carbon 2 of the ring
  - RNA has four nitrogenous bases but the base uracil substitutes for the thymine base of DNA; if a strand of RNA formed from a DNA template, adenine on the DNA would pair with uracil on the RNA; all other bases are the same as DNA bases and pairing
  - RNA can exist in many forms and roles and is found with protein complexes such as ribonucleoprotein.
- Table 3.2 below summarises the main forms of RNA in polypeptide synthesis, along with their roles.

metabolic pathways: sequences of reaction steps; each step depends on a specific enzyme and when combined they result in the formation of a specific product from reactant(s)

Table 3.2 Summary of the main forms and roles of RNA in polypeptide synthesis

Form of RNA	Description and role
messenger RNA (mRNA)	single strand of RNA that complements the base sequence of the DNA which is the template triplets of bases are codons that link to a specific amino acid mRNA is bound to proteins in a complex called messenger RNP (ribonucleoprotein) throughout its processing and transport from the nucleus transfers the genetic code from DNA in the nucleus to the ribosome in the cytoplasm
transfer RNA (tRNA)	this convoluted and relatively short segment of RNA contains a bond to an amino acid and, at the other end, an anticodon triplet of bases that slots into mRNA at the ribosome 'reads' the genetic code to insert amino acids into the polypeptide chain
ribosomal RNA (rRNA)	forms a complex with protein to make up the large and small ribosomal subunits that come together during translation and then disband holds mRNA in place for the insertion of tRNA and acts like a conveyer to assist elongation of the polypeptide chain

Transcription

- The following diagrams and descriptions are intended to help you review the processes involved in **transcription** and **translation**.

transcription: the process in which the DNA genetic code is converted into mRNA code and carried to the cytoplasm

translation: the final steps in the synthesis of polypeptides; occurs on the ribosome and converts the code carried by mRNA into a specific sequence of amino acids in a polypeptide chain

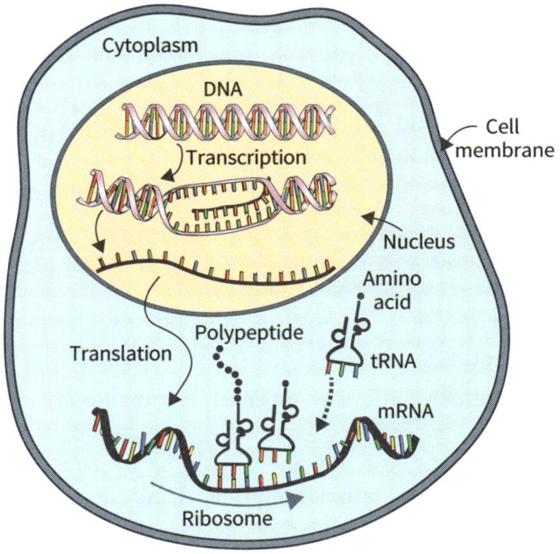

Figure 3.1 Summary of polypeptide synthesis

Source: image adapted from National Human Genome Research Institute, (public domain), via Wikimedia Commons

- DNA is transcribed and then translated into a polypeptide chain. Both these steps involve forms of RNA that carry a sequence of triplets of nitrogenous bases.
- Enzymes play a critical role in facilitating the process and ensuring the accuracy of the conversion of the genetic code, as stored on the DNA, to functional polypeptides and hence proteins.
- The entire process in eukaryotic cells is particularly complex because the genetic code includes start or promoter regions, stop messages, and sections that are deleted in a form of editing.
- As shown in Figure 3.2, the first process in polypeptide synthesis is transcription of DNA into messenger RNA (mRNA). In everyday language, if something is transcribed the meaning remains unchanged but it may be converted from speech to written text, from handwritten to typewritten, and so on. The transcription process is how the genetic code stored in DNA is copied into another form, mRNA, a single-stranded nucleotide polymer.
- Transcription is a rapid process that can occur in very quick succession when polypeptides are needed.

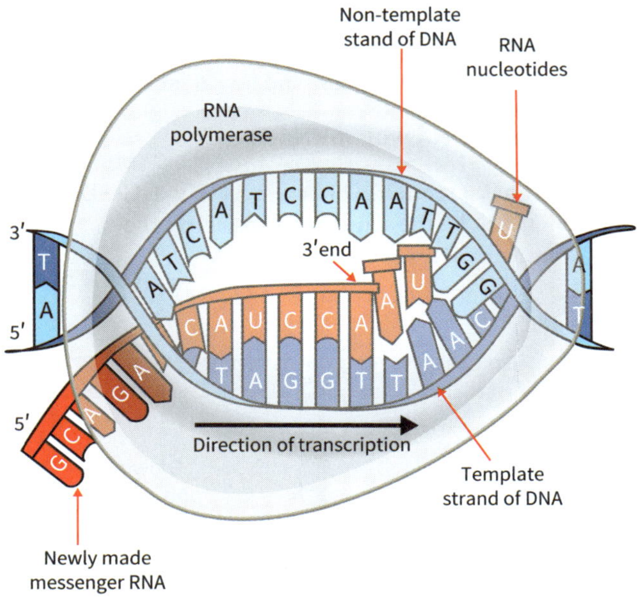

Figure 3.2 Key steps in the transcription of DNA into mRNA

- DNA in the nucleus carries the genetic code. DNA for the gene that is being activated unzips. The starting point of a coding gene, called a promoter, is then identified and one strand acts as the template for the manufacture of one strand of mRNA. This is then processed and carried out through the nuclear pore into the cytoplasm.
- Enzymes such as RNA polymerase assist the manufacture of mRNA by unwinding the DNA, moving the RNA nucleotides into position, and promoting attachment and elongation.
- Each triplet of bases on the mRNA is called a **codon**. Start and stop codons control the part of the DNA template that is 'read' in the process of transcription. A common start codon in eukaryotic mRNA is the base sequence adenine, uracil and guanine. A stop codon at the end of the mRNA will stop the translation process.
- In eukaryotic cells an 'editing' or 'splicing' process occurs to ensure that only the coding template is translated into messenger mRNA. The mRNA is capped and has a tail added to protect it. These additional steps are not shown in Figure 3.2.
- mRNA leaves the nucleus through the pores and enters the cytoplasm. There it attaches to a small subunit of ribosome.
- In Chapters 9 and 10 you will review the role of viruses in disease. Some viruses, such as HIV, are retroviruses. Their genome is single-stranded RNA and, when they enter a host cell, they use a reverse process to transcription to produce DNA that is then inserted into the host's genome.

codon: a triplet of bases on the mRNA that pair up with a triplet of bases on the tRNA (called an anticodon)

→ KEY QUESTIONS

2 Explain the term ***template*** and how it applies to DNA.

3 **Compare DNA with mRNA.**

4 **DNA has a sequence of bases: ATGCCGGGTAC. What will be the sequence of bases on the complementary section of mRNA?**

Answers ➲ p. 50

Translation

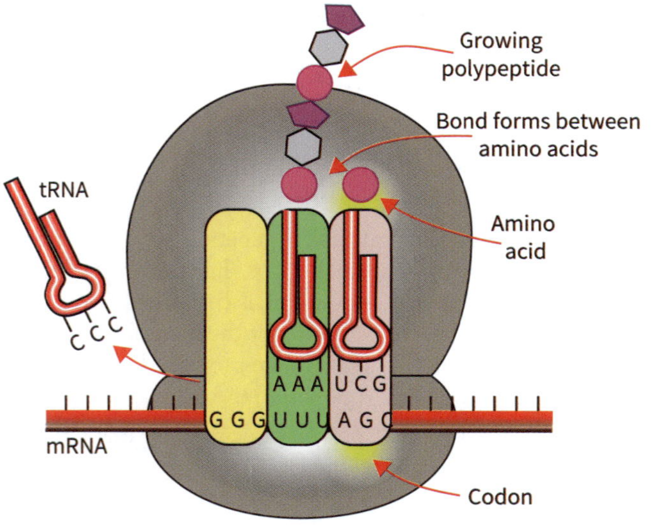

Figure 3.3 Translation (elongation stage)

- The process of translation occurs on the ribosomes. It uses mRNA codons as the template for coding for a sequence of amino acids in a polypeptide chain. tRNA anticodons carry the appropriate amino acid into place. Figure 3.3 shows the mRNA codon UUU matching the tRNA anticodon AAA. Translation is often divided into three stages: initiation, elongation (as shown in Figure 3.3) and termination. These stages are outlined below.
  - Initiation: once in the cytoplasm, a 'start' codon on the mRNA binds between the small and the large ribosome subunits. The start codon on the first part of the mRNA contains the AUG (adenine, guanine, uracil) bases. This matches with the tRNA anticodon UAC (uracil, adenine, cytosine) that carries the amino acid methionine. Methionine is the start of the polypeptide chain being formed.
  - Elongation: the ribosome acts a little like a conveyer belt, moving the mRNA along, and the large subunit has grooves into which fit two tRNA units that each carry an amino acid. When two tRNA units are in place a peptide bond forms, lengthening the polypeptide chain.
  - Termination: a terminal sequence of the mRNA stops the elongation process and releases a polypeptide chain.

The importance of mRNA and tRNA in transcription and translation

- In polypeptide synthesis DNA remains within the nucleus. The ribosomes (the polypeptide making structures) are in the cytoplasm. mRNA is essential to carry the genetic code stored in DNA from the nucleus to the ribosomes.
- Three nucleotides in mRNA form a codon that match with an anticodon on the tRNA.
- mRNA has a start codon, AUG, which initiates most polypeptide translations.
- Three different stop codons in mRNA (UAG, UAA, UGA) each terminate the polypeptide chain as there is no tRNA anticodon that matches.
- tRNA carries a specific amino acid to be joined into the polypeptide chain.
- The correct matching of the codon on the mRNA to the anticodon on the tRNA is critical for the correct sequence of amino acids in the polypeptide chain.

Analysing the function and importance of polypeptide synthesis

- Inheritance and continuity of a species depends on polypeptide synthesis.
- Polypeptide synthesis results in the production of components necessary for protein formation. Proteins are vital to the structure and functioning of cells. For information on examples of proteins and their role and importance see Table 3.5 on page 45.
- Polypeptide synthesis provides the means for the inherited qualities that are carried in the DNA to be mobilised into cellular structure and function.
- Polypeptide synthesis is not the end of the story. The creation of active, functional proteins may involve numerous biochemical processes, including the combination of a number of polypeptides or splitting of polypetide chains, along with the transport or secretion of products. If these fail, cell death or disease may result.

→ KEY QUESTIONS

5 **How important is mRNA in transcription and translation?**

6 **How important is tRNA in transcription and translation?**

7 **What is the function of polypeptide synthesis?**

8 **What is the importance of polypeptide synthesis?**

Answers ➲ p. 50

Assessing how genes and environment affect phenotypic expression

- The total base composition and arrangement of the DNA in an organism is called its **genome**.
- The genetic composition is called the **genotype**. As diploid organisms have two copies of each gene, the genotype for a particular gene is the two forms of the gene or **alleles** that are present.
- The traits of an organism are qualities such as height and eye colour. The specific observable or measurable characteristics of an organism, such as blue eye colour or tall stature, is called its **phenotype**. The phenotype is the result of the interaction of the genotype and the environment. An organism may have a genotype that should produce tall stature but malnutrition may mean the phenotype is only medium stature.
- The trait of blood group is represented in humans in the phenotypes A, B, AB or O types of blood. The genotypes for phenotype A is either AA or AO, phenotype B is either BB or BO, phenotype AB is AB and phenotype O is OO. The three forms of alleles are A, B and O. There is no evidence of any environmental impacts on blood type; therefore, this phenotype is totally inherited. You will review the genetic basis of blood types in Chapter 4.
- For a long time biologists have pondered and researched whether inheritance or environment is the main player in 'who' we are. The debate has been labelled as nature versus nurture, and it has implications for health, child rearing and education, among other factors.
- Some of the earliest studies to assess how genes and environment affect phenotypic expression were carried out on identical twins. Identical twins come from the same zygote, which splits, so they usually have the same genome. Identical twins are often raised in the same environment so early studies focused on identical twins that were separated in early life and brought up in different environments. The Minnesota Twin Family Study was carried out between 1979 and 1999. It compared identical twins with non-identical fraternal twins over time. One of the controversial claims from the Minnesota Twin Family Study was that 70% of intelligence is due to genetics and 30% to environment. For an example of modelling see learn.genetics.utah.edu/content/epigenetics/twins/.
- Table 3.3 provides more examples of the nature versus nurture debate through the study of identical twins in Australia.

genome: the complete set of DNA of an organism and the mapping of gene arrangement
genotype: the genetic make-up of an organism either for a single trait (e.g. eye colour), a set of traits or the entire set of genes of an organism; often represented as a pair of letters for a single trait, such as Tt for height in pea plants

Table 3.3 Summary of some findings and investigations reported by Twins Research Australia

Researcher(s)	Discovery
Associate Professor Jeff Craig, Murdoch Children's Research Institute, 2016	environment in the womb has a large impact on later health
Professor Brian Byrne and Dr Will Coventry, University of New England (since 1999; reported in 2014)	investigated reading ability and literacy development and found impact of teacher not as high as expected as twins in separate classrooms developed similar literacy skills as those in the same classroom
Professors Sam Berkovic and Ingrid Scheffer, the University of Melbourne, the Florey Institute of Neuroscience and Mental Health and the Austin Institute (reported in 2014)	dispelled belief that most forms of epilepsy were due to injury and demonstrated that one serious form was not a result of immunisation but was caused by a gene mutation a diet avoiding glucose was demonstrated to be an effective treatment in some cases
Perminder Sachdev and Julian Trollor at the University of New South Wales Centre for Healthy Brain Aging (since 2015)	the Older Australians Twins Study (OATS) is investigating environmental versus genetic factors that contribute to amyloid plaque development

- You learnt about the ideas of Darwin and Wallace and evolution by natural selection in Year 11. You may also have reviewed earlier ideas of the mechanism of inheritance, as suggested by Lamarck and known as inheritance of acquired characteristics. His ideas were discounted but are now being demonstrated to be at least partially correct. The scientific field of epigenetics has arisen and provided some surprising findings, two of which are listed below.
  - One early study benefitted from excellent health data gathering in the Netherlands. It tracked the children and grandchildren of women who were pregnant in 1944–45 but exposed to extreme malnutrition in an event called the Dutch Hunger Winter, caused by World War II. Women who were malnourished only during the first trimester of pregnancy gave birth to children of normal weight. However, these children later suffered from obesity and other health issues, including a high incidence of mental health problems. A similar pattern of higher obesity levels was also evident in their grandchildren, leading to the conclusion that extreme deprivation had prompted an inherited change in gene expression.

alleles: alternative forms of a gene such as A, B or O blood type alleles, or T and t as height alleles in pea plants
phenotype: the actual observed properties or functioning of an organism that results from gene expression and environmental influences

- An epigenetic experiment has been conducted with offspring from two groups of chickens exposed to different living and feeding conditions. Neither group of offspring had any contact with their parents and were exposed to identical living and feeding conditions. The offspring from the well-fed chickens had less aggressive eating behaviours and were of lower weight than the offspring from deprived chickens. Analysis showed no change in the DNA of either group so the conclusion was that changes occurred in the gene expression and this was passed on to the next generation.

➔ KEY QUESTIONS

9 **What is meant by the term *phenotypic expression*?**

10 **What is the role of genes in phenotypic expression?**

11 **What is the role of the environment in phenotypic expression?**

Answers ➲ p. 50

3 The structure and function of proteins in living things

» Students investigate the structure and function of proteins in living things.

- ➔ Ultimately the sequence of amino acids on the polypeptide chain(s) determines the three-dimensional shape of a protein.
- ➔ Proteins are essential to the structure, function and regulation of the body's cells, tissues and organs.
- ➔ Protein structure is described using four levels: primary, secondary, tertiary and quaternary.
- ➔ Proteins spontaneously fold into the three-dimensional shape that is determined by the sequence of amino acids.
- ➔ Proteins contain functional groups in the amino acids with varying chemical properties. When combined in specific sequences, the functional groups account for the broad range of protein functions.

Table 3.4 Protein structure

Structure level	Key features
primary 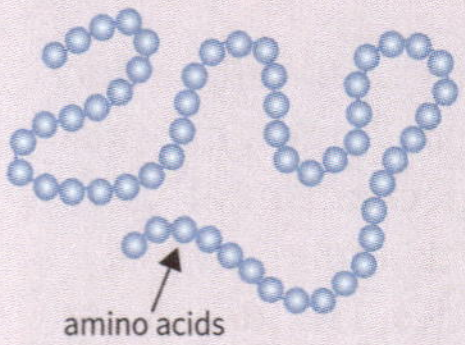	the sequence of amino acids in the polypeptide chain backbone held together by covalent peptide bonds there are 20 different amino acids that form polypeptides the end of the chain that has the unbonded amine group is called the N-terminal and the other end that has the unbonded carboxyl is called the C-terminal the amino acids methionine and cysteine contain sulfur atoms; covalent bonds may form between sulfur atoms (disulfide bonds) in these amino acids and these are also considered part of the primary structure
secondary  	each amino acid has at least one nitrogen atom and one oxygen atom that have the potential to form hydrogen bond attractions (see Chapter 2 where hydrogen bonds between DNA bases were explained) the interactions between the different amino acids in the chain may form regular shapes, such as pleated beta sheets or an alpha helix, as there are local attractions that form hydrogen bonds other regions of polypeptide chains are not regularly arranged or include turns or bends
tertiary 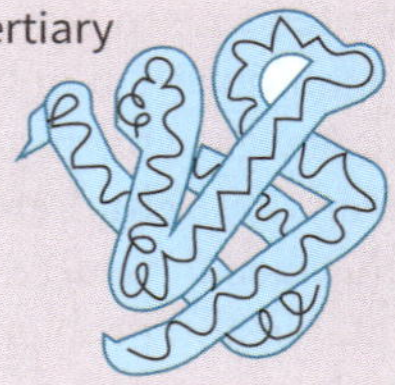	the side-chains on amino acids have differing properties so regions along a chain may be water-attracting or water-repelling these attractions and bridges across more distant amino acids, between alpha helices and beta sheets, and even sulfide bridges between the sulfur atoms, contort the polypeptide into larger specific shapes disulfide bridges are covalent bonds that form between sulfur atoms in amino acids (e.g. cysteine) and are stable in extracellular fluids
quaternary 	proteins may contain more than one polypeptide chain and the attractions between the different chains forms the quaternary structure or conformation disulfide bridges, ionic bonds, hydrogen bonds and other attraction forces hold the chains together each chain is called a subunit and subunits may be the same or different examples of types of configurations include globular proteins, which are the group of proteins that form compact, roughly spherical structures; mostly water soluble another example is fibrous proteins, which have polypeptide chains assembled in rope-like or sheet-like arrangements; these are insoluble

Source: image adapted from NHGRI courtesy of National Human Genome Research Institute, (public domain), via Wikimedia Commons

Table 3.5 Structure and function of proteins in living things

Function	Examples and structural features
catalysts (enzymes): promote and control the chemical reactions and metabolism of organisms, such as the reactions in photosynthesis, respiration, digestion, active transport, assimilation and even polypeptide synthesis may depend on cofactors or coenzymes to carry out their roles some enzymes (e.g. digestive enzymes) are secreted from the cell and processed through the action of the endoplasmic reticulum and the Golgi bodies	pepsin, lactase, catalase the shape of the active site is critical for the formation of the enzyme-substrate complex proteases such as pepsin break down polypeptides by breaking peptide bonds; in some cases they break them in the middle of the chain and in others they attack the end bonds (see Figure 3.4 for an example of structure related to function of a protease)
transport: proteins are integral to the structure and function of the cell membrane, allowing the selective entry or exit of materials proteins may be involved in active transport, in which energy is used to change the shape of the protein so it carries a substance (e.g. sodium ions) across the cell membrane, or in passive transport, such as facilitated diffusion some proteins facilitate transport in body fluids	haemoglobin (carries oxygen), cytochromes (carry electrons) haemoglobin is a globular protein that consists of subunits each of the four subunits contains a haeme molecule which contains a form of iron that can bind with one oxygen molecule serum albumin is a globular protein that carries fatty acids, some hormones and even calcium in blood; it helps maintain the osmotic pressure of blood
storage proteins: some contain molecules that hold metals that help store oxygen, or even the metal, such as iron, itself other proteins store amino acids for the growth and development of embryos, foetuses and infants	ovalbumin (amino acids), casein (globular protein that stores amino acids for the growth and development of mammalian infants) a universal intracellular iron store is the globular protein, ferritin, which contains 24 protein subunits forming a tiny cage with multiple metal–protein interactions it keeps iron in a soluble and non-toxic form; its levels are detected in blood tests for anaemia myoglobin is the disc-shaped protein made up of one polypeptide chain of 153 amino acids wound into eight alpha helices linked together these cradle a haeme molecule that stores oxygen in striated and cardiac muscles; it was the first protein whose structure was determined by X-ray crystallography (in 1958)
mechanical support: proteins provide the strength and varying levels of elasticity needed in many tissues and organs many of these proteins are extracellular	alpha keratin is a rope-like fibrous protein that is responsible for the strength of hair, feathers, hooves, horns, the skin and nails collagen is responsible for the high tensile strength of tendons, cartilage and bones, and the soft, flexible qualities of silk fibrinogen (blood clotting) is a soluble plasma glycoprotein; contains six polypeptide chains and added chains of a small number of simple sugars that becomes a long flexible protein array to form blood clots to stop bleeding
immune protection: the detection and destruction of foreign microbes (e.g. bacteria and viruses) is facilitated by the structure and function of antibody proteins	an antibody (immunoglobulin) is a large, Y-shaped protein produced mainly by plasma cells; used by the immune system (see Chapter 11 for the role of the immune system) the antibody recognises a unique molecule or antigen from the surface of the microbe because of structures on the tip of the 'Y' (see Figure 3.5)
generate movement: cells can change shape because of proteins that interact to use energy to produce force and movement	actin and myosin interact in muscles to cause contraction titin is the largest protein, containing between 27 000 and 33 000 amino acids, and is responsible for resting tension and force transmission in muscles
messengers (hormones): control of growth and differentiation some hormones are secreted from the cell into the blood and, before release, are processed through the action of the endoplasmic reticulum and the Golgi bodies	oxytocin, insulin, somatotropin the human insulin protein is composed of 51 amino acids; it has two parts: an A-chain and a B-chain, which are linked together by sulfur-sulfur bonds (see Figure 3.6) insulin binds with receptor proteins on the cell membrane, promoting changes that allow the entry of glucose, among other substances angiotensin II is a protein hormone made up of a polypeptide of only eight amino acids; its role in humans is to increase blood pressure and activate other hormones

An example of protein structure relating to function

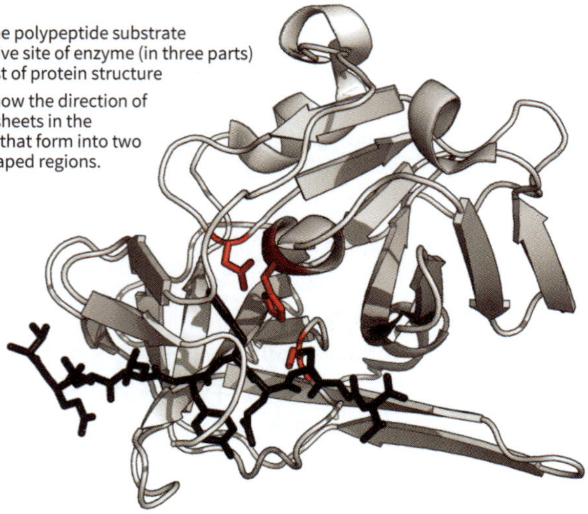

Figure 3.4 Tobacco etch virus protease

Source: image by Thomas Shafee, [CC BY 4.0 (http://creativecommons.org/licenses/by/4.0)], via Wikimedia Commons

- Tobacco etch virus infects some agricultural crops (e.g. capsicum, tomato, tobacco) as well as many weed species. The effects of the virus can be seen by discoloration of leaves and fruit, often giving an etched appearance. Its successful invasion of the host depends on it inserting its own code to take over the steps of the normal host cell's polypeptide synthesis.
- Tobacco etch virus protease is a highly specific enzyme that breaks large protein macromolecules into sections. The protease enzyme structure is shown in Figure 3.5. The active site is a tunnel-like structure into which seven particular amino acids in a chain fit and the protease enzyme cuts the sixth peptide bond of this sequence. This specificity has resulted in biochemists using this protein as a tool for reactions such as separating and purifying proteins.

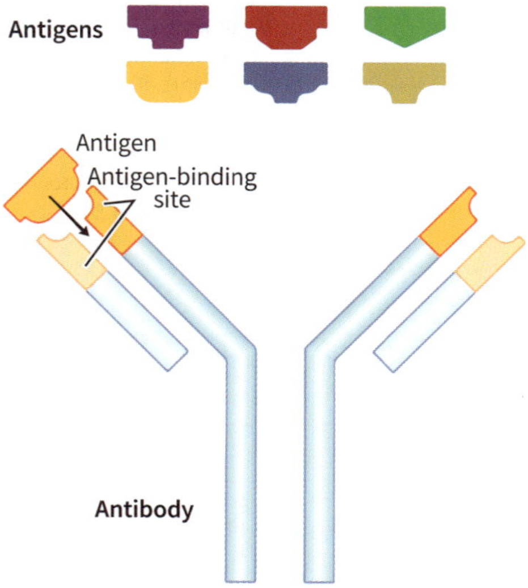

Figure 3.5 Antibody (a protein that binds to a specific antigen)

Source: image by Fvasconcellos, (UTC), (public domain), via Wikimedia Commons

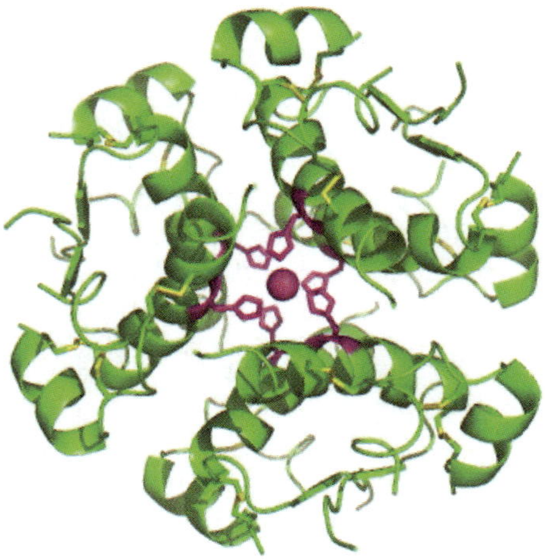

Figure 3.6 Insulin

Source: image by Isaac Yonemoto, [CC BY 2.5 (http://creativecommons.org/licenses/by/2.5)], via Wikimedia Commons

→ KEY QUESTIONS

12 **Distinguish between the four different levels of protein structure.**

13 **Describe five different roles that proteins may play in cells or organisms.**

Answers ➲ p. 50

CHAPTER SYLLABUS CHECKLIST

Are you able to answer these questions from the syllabus for this chapter? Tick each question as you go through the checklist if you are able to answer it. If you cannot answer a question, turn to the relevant page in the study guide to find the answer. For NESA key word meanings, go to www.educationstandards.nsw.edu.au and search 'key words'.

	FOR A COMPLETE UNDERSTANDING OF THIS TOPIC:	PAGE NO.	✓
1	Can I construct appropriate representations to model and compare the forms in which DNA exists in eukaryotes and prokaryotes?	38–39	
2	Can I model the process of polypeptide synthesis, including transcription and translation and assessing the importance of mRNA and tRNA in transcription and translation?	40–42	
3	Can I analyse the function and importance of polypeptide synthesis?	42	
4	Can I assess how genes and environment affect phenotypic expression?	43–44	
5	Can I investigate the structure and function of proteins in living things?	44–46	

Objective-response questions (1 mark each)

1 **Which statement most accurately describes the arrangement of DNA in both prokaryotes and eukaryotes?**
- A DNA is found in chromosomes.
- B DNA is found in the nucleus.
- C DNA is found on plasmids.
- D DNA is the only component of chromosomes.

2 **What is a significant difference between the DNA in eukaryotes compared to that in prokaryotes?**
- A DNA can replicate in eukaryotes but not in prokaryotes.
- B DNA is on linear chromosomes in the nucleus of eukaryotes but not in prokaryotes.
- C DNA is used to make polypeptides in eukaryotes but not in prokaryotes.
- D DNA is associated with protein in eukaryotes but not with prokaryotes.

3 **Which of the following is the correct sequence in protein synthesis?**
- A mRNA binds on ribosome → codon from mRNA pairs with anticodon from tRNA → DNA copied to form mRNA → mRNA leaves nucleus → amino acid carried on tRNA forms a peptide bond to join polypeptide
- B DNA copied to form mRNA → codon from mRNA pairs with anticodon from tRNA → amino acid carried on tRNA forms a peptide bond to join polypeptide → mRNA leaves nucleus → mRNA binds on ribosome
- C DNA copied to form mRNA → mRNA binds on ribosome → mRNA leaves nucleus → codon from mRNA pairs with anticodon from tRNA → amino acid carried on tRNA forms a peptide bond to join polypeptide
- D DNA copied to form mRNA → mRNA leaves nucleus → mRNA binds on ribosome → codon from mRNA pairs with anticodon from tRNA → amino acid carried on tRNA forms a peptide bond to join polypeptide

Use the information below to answer Questions 4 and 5.

Human beta-globin is a polypeptide subunit of the protein haemoglobin. A single mistaken change in the DNA (GAG becomes GTG) at codon 6 results in adenine being replaced by thymine, and hence in the replacement of the amino acid glutamic acid by valine. This results in a misshapen form of haemoglobin leading to the disease sickle-cell anaemia.

4 **Identify the correct mRNA codon for the normal haemoglobin at codon 6 and match the bases of the mRNA codon with the anticodon of tRNA.**
- A mRNA (CTC) codes for tRNA (GAG)
- B mRNA (CUC) codes for tRNA (GAG)
- C mRNA (GAG) codes for tRNA (CTC)
- D mRNA (GAG) codes for tRNA(CUC)

5 **What is a conclusion that can be drawn from the provided information?**
- A The correct sequence of amino acids in a polypeptide is important in the role of the resultant protein.
- B Changing one amino acid in a protein is not important in terms of it carrying out its role.
- C Proteins play a minor role in the functioning of organisms.
- D Every amino acid contained in the protein is essential to its functioning.

Short-answer questions

6 **Analyse the following diagram and compare the process in a with that in b.** (4 marks)

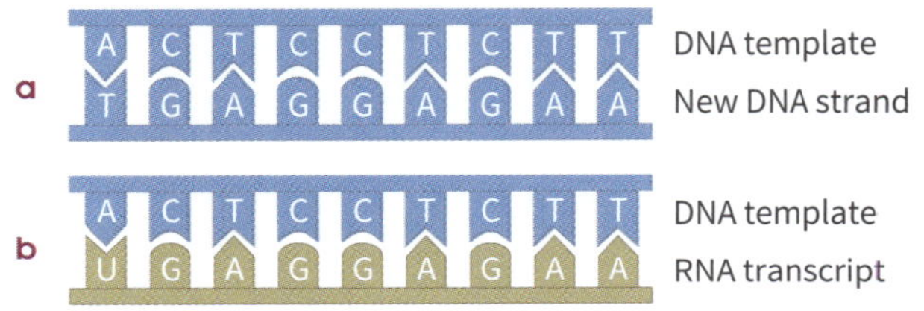

Figure 3.7 Representation of some cellular processes

7 **Identify two key differences in the forms that DNA exists in prokaryotic cells compared to eukaryotic cells and outline two key differences that must occur in polypeptide synthesis as a consequence of these differences.** (4 marks)

8 **Explain the relationship between a phenotype, genotype and environment.** (4 marks)

9 **'The cell is really just a protein factory.' Discuss.** (7 marks)

10 A mRNA segment contains the following codes: AUG, UGG, AAA, CUU, CGU, CAA, CAU, UGU.

a Use the following table to identify the sequence of amino acids in the polypeptide that would be produced. (2 marks)

b Identify one other possible mRNA sequence that would code for the same amino acids. (2 marks)

c Identify the sequences on the anticodons of the tRNA needed to make this polypeptide. (2 marks)

Table A3.1 mRNA codons for specific amino acids

amino acid	mRNA codon
isoleucine	AUU, AUC, AUA
leucine	CUU, CUC, CUA, CUG, UUA, UUG
valine	GUU, GUC, GUA, GUG
phenylalanine v	UUU, UUC
methionine	AUG is also the start codon in mRNA
cysteine	UGU, UGC
alanine	GCU, GCC, GCA, GCG
glycine	GGU, GGC, GGA, GGG
proline	CCU, CCC, CCA, CCG
threonine	ACU, ACC, ACA, ACG
serine	UCU, UCC, UCA, UCG, AGU, AGC
tyrosine	UAU, UAC
tryptophan	UGG
glutamine	CAA, CAG
asparagine	AAU, AAC
histidine	CAU, CAC
glutamic acid	GAA, GAG
aspartic acid	GAU, GAC
lysine	AAA, AAG
arginine	CGU, CGC, CGA, CGG, AGA, AGG
stop codons	UAA, UAG, UGA

ANSWERS

KEY QUESTIONS

Key question ➲ p. 39

1 Both prokaryotes and eukaryotes have chromosomal and extra chromosomal DNA. In prokaryotes, the chromosome is circular, long and intertwined and the DNA is supercoiled. Extrachromosomal DNA is in small circular plasmids that can self-replicate. In eukaryotes, the chromosomal DNA is in the form of chromatin, combined tightly with histone protein and, when dividing, condensed into very dense linear chromosomes that occur in homologous pairs. The extrachromosomal DNA occurs in mitochondria and chloroplasts, where it is circular, and these organelles self-replicate.

There is a lot more DNA in eukaryote cells and it is more densely packed. There is also a lot more non-coding DNA in eukaryotes. Note: There is a theory that mitochondria and chloroplasts were once prokaryote cells that were engulfed by eukaryote cells. The fact that they have circular DNA supports this idea.

Key questions ➲ p. 42

2 In everyday language a template is a rigid object shaped to form a pattern which allows multiple copies to be made. You may use a geometry template to accurately draw shapes or angles. DNA is a template as it stores a code of sequence of bases that can be copied by complementary strands of either DNA in replication or mRNA in polypeptide synthesis.

3 Both DNA and RNA are nucleic acids made of a sugar phosphate backbone with nitrogenous bases attached to the sugar. DNA is double-stranded, with the strands running in opposite directions and coiled into a double helix, with the nitrogenous bases attracting each other internally. mRNA is single-stranded with a backbone of sugar and phosphate, with the bases attached to the sugar. Its simple structure allows it to fit through the pores in the nuclear membrane. The sugars are slightly different with the DNA sugar, deoxyribose, having one less oxygen atom than ribose sugar. Three of the nitrogenous bases are the same in DNA and RNA: these are adenine, cytosine and guanine. The fourth base in DNA is thymine and this is substituted for uracil in RNA.

4 UACGGCCCAUG

5 mRNA formation is essential in transcription as it carries the genetic code from the DNA to the cytoplasm. Precursor mRNA is spliced and edited so that it contains the coding sections of the DNA. Mature mRNA is found on the ribosome and acts as the template for polypeptide synthesis so it is also essential for accurate translation.

6 tRNA plays no role in transcription but is essential to translation as it carries the appropriate amino acid into the polypeptide chain using the mRNA template.

7 The function of polypeptide synthesis is the manufacture of polypeptide chains of the correct composition so that functional proteins can form.

8 Polypeptide synthesis is vital to the functioning of organisms because without proteins of the correct structure and function an organism would not develop or survive.

Key questions ➲ p. 44

9 Phenotypic expression is the total observable characteristics or traits of an organism; it includes appearance (e.g. body size and form), development, behaviour, intelligence and biochemical and physiological properties.

10 Genes code for the production of polypeptides and hence proteins, which control much of the development, appearance, functioning and behaviour of an organism. The environment impacts on gene expression so that even organisms with the same genotype may have differences.

11 Nutrition, lifestyle (e.g. exercise), exposure to chemicals and electromagnetic radiation, levels of nurturing, sensory stimulation and a host of other environmental factors can impact on gene expression and hence phenotype.

Key questions ➲ p. 46

12 Primary structure includes the sequence of amino acids held together by peptide bonds and the positioning of any disulfide bonds that together form the backbone of the polypeptide chains. The secondary structure includes the regular localised arrangements of amino acids into regular structures such as alpha helices or pleated beta sheets. The tertiary structure includes the major bends and folds of the polypeptide chain, with a range of attractions providing stability. The quaternary structure is the combination of polypeptide chains to form a protein molecule. Proteins can consist of one or more polypeptide chains.

13 Proteins may be catalysts, promoting and controlling the reactions that make up cellular metabolism. They lower the activation energy needed for a reaction because of the interaction of their active site and the substrate(s). Proteins can provide a transport role; for example, haemoglobin carrying oxygen around an organism or protein channels allowing entry and exit of materials across a cell membrane. Proteins can 'recognise' foreign substances (antigens) and initiate an immune response to destroy the foreign microbe. Proteins can allow movement; for example, the role of actin and myosin in muscle contraction and cell cytokinesis. Proteins can act as a store; for example, myoglobin, which stores oxygen in muscle tissues, or ferritin, which stores iron.

HSC EXAM-TYPE QUESTIONS

Objective-response questions

1 A. Prokaryotes have a long, circular chromosome; eukaryotes have paired linear chromosomes. **B** is wrong because only eukaryotes have a nucleus. **C** is incorrect because prokaryotes may only have a circular plasmid that contains DNA. **D** is wrong because both groups have proteins associated with chromosomes even though eukaryotes have considerably more protein in the form of histone.

2 B. Chromosomal structures are a major difference. **A**, **C** and **D** are wrong because DNA replicates in both eukaryotes and prokaryotes, makes polypeptides in both, and is associated with proteins in both.

3 D. This option correctly identifies the key steps in transcription and then translation. **A**, **B** and **C** all misrepresent the sequence order.

4 **B**. In RNA C (cytosine) pairs with G (guanine) and A (adenine) pairs with U (uracil). **A** and **C** are wrong because there is no T (thymine) in RNA. **D** is incorrect as mRNA forms complementary bases with DNA.

5 **A**. The amino acid sequence determines the structure and therefore the function of a polypeptide. **B** is wrong as it shows that changing one amino acid can stop a polypeptide from functioning. **D** is incorrect because some changes in an amino acid sequence do not stop a polypeptide from functioning. **C** is incorrect because proteins play a major role in the functioning of cells and organisms.

Short-answer questions

6 **EM** You need to use your understanding of DNA replication from Chapter 2 and apply it to a comparison with a key process in transcription.

Part **a** shows a key process in DNA replication. (✓) DNA unzips and new strands are built around the exposed bases from nucleotides in the nucleus and under the control of enzymes. (✓) Part **b** shows transcription, in which mRNA forms from the unzipped DNA, which acts as a template. (✓) Part **a** is essential for the continuity of a species while part **b** is essential in the activation of the genetic code into functioning polypeptides. (✓)

7 **EM** This is a more difficult question as you have to apply your understanding of the differences in DNA structure and arrangement and then infer differences that would occur in the processes of polypeptide synthesis.

DNA exists as a circular chromosome within a nucleoid region of prokaryotic cells. In eukaryotic cells it is contained within linear chromosomes in a membrane-bound nucleus. (✓) mRNA must travel through the nuclear pores into the cytoplasm and on to the ribosomes in eukaryotic cells. This is not necessary in prokaryotic cells. (✓) DNA in eukaryotic cells also contains large non-coding sections interspersed with the coding sections. The non-coding sections must be edited out of the mRNA so that translation occurs via only using the mRNA that carries code for the amino acids in a particular polypeptide.(✓) Only small amounts of non-coding sections are found in prokaryote DNA so little editing is needed. (✓)

8 **EM** You need to define each term and then explain how they are connected to each other.

Genotype is the genetic make-up of an organism. The term *genotype* can refer to the entire set of genes in an organism. (✓) In relation to particular traits, genotype refers to the identities of alleles carried on each chromosome. The phenotype is the physical expression of the genotype or characteristics (e.g. height, weight, eye colour, leaf shape and size). (✓) In many cases the phenotype is a result of the interaction of the genotype and the environment. (✓) Height will be controlled by genes but also by diet. Many aspects of phenotype (e.g. human intelligence) are influenced by a number of genes in combination and environmental influences. (✓)

9 **EM** This is a challenging question. First, you need to explain the processes of polypeptide synthesis. Second, you need to explain the controls that occur via environmental factors, which influence gene expression, and the differentiation of cells containing the same DNA into many different cell types. Maybe the best conclusion is that even though the cell is to some extent a protein factory, it is a 'living factory' subject to many environmental influences.

To a large extent the cell is a factory that makes and modifies proteins essential to the cell's functioning. The genetic code, stored in the DNA of cells, is transcribed and then translated through the actions of mRNA and tRNA to polypeptide chains. (✓) These chains may be subject to further modification and may be processed and secreted through the endoplasmic reticulum and Golgi bodies (as hormones), or as enzymes to carry out cell metabolism or influence the activity or behaviour of the organism. (✓) In other cases the proteins may be used to build cells (e.g. muscle cells) in which the proteins bring about movement, or secreted to provide structural support or protection (e.g. skin, scales, nails). (✓) Other proteins act as vital parts of the cell membrane to allow the movement of materials into and out of cells. Other transport proteins carry oxygen in body fluids to the parts of multicellular organisms that need it. Other proteins acting on the surface of the cell play a part in recognition of other cells and materials as either belonging to the organism or foreign. (✓)

The cell is more than just a protein factory and environments outside of cells also influence the role of proteins. The quaternary level of protein structure that results from the conformation of a number of polypeptide chains may occur within or outside the cell. (✓) In some cases it is the environment that also influences the final structure and activity of the protein. Structural proteins, such as keratin and collagen, are stable in an extracellular environment where the pH level helps stabilise the structure and the large number of disulfide bonds. Proteins do not occur alone and can be associated with lipids, carbohydrates (glycoproteins) and nucleic acids. These have to be produced and modified into the appropriate structures and while proteins may play a role, the other biomolecules are also important. (✓) Even enzymes often depend on cofactors or coenzymes to carry out their role, so again the cell is more than the interaction of proteins. (✓)

10 **EM** This question requires you to apply skills in interpreting a table and extracting information to come to an understanding of translation in polypeptide synthesis. This understanding needs to demonstrate your knowledge of pairing of bases between the codon in mRNA and the anticodon on tRNA.

- **a** AUG, UGG, AAA, CUU, CGU, CAA, CAU, UGU = methionine, tryptophan, lysine, leucine, (✓) arginine, glutamine, histidine, cysteine (✓)
- **b** AUG, UGG (no alternatives), AAG instead of AAA, CUC or CUA or CUG or UUA or UUG instead of CUU, (✓) CGC or CGA or CGG or AGA or AGG instead of CGU, CAG instead of CAA, CAC instead of CAU and UGC instead of UGU (✓)
- **c** UAC, ACC, UUU, GAA, (✓) GCA, GUU, GUA, ACA (✓)

CHAPTER 4 GENETIC VARIATION

INQUIRY QUESTION:

How can the genetic similarities and differences within and between species be compared?

Close similarity of the genome of species is regarded as an indicator of recent evolutionary connection. In this chapter you will review other ways of comparing genetic similarities and differences between species and also within a species. In addition you will revise some of the sources of genetic variation.

1 Variations in genotypes of offspring

» Students conduct practical investigations to predict variations in the genotype of offspring by modelling meiosis, including the crossing over of homologous chromosomes, fertilisation and mutations.

Predicting variations in the genotype of offspring by modelling meiosis

- Your practical investigation of modelling meiosis will focus on making mathematical predictions based on the causes of variation that occur in the processes of meiosis, particularly the random segregation of homologous chromosomes, crossing over and changes in DNA called **mutations**.
- As part of your modelling of meiosis you will have demonstrated some random events; that is, events that happen by chance alone. You may have used a random event such as tossing a two-sided coin to generate options.
- When you modelled crossing over in meiosis you would have demonstrated its impact on the composition of the resulting chromosomes. You may have tagged positions along the length of the chromosome, with some type of symbol used to represent the different combinations of alleles that resulted.

mutation: permanent alteration of the nucleotide sequence of the DNA in the genome of an organism (types and causes will be reviewed in detail in Chapter 6)

- When modelling mutations in DNA during or just prior to meiosis you would need to use some representation to show that the allele on the chromosome had changed.
- Sexual reproduction depends on fertilisation which, because the gametes are from different gender parents, will also be a source of variation which can be mathematically predicted. Use of the **Punnett square** is an ideal way to model the prediction of genotypes in the zygote from gametes formed in meiosis.
- Diploid organisms possess homologous pairs of chromosomes. The **autosomes** are the chromosomes that carry genes for normal development and appear as identical pairs in a **karyotype**. Usually one pair of chromosomes are the sex chromosomes. In the case of humans they are called the X and Y chromosomes. Females have two X chromosomes and males have an X and a Y sex chromosome. Humans have 23 homologous pairs; that is, 22 autosomal pairs and one pair of sex chromosomes. When modelling meiosis to make predictions about variation of gamete outcomes it is best to keep it simple. Only use small numbers of chromosomes and avoid using sex chromosomes as crossing over gets unnecessarily complicated.

Punnett square: a diagram used to predict the outcome of a cross between two parents of known genotype; also a device used to determine the probability of the possible genotypes of offspring resulting from a particular breeding experiment
autosomes: chromosomes carrying genes whose inheritance pattern is the same for males and females of the species as opposed to the sex chromosomes (in the case of humans, the X and Y chromosomes)
karyotype: the systematic arrangement of the chromosomes of a eukaryotic cell from a light microscope taken during mitosis to show appearance of banding patterns and position of the centromere, ordered from largest to smallest; shown as homologous pairs and numbered for identification

Random segregation

- The practical investigation of modelling crossing over during meiosis to predict variations in genotypes needs to incorporate the random segregation of chromosomes as an additional source of variation.

➔ Figure 4.1 is a simplified diagram of random segregation. Use of colour to delineate the members of each homologous pair of chromosomes that are of maternal origin from those of paternal origin helps demonstrate the impact of random segregation on variation of genotype. Step 1 is shown in Figure 4.2.

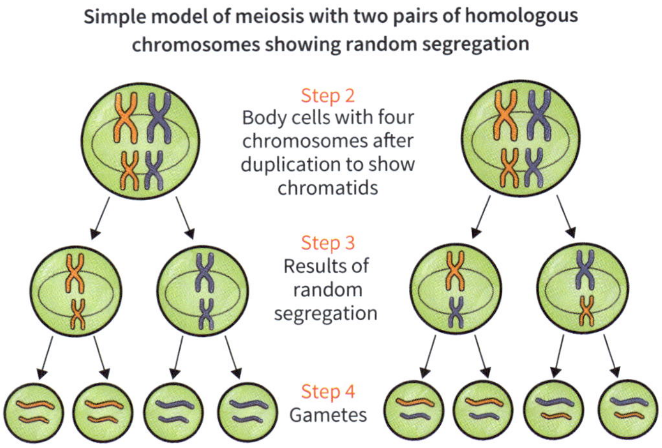

Figure 4.1 Simple model of meiosis with two pairs of homologous chromosomes showing random segregation

➔ KEY QUESTION

1 **What are the origins of the homologous chromosomes in a zygote?**

Answer ➲ p. 65

Crossing over of homologous chromosomes

➔ One of the major sources of variation in meiosis is crossing over. Crossing over involves chromatids breaking at the same point or **locus** and rejoining in two reciprocal non-parental combinations. Combined with the random segregation of homologous chromosomes, many possible combinations of alleles located along the same chromosome can be predicted. For simplicity, modelling of crossing over will make predictions of new combinations without also considering random segregation of chromosomes.

➔ Figure 4.2 uses colour to demonstrate possible new combinations of alleles (**recombinant** chromosomes) along a pair of homologous chromosomes with one crossing over. Figure 4.3 actually labels the alleles along a chromatid.

locus: the position of a specific gene or genetic marker on a chromosome (the plural is loci)

recombinant: of or resulting from new combinations of genetic material; for example, when crossing over occurs, the chromosomes that contain new allele combinations different from those of the parent are the recombinants

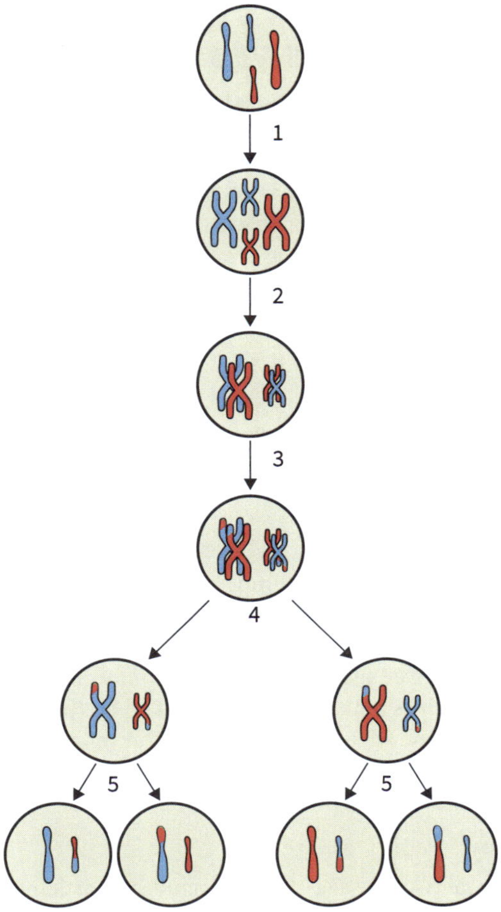

Figure 4.2 Main steps (simplified) in meiosis, showing the results of two pairs of homologous chromosomes crossing over.

Source: image adapted from Peter Coxhead, [CC0], via Wikimedia Commons

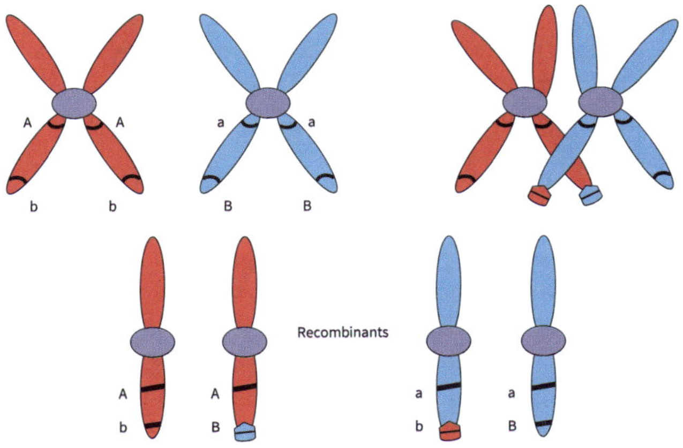

Figure 4.3 Crossing over in homologous chromosomes

Source: image adapted from Abbyprovenzano (own work), [CC BY-SA 3.0 (https://creativecommons.org/licenses/by-sa/3.0)], via Wikimedia Commons

➔ Figure 4.3 above shows two pairs of alleles, Aa and Bb, located on one pair of homologous chromosomes. The middle two chromosomes, resulting from crossing over in meiosis, are the recombinants. The combination of A with B and a with b were not present on the original chromosomes.

FIRSTHAND INVESTIGATION 1

Modelling meiosis, including crossing over of homologous pairs of chromosomes

» Students conduct practical investigations to predict variations in the genotype of offspring by modelling meiosis, including the crossing over of homologous chromosomes.

Aim

to predict the variations in genotype by modelling the crossing over of homologous pairs of chromosomes in meiosis

Materials

two differently coloured lumps of plasticine (one colour represents the chromosomes that were of maternal origin, the other colour those of paternal origin); boiled rice grains, dyed using food dye into four different colours such as pink, yellow, blue and green (pink represents one allele, indicated as A, and yellow another form of the same allele, indicated as a; blue represents allele B and green represents the other form, as b)

Method

1 Create two sausage-shaped lengths of plasticine from each colour. Each colour has one pair of chromatids each of the same length of plasticine. Join the chromatids in the same position (at the centromere) towards one end of the chromatids.
2 Embed pink rice grains near the centromere and on each arm of joined chromatids from maternal origin (similar to A in Figure 4.3). Embed two blue rice grains towards the tip (similar to b on chromatids in Figure 4.3).
3 Embed yellow rice grains in the same position as the pink grains, but this time on the chromatids from the paternal origin (similar to a in Figure 4.3). Embed green rice grains towards the tip (similar to B in Figure 4.3).
4 Line chromosome pairs up (the line represents the metaphase plate). Complete crossing over by overlapping the two adjacent chromatids at the same point (between the two rice grains on the arm of the chromatid), then break them so as to exchange the tips. This is demonstrated in Figure 4.3.
5 Break the centromere between the chromatids into two, making two separate chromosomes, and separate so that each new cell has one member from each homologous pair. You will have four cells, with each containing one chromosome.
6 Tally the resultant alleles.

Results

Table 4.1 Results of allele combinations resulting from crossing over; shaded alleles highlight new combinations

	Crossing over once
Gamete 1	A, b
Gamete 2	A, B
Gamete 3	a, b
Gamete 4	a, B
Result	two new combinations

Conclusion

Chromosomes are formed with new combinations of alleles when crossing over between the loci of the alleles occurs an odd number of times. The greater the variety of alleles and the larger the number of loci included in your modelling will greatly increase the number of various chromosomes.

→ KEY QUESTIONS

2 **How does variation on the allele sequence occur on homologous chromosomes?**

3 **Outline the possible combinations and predict the number of combinations if you modelled crossing over with three sets of alleles along the same chromosome and crossing over up to two times.**

4 **What is the key process you need to demonstrate in your modelling in order to predict the variations that arise from crossing over in meiosis?**

5 **How did you identify the recombinant alleles that resulted from crossing over in meiosis?**

Answers ➲ p. 65

Fertilisation

- Predicting the variation of genotypes that result from fertilisation is demonstrated by analysing a Punnett square diagram.
- Table 4.2 shows a simple Punnett square that demonstrates how in humans there is equal probability of fertilisation resulting in male or female offspring.
- The intersection of the rows and columns that contain the possible genotype of the male and female gametes provides the possible resultant genotypes.
- The probability of each genotype is predicted by the relative number of times it occurs within the Punnett square.

Table 4.2 Punnett square

Male gametes→ Female gametes↓	X	Y
X	XX	XY
X	XX	XY

Mutations

- Mutations as a cause of genotype variation are different to the previously predicted causes of variations because they result in new alleles, not just rearrangement or recombination of the existing alleles.
- Mutation results in changes to the genotype because of changes in DNA. This can be due to **substitution** of one base in a nucleotide for another, the deletion or insertion of a nucleotide, or even rearrangement of sections or whole chromosomes. Ultimately mutations are the source of all variations in genotypes and alleles. Mutations in some sections of DNA may result in no change in function or no new alleles. They may also result in changes that can be beneficial or harmful to the organism (see Chapter 6).
- In this investigation, prediction in variation in genotype will assume that the mutation involves the substitution of an allele for another, such as A for a.
- In the HSC-style questions in Chapter 3 you used an example of a mutation in a polypeptide making up the protein haemoglobin that causes sickle-cell anaemia. The gene in which the mutation occurs is called haemoglobin beta (HBB). The genotype of the normal form of the haemoglobin is written as HbA. The mutated form of the haemoglobin is written as HbS.
- The genotype for a trait on an organism is expressed as letters that represent the alleles. If an individual has both alleles the same for a trait it is said to be **homozygous**. If the alleles are different the organism is said to be **heterozygous**. There are a variety of ways that the alleles may interact to produce the phenotypes in heterozygous genotypes.

substitution (of bases): a type of mutation in which one nitrogenous base is substituted for another, often resulting in a different amino acid in a protein chain if it occurs in a coding section of DNA
homozygous: alleles are homozygous when they are both the same; an organism with the same alleles for a trait is said to be 'true breeding'
heterozygous: alleles are heterozygous when they are both different; an organism with different alleles for a trait is said to be a hybrid; for example, the AB blood type is heterozygous with one A allele and one B allele

→ KEY QUESTIONS

6 How does crossing over produce variation in gametes?

7 How does fertilisation contribute to variation in zygotes?

8 Construct a Punnett square to show the variations possible in the offspring if both parents were heterozygous for the sickle-cell form of haemoglobin; that is, both have sickle-cell trait.

Answers ➲ p. 65

2 The formation of new combinations of genotypes produced during meiosis

» Students model the formation of new combinations of genotypes produced during meiosis, including but not limited to interpreting examples of autosomal, sex-linkage, co-dominance, incomplete dominance and multiple alleles; and constructing and interpreting information and data from pedigrees and Punnett squares.

Autosomal inheritance

- Autosomal inheritance is the inheritance pattern occurring similarly in males and females because the alleles responsible are carried on autosomes.
- The first breakthrough in understanding the pattern of autosomal inheritance was as a result of the work of Gregor Mendel, who conducted experiments on crossbreeding lines of pea plants which were **true breeding**. His results are summarised in Table 4.5. He documented the traits in the offspring (called the **F1 generation**) and then crossed the F1 generation together to produce an **F2 generation**. The experiments are called **monohybrid crosses** because they traced the inheritance of one trait only (hence *mono*) and the parents had alternative forms of the trait (hence *hybrid*).

true breeding: forms of organisms that possess traits that are always carried on to their offspring; that is, offspring show no variation in a particular trait
F1 generation: the direct descendants of a controlled cross
F2 generation: the descendants of a cross between two F1 parents
monohybrid cross: a cross between two pure-breeding parents that differ in the alleles they carry for a particular gene

- Of the seven traits that Mendel crossed, all of the F1 generation possessed only one phenotype. The F2 generation had mixed phenotypes in a ratio approximately of 3:1 (see Table 4.3).

Table 4.3 The results of early genetic experiments with pea plants

Phenotypes crossed	F1 generation results	F2 result
yellow cotyledon vs green cotyledon	all yellow	6022 yellow : 2001 green
axial flower position vs terminal flower position	all axial	651 axial : 207 terminal
round seed vs wrinkled seed	all round	5474 round : 1850 wrinkled
inflated pea pod vs constricted pea pod	all inflated	882 inflated : 299 constricted
grey seed coat vs white seed coat	all grey	705 grey : 224 white
green pea pod vs yellow pea pod	all green	428 green : 152 yellow
long stems vs short stems	all long	787 long : 277 short

- It was apparent that for each phenotype combination, one form of allele 'dominated' the other form in the F1 generation. The term used to describe this relationship was **dominance**. The allele that was apparently carried through the F1 generation and expressed in a lower ratio in the F2 generation was termed **recessive**. These results are now called Mendelian ratios.
- The genotype of a dominant trait is usually written as a capital letter. If true breeding, the tall pea plant would have the genotype TT. If heterozygous, it would be Tt. Because it is recessive the short plant must be homozygous, with the genotype tt.

dominance: a trait that is expressed in a true breeding monohybrid cross in the F1 generation
recessive: the form of genotype that is only expressed when homozygous

- An explanation of Mendel's results can be seen below.

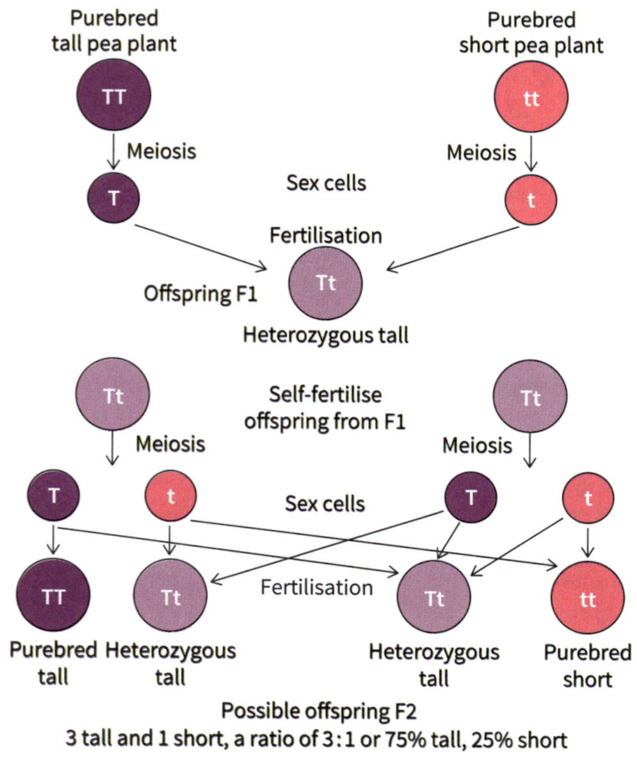

Figure 4.4 Mendel's experiment and results

- Another way of demonstrating the possible genotypes of gametes and combinations resulting from fertilisation is the Punnett square shown in Table 4.4.

Table 4.4 Possible genotypes

Male gametes→ Female gametes↓	T	t
T	TT	Tt
t	Tt	tt

- **Pedigree diagrams** are versions of a family tree that show the inheritance of a characteristic (phenotypes) over at least one generation without showing the genotypes.

pedigree diagrams: representations of inheritance of a trait across a family over generations

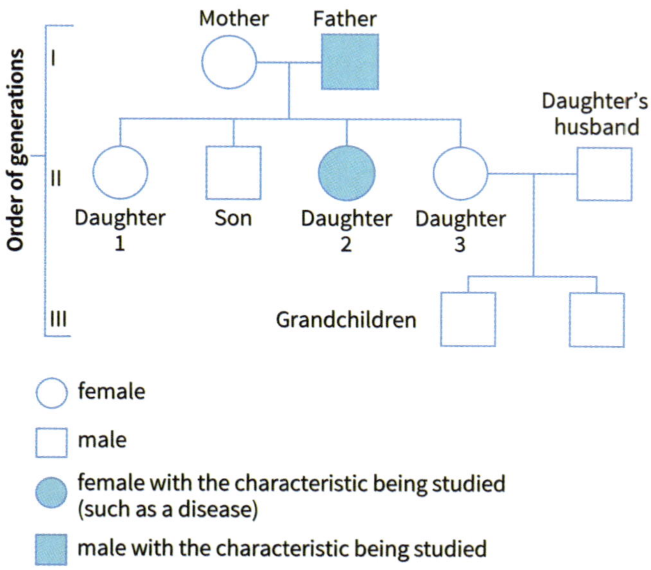

Figure 4.5 A pedigree table or diagram

- Note: In pedigree diagrams circles represent females, squares represent males, shading indicates those expressing the trait, horizontal lines show parents or, if below a vertical line, connect siblings. The key identifies the main symbols used and shows that shading indicates that the phenotype has the characteristic expressed.

- Figure 4.6 below shows the pedigree diagram for the cross between a pure breeding curly-winged fruit fly and a pure-breeding straight-winged (wild type) fruit fly. The green shaded squares and circles show flies that have curly wings. The unshaded squares and circles show flies that have straight wings.

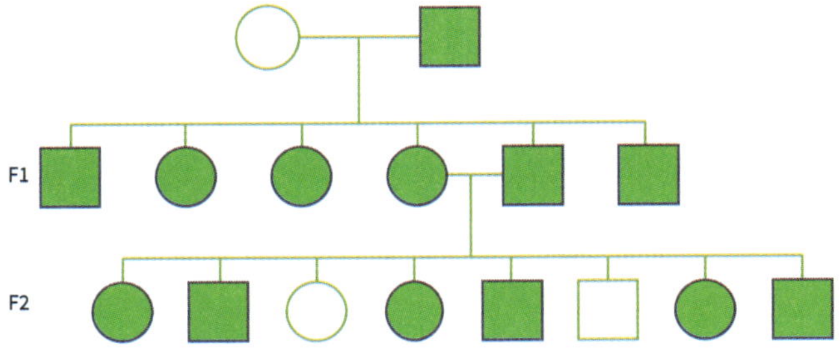

Figure 4.6 Pedigree diagram of a fruit fly cross

- Observations and conclusions include the following.
  - F1 all had curly wings so it is apparent that curly wings are dominant; no difference between males and females in the results confirms the inheritance is autosomal.
  - 75% of F2 were curly winged and 25% straight winged, suggesting that the F1 parents were heterozygous; that is, Cc × Cc.

KEY QUESTIONS

9 **Figure 4.6 is a pedigree diagram showing autosomal dominant inheritance. Explain what is meant by an autosomal dominant pattern of inheritance and how the pedigree diagram shows this pattern?**

10 **Compare the use of Punnett square diagrams with pedigree diagrams in modelling the new combinations of genotypes that result from meiosis and fertilisation.**

Answers ➲ p. 65

Sex-linked inheritance

- Some genes are carried on the sex chromosomes. The pattern of inheritance of these genes is called **sex-linked** and is closely associated with gender. In the case of humans, the male Y chromosome does not carry alleles that are carried on the X chromosome. The male has only one allele for the genes carried only on the X chromosome and so many recessive alleles are more likely to be expressed in the phenotype. The female has two X chromosomes so is less likely to experience a recessive disorder.
- Figure 4.7 shows a different pattern of inheritance to the autosomal dominant pattern. The father has the condition of webbed toes. None of the F1 offspring have this condition but some of the grandsons (F2 or III2, III4, III6 and III7) do.
- No female has the condition but assuming that separate mutations did not occur four times, two daughters (II 4 and II 5) may have been a '**carrier**' to their sons of an allele for the condition. The allele itself is recessive so it is possible for a female to have the condition, provided she has two alleles for the condition (is homozygous).

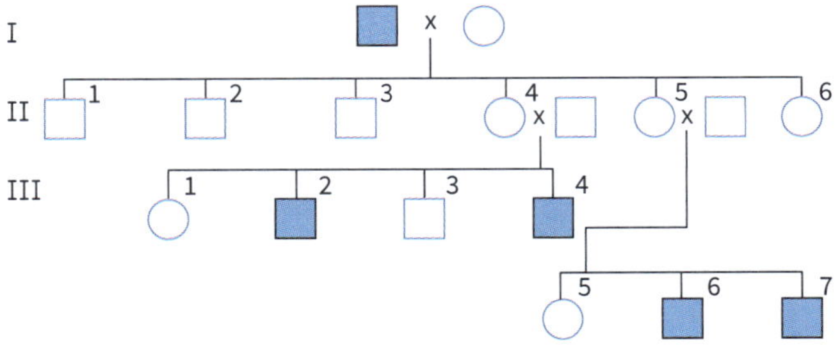

Figure 4.7 Pedigree diagram showing pattern of inheritance of webbed toes

Source: image adapted from Wellcome Images, a website operated by Wellcome Trust, a global charitable foundation based in the United Kingdom, [CC BY 4.0 (http://creativecommons.org/licenses/by/4.0)], via Wikimedia Commons

sex-linked: patterns of inheritance in which each gender has a different numbers of alleles and therefore a different chance of experiencing the condition

carrier: an individual that has an allele that is not expressed in their phenotype but whose offspring may inherit the condition from them

- When Thomas Morgan did crosses between red-eyed and white-eyed fruit flies he found a similar pattern. His (widely accepted) explanation is that some genes are carried on the X chromosome. Fruit flies and humans have females with two X chromosomes and males with one X and one Y chromosome. The Y chromosome carries very few genes (mostly those concerned with conferring 'maleness') so there is only one allele for a number of conditions in males whereas the female has two. A condition that is recessive in the female becomes expressed in the male. Using Table 4.2 it can be seen that males inherit their Y chromosome from their father and must receive their one X chromosome from their mother.
- Thomas Morgan's research on fruit flies was the definitive evidence that genes are carried on chromosomes.
- The convention when using Punnett squares to predict the outcomes of sex-linked inheritance is to show the condition as a superscript on the X chromosome. You may also come across some secondary sources that use bold font to indicate the X chromosome carrying the defective allele. Table 4.5 shows a Punnett square for a cross between a red-eyed male and a white-eyed female fruit fly. Red eyes is dominant over white eyes so the white-eyed female must carry the allele on both X chromosomes. The Y chromosome has no superscript letter as it does not have a locus for the gene.

Table 4.5 Sex-linked inheritance in fruit flies

Male gametes→ Female gametes↓	X^R	Y
X^r	X^rX^R	X^r Y
X^r	$X^r X^R$	X^r Y

Key to table
X^rX^r : white-eyed female
X^RY : red-eyed male
Results of the cross: all red-eyed heterozygous (carrier) females and all white-eyed males

→ KEY QUESTION

11 **In humans the alleles for colour vision are carried on the X chromosome. Full-colour vision is normal (N) and dominant whereas red–green colour blindness is recessive (n). Construct a Punnett square to predict the outcome of offspring from a mother who is heterozygous for colour blindness and a normal-visioned father.**

Answer ⊃ p. 65

Co-dominance

- When both alleles in a genotype are expressed in the phenotype they are described as co-dominant.
- In Chapter 2 you reviewed how genes coded for polypeptides. In some cases one form of allele may contain a mutation that means it does not code for a functional polypeptide. This is independent of the functioning of the other allele. The red-eyed fruit fly has an allele that dominates the white-eye allele as the red-eye allele codes for a polypeptide that produces the red pigment. The white-eye allele is recessive as it lacks the chemical pathways to make pigment. This explains why some alleles are recessive.
- In some cases of **co-dominant inheritance** both alleles function to produce effective polypeptides but they are different polypeptides that have different impacts. In blood type AB, both A and B alleles work to produce antigens so that the red blood cells contain both A and B antigen on their surfaces.
- Genotypes of co-dominant traits are often recorded as both characteristics having a capital letter, with different capital letters reflecting the different genotypes and phenotypes.
- Roan cattle have a mixture of red and white hairs. Red cows have the genotype RR, white cows have the genotype WW, and roan cows have the genotype RW. In some cattle the white hairs dominate and in others the red hairs dominate.

co-dominant inheritance: the genetic expression of both alleles in which the result is the combined impact of both alleles

Incomplete dominance

- In some cases of inheritance the heterozygous form results in a phenotype that is a blend of both allele characteristics. This is called **incomplete dominance**.
- This is the case for the forms of haemoglobin reviewed in the example of sickle-celled anaemia (see page 55). HbA codes for normal haemoglobin. HbS also produces a form of haemoglobin but with one different amino acid, so the haemoglobin molecule changes shape under certain circumstances. The interaction results in a blend of the phenotypes called sickle-cell trait. The incomplete dominance in the sickle-cell trait results in a condition that is not as serious as the homozygous form of sickle-cell anaemia but is still markedly different to people that are homozygous HbA.

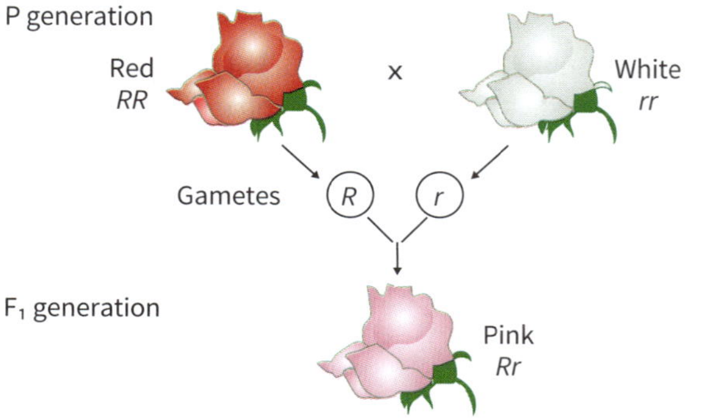

Figure 4.8 Incomplete dominance results in offspring with blended phenotypes.

→ KEY QUESTIONS

12 **Construct a Punnett square diagram to predict the genotype and phenotype outcomes of a cross between two co-dominant roan cattle.**

13 **Distinguish between dominance, co-dominance and incomplete dominance as types of inheritance patterns.**

Answers ⊃ p. 65

Multiple alleles

- Even though diploid organisms have two alleles for each inherited trait, some traits have more than two forms of the allele within the gene pool. This is called **multiple allele inheritance**.
- Multiple allele inheritance is well documented in human blood groups, with alleles to produce A or B blood antigens. Blood group O does not produce any antigens.

incomplete dominance: heterozygous genotype that results in a blend of the alleles in the phenotype
multiple allele inheritance: the existence of three or more alternate forms of alleles that can occupy the one locus

- Multiple alleles result in a number of different phenotypes depending on the interaction of the alleles. In human blood groups, A and B are co-dominant and both are dominant over O.

Table 4.6 Multiple allele inheritance in human blood groups

Blood group	Possible alleles
A	AA or AO
B	BB or BO
AB	AB
O	OO

- Other examples of multiple alleles are coat colour in rabbits at the C locus. Dark rabbits (black, brown or grey) have a dominant allele (C) whereas albino rabbits lack pigment and are recessive (c). Chinchilla (c^{ch}) is dominant over Himalayan (c^{h}) and albino. There is some evidence that when crossed, Himalayan and albino show incomplete dominance.

Figure 4.9 Locus C for coat colour in rabbits has alleles that produce varied colouration from dark to albino.

Constructing and interpreting information and data from pedigrees and Punnett squares

- You constructed and interpreted Punnett squares when predicting the variation in genotypes that can result from meiosis and fertilisation (see pages 54–56).
- You interpreted pedigree diagrams when noting the difference between sex-linked inheritance and autosomal inheritance (see pages 56–57).
- To construct a pedigree diagram you have to display a pattern of inheritance using the conventions of the diagram's form. For example, Figure 4.7 could be constructed from the following information.
  - A man with webbed toes marries a female with normal toes (unaffected). They have three sons followed by three daughters who all have normal toes. The eldest daughter marries an unaffected man and they have one unaffected daughter, followed by an affected son, an unaffected son and another affected son. The second daughter marries a normal man and they have an unaffected daughter followed by two affected sons.

KEY QUESTIONS

14 **Snapdragons have red or white flowers but when crossed produce pink flowers. What type of inheritance is this an example of?**

15 **How would you represent the genotypes in a cross between two pink snapdragons and then predict the outcome?**

16 **Construct a Punnett square to predict the offspring of a cross between two pink snapdragons.**

17 **How do multiple alleles contribute to genetic diversity?**

18 **Use the information above about rabbit coat colour inheritance at locus C to construct a table to predict the phenotypes of all possible hybrid crosses; that is, black × chinchilla, black × Himalayan, black × albino, chinchilla × Himalayan, chinchilla × albino and Himalayan × albino.**

Answers ➲ p. 66

3 Representing frequencies of characteristics in a population

» Students collect, record and present data to represent frequencies of characteristics in a population, in order to identify trends, patterns, relationships and limitations in data, for example: examining frequency data; and analysing single nucleotide polymorphism (SNP).

Examining frequency data

- The following measures of frequencies of characteristics in a population are useful for comparing genetic diversity within a species. Genetic diversity between species is based on comparison of overall genomes.
- The **gene pool** is the stock of different genes in an interbreeding population (e.g. a species). A large gene pool is an indicator of genetic diversity within a species.

gene pool: all of the copies of all the alleles of all the genes in a population

- Allele frequencies indicate the genetic diversity or richness of a gene pool.
- The frequency of an allele, expressed as a proportion or percentage, is called the allele or gene frequency in a population or species. It is calculated by the total number of a particular allele divided by the total number of copies of all alleles for a particular gene on a particular locus in a population.
- The genotype of an organism provides data for calculation of allele frequency. If phenotypes are used to collect data on allele frequency, results may be quite misleading. Dominant alleles will appear more in the phenotype than their frequency might suggest. Many traits are the result of the interaction of environment and alleles or more than one gene. Recessive alleles are often 'carried' without being apparent in the phenotype and so go unrecorded.
- Frequency data is one way that variation within a species can be compared. It is based on determining the proportions of the various alleles located on the one locus of a chromosome.
- It is possible for some alleles in a population to have no variation. This is the case when the existence of the allele in a particular form is essential to survival so that any mutation or variation would automatically be deleted (along with the organism possessing the variation).
- Factors such as genetic isolation, migration, mutation, natural and artificial selection, and chance can affect allele frequencies. Natural change over time of relative allele frequency is called genetic drift, which contributes to microevolution. You will review these concepts in more detail in Chapter 5.
- Population genetics is involved with the study of allele frequencies in populations and how these change over time.
- Many factors influence allele frequency. Earlier in the chapter you looked at the impact of a mutation in the gene for haemoglobin (HbA to HbS) causing sickle-cell shapes in the red blood cells. This mutation is so deleterious in the homozygous form that natural selection probably would have reduced its allele frequency. The fact that in the heterozygous form it provides the benefit of resistance to malaria has allowed the allele frequency to remain relatively high in places where malaria is a threat.

allele frequencies: the proportion or percentage of all possible alleles at a locus of one particular allele

population: a group of potentially interbreeding individuals occurring in a defined space and time (e.g. the total current human population)

FIRSTHAND INVESTIGATION 2

Data collection, recording and presentation to represent frequencies of characteristics in a population

» Students collect, record and present data to represent frequencies of characteristics in a population, in order to identify trends, patterns, relationships and limitations in data.

Aim

to collect, record and present data on frequencies of characteristics in a population and to identify trends, patterns, relationships and limitations in data

Method

1. Identify and select variable human phenotypes on which to collect data. Websites such as www.biologycorner.com/worksheets/geneticsurvey.htm include worksheets for data collection and presentations that explain various characteristics.
2. As an example, collect data within your class, year or school on traits such as hairline (flat or widow's peak), ear lobes (attached or not, Darwin's point or not), tongue-rolling ability, left-handed or not, short big toe or not, curly hair or not, blue eyes or not, and so on.
3. Select one characteristic, such as tongue rolling, and collect data from a family, constructing a pedigree diagram to represent the incidence of the characteristic. Use the diagram to predict whether the alleles for the characteristic are dominant or recessive.
4. Research information about the characteristic surveyed to verify your prediction.

Results

Table of data (see Table 4.7 for a template) and pedigree diagrams (see Figure 4.7 as an example).

Table 4.7 Example of record of data collected on frequencies of characteristics in a population

Trait description	Number and proportion of people with the trait	Number and proportion of people lacking the trait
Blood type O Rh^+	36 of 102 surveyed (35%)	66 of 102 (64%)

Conclusion

Using observations is a very crude way of assessing allele frequency in a population. Many human characteristics do not reflect simple Mendelian ratios or clearly dominant/recessive relationships. Phenotypes may be the result of the interaction of several genes and multiple alleles. Environmental factors may also play a part in gene expression and hence phenotypes.

> **➔ KEY QUESTION**
>
> 19 **In a population of 100 people, 50 have OO blood type, 25 have AO, ten have OB, five have AB, seven have AA and three have BB. Calculate the allele frequencies.**
>
> Answer ➲ p. 66

Analysing single nucleotide polymorphism (SNP)

- Another way to compare similarities and differences in genetic variation within and between species is by analysing **single nucleotide polymorphisms (SNPs)**. SNPs are variations in one nucleotide base, such as thymine being substituted by cytosine on a particular point in the DNA. The variation, when compared with the general population, must occur with at least 1% frequency.
- SNPs may occur in coding or non-coding sections of the genome but are most common in the sections between genes. SNPs account for 90% of naturally occurring human DNA variation and occur on average once every 300 nucleotides.
- **Genetic fingerprinting** does not tend to use SNPs for identification because they are so frequent within the genome, meaning dozens would have to be compared to result in identification. Improvements in technology and automation have increased the potential for the use of SNPs, especially for the identification of degraded DNA.

> **single nucleotide polymorphisms (SNPs):** pronounced as 'snips', SNPs are genomic loci where alternative bases occur in at least 1% of the population
>
> **genetic fingerprinting:** the comparison of DNA to determine identity or familial relationship (e.g. in forensic matching of crime-scene biological evidence)

- When SNPs occur within a gene or a nearby regulatory region they may impact a gene's function and directly cause a disease. Alternatively, they can be used as biological markers to help scientists locate genes associated with disease. The mutation HbA to HbS is an example of an SNP. SNPs have been useful in predicting a patient's response to drugs, susceptibility to toxins, or their risk of developing a particular disease.
- SNPs can be used in sustainable agriculture (e.g. genetic markers have been used to help identify wild stocks of cod fish).
- Many different species share genes (e.g. the gene for haemoglobin). Though it is possible to compare SNPs across species, comparing species genomes has more often been used to identify genetic similarities and differences. SNPs are a way of sampling the genome without the expense and effort of mapping the entire genome.

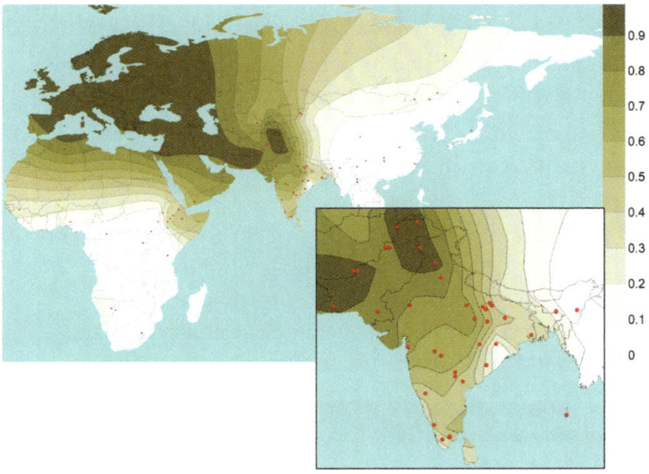

Figure 4.10 Map showing frequency of SNP re142664-A
Source: image by BasuMallick C, Iliescu FM, Möls M, Hill S, Tamang R, Chaubey G, et al, [CC BY 2.5 (http://creativecommons.org/licenses/by/2.5)], via Wikimedia Commons

- Figure 4.10 above shows a map indicating regions where there are variations in the frequency of a SNP. In this case the SNP is on one of the genes that produces a protein associated with the shade of skin colour. Following is an analysis of the findings, which reflect human adaptation and evolution.
  - Skin colour in humans is a diverse trait that is readily observed. Paler skin is more common in humans living away from the Equator, where pale skin absorbs more ultraviolet light to assist in the manufacture of vitamin D. Closer to the Equator, where there is higher light intensity, the ultraviolet light absorption causes mutations that may lead to melanoma. Darker pigmentation provides protection from skin cancer in an environment where there is usually enough light on exposed skin to produce vitamin D.
  - One of the main skin pigmentation genes in humans is SLC24A5.One single nucleotide polymorphism of this gene (SNP ID: rs1426654) explains 25–38% of the skin colour variation between Europeans and West Africans. As seen in Figure 4.10, the ancestral (G) allele of the SNP predominates in African and East Asian populations (93–100%), whereas the derived (A) allele is dominant in Europe (98.7–100%). This derived allele is responsible for lighter skin and is believed to have spread throughout Europe between 6000 and 12 000 years ago. Before that the ancestors of Europeans had comparatively brown skin.
  - The red dots show places where sampling occurred. Data gathering is more than simple observation of phenotypes, as the nucleotide sequence of a particular locus must be obtained, recorded and analysed. A large sample size is important for better reliability.

> **➔ KEY QUESTION**
>
> 20 **Define the term *single nucleotide polymorphism*.**
>
> Answer ➲ p. 66

Identifying trends, patterns and relationships

- Trends are identified by comparing data over time or across different locations or population groups.
- Patterns are detected when there are similar trends, such as an increase in allele frequency over time for several alleles.
- Identification or relationships occur when a trend can be linked to external factors. Changes in human skin pigmentation have been linked to adaptations for protection from skin cancer and the ability to make vitamin D from exposure to ultraviolet radiation.

Limitations in data

- Collecting data on the frequency of characteristics in a population is dependent on observational data, obtaining accurate family records, and decisions about aspects of particular phenotypes. Phenotypes only provide indirect evidence of the genotype and as such are an inaccurate guide to gene frequency.
- Collecting data on SNPs provides an accurate snapshot for specific alleles of genes but the data collection relies on sampling. Determining the base sequences of DNA for a single allele is expensive and requires the use of technology.

→ KEY QUESTIONS

21 **Identify limitations in data gathered to compare genetic similarities and differences within a species.**

22 **Identify limitations in data gathered to compare genetic similarities and differences between species.**

Answers ➲ p. 66

CHAPTER SYLLABUS CHECKLIST

Are you able to answer these questions from the syllabus for this chapter? Tick each question as you go through the checklist if you are able to answer it. If you cannot answer a question, turn to the relevant page in the study guide to find the answer. For NESA key word meanings, go to www.educationstandards.nsw.edu.au and search 'key words'.

	FOR A COMPLETE UNDERSTANDING OF THIS TOPIC:	PAGE NO.	✓
1	Can I conduct practical investigations to predict variations in the genotype of offspring by modelling meiosis, including the crossing over of homologous chromosomes?	52–54	
2	Can I conduct practical investigations to predict variations in the genotype of offspring by modelling meiosis, including fertilisation and mutations?	54–55	
3	Can I model the formation of new combinations of genotypes produced during meiosis, including but not limited to: interpreting examples of autosomal and sex-linkage?	55–58	
4	Can I model the formation of new combinations of genotypes produced during meiosis, including but not limited to: interpreting examples of co-dominance, incomplete dominance and multiple alleles?	58–59	
5	Can I construct and interpret information and data from pedigrees and Punnett squares?	59	
6	Can I collect, record and present data to represent frequencies of characteristics in a population, in order to identify trends, patterns, relationships and limitations in data; for example, examining frequency data?	59–60	
7	Can I collect, record and present data to represent frequencies of characteristics in a population, in order to identify trends, patterns, relationships and limitations in data; for example, analysing single nucleotide polymorphism (SNP)?	61–62	

HSC EXAM-TYPE QUESTIONS

Objective-response questions (1 mark each)

1 A cell has six chromosomes arranged in three pairs. How many different possible combinations of chromosomes would result in the random segregation stage of meiosis?

A 2
B 4
C 6
D 8

2 Observe Figure 4.11 below.

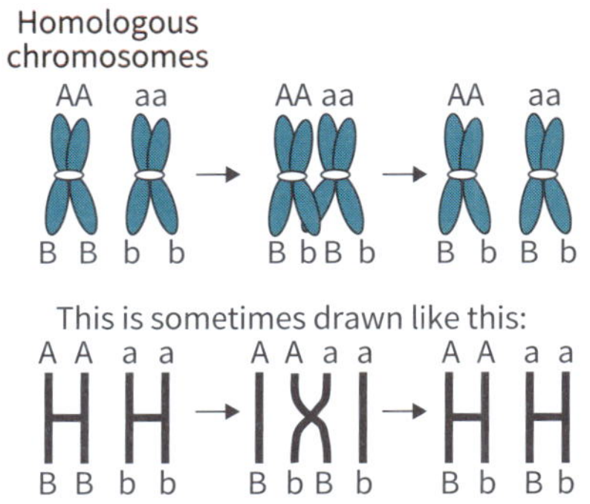

Figure 4.11 Meiosis event

What does this figure represent?

A mutation
B crossing over
C fertilisation
D random segregation

3 The following pedigree diagram shows the occurrence of albinism in three generations. Albinos are individuals with little or no pigment in the skin or hair.

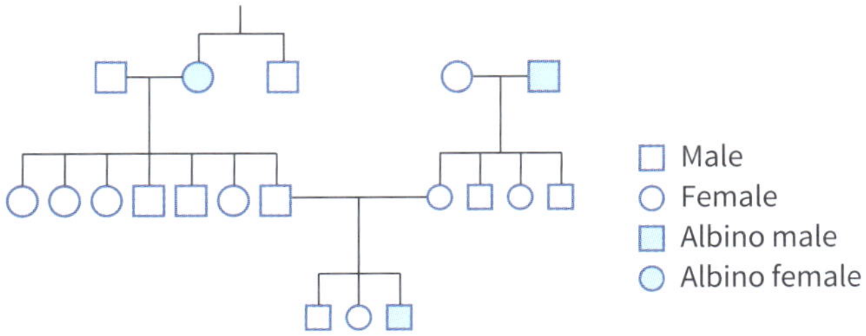

Figure 4.12 Pedigree diagram

From this family tree, it appears albinism is:

A dominant.
B recessive.
C sex-linked.
D co-dominant.

4 Carefully observe Figure 4.13 below.

Figure 4.13 Cattle dog
Source: image by Eva HoldereggerWalser (own work), [CC BY-SA 3.0 (https://creativecommons.org/licenses/by-sa/3.0)], via Wikimedia Commons

Assume this male dog is the result of a cross between a red bitch and a blue cattle dog. What type of inheritance is most likely to have contributed to the coat colour?

A co-dominance
B incomplete dominance
C multiple alleles
D sex-linked inheritance

5 You are asked to collect data on the frequency of the allele for brown eyes in humans. What is NOT one limitation of conducting an eye-colour survey of students at your school?

A The sample of students may be too small to represent the human population.
B The sample of students at your school may not be a fair representation of the variation in the whole population.
C You cannot tell from the phenotype what the actual genotype for brown eyes is.
D It is very difficult to obtain accurate data on the phenotype frequency of brown eye colour.

Short-answer questions

6 An individual has the base sequence of their genome screened and is found to have a unique mutation at one nucleotide. Can this mutation be classified as a single nucleotide polymorphism? Explain your answer. (2 marks)

7 Carefully study the following Punnett square and pedigree diagrams. Compare the information about inheritance patterns you can obtain when interpreting a Punnett square diagram with what you can obtain from a pedigree diagram. (4 marks)

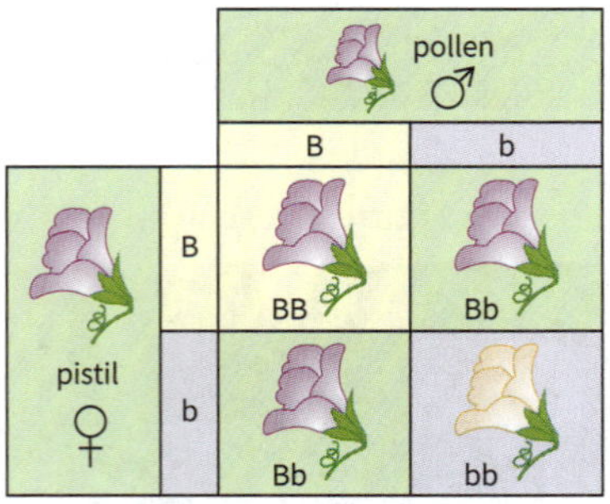

Figure 4.14 Punnett square diagram of a genetic cross between pea plants with mauve coloured flowers

Source: image adapted from Madeline Price Ball (Madprime, own work), [CC0, GFDL (http://www.gnu.org/copyleft/fdl.html)], via Wikimedia Commons

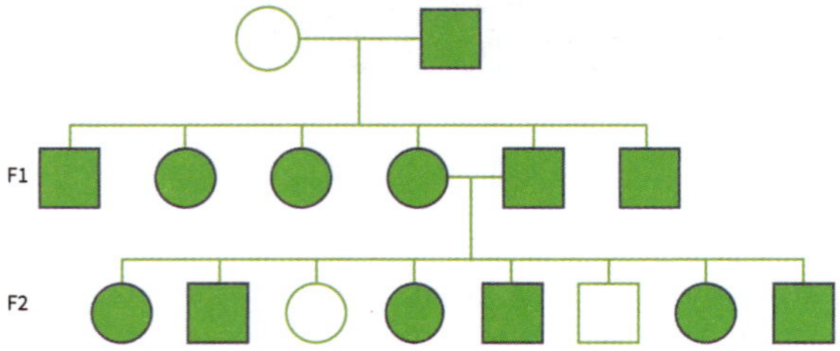

Figure 4.15 Pedigree diagram of crosses between curly-winged flies (shaded green) and wild type (normal) flies (unshaded)

8 **There is a set of multiple alleles of a gene for coat colour in mice. Four of those alleles are as shown in Table 4.8.**

Table 4.8 Inheritance of coat colour in mice

Allele	Phenotype
C	full colour (wild)
cch	chinchilla
cd	dilution
c	albino

The gene locus is not sex linked and each allele is dominant to those lower in the table (i.e. cch is dominant over cd). Construct Punnett squares for the crosses indicated below and give the phenotypic ratios expected from each. (4 marks)

- **a** **wild (heterozygous for dilution) × chinchilla (heterozygous for albino)**
- **b** **chinchilla (heterozygous for albino) × albino**

9 **Evaluate sources of genetic (genotype and phenotype) variation within a species.** (5 marks)

10 **How are genetic (genotype and phenotype) similarities and differences within a species compared?** (6 marks))

ANSWERS

KEY QUESTIONS

Key question ➲ p. 53

1 One member of each homologous pair of chromosomes comes from the maternal parent and the other from the paternal parent.

Key questions ➲ p. 54

2 Variation of the allele sequence occurs if crossing over has occurred between chromatids of homologous chromosomes. The chromosome from the maternal parent may be a combination of both of her parents' alleles if crossing over occurred. The other homologous chromosome from the paternal parent may be a copy of a chromosome from his mother, father or a combination of both if crossing over occurred.

3 Alleles Aa, Bb and Cc could produce ABC, abc with no crossing over, ABc and abC with one crossing over between B and C, Abc and aBC with one crossing over between A and B and AbC and aBc with crossing over occurring twice, once between each allele. That is, eight variations of alleles along a chromosome in the gamete.

4 The key process you need to demonstrate in crossing over in meiosis is the entangling of two adjacent chromatids from homologous pairs, their breaking at the same point (locus), and the exchange of chromatid material from one to the other.

5 The alleles needed to be marked on the arms of chromatids associated with the crossing over. If this was done using colour-coding for the forms of alleles, the resulting four cells would each contain a chromosome with at least two alleles marked. If the colour combination was different from that on the parent chromosomes, it was a recombinant.

Key questions ➲ p. 55

6 Crossing over produces chromosomes with new combinations of alleles. Gametes produced after crossing over has occurred will contain new combinations of maternally and paternally derived alleles.

7 The combination of two sets of quite different haploid sets of chromosomes will contain a variety of alleles.

8

Table A4.1 Punnett square of cross between parents with sickle-cell trait

Male gametes→ Female gametes↓	HbS	HbA
HbS	HbSHbS	HbSHbA
HbA	HbAHbS	HbAHbA

Results: 25% HbAHbA—homozygous for the normal form of haemoglobin.

50% HbSHbA—heterozygous; the phenotype is called the sickle-cell trait in which individuals usually have sufficient normal haemoglobin to maintain a normal life but they do experience the benefit of resistance to the malarial parasite.

25% HbSHbS—homozygous for sickle-cell anaemia; this condition is potentially life-threatening as the sickle-shaped red blood cells clog blood vessels. The red blood cells survive for only ten to 20 days, meaning that anaemia is often experienced as the body struggles to produce enough red blood cells.

Key questions ➲ p. 57

9 Autosomal dominant inheritance is a trait carried on normal chromosomes so both males and females may be affected equally. It is expressed in the phenotype if in the heterozygous form. Figure 4.6 shows this pattern because all the F1 generation (hybrids) have curly wings (dominant) but some of the F2 generation have the normal (recessive) condition.

10 Punnett square diagrams are designed to outline the possible genotypes of gametes and, at the intersection of rows and columns, the possible genotype outcomes of the zygotes resulting from fertilisation. The genotypes of zygotes can then be used to predict the phenotypes of offspring. Pedigree diagrams have a different purpose as they plot the family history of particular phenotypes of a trait over at least one generation. It may be possible to deduce aspects of the inheritance pattern, such as whether it is a dominant or recessive genotype, from the data or whether the trait is autosomal.

Key questions ➲ p. 58

11 The male will have the genotype X^NY and the female will have the genotype X^NX^n.

Table A4.2 Punnett square of inheritance of colour vision

Male gametes→ Female gametes↓	X^N	Y
X^N	X^NX^N	X^NY
X^n	X^NX^n	X^nY

Result: Normal vision in daughters but half will be carriers for the colour blindness. Half the sons will be colour blind and the other half will have normal colour vision.

12 The roan parents would each have the genotype RW. The predicted outcome would be one red (RR) : two roan (RW) : one white (WW).

Table A4.3 Punnett square for a hybrid cross of co-dominance in coat colour of cattle.

Male gametes→ Female gametes↓	R	W
R	RR	RW
W	RW	WW

13 The F1 generation of a monohybrid cross will all have a phenotype of a dominant allele in dominant/recessive inheritance; that is, one allele completely 'overruns' the other allele in the way it is expressed. In co-dominance and incomplete dominance both alleles are classed as dominant and both have hybrids that are not the same as the pure breeding parents. In co-dominance the F1 generation will have a mixture of both alleles being expressed, such as roan cattle having some white hairs and some red hairs. In incomplete dominance the resulting phenotype is a blend of the characteristics of both parents.

Key questions ➲ p. 59

14 This is an example of incomplete dominance as the heterozygous hybrid has a phenotype that is a blend of the colour of the flowers of the parents.

15 A cross between two pink snapdragons would be Rr × Rr.

16 In a Punnett square the cross between two pink snapdragons is as shown in the table below. There would be offspring in the ratio of one red (RR) : two pink (Rr) : one white (rr).

Table A4.4 Punnett square of a cross between two incompletely dominant pink snapdragons

Male gametes→ Female gametes↓	R	r
R	RR	Rr
r	Rr	rr

17 Multiple alleles generally refers to the presence of more than two possible alleles for a gene. While each diploid individual will still only possess two of the possible total number of multiple alleles, there is more chance of variation in the zygote as a result of fertilisation.

18

Table A4.5 Results of hybrid rabbit crosses

Cross	Predicted phenotype
black × chinchilla	black
black × Himalayan	black
black × albino	black
chinchilla × Himalayan	chinchilla
chinchilla × albino	chinchilla
Himalayan × albino	blended Himalayan/albino

Key questions ➲ p. 61

19 There are 200 alleles in the whole population.
A frequency = 44 ÷ 200 × 100 = 22%,
B frequency = 21 ÷ 200 × 100 = 10.5%,
O frequency = 135 ÷ 200 × 100 = 67.5%.

20 A single nucleotide polymorphism is a genomic locus where alternative bases occur in at least 1% of the population so it is not a one-off mutation. It may occur in a region of DNA that codes for a protein, such as the SNP resulting in HbS rather than the more common HbA form of haemoglobin. It may also occur in non-coding sections of DNA, such as parts of the DNA involved in gene expression.

Key questions ➲ p. 62

21 One of the main limitations of data gathered to compare genetic similarities and differences within humans is that the sampling may be insufficient. Methods include observations of phenotypes (e.g. hair colour) but many aspects of phenotypes are not able to be observed. You cannot project genotype frequency directly from phenotype frequency. SNPs are used to compare human DNA nucleotide base changes and they account for 90% of human variation, both in coding and non-coding sections of DNA.

22 Sampling across the geographic range of all species is necessary to ensure a fair comparison of SNPs or entire genomes. Data gathering is demanding of equipment and time and is expensive for genetic comparisons.

HSC EXAM-TYPE QUESTIONS

Objective-response questions

1 **B.** If you label the chromosomes Xx, Yy, Zz the combinations are 1 : X, Y, Z with x, y, z, 2 : X, y, Z with x, Y, z, 3 : X, Y, z with x, y, Z and 4 : X, y, z with x, Y, Z. **A** is wrong as it has not considered all combinations and **C** and **D** are wrong because some combinations have been counted twice.

2 **B.** This correctly identifies the diagram showing the chromosomes crossing over. Mutation, fertilisation and random segregation are not represented so **A**, **C** and **D** are wrong.

3 **B.** The gene would not be transmitted to the third generation without being expressed if it were dominant, making **A** wrong. **C** is incorrect because if it were sex-linked, the affected female should have passed the condition on to all of her sons. **D** is incorrect because if it were co-dominant, the parents of the albino male in the third generation would have exhibited some evidence of carrying the gene.

4 **A.** The dappled appearance suggests that both alleles for coat colour are being expressed. **B** is wrong because the coat colour is not a blend. **C** is wrong because no matter how many alleles exist for coat colour, each individual will only have two. **D** is incorrect because if coat colour was sex-linked the male would have only one allele and would not likely be a mixture of colours.

5 **D.** It is generally not difficult to decide if eye colour is brown. Answers **A** and **B** are all significant limitations of collecting data on the frequencies of characteristics in a population. **C** is a statement that reflects the limitation of using phenotypes to infer information about genotypes.

Short-answer questions

6 **EM** Your explanation needs to outline the similarity and the difference between the SNP and a single mutation.

A single nucleotide polymorphism is a different form of a nucleotide on a piece of DNA, which results from a different nitrogenous base change in one nucleotide. It has an allele frequency of at least 1% of a population. (✓) Its origin may have been a one-off mutation but it has continued in the population and its frequency has increased. The SNP HbS is a form of haemoglobin that has provided resistance to the pathogen that causes malaria so its frequency is high in populations exposed to the disease. (✓)

7 **EM** Maximise marks by making full use of the diagrams to stimulate analysis and comparison.

Punnett square diagrams serve a specific purpose of predicting the genotype and phenotypes that result from gametes combining in fertilisation to produce the zygote. (✓) They rely on applying understanding of the relationship between alleles, such as dominance or incomplete dominance. They can be adapted by showing alleles on the sex chromosomes to predict the outcomes of sex-linked inheritance. (✓)

Pedigree diagrams are really family trees that reflect the inheritance of a characteristic over several generations. (✓) In some cases pedigree diagrams can be used to determine the relationship between alleles or whether they are carried on the sex chromosomes. In other cases a pedigree diagram can show the incidence of a mutation that results in a new allele. (✓)

8 **EM** You need to construct two Punnett square diagrams and interpret the result of the crosses with the stated genotypes to predict the phenotype outcomes.

a Wild (heterozygous for dilution or C cd) × chinchilla (heterozygous for albino or cch c)

Table A4.6 Punnett square diagram

Wild gametes→ Chinchilla gametes↓	C	cd
cch	C cch	cch cd
c	C c	cd c

Construction of Punnett square (✓)

Predicted phenotype ratio: 2 wild type (C cch and C c) : 1 chinchilla (cch cd) : 1 dilution or twice as many wild type as chinchilla and dilution. (✓)

b Chinchilla (heterozygous for albino) × albino

Table A4.7 Punnett square diagram

Chinchilla gametes→ Albino gametes↓	cch	c
c	cch c	c c
c	cch c	c c

Punnett square constructed correctly (✓)

Predicted phenotypic ratio: 2 chinchilla (cch c) : 2 albino (c c) or equal numbers of chinchilla and albinos. (✓)

9 **EM** You need to identify the sources of variation, outline their role, and make a judgement about how important they are in differences in genotypes and phenotypes in a species. You need to draw on your knowledge from Chapter 3 on the roles of genes and the environment on phenotype expression.

Processes that occur in meiosis such as crossing over and segregation of homologous chromosomes 'reshuffle' alleles on chromosomes. (✓) Fertilisation also results in new combinations of alleles in the zygote. The ultimate (and most important) source of variation is mutation, which is the source of new alleles. Mutation may result in multiple alleles, which gives increased potential for variation in gametes and zygotes, and changes in allele frequency. (✓)

Variations in inheritance patterns include sex-linked inheritance, in which alleles are carried on sex chromosomes so different patterns of phenotypes and genotypes are experienced by either gender. (✓) Interactions between alleles such as dominance, recessiveness, incomplete dominance and co-dominance result in variations of phenotypes for particular heterozygous genotypes. (✓) Incomplete dominance, for example, results in a blended phenotype in the hybrid but co-dominance results from both alleles being expressed, so both A and B antigen or both red and white hairs in a hybrid roan cow. Finally, the environment (e.g. diet) interacts with genes to result in the final phenotype of an individual. (✓)

10 **EM** Part of your revision should be an analysis of the Inquiry Question for each chapter and 'marrying it' with the chapter content. Again, it is a matter of deconstructing each part of a question and addressing all parts. In this case you need to apply your understanding of the sources of variation in Section 1 of the chapter and the ways of collecting and processing data about genetic variation in Section 2.

Genotype variations require data of the genome to be collected and analysed. (✓) This can be done by selecting and comparing single nucleotide polymorphisms that, in the case of humans, account for 90% of genetic variation. (✓) Gene expression/regulation and the environment may influence the way the phenotype reflects the genotype. Because of this, to effectively compare genetic similarities and differences the genotypes need to be compared with the phenotypes. (✓)

Phenotype similarities and differences are compared by surveying or otherwise collecting sampled data about phenotypes such as blood types, hair or eye colour, or the presence of certain genetic conditions such as genetic diseases. (✓) The particular phenotypes may depend on the species whose population variation is being assessed. It may be coat colour in some mammals or the presence of various chemical defences in eucalypts. (✓) Sampling needs to take account of the geographic distribution of the species, as variations such as skin colour in humans can be linked to distance from the Equator. (✓)

CHAPTER 5 INHERITANCE PATTERNS IN A POPULATION

INQUIRY QUESTION:

Can population genetic patterns be predicted with any accuracy?

In Chapter 4 you investigated frequencies of characteristics of populations. In this chapter you will first review the technologies that have expanded the range of genetic studies of populations, and then examine some key applications of these technologies that have changed our understanding of human evolution, conservation practices and disease management.

Before the development of these technologies scientists bred organisms and studied the traits of offspring, cells and cell theory, and made inferences about population genetic patterns. Technologies now allow DNA to be studied directly.

1 The use of technologies to determine inheritance patterns in a population

» Students investigate the use of technologies to determine inheritance patterns in a population using, for example, DNA sequencing and profiling.

- In this section you review the technologies that are used to investigate inheritance patterns and changes that occur across populations. These technologies include **DNA sequencing** and **DNA profiling**. Both of these technologies depend on a variety of processes which involve the manipulation of DNA.
- DNA sequencing is a complex set of technology-dependent processes used to determine the sequence of nucleotide bases in the genome. This has led to the study of inheritance patterns which has improved our understanding of:
  - human evolution
  - inheritance of certain diseases or disorders
  - levels of genetic diversity within species, especially in relation to conservation management.
- DNA profiling uses technology to determine human identity and familial relationships based on comparison of a small but highly variable fraction of the total genome.

DNA sequencing

- DNA sequencing is used to identify the sequences of nitrogenous bases in a DNA molecule. It is possible to discover genetic patterns and changes in populations, such as mutations, single nucleotide polymorphisms and other changes in DNA nucleotide sequences, when comparisons are made between DNA molecules. These technologies have been refined since the late 1970s and improved with the assistance of processes that produced the data used in the Human Genome Project (see Table 5.1).
- DNA sequencing involves many of the steps outlined below.
  - The DNA is denatured into separate strands.
  - Primers and specific enzymes (see Chapter 2 on DNA replication) are used to make copies of the labelled strands for analysis: this process is used to amplify the DNA and traditionally involved **polymerase chain reaction (PCR)**.
  - The DNA strand is broken into labelled pieces by using particular enzymes called nucleases in a process known as hydrolysis: an alternative method is to synthesise pieces of varied lengths of DNA containing labelled nucleotides using the original DNA template. Polymerase enzymes form pieces of complementary strands that are labelled in some way. This is called sequence by synthesise (SBS) because the direct action of DNA polymerase produces the observable output.
  - The pieces of DNA are separated so the labelled bases can be identified in order: one traditional method of separation of the fragments is electrophoresis.
  - Detecting the base sequence occurs by either recognising overlapping sequences so the entire sequence can be compiled, or identifying labelled nucleotides in sequence using detection methods such as radioactive labels, fluorescent dyes or reactions that produce light or changes in pH.

DNA sequencing: the processes used to determine the order of nucleotides in DNA

DNA profiling: the process used to determine an individual's DNA characteristics, often for the purposes of paternity testing, genealogy, identification of remains or criminal investigations

polymerase chain reaction (PCR): a form of artificial DNA replication (molecular photocopying) that uses specific enzymes and heating and cooling to make unlimited copies of a specific DNA sequence

➔ Figure 5.1 gives an example of one SBS method of DNA sequencing. In the example fragments of DNA have been synthesised from an original DNA strand but special bases containing dideoxyribose instead of deoxyribose sugar were added for each of the four DNA nitrogenous bases. When they were added at random, they terminated the sequence being formed.

➔ Table 5.1 below summarises some of the early major advances in technology that resulted in dramatic improvements in DNA sequencing.

➔ Technological advances continue to be developed and applied to each of the steps. These technologies have enabled innovations in DNA sequencing and are often referred to as second and third generation sequencing. Some of these advances are outlined below.
- Improved automation in the process makes it possible to speed up the rate of DNA sequencing; for example, sequencers such as ABI PRISM allow simultaneous sequencing of hundreds of samples.
- Other methods of DNA **clonal** amplification have been developed, including bridge, bead or rolling circle amplification.
- There have been improvements in labelling and safety along with faster computer analysis.

clonal: making multiple identical copies

A: DNA sequence	ATGCAGCGTTACCATG
B: fragments synthesised when terminal A dideoxyribose bases added	A ATGCA ATGCAGCGTTA ATGCAGCGTTACCA
C: fragments synthesised when terminal T dideoxyribose bases added	AT ATGCAGCGT ATGCAGCGTT ATGCAGCGTACCAT
D: fragments synthesised when terminal G dideoxyribose bases added	ATG ATGCAG ATGCAGCG ATGCAGCGTTACCATG
E: fragments synthesised when terminal C dideoxyribose bases added	ATGC ATGCAGC ATGCAGCGTTAC ATGCAGCGTTACC

F: DNA fragments are separated based on their size (the shorter fragments progress further in the electrophoresis gel)

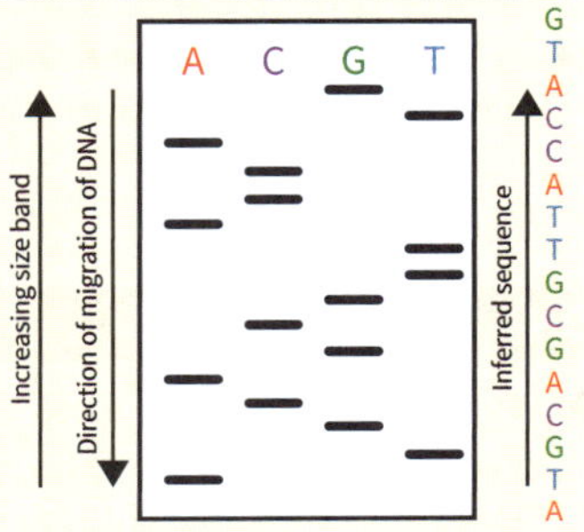

Figure 5.1 Determining DNA base sequence by interpreting the pattern of DNA fragments separated by electrophoresis

Table 5.1 Major advances in technology used in DNA sequencing

Date/Researchers	Technology	Advantages/Disadvantages
1977: Maxam-Gilbert's sequencing was the first widely adopted technique	produced single-stranded DNA that was chemically digested at specific bases in sequence and radioactively labelled at one end of the DNA strand fragments were separated by a process of electrophoresis and identified using autoradiography	safety issues via the use of radioactive labels and toxic chemicals a slow, laborious process discovered the different gene organisation in eukaryotes to prokaryotes; discovered DNA included 'non-coding' as well as 'coding' sequences
1977: Sanger; the widely adopted breakthrough change	used a third form of the DNA sugar (dideoxyribose) that would cause selective chain termination when incorporated into the four different nucleotides used DNA polymerase enzymes rather than chemicals to produce the fragments and a radioactive RNA primer	more efficient, reliable, safer and better suited to automation used in the Human Genome Project aspects of this method are still used when relatively short DNA sequences are analysed
1986: Leroy Hood and co-workers; commercialisation of DNA sequencing	introduced fluorescent dideoxy-terminators, laser-induced fluorescent detection and computerised base labelling	reduced number of reaction vessels needed; contributed to the first draft of the Human Genome Project, which was completed ahead of schedule
1990s: Harold Swerdlow; increased automation	gel electrophoresis was replaced by 50 μm capillary tubes which allow higher voltages	the quicker heat dissipation resulted in shorter run times
1990s: Pål Nyrén; pyrosequencing method of DNA sequencing was the first of the next generation sequencing technologies	introduced a two-enzyme process which used natural nucleotides and emitted light when detecting the sequence reduced the need for lengthy washing steps	DNA sequencing results could be observed in real time; the process allowed mass parallelisation of sequencing reactions, improving time and efficiency; levels of brightness when there were multiple sequences of the same nucleotide caused problems

- The replacement of sequence by synthesise (SBS) with sequencing using **DNA ligase**: this process is called sequencing by **oligonucleotide** ligation and detection (SOLiD).

➔ A more recent advance, nanopore technology, means that DNA replication is no longer necessary as single molecules of DNA can be sequenced. Nanopore technology processes single strands of DNA through a biological nanopore embedded in a synthetic membrane.

➔ KEY QUESTIONS

1 **What is DNA sequencing?**

2 **Outline some of the technologies that are used in DNA sequencing.**

3 **Explain the role of the polymerase chain reaction.**

4 **How can DNA sequencing be used to identify inheritance patterns in a population?**

Answers ➲ p. 78

DNA profiling

➔ DNA profiling, also known as DNA fingerprinting, was developed in the 1980s and is used to analyse an individual's DNA. It is often used for identification based on comparing small, specially selected DNA segments. It is a much simpler and more selective process than whole genome sequencing. DNA profiling is used in forensics and in the identification of family relationships. Criminal identification relies on the purity of DNA samples.

➔ The process usually focuses on analysis of nuclear DNA. **Mitochondrial DNA (mtDNA)** has also been analysed in samples of bone, teeth or old hair shafts. Because mtDNA is inherited purely through the maternal line it can be useful for identifying unidentified remains or missing persons.

➔ The human Y chromosomes are also used in DNA profiling to assist in the investigation of crimes that may have involved a number of males.

➔ DNA profiling can take a minimum of three days though a portable device developed in 2017 claimed to be able to match biological samples to a national DNA database in one hour.

DNA ligase: a specific type of enzyme that breaks up DNA molecules
oligonucleotide: a short DNA or RNA molecule
mitochondrial DNA (mtDNA): DNA carried by the mitochondria that self-replicates independently of the nucleus and mitosis; mitochondrial DNA is passed to offspring through the maternal lineage

➔ DNA profiling depends on a range of technologies and methods. The earlier process required relatively large samples of DNA. Steps included are outlined below.
- The collection of biological samples, DNA extraction and purification is required.
- **Restriction fragment length polymorphism (RFLP)** analysis is a process which uses enzymes, called restriction enzymes, to cut DNA at specific locations into fragments that are separated and compared. This is applied to large samples of DNA. The fragments often vary in length between individuals because they contain **variable number tandem repeats (VNTRs)**. These are sometimes called mini-satellites.
- Fragments are separated through the use of electrolysis. This is followed by a transfer of the fragments to a membrane in a process called blotting. On the membrane, specific radioactive DNA probes hybridise with complementary DNA fragments. Fluorescent dyes may replace radioactive probes. The membrane is washed free of excess probe.
- X-ray film is used to detect a radioactive pattern, followed by comparison of the DNA band pattern to other DNA samples.

➔ Polymerase chain reaction (PCR) analysis has largely replaced the RFLP method because it is faster, more reliable, and can be used on degraded or small DNA samples. The steps are outlined below.
- The collection of biological samples, DNA extraction and purification is required.
- **Short tandem repeat (STR)** loci are targeted with sequence-specific primers and amplified using the polymerase chain reaction. Generally about four loci are targeted and, because they are highly variable, four are sufficient for DNA comparison.
- The DNA fragments produced are separated by gel or capillary electrophoresis.
- Fluorescent dye reveals the DNA bands that can be used for comparison of samples.

➔ DNA profiling can have limited application because of the factors below.

restriction fragment length polymorphism (RFLP): lengths of DNA that have been cut by restriction enzymes and whose lengths vary, often depending on how many repeats of VNTR they enclose
variable number tandem repeats (VNTRs): a sequence of between eight and 50 nucleotide bases that repeat in tandem; the number of repeats (not the actual sequence) tends to be variable between individuals, making this technology useful for DNA profiling
short tandem repeats (STRs): between two and ten nucleotide bases that repeat in tandem sequence, usually found in non-coding sections of DNA; the number of repeats is found to vary significantly between individuals

- Evidence could be contaminated and small amounts of contamination are highly magnified in PCR analysis of STRs. The RFLP method is also prone to error.
- The STRs that are analysed are only a sample; therefore the results are quoted as a probability that there is a perfect match.
- Errors in DNA profiling may result from 'planted evidence'. This issue is often countered with the argument that DNA profiles can be used to prove a suspect's innocence.

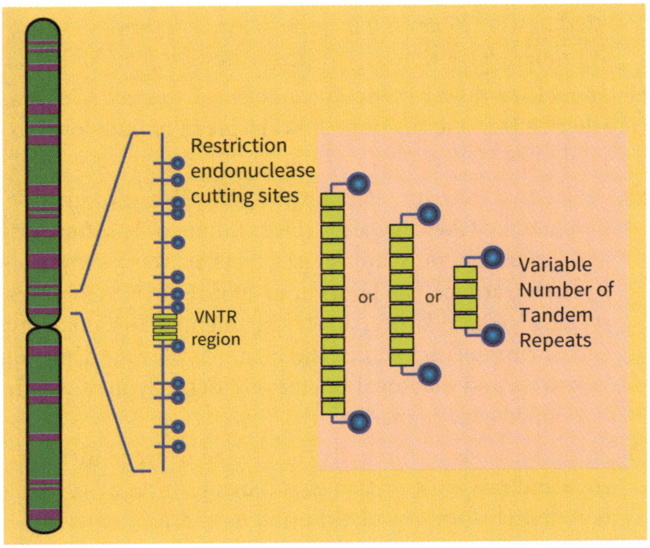

Figure 5.2 Human chromosome showing a region containing VNTRs

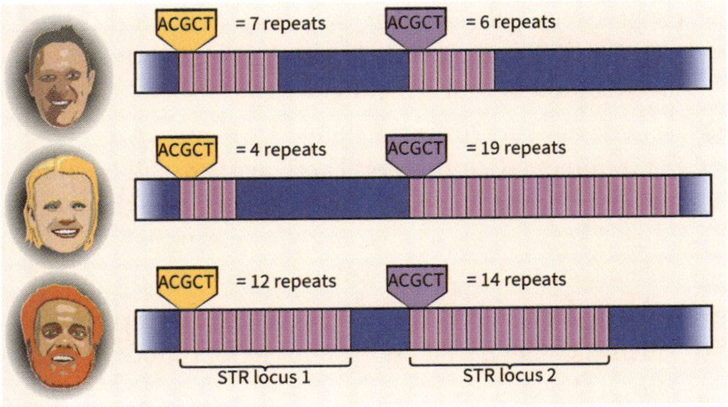

Figure 5.3 Comparative STR lengths at two specific loci

→ KEY QUESTIONS

5 **Distinguish DNA sequencing from DNA profiling.**

6 **Outline some of the technologies used in DNA profiling.**

7 **How is DNA profiling used to determine inheritance patterns in a population?**

8 **What is the relationship between short tandem repeats (STRs) and variable number tandem repeats (VNTRs)?**

Answers ⊃ p. 78

2 The use of data analysis from a large-scale collaborative project

» Students investigate the use of data analysis from a large-scale collaborative project to identify trends, patterns and relationships; for example, the use of population genetics data in conservation management; population genetics studies used to determine the inheritance of a disease or disorder; population genetics relating to human evolution.

Population genetics

→ **Population genetics** is the study of the frequencies of variants of alleles, the processes bringing about changes in alleles, and population structures and divisions.

→ Many genes within a population are **polymorphic**; that is, they occur with multiple variants or alleles. Single nucleotide polymorphisms are the most common genetic variation among humans (see page 61).

→ It is important that data analysis in population genetics takes sample sizes into account. For example, the Human Genome Project determined the entire sequence of human DNA but only a sample of human DNA was analysed.

→ Single nucleotide polymorphisms are another way of sampling genetic variation.

→ Data analysis of information about allele frequency and changes contributes to decisions about conserving endangered species, provides evidence for the genetic component cause of some diseases, and provides evidence of human evolution and migration.

→ The levels, origins and patterns of genetic variation help to explain the areas of adaptation, speciation and microevolution studied in Year 11.

→ The diversity within a gene pool may result from population size, mutation, **genetic drift**, natural selection, diverse environments and population movements resulting in **gene flow**. You will review these concepts in Chapter 6.

population genetics: the study of genetic characteristics of populations and how and why the allele frequencies change over time and space
polymorphic: exists in more than one form
genetic drift: changes in allele frequency in populations, particularly if the population is small, which result from random or chance events (also called sampling error)
gene flow: the movement of genes between populations, resulting in an increase in genetic diversity because of migrations

→ KEY QUESTIONS

9 What is the purpose of population genetics?

10 Distinguish between gene flow and genetic drift.

11 How can the study of single nucleotide polymorphisms assist population genetics?

Answers ➲ p. 78

Conservation management

- In Year 11 you learnt that functional ecosystems depend on biological diversity, including genetic variation.
- Concerns in conservation genetics include small, fragmented or **inbreeding populations**. Endangered and vulnerable species are also at increased risk because of a lack of genetic diversity.
- Size and isolation of populations have an impact on genetic diversity. Populations restricted to island habitats are well-known for their vulnerabilities. One feature of island and small population habitats is the impact of the initial migrants or founders on genetic characteristics. The **founder effect** is the term used to describe this factor in genetic diversity.
- You may have investigated well-publicised examples of the Wollemi pine or the Tasmanian devil (*Sarcophilius harrisii*) as examples of endangered species that lack genetic diversity.

inbreeding populations: populations that are sufficiently small so breeding occurs between genetically related individuals; as a result there is reduced fitness of individuals and of the long-term adaptability of the population

founder effect: the isolation of small groups that lack genetic variation, which can result in changes to the genetic make-up of that group

- Lack of genetic diversity is one factor contributing to a number of species in Figure 5.4 being classified as endangered. Other factors contributing to declining populations or threats include the need for specific requirements, such as bamboo by the panda.

Population genetics used to help conserve the critically endangered Rice-flower shrub

- Rice-flower is a small, spreading shrub found in grasslands and open shrubland. It comprises two subspecies, Spiny Rice-flower (*Pimelea spinescens spinescens*) and Wimmera Rice-flower (*Pimelea spinescens pubiflora*).
- Spiny Rice-flower is endemic to central-western Victoria, where it is found in about 20 widespread populations totalling 12 000 plants.
- A National Recovery Plan for the Spiny Rice-flower was prepared by Oberon Carter (Department of Sustainability and Environment, Victoria) and Neville Walsh (Royal Botanic Gardens, Melbourne) in 2006. Other researchers include Victoria University, Department of Sustainability and Environment, Victoria, the Centre for Stress Adaptation Research, and the Royal Botanic Gardens, Sydney, which developed genetic markers.
- Spiny Rice-flower is mainly threatened by weed invasion, road and rail works, grazing, and inappropriate fire regimes. Historically, distribution has been fragmented by land clearing for settlement, agriculture and industry.
- Wimmera Rice-flower is the subject of an integrated conservation program that includes genetic studies, translocation and reintroduction of plants, and investigation of root structure.
- Genetic markers (microsatellites) are being developed for Wimmera Rice-flowers. These markers will provide further information on genetic variation across the species range. Leaf samples have been collected and DNA isolated for assessment of genetic variability.
- For long-term survival it is necessary to employ methods which encourage wild populations to reproduce in their natural habitats.

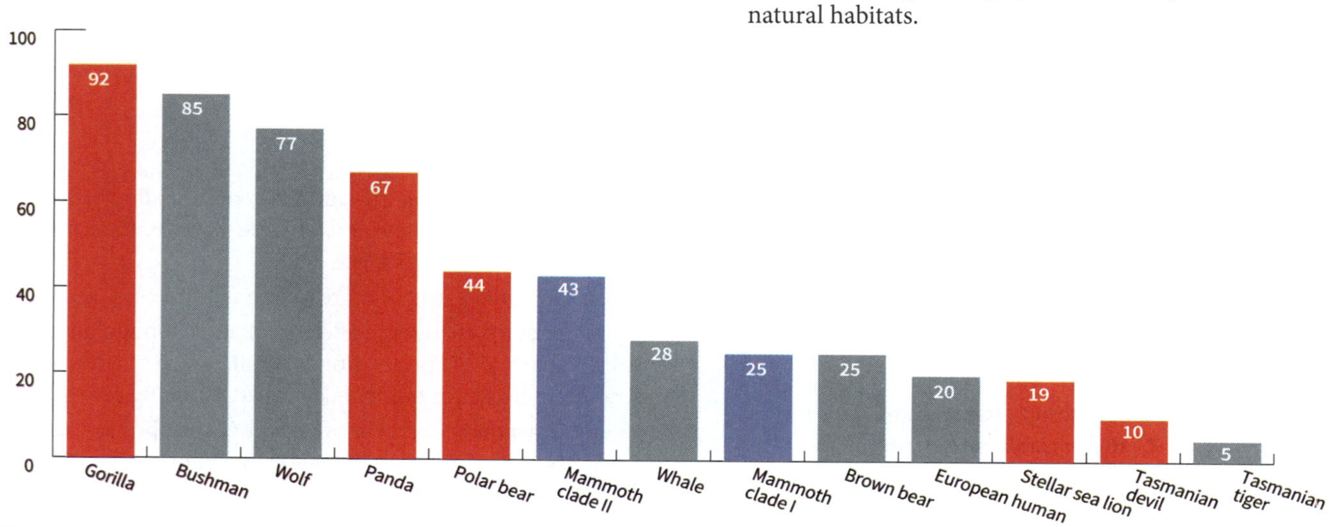

Figure 5.4 Genetic diversity of endangered (red), extinct (grey) and other species (blue) measured using SNP data

Population genetics used to help conserve the endangered Tasmanian devil

- The population of Tasmanian devils has been decimated by the rapid spread of Devil Facial Tumour Disease (DFTD), a contagious cancer.
- An unstable climate between three to five thousand years ago caused a dramatic reduction in population, distribution and diversity.
- Research has been funded by the Australian and Tasmanian governments, Sydney University, San Diego Zoo Global and the Zoo and Aquarium Association, Mosman. These efforts have been supported by considerable public donations and by other groups and zoos housing refuge populations.
- Lack of genetic diversity (devils have approximately one million SNPs, which is relatively low) is believed to be a factor in susceptibility to the disease. In particular, devils have low diversity in the genes associated with the immune response called the Major Histocompatability Complex.
- Genetic studies of cancer cells suggest the cells have properties that enable them to avoid detection and destruction by the devil's immune system. The genome of healthy Tasmanian devils' cells and the cancer cells have been compared, with 20 000 mutations found in cancer cells.
- Devils in northwest Tasmania seem to be more resistant to DFTD so there is potential to discover if there is a genetic basis to this resistance.
- Genetic investigations are now focusing on behavioural adaptations. These include younger breeding ages in females and shorter dispersal distances. Researchers are investigating whether these adaptations have a genetic basis (called a genetic footprint). To date the evidence is that the variations detected in devil genome are random and are the result of 'genetic drift' rather than natural selection.
- Cancer-free preserves of the population are being established on the mainland and other islands. Current predictions are that the devils will be extinct in the wild within 35 years.

KEY QUESTIONS

12 **What factors within population genetics influence the chances of survival of an endangered species?**

13 **What practices in conservation management can help alleviate lack of genetic diversity and/or small populations?**

Answers p. 78

The inheritance of a disease or disorder

- It is thought that almost all human non-infectious disease, except conditions caused by trauma, has a genetic component. In some cases (e.g. Huntington's disease and cystic fibrosis) the genetic component is large and the inheritance and cause of the disorder is understood. Other diseases (e.g. adult onset diabetes) have both genetic and environmental components and there is still a lot to be understood about the triggers to the disease.
- Diseases with environmental and genetic components include cancers, coronary heart disease, schizophrenia, autism and multiple sclerosis. Recent successful investigations on the genetic aspects of these diseases resulted from the **epidemiological** study of large sample sizes for an individual trait. These investigations involved international collaboration on an unprecedented scale. In Chapter 15 you will review epidemiology and human diseases.
- Genetically isolated populations (or 'population isolates') such as Iceland and Finland have already proved useful in the identification of rare recessive disease genes. Such genes are only detectable in isolated populations with a limited number of founding members. In these populations, rare disease alleles are enriched, resulting in homozygote individuals affected by the disease.
- Iceland's population is not a conventional 'population isolate' but genetic drift over generations has resulted in relatively small genetic variation. Researchers used data from well-known genealogy, large extended families, and the more than 50% of the adult population that have volunteered their medical and genetic information. This, among other benefits of a geographically isolated population, has enabled the identification by means of linkage and, more recently by **genome-wide association (GWA) studies**, of an impressive number of variants contributing to the development of common yet complex diseases. Among these are gene loci for myocardial infarction and stroke, type 2 diabetes, atrial fibrillation and prostate cancer.

epidemiological: the study of the distribution and determinants of health-related status or disease occurrence and the application of this study to the control of diseases and other health problems

genome-wide association (GWA) studies: an approach that involves rapidly scanning markers across the complete sets of DNA, or genomes, of many people in order to find genetic variations associated with a particular disease

Table 5.2 Large-scale population genetics study of disease

Description of the project	Identified trends, patterns and relationships	Strategies/Goals/Outcomes/Data analysis
the Wellcome Trust Case Control Consortium (WTCCC) conducted a large series of case-control studies, combined with research from Oxford University, to map variation in human genotypes across the UK	compared the DNA of people with tuberculosis, coronary heart disease, type 1 diabetes, type 2 diabetes, rheumatoid arthritis, Crohn's disease (a type of inflammatory bowel disease), bipolar disorder and hypertension (2000 for each disease group) with that of the DNA from 3000 healthy controls	compared 500 000 SNPs of people with a single disease findings included a clear genetic link to obesity and three new genes linked to type 2 diabetes a major gene region on chromosome 9, identified by independent studies on coronary heart disease, was verified four chromosome regions containing genes that can predispose people to type 1 diabetes were discovered along with three new genes for Crohn's disease and a gene linking the two autoimmune diseases: type 1 diabetes and Crohn's disease

→ KEY QUESTIONS

14 **Sickle-cell anaemia is a genetic disease resulting from a single nucleotide polymorphism of the gene responsible for the production of the haemoglobin protein. How would you use population genetics to investigate this disease?**

15 **Why are population isolates useful in determining the population genetics of a genetic disorder?**

Answers ➲ p. 78

Human evolution

- Population genetics relying on data from DNA has contributed to the discussion about modern human evolution. Modern humans share a relatively recent link with two ancestral groups: Neanderthals and Denisovans. Table 5.4 provides a case study of analysis of data to deduce trends, patterns and relationships between these three groups.
- Analysis of data about the similarities and differences in human alleles includes efforts to describe population structure, including diversity, within and between populations; to reconstruct population history, migration, relationships and changes; and to understand adaptation to local environments.
- Factors in human evolution include mutation, natural selection, genetic drift and gene flow. These factors have combined in a way that has resulted in humans having relatively low genetic diversity. Globally, representative samples of humans have variation in only about 0.08% of DNA base pairs in the nuclear genome.
- Genetic variation within populations is compared with global variations and then analysed. Results have demonstrated that genetic variation within local human populations is as significant as variations between discrete population clusters. The concept of biological race is no longer regarded as a valid concept.
- Sudden reductions in population sizes because of environmental events (e.g. earthquakes, drought, floods, fires or disease) or wars are called **population bottlenecks**. For example, about 70 000 years ago the Toba super-volcano erupted in Indonesia and caused major environmental changes. There is evidence that this caused population bottlenecks in a number of primates, including humans.
- **Coalescent theory** has been used to reconstruct the ancestral state of the population. It has been suggested that at the time of the Toba super-volcano the human population was between 10 000 and 30 000. Coalescent theory makes assumptions about the rate of change of alleles and so is not beyond refute. Some examples using coalescent theory to trace human ancestry include:
  - There is extremely low global genetic variation on the Y chromosome. Y chromosome inheritance occurs along the paternal line (from father to son) and has relatively low amounts of recombination from crossing over. It has been suggested that the global diversity found on the Y chromosome may have all been generated within the last 60 000 years.
  - Mitochondrial DNA (mtDNA) or non-nuclear DNA, passed from mother to offspring, has been used in coalescent theory. Studies based on an assumed consistent rate of mutation of mtDNA have been used to develop a 'molecular clock'. The term coined for the last common ancestor is 'mitochondrial Eve'. This has attracted considerable publicity and debate.

population bottlenecks: sudden reductions in population size which severely impact diversity

coalescent theory: the application of population genetic studies of particular inheritance patterns, such as genes on the Y chromosome or mtDNA, in order to estimate a time of shared ancestry in the past

- Fossil evidence has resulted in the widespread belief that the first modern humans evolved in Africa. Two theories about human evolution and migration remain but both theories accept that *Homo erectus* populations migrated out of Africa about one million years ago.
- The 'Out of Africa' theory suggests that there was a second wave of migration of *Homo sapiens*, who had evolved from *Homo erectus* stock between 130 000 and 200 000 years ago, and this group replaced all other human ancestral groups around the world.
- The alternative, multi-regional hypothesis (also called the theory of regional continuity) suggests that from the initial migration the *Homo erectus* populations evolved into a new species and continued gene flow maintained the existence of only one species, which eventually became *Homo sapiens*.
- Genetic evidence has suggested that there has also been an ancient contribution to the human genome from southern Asia. This might suggest that human evolution is 'mostly out of Africa'.
- Investigation of the relationship between Neanderthals, Denisovans and modern humans has taken several approaches as shown in Table 5.3 below. The fact that only a few specimens have provided DNA (either mitochondrial or nuclear) has meant that conclusions are tentative, not always consistent, and prone to change or dispute.
- Some researchers believe that around 450 000 years ago, ancestors of the Neanderthals and Denisovans migrated from Africa to Europe and Asia. Those who travelled east became the Denisovans. Those who stayed in the west were the Neanderthals.

Table 5.3 Evidence of human evolution from ancestral human fossils and population genetics

Location	Date	DNA finding	Conclusion
Neanderthal specimens collected at Feldhofer cave in the Neander Valley, Germany	40 000 years ago	modern human mtDNA varies on average at eight sites but Neanderthal specimens differ at 27 mtDNA sites on the one gene locus	not clear whether Neanderthals were a separate human race or separate species
the Mezmaiskaya cave in Russia	29 000 years ago (most recent Neanderthal find in Eurasia)	modern human mtDNA varies on average at eight sites but this Neanderthal specimen (a two-week old infant) differs at 23 mtDNA sites on the one gene locus and shows a 3.48% divergence from the Feldhofer cave specimen	Neanderthal extinction paralleled the arrival of the first modern humans the Feldhofer cave finds and this specimen belong in a **clade** distinct from modern humans, suggesting their mtDNA types have not contributed to the modern human mtDNA pool
Cueva Mayor–Cueva del Silo cave system in north-central Spain (includes Sima de los Huesos, or the 'pit of bones')	430 000 years ago (oldest DNA ever)	the mtDNA of the femur of one individual found in the pit confirmed that the owner was more closely related to the Denisovans other bones show some similarities to those of Neanderthals but they also appear to be connected to ancient humans	modern humans split from the ancestors of Neanderthals hundreds of thousands of years earlier than thought nuclear DNA evidence from the site suggests that Neanderthals and the Denisovans had diverged from their common ancestor by 765 000 years ago, which split from the modern human lineage much earlier than expected (previously believed to be between 315 000 and 540 000 years ago) *Homo heidelbergensis* was believed to be the common ancestor of modern humans, Neanderthals and Densovians, but now *Homo antecessor* (dating to 900 000 years ago) may be the common ancestor; this requires confirmation via *Homo antecessor* specimens being found in Africa or the Middle East

clade: a group consisting of a common ancestor and their lineal descendants; sometimes called a 'branch' in the 'tree of life'; clades can be nested within larger clades

→ KEY QUESTION

16 **What are three problems that arise from using data analysis of DNA to determine the course of human evolution?**

Answer ➲ p. 78

CHAPTER SYLLABUS CHECKLIST

Are you able to answer these questions from the syllabus for this chapter? Tick each question as you go through the checklist if you are able to answer it. If you cannot answer a question, turn to the relevant page in the study guide to find the answer. For NESA key word meanings, go to www.educationstandards.nsw.edu.au and search 'key words'.

FOR A COMPLETE UNDERSTANDING OF THIS TOPIC:		PAGE NO.	✓
1	Can I investigate the use of technologies to determine inheritance patterns in a population using, for example, DNA sequencing?	68–70	
2	Can I investigate the use of technologies to determine inheritance patterns in a population using, for example, DNA profiling?	70–71	
3	Can I investigate the use of data analysis from a large-scale collaborative project to identify trends, patterns and relationships in the use of population genetics data in conservation management?	71–73	
4	Can I investigate the use of data analysis from a large-scale collaborative project to identify trends, patterns and relationships in the use of population genetics studies used to determine the inheritance of a disease or disorder?	73–74	
5	Can I investigate the use of data analysis from a large-scale collaborative project to identify trends, patterns and relationships in the use of population genetics relating to human evolution?	74–75	

HSC EXAM-TYPE QUESTIONS

Objective-response questions (1 mark each)

1 **What does DNA sequencing determine?**
- A the number of nucleotides in a DNA molecule
- B the sequence of sugars in a DNA molecule
- C the number of bases in a DNA molecule
- D the sequence of DNA bases in a DNA molecule

2 **What can the comparison of DNA sequences be used to identify?**
- A the perpetrator of a crime
- B the father of a child
- C the existence of a single nucleotide polymorphism
- D the chances of contracting a disease

3 **What does DNA profiling compare?**
- A the sequence of bases of DNA
- B the length and extent of highly variable tandem repeating sequences of DNA
- C the length and extent of single nucleotide polymorphisms
- D the blood groups of crime suspects and blood found at crime scenes

4 **What is a useful application of the study of population genetics?**
- A conservation management of an endangered species
- B treatment of genetic disease
- C identification of a crime suspect
- D understanding racial differences

5 **What investigation assists our understanding of human evolution?**
- A comparison of the DNA of all human ancestors
- B DNA evidence of the causes of population bottlenecks
- C analysis of variations in mitochondrial and Y chromosome DNA
- D identification of variable number tandem repeats (VNTRs) in the human genome

Short-answer questions

6 **Compare the technologies involved in DNA sequencing and DNA profiling.** (5 marks)

7 **Explain three issues or problems in data analysis used to identify trends, patterns or relationships that have contributed to conservation management, understanding of human evolution, and the inheritance of a disease or disorder.** (6 marks)

8 **Cystic fibrosis is a genetic disease resulting from a single nucleotide polymorphism. Explain how data analysis and population genetics could help to determine the inheritance of this disorder.** (4 marks)

9 **The genome of a Tasmanian tiger preserved as a four-week-old pup in 1909 has been sequenced by Melbourne University. Data suggested that the Thylacine was in genetic decline before Europeans hunted it to extinction. Assuming you possessed current understanding of population genetics and had access to current technology, what are some recommendations you would make to help to preserve the species if you were responsible for Thylacine conservation management in 1850?** (5 marks)

10 **Explain, using an example, how population genetics has assisted our understanding of human evolution.** (4 marks)

ANSWERS

KEY QUESTIONS

Key questions ➲ p. 70

1 DNA sequencing includes the use of technologies to determine the sequence of nitrogenous bases in the genome of an organism, population or species.

2 Technologies that are used in DNA sequencing include the use of enzymes to either fragment DNA (nucleases) or make copies of pieces of DNA (polymerases), the separation of DNA fragments by using gel electrophoresis or nanopores, and labelling bases for identification by using radioisotopes or fluorescent dyes.

3 The polymerase chain reaction produces multiple copies of DNA from a single source in order to provide sufficient DNA for sequencing or profiling.

4 Once the sequence of bases is known for the entire genome, the extent, location and role of variations in the base sequence can be determined. A single nucleotide polymorphism may be detected in individuals suffering a specific disease so the mutation that causes the disease may be identified.

Key questions ➲ p. 71

5 DNA sequencing includes the use of technologies to determine the sequence of nitrogenous bases in the genome of an organism, population or species. DNA profiling is the use of parts of the DNA of an individual that are highly variable to match them for the purposes of determining identity or relationships.

6 Technologies used in DNA profiling include the use of nucleases to break DNA into useful comparable segments, gel electrophoresis, the polymerase chain reaction, and labelling to detect and compare the progression of separated DNA segments.

7 Offspring obtain half their DNA from each parent. If one parent (e.g. the mother) is known, then the half of the DNA segments that do not match with the mother must match with the potential father for paternity to be confirmed.

8 Both are sequences of DNA bases that repeat themselves one after the other (i.e. in tandem). In the case of STRs only between two and ten nucleotides are repeated whereas in VNTRs, between eight and 50 nucleotide bases are repeated. Both help to identify individuals because the number, not the actual base sequence, is highly variable.

Key questions ➲ p. 72

9 Studying the genetic characteristics of a population can assist understanding of how species have evolved, how to manage the population to maintain diversity and population levels, and how diseases or disorders may be inherited.

10 Gene flow results in an increase in genetic diversity because of population mobility whereas genetic drift results in changes in allele frequency, either positive or negative, from random events as opposed to natural selection.

11 Single nucleotide polymorphisms are one measure of genetic variation. They are used to investigate the genetic diversity within a species and to compare diversity between species.

Key questions ➲ p. 73

12 Size and dispersal of a population and genetic diversity have an impact on the chances of survival of an endangered species.

13 Practices in conservation management that help to overcome lack of genetic diversity include the establishment of wildlife corridors to ensure movement of population members. This aims to increase the opportunity for diverse members to interbreed. DNA profiling can be used to identify more diverse members of an endangered population. This means breeding can be manipulated to maximise the opportunities for diverse offspring.

Key questions ➲ p. 74

14 You would sample populations to determine the frequency and distribution of the allele for the defective form of haemoglobin. You would then search for patterns and relationships in the data to identify populations and places where it is more frequent and attempt to correlate the increased frequency with external factors such as climate and exposure to mutagens.

15 Large samples are usually required in order to obtain sufficient data to analyse trends in genetic disorders. Some population isolates have a high incidence of a particular disease, e.g. Tasmania has an unexpectedly high incidence of Huntington's disease. These groups can be studied to try to determine why natural selection has not diminished the frequency of the harmful mutation.

Key question ➲ p. 75

16 Problems include the degraded nature of DNA from fossilised human remains, small numbers of fossils useful for DNA evidence, and inconsistencies between evidence from nuclear and mitochondrial DNA.

HSC EXAM-TYPE QUESTIONS

Objective-response questions

1 D. DNA sequencing involves the determination of the sequences of the bases that are attached to the nucleotides for the whole genome. **A** and **C** are wrong because DNA sequencing is about the sequence, not the numbers. **B** is incorrect because all DNA nucleotides have the same sugar (deoxyribose).

2 C. Comparing sequences of DNA bases will identify single nucleotide polymorphisms. **A** and **B** are incorrect because DNA profiling, not DNA sequencing, can identify perpetrators of crimes or familial relationships. **D** is wrong because while some diseases are caused by gene defects, many diseases have both genetic and environmental components. This means DNA sequencing is not always a reliable predictor of disease occurrence.

3 B. DNA profiling focuses on comparing the highly variable parts of the genome (the VNTRs or STRs) for the purposes of identification. **A** is wrong because that is the result of DNA sequencing. **C** is wrong because SNPs result from a single change in one nucleotide in the DNA sequence so they do not vary in length. **D** is incorrect because comparing blood types does not require DNA profiling.

4 A. Population genetics has been used to help conservation management of species such as the Tasmanian devil. **B** is incorrect because population genetics helps identify but not treat certain

genetic diseases. **C** is incorrect because it is the result of DNA profiling. **D** is wrong because population genetics has revealed insignificant variation across subgroups of human populations.

5 **C**. Comparing mtDNA has been used as a biological clock to determine the time elapsed since there was a shared common human ancestor while Y chromosome variation has revealed information about human male migration rates. **A** is wrong because only the DNA of recent human ancestors can be compared. **B** is incorrect because the causes of population bottlenecks are volcanic eruptions or other catastrophes. **D** is wrong because the VNTRs are too variable to give comparative information about human evolution.

Short-answer questions

6 **EM** You need to compare the processes of DNA sequencing and profiling and then draw out the differences and similarities in the technologies involved in each.

DNA sequencing is a very demanding process as the sequence of nitrogenous bases in each component nucleotide has to be determined. In DNA profiling the use of highly variable parts of DNA are compared to determine a person's identity. (✓) Both technologies may use the polymerase chain reaction to make many identical copies of the DNA being investigated. (✓) Both technologies separate fragments of DNA in processes such as electrophoresis. (✓) DNA sequencing is more complex as DNA nucleotides have to be labelled and detected in sequence through techniques such as radioisotope labelling or the use of fluorescent dyes. (✓) DNA profiling uses comparison of the lengths of highly variable fragments, making identification after separation a much easier process, though one that relies heavily on the use of the correct enzymes to cut the appropriate fragments of DNA. (✓)

7 **EM** You need to identify three issues or problems in obtaining or interpreting and analysing data for each of the specific purposes of population genetics listed. You also need to outline the causes and effects of these issues.

Problems using population genetics to investigate human evolution arise from a lack of data (human fossils are difficult to find and often contain only degraded DNA). (✓) Only rough estimates of their age can be made because of the limitations of radiocarbon dating. (✓)

Problems using population genetics to determine if there is a genetic component to diseases (e.g. type 1 diabetes) include the need for large samples to determine a relationship. There is also a need for control population groups for comparison purposes. (✓) The process is time consuming, expensive and labour intensive, and dependent on extensive use of DNA analysis, such as DNA sequencing, in order to identify single nucleotide polymorphisms. (✓)

An issue with using data analysis to help better manage endangered species is that if a lack of genetic diversity of a very small population is determined to be a major factor, little can be done to enrich genetic diversity apart from promoting the breeding of more diverse members of the species. (✓)

One example of the complicated nature of using population genetics to help maintain a population is the Tasmanian devil. Lack of genetic diversity in the devil's genes responsible for the immune response was believed to be responsible for the initial rapid spread of the disease. It has since been found that the mutations that cause the cancer are highly diverse and adaptable, reducing the ability of the animals to counter the cancer threat. (✓)

8 **EM** You need to identify what data could be collected and explain how it should be analysed to help investigate the inheritance of the disorder. You may not know this specific example but you could apply the concepts learnt from the example you studied.

DNA sequencing can be used to detect single nucleotide polymorphisms. (✓) This could then be used to identify carriers of the disease and explain the risks of their offspring contracting it. (✓) Many diseases are experienced at different levels of severity, so research could also help understand if environmental factors (e.g. pollution or diet) can help in disease management if the presence of the SNP is detected. (✓) Population genetics data can be used by researchers to better understand factors such as genetic drift or flow, which may impact on the incidence of the disease in a particular population group, e.g. a population isolate. (✓)

9 **EM** This question challenges you to apply your understanding of the role of population genetics in conservation management to a different example.

Strategies could include those outlined below.

Research in detail the population genetics of the Thylacine. Identify, using single nucleotide polymorphisms, the level of genetic diversity. (✓)

Research the population status of the Thylacine. Determine if there is similar genetic diversity within population groups to the diversity level across the whole population. (✓)

Wildlife corridors allow movement of existing populations of Thylacine so populations could continue to interbreed. Another approach would be deliberately introducing genetically diverse individuals into a population in order to create gene flow. (✓)

Establish populations of Thylacine in zoos and reserves and possibly even in areas on mainland Australia. Compare the genomes of the populations to establish the population genetics of these reserve populations and control breeding in order to maintain as much diversity as possible in these populations. (✓)

Protect Thylacines from hunting and subsidise or otherwise assist farmers with other methods of ensuring stock were not subject to Thylacine predation. (✓)

10 **EM** This question requires you to select an appropriate example to show how the analysis of DNA caused an improvement in understanding of human evolution.

Two ancestral groups, the Neanderthals and Denisovans, have provided nuclear and mitochondrial DNA that can be compared with that of modern humans. One group of Neanderthals, found in the Hohlenstein-Stadel cave in Germany, had mitochondrial DNA that could be used to date the fossils to about 124 000 years ago. This was a situation where it was not possible to accurately use radiocarbon dating. The calculation of the date is the result of using differences in mtDNA as a molecular clock, based on the assumption that the rate of mutation of this DNA has been constant. (✓)

The fact that the DNA of this group was very different to DNA from groups of Neanderthals found at other locations suggests they were more diverse and diverged earlier than originally thought (270 000 years ago). (✓) Nuclear DNA of Denisovans and Neanderthals matches more closely than nuclear DNA of modern humans. Neanderthals and modern humans have more similar mtDNA. This suggests that interbreeding between the Neanderthals and the mobile modern humans was relatively minor (✓) and the link and interbreeding between Denisovans and Neanderthals was relatively stronger. (✓)

CHAPTER 6 MUTATION

INQUIRY QUESTION:

How does mutation introduce new alleles into a population?

In Module 5 you reviewed patterns of inheritance, including causes of genetic variation and its impact on populations. In this module you will review the natural- and human-induced causes and effects of genetic change. This includes biotechnology and gene technology. In this chapter you review causes of mutations.

1 How a range of mutagens operate

» Students explain how a range of mutagens operate, including but not limited to electromagnetic radiation sources, chemicals and naturally occurring mutagens.

Mutations

- Mutations are permanent changes in the DNA or RNA of a cell and include changes to the nucleotide sequence of DNA or RNA. DNA replication and polypeptide synthesis have built-in checks and repair pathways. Repair systems include DNA repair or removal, or tolerance of the change. A mutation makes a permanent change to the genome when there is a failure in correction processes.
- Mutations in the genome of an organism may harm cells and cause certain diseases, such as cancer (see Chapter 14).
- Mutations may arise spontaneously though research has provided evidence that mutagens increase the rate of mutation. Mutagens are chemicals or energy forms, such as radiation, that promote changes in the genetic material (DNA) and hence mutations. In Chapter 1 you reviewed the use of mutagens in plant agriculture to produce plant mutants that would hopefully possess beneficial characteristics. In this chapter you review some of the harmful impacts that occur due to exposure to mutagens.
- Mutagens are often classed as physical, such as electromagnetic radiation, chemical, such as cigarette smoke, or natural, such as some viruses.

Electromagnetic radiation sources

- **Electromagnetic radiation** refers to a range of photons carrying energy in the form of electromagnetic fields through time and space. The range extends in a spectrum that includes high energy, short wavelength, high frequency gamma rays, X-rays and ultraviolet light. Gamma rays, X-rays and ultraviolet or UV light are mutagens.
- Penetration distances vary with the energy levels of electromagnetic radiation, hence the use of X-rays for diagnostic imaging. This also determines the depth of tissue damage that can result from exposure.
- The strength of electromagnetic radiation exponentially decreases from its source. Mobile phones emit microwaves, which are generally considered harmless, but the frequency of use and proximity to the brain of microwaves has worried some people.
- Radiation can damage DNA at any time in the cell cycle.
- X-rays and gamma rays are classed as ionising radiation because when they strike atoms and molecules they 'bump off' an electron, leaving an ion or free radical behind. Therefore ionising radiation can break DNA molecules, damage bases, and cause base and large chromosomal deletions. It also causes cross-linking of DNA to itself or to proteins. Most of this damage can be repaired by the DNA-repair pathway (enzymes) though continued exposure increases the risk that the damage cannot be repaired. Experiments showed a linear direct relationship between the amount of X-ray exposure and the number of induced mutations, no 'safe' level of X-rays and a cumulative effect.
- Cells that have been exposed to electromagnetic radiation may release **reactive oxygen species (ROS)**, which directly damages nuclear DNA, modifies bases or deactivates repair.
- Ultraviolet light above 260 nm in wavelength is absorbed by DNA bases, causing pyrimidine dimers to form (see the 'after' image in Figure 6.1) and the possibility of replication and transcription errors. It may also stimulate mutations and chromosomal rearrangements.

electromagnetic radiation: oscillating electric and magnetic fields, which travel at right angles to each other in a bundle of energy called a photon

reactive oxygen species (ROS): peroxides, superoxides and hydroxyl radicals that release reactive forms of oxygen

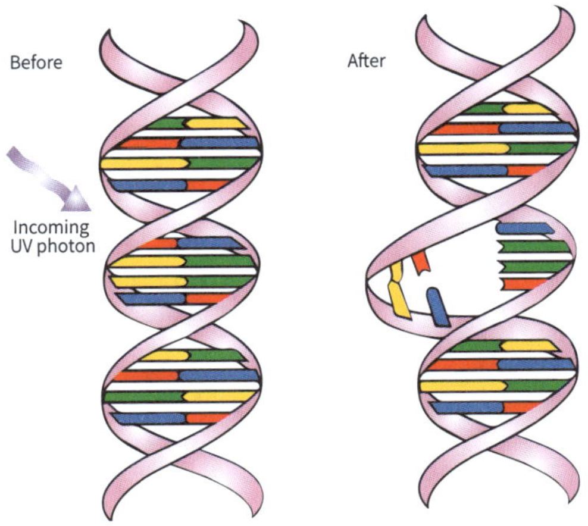

Figure 6.1 Ultraviolet light damaging DNA, causing pyrimidine dimers to form

Chemicals

- Chemical mutagens cause changes to the DNA nucleotide bases or to the three-dimensional structure and length. In many cases the mutagen was discovered because of its role as a carcinogen.
- Some chemotherapy drugs are also chemical mutagens. They work because cancer cells divide more frequently than most other body cells and the DNA damage is sufficient to stop their cell division.
- The first chemical mutagen, nitrogen mustard (a component of mustard gas), was detected in 1942.
- Cigarette smoke is a well-publicised carcinogen. It contains compounds such as polyaromatics, acrolein, 1,3 butadiene, acetaldehyde, n-nitrosamines, benzene and aromatic amines which have been shown to be chemical mutagens.

Table 6.1 Chemical mutagens

Chemical mutagen	What it does	Possible source
5-bromo-deoxyuridine	mimics nitrogenous bases	exposure results from it being used to detect actively dividing cancer cells in living organisms
nitrogen mustards (from World War I and II), nitrosoureas, alkyl sulfonates, triazines and ethylenimines	alkylating agents produce mutations by adding small hydrocarbon groups, such as methyl groups, to either the bases or the backbone phosphate groups of DNA	exposure to mustard gas used as a chemical weapon or a component (and side effect) of chemotherapy
deaminating agents (e.g. dimethyl nitrosamine)	removes amine groups from bases, such as converting cytosine bases to uracil, producing mutations	produced in an animal's stomach after eating foods containing nitrous acid or nitrite
intercalating agents such as acridine orange, proflavin and ethidium bromide	insert themselves into DNA, 'stretching' it and causing **frameshift mutations**	used in laboratories as dyes or mutagens
DNA adducts such as polycyclic aromatic hydrocarbon (PAH)	DNA adducts are a form of DNA damage in which the chemical mutagen attaches to the DNA	found in coal and tar deposits and released by the incomplete combustion of organic matter

Naturally occurring mutagens

- Naturally occurring mutagens may physically or chemically promote mutations. Historically, naturally occurring mutagens have often been detected and investigated because they correlate strongly with increased incidences of cancer.
- Table 6.2 lists some naturally occurring mutagens including oncoviruses. In Chapter 9 you will review non-cellular pathogens such as viruses.

frameshift mutations: mutations that result in changes in the reading frame of codons in mRNA

Table 6.2 Naturally occurring mutagens

Natural mutagen	What it does	Possible source
viruses have been demonstrated to be the cause of 'contagious' or transmissible cancers, such as Tasmanian devil facial tumour disease and cervical cancer, caused by human papilloma virus type 16 and 18 (HPV)	viruses take over their host's genome and, in some cases, cause mutations viral genomes of the two types of HPV have been detected in human cancerous cells; the viruses may also override the cell's defence mechanisms some viruses operate in a 'hit and run' manner and leave no genetic evidence of their role in promoting cancer-causing mutations	viruses that can cause cancer are called oncoviruses HPV accidentally integrates into the host's genome to cause cancer others may cause cancer indirectly because they result in long-term inflammation in the host's cells; some may enhance the action of oncogenes that pre-exist in the host
aflatoxin (a mycotoxin)	fat-soluble molecule that is rapidly absorbed and processed by enzymes in the liver, which convert it to a reactive mutagen that damages DNA	produced by moulds and resulting from contaminated food, especially grain in hot, humid conditions

Natural mutagen	What it does	Possible source
flavonoids and compounds such as pyrrolizine alkaloids, ally isothiocyanate or cycasin	investigated because of toxic effects in livestock and poultry, and liver and other cancers in inhabitants of certain localities some alkaloids are not very reactive but are turned into mutagens by the liver	found in many leafy green plants
reactive oxygen species (ROS)	result in mutagens attached to nitrogenous bases, as well as DNA strand breaks and crosslinks	produced in mitochondria and chloroplasts as part of cellular metabolism ROS also result from exposure to pollutants, tobacco, smoke or radiation and plants produce ROS in response to environmental stress
metals such as mercury and cadmium and metal compounds	compounds containing some metals (e.g. aluminium, arsenic, copper, lead, mercury, nickel) can induce chromosomal aberrations or abnormal cell divisions in animal or plant cells	found in rocks and soil and in high concentrations in ores
background radiation	some background radiation is naturally occurring electromagnetic radiation except for alpha and beta rays, which are energy-carrying particles that are also ionising	radioactive decay from naturally occurring radioactive isotopes in soil, rock and wood; cosmic rays from sun and space; radon in the atmosphere radioactive isotopes can emit alpha and beta particles in addition to gamma rays

→ KEY QUESTIONS

1 How do electromagnetic sources operate as a mutagen?

2 How do chemicals operate as mutagens?

3 How do naturally occurring mutagens operate?

Answers ⊃ p. 88

2 The causes, processes and effects of different types of mutation

» Students compare the causes, processes and effects of different types of mutation, including but not limited to point mutation and chromosomal mutation.

Point mutation

- A change in one nucleotide in a DNA sequence is called a **point mutation**. When a point mutation, in which one base is substituted for another, occurs with sufficient frequency in the population it will result in a single nucleotide polymorphism (SNP) (see Chapter 4). Impacts of SNPs on humans includes a range of conditions such as curly hair, sickle-cell anaemia, variations in responses to drugs, and increased risks of suffering diabetes, obesity, hypertension and some psychiatric disorders.

point mutation: a change in one nucleotide in DNA, such as substitution of one nucleotide for another; an insertion or deletion of a nucleotide

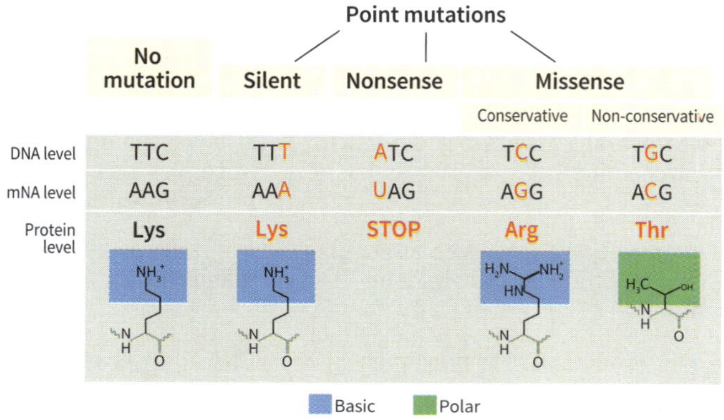

Figure 6.2 Point mutation

Source: image from Jonsta247 (own work) via Wikimedia Commons

- If substitution of a nitrogenous base occurs in a 'coding' part of DNA it could result in the substitution of one amino acid for another in polypeptide synthesis. Substitution in part of the 'non-coding' DNA could alter the expression of a gene. Some point mutations may have no observable impact on phenotype or gene expression.
- Other changes that result in point mutations are the insertion of an extra nucleotide or deletions of a nucleotide. These may result in frameshift mutations. Insertions and deletions change the following triplets and hence the codons and anticodons in the process of polypeptide synthesis.
- Point mutations occur spontaneously, usually during DNA replication when a cell is undergoing mitosis or meiosis. Cellular metabolism produces free radicals and active oxygen molecules that can harm DNA by breaking either single or both DNA strands. DNA bonds themselves can degrade. The rate of mutations increases in the presence of mutagens.

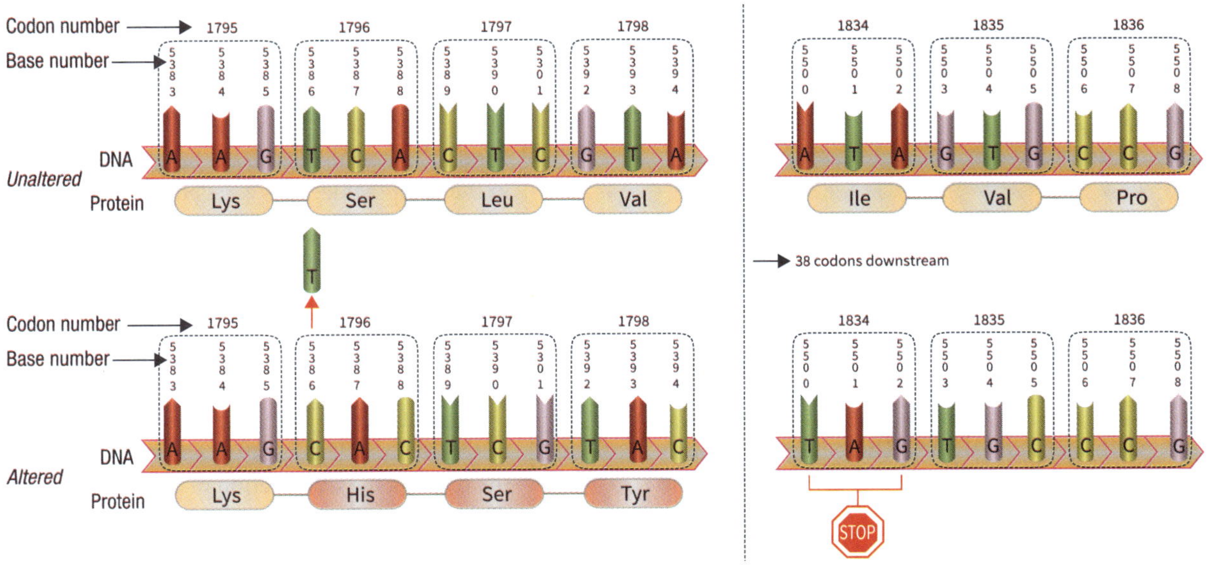

Figure 6.3 Frameshift mutation

Source: image by Genomics Education Programme, [CC BY 2.0 (http://creativecommons.org/licenses/by/2.0)], via Wikimedia Commons

Chromosomal mutation

- **Chromosomal mutation** occurs when a cell is undergoing cell division. It can also occur as a result of exposure to ionising radiation and some chemical mutagens. Chromosome mutations occur by chromosome rearrangement and changes in chromosome numbers. The changes are large enough to be observed through karyotype comparison.
- The four main types of chromosomal rearrangement, as shown in Figure 6.4, are outlined below.
  - Deletion occurs when a chromosome breaks and a section of chromosome is lost.
  - Duplication occurs when a section of chromosome is copied and added.
  - Inversion occurs when a section of chromosome breaks free, reorients and re-joins in the reverse order.
  - Translocation occurs when a section of chromosome breaks off but re-joins a different, non-homologous chromosome.
- Chromosome numbers change when whole chromosomes or even sets of chromosomes are lost, gained or rearranged. **Non-disjunction** occurs when homologous chromosome pairs fail to segregate during cell division. Non-disjunction is a cause of **aneuploidy**. If non-disjunction occurs in one chromosome pair, one of the daughter cells has an extra chromosome and the other daughter cell has one missing chromosome.

chromosomal mutation: a chromosome rearrangement resulting in a change in chromosome structure or number
non-disjunction: the failure of homologous chromosomes to separate during either mitosis or meiosis
aneuploidy: the condition in a cell where the chromosome number is not a multiple of the haploid number

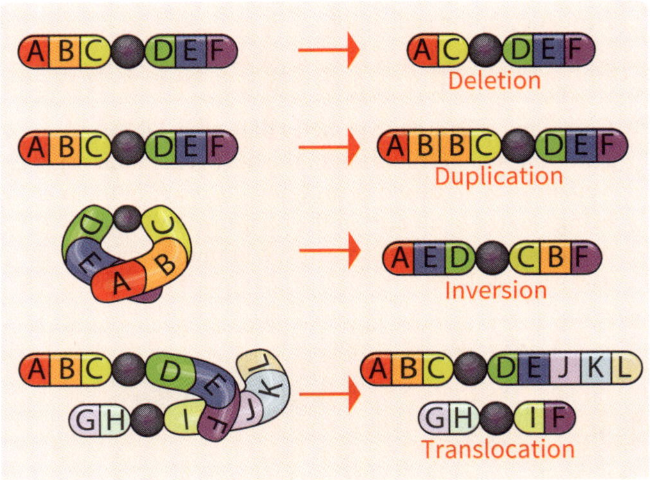

Figure 6.4 Chromosome mutation

- Chromosomal mutations in germ-line cells in humans result in a substantial proportion of congenital abnormalities and inherited diseases; for example, Kleinfelter's syndrome. Chromosomal mutations in somatic cells may result in cell death or the loss or reduction of cell function. They may also contribute to diseases such as cancer. These are reviewed in detail in Section 3.

→ KEY QUESTIONS

4 What are the causes and processes of point mutation?

5 What are the causes and processes of chromosomal mutation?

6 What are the effects of point mutation?

7 What are the effects of chromosomal mutation?

Answers ➲ p. 88

3 Somatic and germ-line mutations

» Students distinguish between somatic mutations and germ-line mutations and their effect on an organism.

Somatic mutations

- Somatic mutations are mutations in the somatic or normal body cells that occur in an individual after conception but are not passed on to offspring.
- The mutations are carried in the daughter cells of the original mutant cell and therefore the initial impact is only on specific cells and tissues. These mutations may have varying effects on the individual depending on the part of the DNA where they occur.
- Mutations can contribute to the development of diseases or disorders in individuals. A series of mutations can result in cancer.
- Somatic mutations do not directly change the allele frequency in the gene pool.

Germ-line mutations

- Germ-line cells are the gametes that carry DNA and genes through fertilisation to the next generation zygote. Mutations in fertilised germ-line cells result in them being carried to all cells in the offspring. Inherited diseases and disorders are a result of mutations in germ-line cells.
- Germ-line mutations have the potential to change allele frequency in the gene pool.

KEY QUESTION

8 **Distinguish between somatic mutations and germ-line mutations and their effect on an organism.**

Answer ➲ p. 88

4 The significance of 'coding' and 'non-coding' DNA segments in the process of mutation

» Students assess the significance of 'coding' and 'non-coding' DNA segments in the process of mutation.

Mutation in 'coding' DNA segments

- 'Coding' sections of DNA segments comprise the genes that are templates for the formation of polypeptides. If a mutation occurs in the coding then the consequences can include the outcomes outlined below.
  - A different mRNA codon but the same amino acid is inserted into the polypeptide so there is no change in phenotype. This is called a **silent mutation**, as shown in Figure 6.2, which illustrates a point mutation.
  - A point mutation may result in a different amino acid in the polypeptide. This is called a **missense mutation**. If the substituted amino acid has similar properties to the original amino acid in the polypeptide chain there is no change to the function of the protein. If the substituted amino acid has different properties this results in a change in the polypeptide structure and a subsequent change to the function and phenotype.
  - The mutation results in the substitution of a 'stop' mRNA codon. This is called a **nonsense mutation** and the result is the termination of the synthesis of the rest of the polypeptide chain. The polypeptide becomes non-functional. Figure 6.2 shows a frameshift mutation resulting from the deletion of the nucleotide containing thymine. Subsequent amino acids are changed as a result in the change of the triplets until a stop codon is created.
- Not all genes are equal when it comes to the effect of a mutation making the gene non-functional. Some genes cause a cascade of impacts. For example, homeobox genes control the development of most organ systems. Scientists working on a mutation of fruit flies in which a leg grew in the place where antennae normally grew found the cause to be a single mutation in a homeobox gene.

silent mutation: a base substitution mutation that results in the same amino acid in the resulting polypeptide and thus no observable impact
missense mutation: a base substitution mutation that results in one changed amino acid in the polypeptide sequence; depending on the position of the amino acid and its chemical properties it may (if non-conservative) or may not (if conservative) change the functioning of the resulting protein
nonsense mutation: a mutation that results in a stop codon being prematurely inserted into a polypeptide so its synthesis is terminated and it becomes non-functional

Mutation in 'non-coding' DNA segments

- 'Non-coding' DNA segments in the genome have a range of functions, including regulation of gene expression. The impact of mutations in 'non-coding' segments can range from having no effect to preventing polypeptide synthesis and hence some protein functions.
- If the 'non-coding' DNA that experiences a mutation is a promoter sequence, DNA polymerase may not be able to bind, thereby preventing the transcription of a polypeptide.

- Huntington's disease is an example of a disease in which population genetics contributed to our understanding. The disease is a result of mutation in the 'non-coding' DNA segment that influences gene expression. More than 40 short tandem repeats of the base sequence CAG results in the disease while less than 35 repeats results in the normal condition. Between 35 and 40 is intermediate and some aspects of the disease may appear. Mutations in germ-line cells that already have a high number of repeats are more likely to result in the offspring experiencing the disease.

→ KEY QUESTIONS

9 **How significant are DNA mutations in the 'coding' segments of DNA?**

10 **How significant are DNA mutations in the 'non-coding' segments of DNA?**

Answers ➲ p. 88

5 The causes of genetic variation relating to the processes of fertilisation, meiosis and mutation

» Students investigate the causes of genetic variation relating to the processes of fertilisation, meiosis and mutation.

- Mutations are the ultimate source of new alleles and genetic change.
- As a form of revision, watch this five-minute video on modelling of mutations at learn.genetics.utah.edu/content/selection/sources/.
- Table 6.3 compares the role of fertilisation, meiosis and mutation in causing variation.
- In Chapter 9 you will review how humans use artificial insemination and artificial pollination to produce more productive crops and livestock by reducing variation in fertilisation.

Table 6.3 Comparing sources of variation

	Fertilisation	Meiosis	Mutation
How this causes genetic variation	allows new combinations of alleles	random segregation causes new combinations of chromosomes and hence alleles crossing over gives new combinations of alleles along chromosomes	produces new alleles and changes in gene regulation and expression or chromosome number and arrangement
Impact on gene pool	natural selection may play a role by increasing the chances of more 'fit' parents producing the gametes	the processes in meiosis result in variation in combinations of alleles but not allele frequency	potential to change allele frequency in gene pool

- Fertilisation and meiosis are the key events that facilitate genetic variation in sexual reproduction.
- In-vitro fertilisation is used, along with genetic screening, to reduce the chance of genetic disease in some humans.
- In Chapter 1 you reviewed how mutagens were used to promote variation to assist the selective breeding of crops.
- In Chapter 8 you will review how meiosis has been manipulated to cause chromosome engineering for plant breeding and research.

→ KEY QUESTIONS

11 **How does fertilisation cause genetic variation?**

12 **How does meiosis cause genetic variation?**

13 **How does mutation cause genetic variation?**

Answers ➲ p. 88

6 The effect of mutation, gene flow and genetic drift on the gene pool of populations

» Students evaluate the effect of mutation, gene flow and genetic drift on the gene pool of populations.

- The diversity within a gene pool may result from population size, mutation, **genetic drift**, natural selection, diverse environments and population movements resulting in **gene flow**.
- The effect of mutations, genetic drift and gene flow on the gene pool are much greater in small populations.
- One example is the study of Tasmanian devils described in Chapter 5.

genetic drift: random chance effects on allele frequency in gametes and populations, particularly if the population is small
gene flow: the movement of genes between populations, resulting in an increase in genetic diversity because of migration

Table 6.4 Effects of mutation, gene flow and genetic drift on the gene pool

Cause	Effect	Evaluation
mutation	initially mutation increases the variety of alleles in the gene pool	most significant cause of change to gene pool mutation impact may be positive, negative or neutral
genetic drift	random changes in allele frequency due to a range of factors in small populations random events might lead to increases or decreases of allele frequency	least important and less likely to alter gene pool composition unless the population is very small
gene flow	migration into a population can increase the genetic diversity due to gene flow conservation management practices include the maintenance of wildlife corridors to ensure gene flow	generally positive impact by increasing the diversity of the gene pool

→ KEY QUESTIONS

14 What effect does mutation have on the gene pool of a population?

15 How important is the effect of mutation on the gene pool of a population?

16 What effect does genetic drift have on the gene pool of a population?

17 How important is the effect of genetic drift on the gene pool of a population?

18 What effect does gene flow have on the gene pool of a population?

19 How important is the effect of gene flow on the gene pool of a population?

Answers ➲ p. 88

CHAPTER SYLLABUS CHECKLIST

Are you able to answer these questions from the syllabus for this chapter? Tick each question as you go through the checklist if you are able to answer it. If you cannot answer a question, turn to the relevant page in the study guide to find the answer. For NESA key word meanings, go to www.educationstandards.nsw.edu.au and search 'key words'.

	FOR A COMPLETE UNDERSTANDING OF THIS TOPIC:	PAGE NO.	✓
1	Can I explain how a range of mutagens operate, including but not limited to electromagnetic radiation sources, chemicals, naturally occurring mutagens?	80–82	
2	Can I compare the causes, processes and effects of different types of mutation, including but not limited to point mutation?	82–83	
3	Can I compare the causes, processes and effects of different types of mutation, including but not limited to chromosomal mutation?	83	
4	Can I distinguish between somatic mutations and germ-line mutations and their effect on an organism?	84	
5	Can I assess the significance of 'coding' and 'non-coding' DNA segments in the process of mutation?	84–85	
6	Can I investigate the causes of genetic variation relating to the processes of fertilisation?	85	
7	Can I investigate the causes of genetic variation relating to the processes of meiosis?	85	
8	Can I investigate the causes of genetic variation relating to the processes of mutation?	85	
9	Can I evaluate the effect of mutation on the gene pool of populations?	85–86	
10	Can I evaluate the effect of gene flow on the gene pool of populations?	85–86	
11	Can I evaluate the effect of genetic drift on the gene pool of populations?	85–86	

HSC EXAM-TYPE QUESTIONS

Objective-response questions

(1 mark each)

1 **Which of the following contains all mutagens?**
- A radio waves, chlorophyll, mustard gas
- B X-rays, alkylating agents, reactive oxygen species (ROS)
- C microwaves, mRNA, polymerase
- D infrared light, deaminating agents, haemoglobin

2 **What is the key feature of a point mutation?**
- A Point mutation only influences one nucleotide.
- B Point mutation only changes one amino acid.
- C In point mutation, one nucleotide is inserted, deleted or substituted.
- D In point mutation, one chromosome experiences an insertion, deletion or substitution.

3 **What is the difference between somatic and germ-line mutations?**
- A Somatic mutations are transmitted to offspring.
- B Germ-line mutations are transmitted to offspring.
- C Somatic mutations occur only in 'coding' DNA segments.
- D Germ-line mutations occur only in 'coding' DNA segments.

4 **Which of the following contains causes of genetic variation in offspring?**
- A random segregation, germ-line mutation, crossing over
- B DNA replication, genetic drift, gene flow
- C fertilisation, mitosis, somatic mutation
- D germ-line mutation, cytokinesis, genetic drift

5 **Which is a likely cause of a chromosome rearrangement?**
- A an alkylating agent
- B a silent point mutation
- C translocation
- D non-disjunction

Short-answer questions

6 **A student wrote the following sentence in the answer to an examination question: 'Mutations are harmful changes to alleles in a population'. Evaluate this statement.** (5 marks)

7 **Compare the impact of gene flow and genetic drift on the gene pool and biodiversity.** (4 marks)

8 **How do mutations introduce new alleles into a population?** (6 marks)

9 **Compare the effects of point mutations with chromosomal mutations.** (8 marks)

10 **How are mutations caused?** (5 marks)

ANSWERS

KEY QUESTIONS

Key questions ⊃ p. 82

1 High-energy sources of electromagnetic radiation such as gamma rays, X-rays and ultraviolet light deliver energy that changes atoms and molecules within DNA and even breaks chromosomes.

2 Chemicals operate as mutagens in a variety of ways. They can cause nucleotide base substitutions, add alkyl groups, remove amine groups and often prevent DNA from replicating.

3 There are naturally occurring forms of electromagnetic radiation and chemicals from plants, moulds and in the soil, which can damage DNA and chromosomes. Some produce reactive oxygen species (ROS) which can form DNA cross links and breaks.

Key questions ⊃ p. 83

4 Point mutations involve a substitution, deletion or insertion of one nucleotide base. Deletions and insertions result in a frameshift of triplets so polypeptide chains are significantly disrupted or terminated.

5 Chromosomal mutations result from electromagnetic radiation and some chemicals and result in rearrangements of chromosome pieces such as inversions, translocations, duplications and deletions of chromosome parts. Non-disjunction of homologous chromosomes can result in an entire missing or additional chromosome.

6 Point mutations cause new alleles in germ-line and somatic cells.

7 Chromosomal mutations cause congenital fatalities and abnormalities in germ-line cells, and cell death, malfunction and cancer in somatic cells.

Key question ⊃ p. 84

8 Somatic mutations are changes in DNA in body cells that can result in diseases such as cancer. The mutations are carried in the daughter cells of the original mutant cell so they initially only impact specific cells and tissues. These mutations may have varying effects on the individual depending on the part of the DNA where they occur. Mutations that result in aggressive cancers have a disproportionate impact because those cells replicate excessively and can migrate to other parts of the body.

Germ-line mutations are passed on through generations and potentially impact all body cells in the offspring. They can cause inherited diseases such as haemophilia, cystic fibrosis, muscular dystrophy and Huntington's disease.

Key questions ⊃ p. 85

9 Mutations in the 'coding' segments of DNA may have no impact or result in the failure to produce a functional polypeptide and possibly a genetic disease.

10 Mutations in the 'non-coding' segments of DNA may have no impact or result in changes in gene expressions, similar to the impact in some 'coding' DNA mutations. Huntington's disease is a result of mutations in 'non-coding' DNA.

11 Fertilisation combines alleles from two different parents to create a new combination of alleles in the zygote.

12 Random segregation in meiosis results in gametes with new combinations of chromosomes. Crossing over in meiosis results in chromosomes with new combinations of alleles along their length.

13 Mutation creates new alleles, which adds to genetic variation. Mutation can also change the frequency of alleles in the gene pool.

Key questions ⊃ p. 86

14 Mutation increases the variety of alleles in a gene pool.

15 Mutation is important because diversity increases survival chances, especially in a changing environment.

16 Genetic drift causes random changes to the gene pool, which can have a large effect on small populations.

17 Genetic drift is not important in large populations but can have a significant impact on small populations.

18 Gene flow increases genetic diversity in a population.

19 Gene flow is important because diversity helps survival in changing environments. It is especially beneficial for small populations.

HSC EXAM-TYPE QUESTIONS

Objective-response questions

1 **B**. This option contains three mutagens. **A** only contains one (mustard gas), **C** contains none and **D** contains one (deaminating agents) so these options are incorrect.

2 **C**. This option correctly explains possible changes that result in a point mutation. **A** is incorrect because it could cause a frameshift alteration. **B** is incorrect because some point mutations involve no change to amino acids in the polypeptide sequence. **D** is wrong because chromosomal mutations involve major changes in gene sequences.

3 **B**. Germ-line mutations occur in the production of gametes. **A** is incorrect because somatic cells are body cells not directly involved in reproduction. **C** and **D** are incorrect because somatic and germ-line mutations can occur throughout 'coding' and 'non-coding' DNA segments.

4 **A**. This option correctly lists causes of variation that could be passed on to offspring. **B** is wrong because gene flow and genetic drift result in changes to gene pools of populations, not individuals, and DNA replication does not cause variation. **C** is wrong because mitosis and somatic mutations produce cells that are not involved in reproduction. **D** is incorrect because cytokinesis is the separation of cytoplasm during cell division (so it is not a cause of variation).

5 **C**. This option correctly identifies translocation as a cause of change in whole chromosomes. **A** is incorrect because alkylating agents stop DNA replication. **B** is incorrect because silent point mutations result from only one nucleotide substitution. **D** is wrong because non-disjunction results in change in chromosome number, not rearrangement.

Short-answer questions

6 EM In order to answer this question you have to deconstruct the sentence into separate components. Are mutations harmful changes to alleles? Do mutations damage the gene pool of a population?

Mutations cause changes to more than alleles and they can also be neutral or beneficial to individuals. (✓) Mutations occur in both 'coding' and 'non-coding' DNA segments and also cause changes in the numbers and rearrangements of chromosomes. (✓) Mutations may have no impact on phenotype or be beneficial or harmful to individuals. (✓) Mutations are the source of variation in alleles that contributes positively to biodiversity. (✓) Biodiversity has potential benefits for the survival and evolution of a species. (✓)

7 EM Start your answer by distinguishing between gene flow and genetic drift. Then, relate each to their different impacts on variation in the gene pool.

Genetic drift is the random changes in allele frequency in the gene pool that have more impact when the gene pool and population are small. (✓) Gene flow is the result of immigration, when new members entering the population add to the variety and frequency of alleles in the gene pool. (✓) Gene flow is particularly advantageous in small populations and gene pools. (✓) Because genetic drift is random and unrelated to natural selection, its impact on biodiversity could be neutral, positive or negative. (✓)

8 EM You need to address the causes and processes of mutations that relate to the production of new alleles and then identify and explain the factors that influence the incorporation of new alleles into a population.

Point mutations in 'coding' DNA segments (✓) have the potential to create new alleles if they occur in germ-line cells. (✓) Substitution of bases (✓) may result in a different amino acid in the resultant polypeptide chain that, if it is missense non-conservative, could alter the structure and function of the protein that the polypeptide comprises. (✓) This would result in a mutation that produces a change in phenotype and, as such, this could be identified as a new allele. Other point mutations such as silent and conservative missense probably have no impact on phenotype and so a new allele would not be identified. (✓) Gametes resulting from mutated germ-line cells need to be involved in fertilisation to be carried on to offspring and thereby introduced into the population. (✓)

9 EM To answer this question, identify the key similarities and differences between the effects of point and chromosomal mutations.

Point mutations involve a change in one nucleotide base in a DNA segment. They may occur when one base is substituted for another and so have the potential to cause a change in one amino acid in a polypeptide chain or produce a stop codon. Point mutations in 'coding' DNA segments may result in a new allele or a non-functional polypeptide. (✓) If the point mutation is a result of the insertion or deletion of a nucleotide, the result can be a frameshift that changes the whole sequence of amino acids following. (✓) Additionally, the frameshift may result in the premature insertion of a stop codon which means that the functional polypeptide will not be synthesised. Point mutations in 'non-coding' DNA segments may result in changes to gene expression and regulation. (✓) Point mutations in somatic cells may result in diseases such as cancer. (✓) In germ-line cells, point mutations may result in genetic disease or disorder in the offspring. (✓)

Chromosomal mutations usually cause a major disruption to both 'coding' and 'non-coding' DNA segments. (✓) Chromosomal mutations in germ-line cells are often fatal or lead to congenital abnormality in the development of the zygote. One example in humans is trisomy 21, which results from fertilisation of a gamete formed by the non-disjunction of chromosome 21. (✓) The condition that results is Down's syndrome, which has major health implications for the offspring. Chromosomal mutations in somatic cells can result in the development of cancers, cell death, or loss of cell functionality. (✓)

10 EM This open question relies on your syllabus knowledge to ensure you address aspects of spontaneous mutations and those caused by mutagens, such as electromagnetic radiation, chemicals and naturally occurring mutagens.

Mutations may occur spontaneously in cells during the processes of DNA duplication or polypeptide synthesis. (✓) Cells have processes to monitor and repair DNA damage but occasionally these fail and the change is passed on to either somatic or germ-line cells.

Mutagens are forms of energy or chemicals that promote changes in DNA or chromosomes. (✓) Electromagnetic radiation can physically break DNA and chromosomes (✓) but also interact with chemicals such as reactive oxygen species (ROS). (✓) Some organisms, including moulds (aflatoxin) and plants (flavonoids and alkaloids), also produce chemicals that react with and change DNA. (✓)

CHAPTER 7 BIOTECHNOLOGY

INQUIRY QUESTION:

How do genetic techniques affect Earth's biodiversity?

In Chapter 1 you reviewed how scientific understanding of reproduction had been applied to plant and animal agriculture. You reviewed a small component of the role of **biotechnology** as well as some examples of gene technology applied to agriculture. The content of this chapter adds to the Chapter 1 information. You will review examples of the application of biotechnology to industry, bioremediation, research, therapeutics and medicine. In Chapter 8 you will focus in more detail on the use of gene technologies.

1 Investigating the past, present and future uses and applications of biotechnology

» Students investigate the uses and applications of biotechnology (past, present and future), including analysing the social implications and ethical uses of biotechnology, including plant and animal examples; researching future directions of the use of biotechnology; evaluating the potential benefits for society of research using genetic technologies; and evaluating the changes to the Earth's biodiversity due to genetic techniques.

➔ Examples of **genetic techniques** include the polymerase chain reaction (PCR) and cloning, and the use of probes, electrophoresis and next generation sequencing methods.

➔ Examples of traditional biotechnology include:
- growing crops as a result of seed germination and asexual plant propagation techniques (see Chapter 1)
- raising livestock in a range of free range and confined accommodation types

biotechnology: the use of living systems and organisms to make products, help solve human problems or satisfy needs or wants

genetic techniques: the methods, tools and skills used to study genetic phenomena, such as heredity, variation, and DNA structure and function

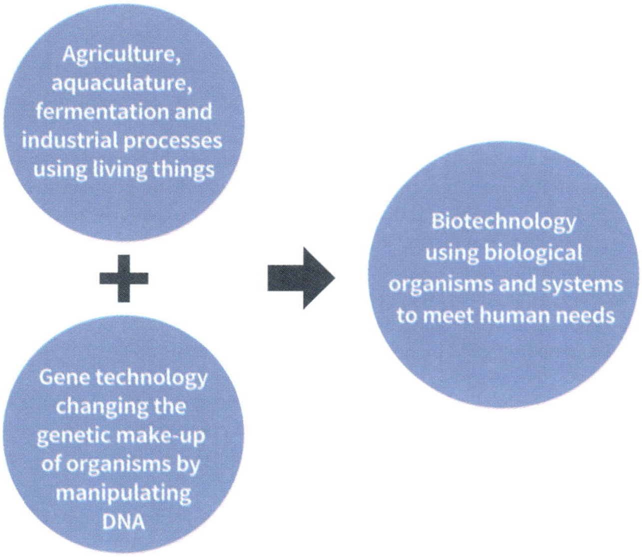

Figure 7.1 The relationship between gene technology and biotechnology

- fishing and aquaculture of fish, shellfish and molluscs, including producing effective and sustainable food sources for aquaculture
- selective breeding of plants, animals and microorganisms to improve output or economics
- the use of microorganisms for the production or preservation of food and pharmaceuticals
- a variety of food-growing activities from subsistence farming or permaculture to large-scale mechanisation and monocultures
- laboratory production of microorganisms and their products
- the use of animals such as mice to produce **monoclonal antibodies (MAbs)**; these are used to detect or diagnose certain diseases, in pregnancy-testing kits, and in the treatment of cancer, bacterial poisoning and snake bite (see Figure 7.2 for a summary of the process of making MAbs)

monoclonal antibodies (MAbs): proteins (immunoglobulin Ig) artificially produced by identical immune cells and capable of interacting with a specific target macromolecule; used for research and diagnosis or to treat cancer and other diseases

- the production or extraction of enzymes for a variety of uses; a recent example is **ELISA (enzyme-linked immunosorbent assay)**, which links enzymes to MAbs to give a quantitative estimate of the amount or concentration of a macromolecule in a solution (e.g. plasma, serum or cerebrospinal fluid).

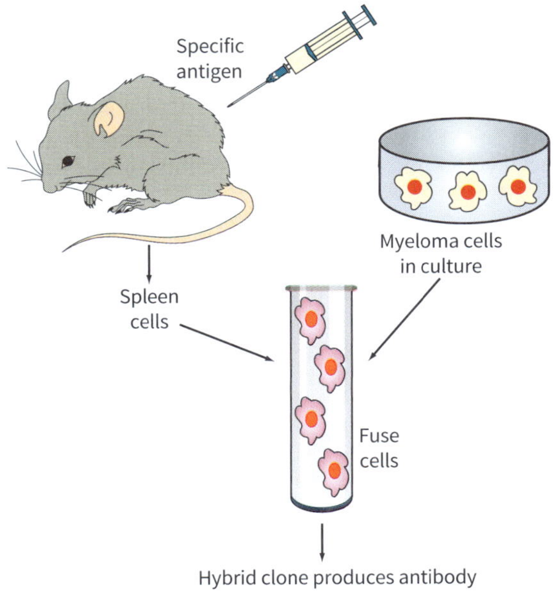

Figure 7.2 Making monoclonal antibodies (MAbs)

Examples of gene technologies

- Gene technologies can involve the production of organisms that contain genes from other species; that is, transgenic crops or animals or genetically modified organisms (GMOs). Gene technology is the focus of Chapter 8.
- **RNA interference (RNAi)** or gene silencing involves impairing the polypeptide synthesis process by the interference of production of mRNA to render specific genes non-functional (see Chapter 3 for information on polypeptide synthesis).
- **Marker assisted breeding** is used to detect desirable genetic traits (e.g. disease resistance, size or high yield) and link these to other gene markers so that selective breeding is more efficient and directed.

ELISA (enzyme-linked immunosorbent assay): enzymes attached to MAbs; used to determine the concentration of a targeted substance in body fluids
RNA interference (RNAi): a precise, efficient and stable way of inactivating a specific gene; useful in research and disease control
marker assisted breeding: selective breeding based on improved ways of selecting desirable organisms for breeding by tracking DNA markers associated with the desirable traits; involves no production of transgenic organisms

- The **CRISPR-Cas9** (Clustered Regularly Interspaced Short Palindromic Repeats) gene editing tool gives precision to gene technology. A RNA guide of about 20 nucleotides that match the specific bases of the target DNA locate the correct locus and then the Cas9 enzyme cuts both DNA strands. As the cell attempts to repair the break a chain of nucleotides carried by CRISPR are inserted to render the desirable change. A main focus of this research is to prevent vector mosquitoes from being able to transmit diseases such as malaria. Figure 7.3 attempts to demonstrate the precision of the CRISPR-Cas9 process of DNA manipulation (see also Chapter 12).
- **Gene drives** use 'selfish' genes that can increase the frequency of certain alleles in the population.
- Chromosome engineering is the deliberate alteration of the pattern of inheritance of chromosomes. Methods include causing chromosome deletions, inversions or translocations, and inducing extra or missing single chromosomes or extra whole sets of chromosomes (polyploidy) for a desired genetic effect. Chromosome engineering is used as a research tool to better understand genetic disease and in plant breeding. An example is the deliberate production of triploid fish in aquaculture as described in Chapter 1. One process in chromosome engineering is to prevent crossing over during meiosis. This maintains intact parental chromosomes so that homozygous lines can be bred from hybrids. Recent improvements have resulted in the manipulation of chromosome segments that carry the target gene only and not additional unwanted genes.

CRISPR-Cas9: a combination of RNA sequence and a restriction nuclease that has enabled precise gene editing by inserting, replacing, deleting or regulating genes
gene drives: a genetic system that builds on CRISPR-Cas9 to greatly increase the chance of a linked gene being inherited

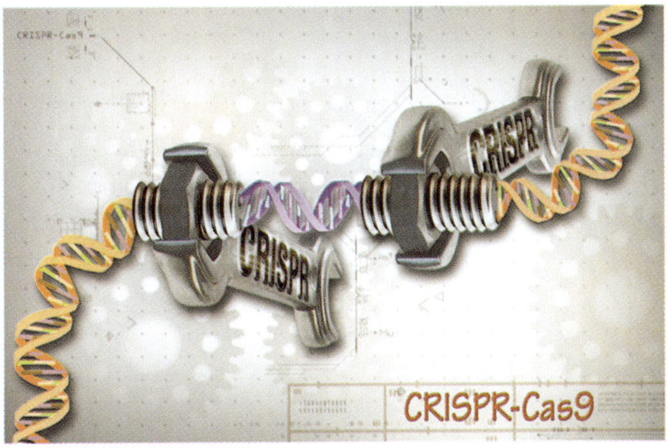

Figure 7.3 Model of the CRISPR-Cas9 gene editing tool
Source: image by National Human Genome Research Institute (NHGRI), [CC BY 2.0 (http://creativecommons.org/licenses/by/2.0)], via Wikimedia Commons

➔ KEY QUESTIONS

1 **Compare past and present uses of biotechnology.**

2 **What are some possible future major directions in biotechnology?**

Answers ➲ p. 101

Past uses of biotechnology

➔ Table 7.1 below provides a wide variety of examples of the past uses of biotechnology and the social and ethical issues that have arisen. Many of these methods have continued, with refinements and better understanding of the processes. Trends include a transition towards increased use of gene technologies and a more refined and efficient use of traditional biotechnologies.

➔ A resource which adds to the table below can be found at www.industry.gov.au/industry/IndustrySectors/nanotechnology/Publications/Documents/Agricultural_Biotechnology_Developments2008.pdf. It is dated at 2008 but contains timelines that are useful for identifying past and present uses of biotechnology as well as discussion of social issues and ethics and projections for the future.

Table 7.1 Examples of past uses of biotechnology

Applications	Past	Social implications	Ethical implications
agricultural uses and applications	selective breeding, fermentation of food and beverages for food preservation and maintaining nutrition, simple aquaculture to improve availability of protein sources genetic modification of soybeans started in 1988 and is now the most widely used GMO **biological controls** were used to control some pests (e.g. in 1925 *Cactoblastis* moth larvae was used to control prickly pear in NSW and Queensland)	improvements in health and nutrition some social groups experienced improvements in productivity, wealth and increased leisure and autonomy countries responded differently to the acceptance of GM organisms as food or food ingredients and legislated and regulated accordingly	habitat destruction for many species, loss of **biodiversity** animal welfare (e.g. free-range versus penned animals, live animal transport and export) soil erosion, degradation and salinity irrigation impacts on river systems, increased salinity
industrial and environmental uses and applications	from 1920s bacterial sludge was used in sewage treatment anthrax was used in 2001 against 22 people in the USA in what is regarded as **bioterrorism** fermentation was used for production of useful substances such as citric and lactic acid in 1989 bacteria were used to help clean up the Exxon oil spill in Alaska	better waste disposal and reduced contamination of water and soil reduced security because of increased bioterrorism threat; the Biological Weapons Convention (1972) was co-signed by 103 nations but there has been substantial evidence that not all signatories have complied less dependence on imports, improved access to a range of products reduced damage from accidental environmental spills	international conventions require scrutiny and there is an increased need for nations to be defensively prepared for the use of biological weapons
medical uses and applications	production of vaccines, anti-venoms, antibiotics (e.g. penicillin), prebiotics and probiotics, dietary supplements and herbal remedies from 1920s insulin was extracted from pigs and cattle for the treatment of diabetes	need to ensure quality and efficacy of many herbal remedies and pharmaceuticals some supplements are not subject to regulation and labelling restrictions in some countries treatments are more safe and predictable	precedence set for the use of animal models for research and testing of products the discussion of benefits versus harm caused is raised in production of MAbs which are used to fight, diagnose and research diseases as well as to test new drugs

biological control: the use of organisms to control the population size and distribution of a pest species
bioterrorism: the use of infectious agents, such as anthrax, or other harmful biological substances, such as toxins, for terrorism
biodiversity: diversity within species (including genetic diversity) and between species and ecosystems

Applications	Past	Social implications	Ethical implications
medical uses and applications	since 1978 recombinant DNA has been used to produce human insulin and growth hormone animals were (and are) used for testing products and drugs, and in the production of monoclonal antibodies (MAbs) genetically engineered animals such as the OncoMouse were (and are) used for cancer research	evolution of resistant strains of pathogens resulting from inappropriate use of antibiotics (see also Chapter 12) increased confidence in medical diagnoses and early diagnoses of certain diseases and infections	methods for producing MAbs in animals (particularly the ascites method which is banned in Australia) cause pain, distress or discomfort and final euthanasia of the animal in-vitro methods use **bioreactors** and reduce the levels of distress but some animals still need to be euthanised

Present uses of biotechnology

→ Examples of agricultural uses and applications are outlined in the points below:
- Transgenic crops are designed to counter abiotic and biotic factors. Examples include drought resistance (abiotic) and Bt cotton to prevent insect attacks (biotic). Cereal varieties have been developed using RNAi to provide resistance to Barley Yellow Dwarf virus, Rice Stripe virus and Western Corn rootworm.
- Crops with enhanced nutritional benefits (biofortification) have been developed; for example, having increased vitamin content. This is achieved through selective breeding and genetic modification, which has prompted much attention and debate. These methods have seen the introduction of nutraceuticals; that is, foods that have nutritional and therapeutic benefits, such as forms of bread, milk and margarine that lower cholesterol. Other nutraceuticals are dietary supplements and medical foods prescribed for patients by a doctor for a specific purpose. The CSIRO has developed DHA canola that contains high quality in-demand oils rich in omega-3 DHA.
- The CSIRO has used RNAi to simultaneously reduce the function of three genes in cottonseed to retain high levels of oleic acid and reduce undesirable fatty acids such as cholesterol raising trans-fatty acids. The product is a healthier oil for frying.
- GM canola produces an oil resembling that from coconuts or palms.
- Improved post-harvest processing of timber such as eucalypts is possible because of changed levels of lignin in the wood.
- Aquaculture has been expanded to increase seafood production. Prawn aquaculture now uses a natural food source produced by marine microbes that is cost-effective and more sustainable. CSIRO researchers have learnt how to manipulate marine microbial organisms and their environment in order to maximise their productivity and stimulate them to produce a novel **bioactive** product which enhances prawn growth.

bioreactors: apparatus that carry out biological reactions on an industrial scale in order to produce useful substances
bioactive: a substance having a biological effect

→ Examples of industrial and environmental uses and applications are outlined below:
- Products and processes (e.g. **bioremediation**) that degrade toxic or harmful chemicals and wastes by the use of bacteria or other microorganisms have been developed.
- Diagnostic tests (e.g. testing seafood to identify species composition) are used to detect and eliminate the illegal use of endangered species. **Biosensors** quickly and accurately detect water contamination from pollution or pathogens such as *E. coli*, *Giardia* and *Legionella*.
- Biofuels such as biodiesel, ethanol and methane are made from plant materials and animal waste (renewable biomass). These replace non-renewable fossil fuels with renewable resources. The CSIRO has used RNAi to produce safflower seed oil that contains a valuable fatty acid used for industrial chemical production and fuel.
- Biodegradable plastics are made from petrochemicals but genetic engineering is used to help them break down more quickly.
- Bioplastics have been made from corn starch. An example is polylactide acid, used for food containers, that looks and behaves like polyethylene and polypropylene but is compostable.

→ Examples of medical uses and applications are outlined below:
- **Biopharming** produces proteins, vaccines, hormones and blood clotting and blood thinning agents. Potatoes are grown to produce the vaccine for human papilloma virus (HPV).
- RNAi gene silencing is used to improve the production of antibodies for treating human diseases (e.g. Non-Hodgkin lymphoma) and inflammatory conditions (e.g. arthritis).

bioremediation: the use of plants and microorganisms to consume or otherwise help remove materials (e.g. toxic chemical wastes and metals) from contaminated sites, especially soil and water
biosensors: devices which use living organisms or biological molecules (e.g. enzymes or antibodies) to detect the presence of chemicals or pathogens
biopharming: the use of genetically modified organisms to make medications

- Milk-derived chewing gum releases calcium and phosphate in a form that re-mineralises tooth enamel.
- Patients with faulty heart valves that cannot be repaired need valve replacements. Mechanical valves can be used for younger people but the patients require ongoing anti-coagulation medication. Currently human (from deceased donors), pig, horse and cow heart valves are the choice for transplants in patients over 70 years.
- Knockout genes are genes which have been made inoperable (see Table 8.1. in Chapter 8). Knockout organisms are those with knockout genes (e.g. knockout mice) that are used to research the role of genes where the sequence but not the function is known. These are used for studying and modelling human diseases such as cancer. Knockout mice are also used in the development and testing of drugs and other therapies.

➔ Examples of social implications are outlined below:
- There have been rapid improvements in human life expectancy. This has led to increasing concerns about the costs of aged care.
- An increased awareness and concern over additives in food, such as pesticides and hormones, has been accompanied by a growing trend for the promotion of 'organic' foods.
- The benefits of the use of biotechnologies are not available to many poorer people, with some arguing that the technologies widen the gap between wealthy, resource-rich communities and those who are less well-off.
- There has been an increase in the influence on societies of companies involved in modern biotechnologies.
- Competition between agricultural land for food and land for biofuels or other GM products has increased. There has also been an increase in public acceptance of gene technologies. Currently no fresh GM foods are sold in Australia but derived products (e.g. oils) are and they do not have to be labelled as GM.
- Some developing countries may lose important export crops, such as palm and coconut oils and farmed prawns, and this will likely impact their economies.
- The development of more drought-resistant sources of protein in Australia has resulted in less dependency on imported foods.
- Golden Rice could help to alleviate vitamin A deficiency blindness in developing countries.
- Cheaper products that use fewer resources or produce fewer pollutants have been developed. There has also been better management of wild food resources.
- Better preventative medical practices, diagnoses and treatment, particularly of diseases such as genetic disorders and cancer, have been developed. This has been accompanied by the elimination of some pathogens or vectors.

➔ Examples of ethical uses are outlined below:
- There is the potential to reduce humans' ecological footprint and improve environmental management but there is also the possibility of irreversibly changing species and ecosystems. In some cases this problem can be addressed by a different ecosystem providing the resource (e.g. terrestrial instead of marine). For example, ocean-based algae and fish are currently the main source of omega-3 DHA but DHA canola could replace this source which would help conserve fish stocks.
- Human health has experienced vast improvements but this has been to the detriment of animal models used for research. There is also concern about access to the best medical care because of patents making some treatments more expensive.
- A 2010 case saw a farmer in Kojonup, Western Australia lose 'organic farm' status because of contamination from a neighbour's GM canola crop, which blew in to his farm. The protracted court case attracted much publicity and raised questions of ethics.
- There is the potential to reduce the emissions that are contributing to global warming.

Possible future uses of biotechnology

➔ Examples of possible agricultural uses and applications are outlined below:
- The isolation and insertion of genes from nitrogen-fixing bacteria has reduced the need for nitrogenous fertilisers, expanded areas where crops can be grown and eliminated problems associated with fertiliser run-off.
- There has been an expansion in the ability to control the gender of animals; for example, producing only male fish, which do not spend energy in producing eggs. The CSIRO is using gene technology to put a biological marker on the chicken sex chromosome to enable pre-hatch determination of the sex of fertilised eggs. Male hatchlings are currently culled in poultry farming and some European countries are looking at banning male culling for ethical reasons. Male-producing fertilised eggs could be used for growing influenza vaccines.
- The CSIRO is using marker-assisted breeding to try and breed hornless cattle in order to end the painful practice of dehorning beef cattle.
- CSIRO scientists have identified two genes resistant to powdery and downy mildew, providing an alternative to fungicides in the battle against these grape pathogens. Gene silencing is being used to eliminate pathogens such as varroa mites in bee colonies or viruses in tobacco plants. Advances such as these could result in the reduced use of pesticides.

- Examples of possible industrial and environmental uses and applications are outlined below:
  - There is the potential to solve problems such as climate change, food and energy security and waste disposal. There is also the potential to eliminate waste through processes such as the use of plants to remediate contamination.
  - Biomarkers have been developed that act as environmental indicators and are used to assess the level of exposure and the impact of pollution or damage caused by pollutants.
  - New materials such as very strong spider silk protein have been developed. Transgenic goats currently produce this in their milk but a means of extraction still needs to be developed.
  - There is the possibility of eliminating some exotic pests (e.g. carp, cane toads, foxes) through the incorporation of a 'daughterless' gene.
- Examples of possible medical uses and applications are outlined below:
  - There is the possibility of developing cultures of human stem cells for use in drug trials. There is also the potential for reducing the time and costs for the development of effective and even personalised medicines.
  - The development of virus-free **xenotransplantation** of animal organs and tissues in humans would allow for more widespread use.
  - The elimination of certain infectious diseases by genetically altering animal vectors such as mosquitoes through the use of CRISPR-Cas9 and gene drives may be achievable.
  - The CSIRO is using RNAi technology to find ways to 'silence' viruses. An example is limiting the hepatitis virus's ability to reproduce itself.
  - Recombinant monoclonal antibodies (rAbs) are produced more quickly and are more effective than MAbs antibodies. rAbs rely on the equivalent of an in-vitro human immune system; that is, an antibody 'gene library'. Viruses that attack bacteria (bacteriophages) have genes inserted in a way so the bacteriophage produces and displays the protein product of the gene on its outside coat. These are easily screened to produce a 'library' of these antibody proteins that are then made available by biotech companies.

xenotransplantation: the transfer of cells, tissues or organs from one species to another, including from animals to humans

- Examples of possible social implications include concerns over food security and political freedom. In addition, social and economic equity needs to be rethought and new social, economic and political systems put in place.
- Examples of possible ethical uses are outlined below:
  - There have been advances in environmental restoration and the management of endangered species and ecosystems; for example, the collection of coral sperm and eggs at Heron Island to help regenerate coral reefs.
  - Stem cells may eliminate the use of animal models for disease research.
  - There is the potential for 'artificially produced food', with microorganisms used to reduce our dependence on animal sources of protein. Such foods might be more acceptable to vegetarians and vegans.
  - The ethical concerns of using MAbs may be addressed by the production of rAbs. However, intellectual property laws have allowed commercial interests to restrict access to improved methods, increasing costs and accessibility to rAbs.
  - There are ethical concerns about the increased use of animals for xenotransplantation.

The social implications of biotechnology

- Some of the concerns for societies are outlined in the points below:
  - There is a need to consider possible impacts on developing countries through losing traditional markets and products (e.g. palm oil) with a subsequent loss of foreign exchange. The biosynthesis of high-value ingredients such as vanilla, pyrethrum and rubber could ultimately transfer production into industrial bioreactors, again impacting on countries which produce these raw materials.
  - There are issues related to food safety and security, combined with the need for regulations and labelling, to ensure consumer equity and knowledge.
  - There is the potential for environmental damage via an increase in monocultures, loss of habitats and biodiversity. Concerns include some GM crops that do not biodegrade as readily so soil structure and fertility degrades over time. Impacts on non-target species, such as the use of insecticide-containing GMOs, impact non-harmful insects and the food chain. If herbicide-tolerant genes escape into weed populations there is potential for weed problems to escalate.
  - Loss of autonomy is a major social implication. Many unforeseen problems (e.g. environmental damage) impacts people and groups who have elected not to use modern biotechnology. The power and influence of some companies who patent or otherwise control the new biotechnologies may become unacceptable.
  - There is the potential for bioterrorism or the use of biological weapons in warfare.
  - Laboratory safety and security, including the risk of the release of genetically modified plants or genes, is a concern.
  - There are ethical concerns about future directions in biotechnology, such as the potential to create life.

- Biotechnology also raises discussion of potentially positive social implications. These include:
  - less starvation and undernutrition
  - less environmental damage because of more efficient food production
  - increased life expectancy and quality of life
  - personalised medicine and management of health risks
  - more resources and environmentally friendly and reliable energy supplies.

The ethical uses of biotechnology

- The ethics of the application of biotechnology often becomes a discussion about beneficial consequences outweighing harmful consequences. This is a perplexing issue because problems and harmful consequences may not be immediately apparent. Other aspects of ethics revolve around individual and societal rights and responsibilities, autonomy, different cultural, spiritual and religious perspectives, and upholding the virtues that communities commonly accept as good.
- Some particular ethical issues are about the right of humans to:
  - use animal parts (e.g. pig hearts) for xenotransplantation
  - deliberately breed animals for research into disease; for example, the OncoMouse and zebra fish; OncoMouse is a laboratory mouse that has genes inserted to promote the development of tumours with the mice then used as models of human cancer for researching and trialling treatments; zebra fish are useful research organisms because they share 70% of genes with humans and produce nearly-transparent embryos that can be used to study embryological development
  - change the genomes of other organisms and potentially change ecosystems and biodiversity (see the case study on page 97 for more information on this topic).

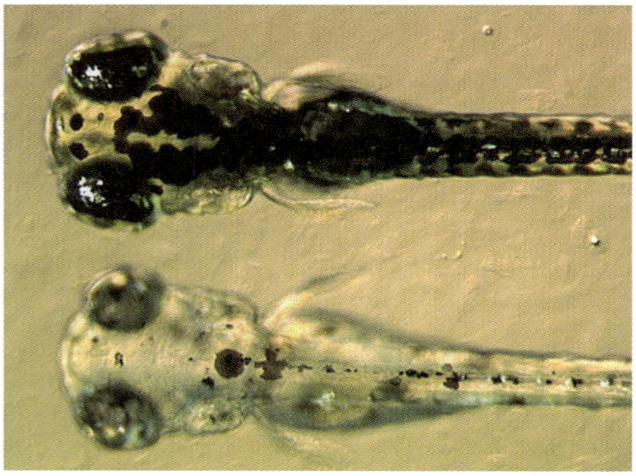

Figure 7.4 Zebra fish embryos (the lower embryo is a mutant and lacks the pigment melanin)

Source: image by Adam Amsterdam, Massachusetts Institute of Technology, Boston, [CC BY 2.5 (http://creativecommons.org/licenses/by/2.5)], via Wikimedia Commons

- Some resources that might assist your review of ethical uses are an online discussion of ethics at Biotechnology online (www.archive.industry.gov.au/Biotechnologyonline.gov.au/biotec/ethics.html) and TED talks on gene editing (www.ted.com/talks/jennifer_kahn_gene_editing_can_now_change_an_entire_species_forever) and reprogramming life (www.ted.com/talks/juan_enriquez_we_can_reprogram_life_how_to_do_it_wisely#t-3294).

→ KEY QUESTIONS

3 **Compare the social implications of the use of biotechnology as it has changed from the past to the present and potentially into the future.**

4 **What ethical issues have arisen in recent times that were not a concern in past uses of biotechnology?**

Answers ➲ p. 101

Researching future directions of the use of biotechnology

- Your research into future directions of the use of biotechnology may have reviewed specific techniques, such as gene drives, or focused on one industry or project. Some resources for examining future directions in biotechnology are outlined below.
  - Read the occasional paper from the Office of the Chief Scientist of Australia at www.chiefscientist.gov.au/wp-content/uploads/OCS-Gene-Editing-and-CRISPR.pdf.
  - Examine the use of bio-computers, which involves a computer that is programmed to control gene expression being inserted into a living cell. The bio-computer would consist of DNA and enzymes.
  - Gene therapy involves attempting to correct faulty genes in human cells. One problem in this field is the process of transferring genes into recipient human cells via suitable vectors. Currently bacterial plasmids are popular vectors and other techniques include the use of microinjection, direct DNA uptake, electroporation or using viruses or transposable elements as vectors. Few of these vectors can be applied when trying to correct faulty human genes. See ghr.nlm.nih.gov/primer/therapy/procedures to learn more about gene therapy and vectors.
  - Read the research summaries of biotechnology briefs at www.agriculture.gov.au/SiteCollectionDocuments/ag-food/biotech/value-of-biotechnology-applications.pdf.
  - Examine the potential of biotechnology to help Australia address climate change at www.agriculture.gov.au/SiteCollectionDocuments/ag-food/biotech/climate-change-and-biotechnology.pdf.

Case study into coral reefs

- A case study in the ethical use of biotechnology and future directions revolves around the work of scientists to help prevent the devastation of coral reefs, such as the Great Barrier Reef, due to global warming causing increased water temperatures and coral bleaching.
- Techniques being considered include selective breeding of corals, investigating genomes, especially of the coral symbionts, and detection, cloning and insertion of genes from heat-resistant corals into other species. Some of this research is summarised in the points below.
  - The Townsville-based Australian Institute of Marine Science is identifying 'super coral' survivors of heat events. Other strategies include developing new strains of the symbiotic algae that corals rely on and testing inoculations of protective bacteria. The hope is that the bacteria may remove reactive oxygen molecules that start the bleaching process when released by heat-stressed corals. This raises the possibility of spraying probiotics on a reef to strengthen it. The alternative—growing and planting corals—is a demanding and lengthy process.
  - The University of Rio de Janeiro in Brazil has shown that inoculating corals with a mixture of ten hydrocarbon-eating bacteria helps them survive oil spills.
  - Researchers in Florida are investigating techniques to rapidly re-establish reefs killed by heat stress while in Hawaii they are studying the biology of corals that survived sewerage pollution.
  - Caribbean nations are banding together to create a genetic storage bank for corals as a 'back-up plan' if all of today's reefs die.

→ KEY QUESTIONS

5 **How can you determine future directions in the use of biotechnology?**

6 **Identify an area in each of agriculture, medicine, industry and environmental management where biotechnology may be used.**

7 **What biotechnologies are likely to be used in the future?**

Answers ➲ p. 101

Evaluating the potential benefits for society of research using genetic technologies

- As part of your evaluation of potential benefits for society of research using genetic technologies you may have asked critical questions and developed skills in detecting and critically analysing biased arguments. Some examples of the areas you might have examined and the questions you may have asked are provided in Table 7.2 below.

→ KEY QUESTIONS

8 **What are five potential benefits for society involving research using genetic technologies?**

9 **What criteria are useful in evaluating the benefits for society from research using genetic technologies?**

Answers ➲ p. 101

Table 7.2 Issues and questions on the benefits to society of using genetic technology

Less starvation and undernutrition	Will this simply result in an increase in human population?	How will this impact on the current obesity epidemic in developed nations?	How will the increased food supplies be equitably distributed?
Less environmental damage because of more efficient food production	Is it possible to recreate complex ecosystems that have been previously destroyed?	Do we need to change expectations and demand to genuinely reduce environmental damage?	
Increased life expectancy and quality of life, personalised medicine and management of health risks	Does this create more implications for aged care?	Does more people living longer and better lives increase demand for resources?	Is it possible to equitably improve health care for all groups and societies?
More resources and environmentally friendly and reliable energy supplies, including better waste management or elimination and less pollution	Is this possible without changes in human behaviour, such as less use of motor transport?		

Evaluating the changes to the Earth's biodiversity due to genetic techniques

- Genetic techniques include the tools of genetic research. This research has enabled the applications that include gene technology (see Chapter 8) and its applications to agriculture, industry and medicine.
- Changes to Earth's biodiversity as a result of genetic techniques can be direct or indirect, and include the points outlined below:
  - There is an initial increase in genetic diversity as a result of new genetic combinations that are not possible naturally.

- The uptake of the use of genetically identical or similar organisms in agriculture results in a reduction of overall genetic diversity because of habitat destruction and degradation, along with declining populations and pollution.
- There may be an increase in the rate of species extinction along with proliferation of some species or breeds such as pesticide-resistant weeds, insects or microorganisms.
- There is an increase in the life expectancy and size of the human population that, in turn, increases the impact on the environment and biodiversity.
- The potential elimination of species considered as pests to humans, such as the malaria vector mosquito along with non-target species, reduces biodiversity.

➔ Awareness of some of the problems of reducing biodiversity has prompted a number of positive strategies, as outlined below:

- Seed and gene banks have been established to preserve the genes of wild and endangered species.
- Captive breeding programs for endangered species (e.g. the Devil Ark at Barrington Tops) have been established to help nurture disease-free populations of Tasmanian devils.
- Many human environmental impacts have been negated through the development of biofuels, bioremediation and ecosystem restoration.
- More humane ways of controlling feral or exotic species have been developed.

➔ Given the reality of the current size and impact of the human population, the evaluation of changes to the Earth's biodiversity could be based on questions such as the following:

- Do genetic techniques change the Earth's biodiversity more than current human activities?
- Do genetic techniques change the Earth's biodiversity more than is acceptable for sustainability?

Table 7.3 Positive and negative changes to Earth's biodiversity due to genetic techniques

Potentially positive changes to biodiversity as a result of genetic techniques	Negative and potentially negative impacts on biodiversity as a result of genetic techniques
metabarcoding promises quicker, cheaper and more accurate assessments of biodiversity in groups that are difficult to assess by traditional methods sequencing low-quality DNA extends the range of useable materials to include museum specimens, archaeological remains and environmental samples see the history of DNA bar coding at dnalc.org/resources/dnatoday/111103-history-dna-barcoding.html	the introduction of genetically modified organisms (GMOs) into natural ecosystems is likely to cause the loss of species and habitat diversity gene drives have the potential to eliminate species or proliferate them
genomic and transcriptomic data can be used to assess the potential of populations to adapt to new challenges the development of powerful new tools for critical assessment of biodiversity, especially genetic diversity, and consequently the identification of potential bio-resources	adverse biological effects on non-target populations and ecological and evolutionary disruption may be either the direct result of the introduced transgene(s) or, alternatively, the indirect result of socio-economic conditions related to the application of recombinant DNA technologies
in the near future gene-editing tools may help endangered species cope with change, while gene drives might control unwanted species and help wanted species	several direct non-target effects on beneficial and native organisms by GMOs have been reported; an example is the transgenic Bt cotton plant which affects a wide array of non-target insects such as butterflies, moths and beetles
de-extinction has become a serious prospect	some GM crops have been shown to affect soil ecosystems by decreasing the rate of decomposition of organic wastes, affecting carbon and nitrogen levels and decreasing the diversity of soil microbial populations
four different transgenic crops (oilseed rape, potato, maize and sugar beet), grown in 12 different habitats and monitored over a period of ten years, showed no cases where transgenic plants were found to be more invasive or persistent than their conventional counterparts	**genetic pollution** is gene flow into wild populations; scientists want to avoid this in order to maintain the diversity of native populations the idea of destroying natural gene pools by adding genes to and from organisms is unacceptable and seen as sabotaging billions of years worth of natural selection created by nature

metabarcoding: a rapid method of biodiversity assessment that combines DNA-based identification and high-throughput DNA sequencing
genetic pollution: the uncontrollable and undesirable flow of genes into wild populations

Potentially positive changes to biodiversity as a result of genetic techniques	Negative and potentially negative impacts on biodiversity as a result of genetic techniques
there is wiser and more efficient utilisation of bio-resources, both as a genetic resource for production and in the remediation of altered or degraded ecosystems	episodic **genetic erosion** could threaten the genetic diversity on which gene technology depends; for example, micro-propagation and the consequent production of identical clones discourages genetic diversity through evolutionary adaptations
ongoing research may assist in saving whole ecosystems (e.g. coral reefs) rapid propagation or breeding of a newly discovered species may prevent overexploitation from its natural habitat	biotechnological methods leading to the identification of a plant material for an important pharmaceutical use would raise the value of the material, resulting in increased collection pressures on the plant, which in turn would lead to overexploitation and species loss
inbuilt sterility in GMOs reduces the risk of escape into natural habitats	adverse impacts on biodiversity through the introduction of GMOs may also result from disturbance of the dynamic population equilibrium of ecosystems the population size of native taxa may be reduced by the enhanced ability of GMOs to invade the natural habitats of native species
	another possible direct impact of GMOs specifically raised for conferring viral resistance is the likely emergence of new viruses with new biological characteristics through recombination; similarly there is the potential for the emergence of resistant insect pests, other pathogens and weeds
	increased practice of monoculture as a result of biotechnology; monoculture usually involves vast fields planted with a single high-yielding variety, using inputs such as hi-tech irrigation systems, along with fertilisers and pesticides to maximise production monoculture results in low genetic diversity, making crops vulnerable to pests or disease which can quickly spread associated land management practices may also degrade soils and pollute waterways

genetic erosion: the general loss of genetic diversity (e.g. varieties, breeds or crop species) or within a species (e.g. loss of genes or alleles)

→ KEY QUESTIONS

10 What are five potentially positive changes to biodiversity that result from research using genetic technologies?

11 What are five potentially negative changes to biodiversity that result from research using genetic technologies?

12 What criteria can be used to evaluate the changes to Earth's biodiversity as a result of using genetic technologies?

Answers ➲ p. 101

CHAPTER SYLLABUS CHECKLIST

Are you able to answer these questions from the syllabus for this chapter? Tick each question as you go through the checklist if you are able to answer it. If you cannot answer a question, turn to the relevant page in the study guide to find the answer. For NESA key word meanings, go to www.educationstandards.nsw.edu.au and search 'key words'.

	FOR A COMPLETE UNDERSTANDING OF THIS TOPIC:	PAGE NO.	✓
1	Can I investigate the uses and applications of biotechnology (past, present and future), including analysing the social implications and ethical uses of biotechnology, including plant and animal examples?	90–94	
2	Can I investigate the uses and applications of biotechnology (past, present and future), including researching future directions of the use of biotechnology?	94–95	
3	Can I investigate the uses and applications of biotechnology (past, present and future), including evaluating the potential benefits for society of research using genetic technologies?	95–96	
4	Can I investigate the uses and applications of biotechnology (past, present and future), including evaluating the changes to the Earth's biodiversity due to genetic techniques?	97–99	

HSC EXAM-TYPE QUESTIONS

Objective-response questions (1 mark each)

1 **Which of the following is an example of biotechnology?**
- A having a pet dog
- B using plants in gardens to attract birds
- C keeping chickens for their eggs
- D filling your car with petrol

2 **Which of the following is a recent use of biotechnology?**
- A selective breeding of wheat
- B fermenting grape juice to make wine
- C 'daughterless' genes inserted into cane toads to prevent the next generation from breeding
- D the use of enzyme-linked immunosorbent assay (ELISA) to detect the human immunodeficiency virus (HIV)

3 **A social implication of the biotechnological use of plants is:**
- A reduced genetic diversity because of tissue culture cloning of identical ferns.
- B reduced blindness due to vitamin A deficiency because of the development of Golden Rice.
- C the use of genetically modified cotton seed oil to produce biofuel.
- D the evolution of insects resistant to the Bt toxin because of the widespread planting of Bt cotton.

4 **Which of the following examples describe an ethical use of animals in biotechnology?**
- A breeding 'daughterless' rabbits to replace the use of calicivirus in pest control
- B the genetic modification of cows to produce human insulin in their milk
- C breeding virus-free pigs for safer use in xenotransplantation
- D using animals such as horses and mice to produce anti-venoms or antibodies

5 **Which of the following is an application of biotechnology that could improve biodiversity?**
- A gene technology used on rice to include a precursor to vitamin A to reduce vitamin A deficiency blindness
- B selective breeding of sheep to improve the quality of the wool
- C the use of bioreactors to produce pharmaceuticals
- D the use of metabarcoding to detect the catch of protected species of seafood

Short-answer questions

6 **Discuss the social implications of biotechnological use of a named plant and animal.** (8 marks)

7 **Choose the area of research using genetic technologies that you believe has the greatest potential benefit for society and justify your choice.** (6 marks)

8 **From your consideration of future directions in biotechnology analyse the prospect that future food may come in pill form.** (4 marks)

9 **How do genetic techniques affect Earth's biodiversity?** (5 marks)

10 **Compare the past applications and uses of biotechnology with current applications and uses.** (4 marks)

ANSWERS

KEY QUESTIONS

Key questions ➲ p. 92

1 Past uses of biotechnology included selective breeding of plants and animals and the use of microorganisms to produce food, beverages and medicines. In addition to these practices, current use involves the genetic manipulation of organisms and the expanded applications of biotechnology to industrial and environmental fields along with an increased range of applications to the health of humans and other useful organisms.

2 Some possible major directions in biotechnology include increased uses for renewable energy, new materials (e.g. biodegradable plastics), organs for transplantation, better waste management and recycling, and new tools for research.

Key questions ➲ p. 96

3 Past social implications of the use of biotechnology included greater food security and safety and better human health, including increased life expectancy and population size. At present and potentially into the future biotechnology may reduce global food shortages and dramatically increase life expectancy, medical costs and inequity in society.

4 New and potential ethical issues are the increased use of animals for research and replacement of human organs and tissues, along with genetic pollution. Genetic erosion is occurring on an increased scale. Other issues include the rights and responsibilities of farmers in terms of preventing genetic pollution. Ethical debates about harm versus benefits are more complex when the potential harm may not be known.

Key questions ➲ p. 97

5 Identifying future directions in the use of biotechnology involves projecting current trends or identifying problems along with brainstorming and mind-mapping possible solutions that could arise by using living organisms.

6 Plants could be engineered so that every part had a use (agriculture); cells could be engineered to repair themselves (medicine); organisms could be developed that extract and shape metals into useful items directly from their ores (industry); organisms could be developed that better store carbon to reduce the levels of greenhouse gases (environmental management). Note that these are examples and there are many possible options for this answer.

7 Biotechnologies likely to be used in the future include traditional practices such as selective breeding and fermentation but with refinements, such as the use of marker-assisted breeding. The use of some gene technologies, such as cloning desirable organisms, gene silencing and genetic modification to produce medicines and some food ingredients, is also likely.

8 Five potential benefits are more nutritious food, especially in developing countries, better biofuels, better medicines, better diagnoses and less pollution.

9 Useful evaluation criteria include better health (e.g. life expectancy, nutrition and life quality), improved equity of access to resources and medicine, and better quality of the environment and biodiversity.

Key questions ➲ p. 99

10 Five potentially positive changes include the return to natural ecosystems because of elimination or reduction of exotic species and pests, bioremediation of land and water that has been contaminated by human land use, reduced pressure on natural populations because of the increased ability to produce alternative resources, more resilient species and ecosystems, the potential to reduce climate change via the production of biofuels, and better detection and monitoring of biodiversity (using metabarcoding).

11 Five potentially negative changes are species extinction, loss of wild types or breeds, decline of non-target species such as beneficial insects, dominance of monocultures of preferred varieties of crops or livestock, and genetic pollution (e.g. resulting from the spreading of genes or the inadvertent development of resistant pathogens such as viruses).

12 Evaluation criteria include reduction of diversity within species, reduction of the diversity of species, and reduction of the diversity of ecosystems.

HSC EXAM-TYPE QUESTIONS

Objective-response questions

1 C. The use of living organisms (chickens) to produce products which we need (eggs) is an example of biotechnology. **A** and **B** are incorrect as companion animals or plants to attract birds are not using organisms for products. **D** is wrong because petroleum is a fossil fuel.

2 D. ELISA is a currently used biotechnology. **A** and **B** are wrong because they are traditional biotechnologies. **C** is incorrect as this technology is still under development.

3 B. An improved diet implies improved vision for communities at risk of vitamin A deficiency. **A**, **C** and **D** are logical consequences of use of biotechnology on plants but the issues raised are environmental (A), associated with genetic erosion (C), or the evolution of resistant animals (D).

4 A. This option is correct because it is a much more humane way of reducing rabbit numbers. **B**, **C** and **D** are wrong because their purpose is to use animals to improve human health, often to the detriment of the animal.

5 D. Better policing of protected species should help conserve the species and maintain biodiversity. **A** is wrong because it will tend to result in more use of monocultures and **B** will result in less variety of sheep breeds. **C** is incorrect because many of the new pharmaceuticals are designed to totally eliminate the pathogen.

Short-answer questions

6 EM You have to apply understanding of the social implications of biotechnology to a plant and an animal example. Careful selection of your examples might make answering the question more straightforward.

Golden Rice has been in development for over 20 years but is still not grown commercially. The result of this is that the projected social implication of a reduction in vitamin A deficiency blindness and improved immunity, especially in children, has not occurred. (✓) Economic factors play a role as even though companies producing the rice have waived the costs associated with patents they set a condition that the rice would not be sold at a premium price.

This was to ensure accessibility for the poor but may have had a negative impact on willingness to take up cultivation of the rice. (✓)

Greenpeace has conducted an ongoing anti-GM campaign and has used terms such as 'contamination' when applied to GM crops. This, along with concerns about genetic pollution, has had a negative impact. (✓) Countries such as Thailand which export rice to Europe, where there is strong opposition to GM, have categorically turned down the use of Golden Rice. (✓)

Insulin-dependent diabetics (type 1 and some type 2 diabetics) rely on injections of insulin or insulin pumps to ensure their blood glucose levels are controlled. See Table 3.5 and Figure 3.6 for an outline of the structure and function of insulin.

Prior to the discovery of insulin a diagnosis of diabetes meant a dramatically shortened life expectancy. Over decades of incremental improvements, traditional biotechnology greatly improved the length and quality of life of type 1 diabetics. (✓) Diabetics were treated with insulin mainly extracted from cow or pig pancreases or a mixture of these. Improvements in extraction and purification techniques since the first extractions in the early 1920s has greatly improved patient outcomes. (✓)

Gene technology has now been applied with microorganisms such as *E. coli* having human insulin genes inserted. The initial impetus for the change from traditional methods was a perceived shortage of insulin from animal sources and because human insulin is slightly more soluble and rapidly absorbed. Despite some trials having shown no difference in safety and efficacy and GM insulin being more expensive there has been a transition to GM insulin since its introduction in the 1980s. (✓) Another change is the development of a new form of insulin (analogue insulin), which has increased the price 100-fold with questionable benefits except economic ones to the company producing it. (✓)

7 **EM** Possible choices for your answer include CRISPR-Cas9, gene drives, ELISA, RNAi (gene silencing), rAbs or marker-assisted breeding. You need to demonstrate thorough understanding of the gene technology and justify its potential for positive impacts on society when compared with other possible choices.

RNAi or gene silencing has the potential to bring the greatest benefits to society. It is more acceptable than other gene technologies because it influences gene expression. GM technologies that change DNA are still not accepted by many societies and countries. (✓) RNAi has a diverse range of applications for agriculture, human nutrition and health, and better biofuels. It is an invaluable tool for genetic research in terms of determining gene function and linking phenotypes to genes. (✓)

Gene silencing has the potential to greatly improve food production by eliminating pathogens of many plant crops and animals without the need for pesticides. Cereal varieties developed using RNAi have resistance to Barley Yellow Dwarf virus, Rice Stripe virus or Western Corn Rootworm. Gene silencing has been used experimentally to prevent prawn death from Gill-associated virus. There can also be an improvement in the nutritional value of crops. (✓)

Gene silencing is currently used to improve production of antibodies used to treat human diseases (e.g. Non-Hodgkin lymphoma) and inflammatory conditions (e.g. arthritis). It also has the potential to treat diseases arising from faulty gene expression. Diseases caused by viruses such as the human immunodeficiency virus (HIV), human papilloma virus (HPV), hepatitis and influenza, may be eliminated by using RNAi. (✓)

RNAi has great potential for expanded use in genetic research. It can be used to identify the function of a gene and determine the gene responsible for a specific phenotype. RNAi could add precision to marker-assisted breeding by identifying genes of interest. It has assisted the CSIRO in breeding improved wheat varieties and marine stock for aquaculture. (✓)

RNAi has the potential to help society by contributing to a reduction in global warming via the more efficient production of biofuels. (✓)

8 **EM** This question makes a prediction about the development of food substitutes. You need to use your understanding of possible future directions in gene technologies along with barriers to the acceptance of these technologies in order to analyse future prospects.

Current trends in the use of biotechnologies in food production include gene technologies which increase the nutritional content of staple foods such as bananas. This has the potential to overcome vitamin A deficiency and its harmful impact on both the immune system and eyesight. (✓) Despite this there are many hurdles to overcome, with farmers and consumers proving resistant to GM foods. Because of this many of the benefits have yet to be realised. (✓)

The increased use of dietary supplements and other nutraceuticals suggests that humans have accepted that their nutritional intake may not be derived from food alone. It is a major leap in terms of social implications to acceptance of food in the form of a pill produced by bioreactors. For example, what would be the impact on farmers? (✓) A major catastrophe such as nuclear, chemical or biological warfare may result in conventional food production being abandoned, leading to public acceptance of the 'food pill' not being a matter of choice. (✓)

9 **EM** This inquiry question requires you to identify the genetic techniques or their applications and uses and determine the impact on population genetics, species diversity and ecosystem diversity.

Initially genetic techniques such as genetic modification of a species produces new genetic combinations not possible in nature, leading to an increase in biodiversity. (✓) Some applications of GM result in a reduction of biodiversity. Herbicide-resistant crops allow farmers to more easily control weeds with the use of herbicides. While these herbicides reduce weeds, they also reduce other species that might depend on the weeds (e.g. seed-eating birds). Crops genetically modified to contain insecticides from bacteria (e.g. Bt cotton) reduce non-target insect species as well as those that are harmful. (✓) The costs of application of genetic techniques promote the need for monocultures and large-scale aquaculture which can reduce ecosystem diversity because of factors such as habitat destruction, water and soil pollution, and increased salinity of soils. (✓) Techniques such as RNAi have the potential to reduce biodiversity by eliminating some pathogens. (✓) Genetic techniques that improve human health and life expectancy may increase the rate of growth of the human population, placing far greater stresses on biodiversity. (✓)

10 **EM** You need to identify the similarities and differences between traditional biotechnologies with current biotechnologies that additionally use genetic technologies.

Past uses of biotechnology included production and preservation of foods by fermentation, growing crops and raising livestock through agriculture and aquaculture. Selective breeding of strains or breeds was a key to improvement. (✓) Traditional biotechnology also included the production and use of monoclonal antibodies, antibiotics and anti-venoms for medical diagnoses and treatments. Biological control of pests required substantial research and resulted in varying levels of success. (✓)

Present uses of biotechnology include more effective selective breeding by using marker-assisted features to rapidly breed more productive organisms. RNAi is used to silence gene expression to improve resistance to diseases and better understand and control the role of some genes. (✓) Improved crops and animals have enhanced agriculture and aquaculture. Biological control can be combined with CRISP-Cas 9 and gene drives to eliminate disease vectors or pathogens or render female pests only capable of producing male offspring. (✓)

CHAPTER 8 GENETIC TECHNOLOGIES

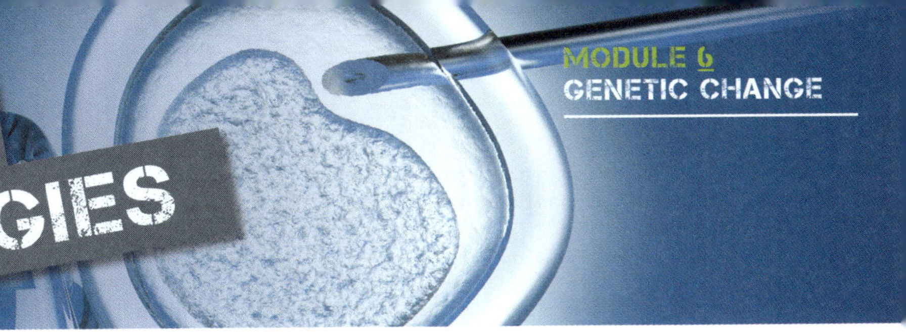

INQUIRY QUESTION:

Does artificial manipulation of DNA have the potential to change populations forever?

In Chapter 7 you reviewed biotechnology and its past, current and future uses and impacts. Future directions are closely linked with the component of biotechnology that uses genetic technologies. In this chapter you will review research in various fields of science, including agriculture, industry and medicine. You will also review the impact of gene and reproductive technologies, including cloning, on biological diversity.

1 Investigating the uses and advantages of current genetic technologies that induce genetic change

» Students investigate the uses and advantages of current genetic technologies that induce genetic change.

➔ In Chapter 5 you reviewed how genetic change can be a result of changes in allele frequency and in Chapter 6 how changes in the alleles themselves resulted from mutation. In Chapter 7 you reviewed how gene editing or the artificial introduction of a gene from one species into another produced genetic change. If the change occurs in the germ-line or embryonic stem cells (as in knockout mice in Table 8.1) the change may be ongoing. Temporary genetic change can result from the alteration of the expression of genes such as RNAi (gene silencing). The incorporation of gene drives along with transgenic interventions can mean gene frequency can change rapidly within a population.

➔ Uses of current genetic technologies are summarised below and include:

- rapid improvement in agricultural outputs with its attendant economic and social benefits
- improved drugs and medicines
- a transition to more renewable fuels and energy sources
- more recycling of nutrients and better waste management
- better understanding of genetics and diverse research tools.

Table 8.1 Uses and advantages of current genetic technologies

Technology	Uses	Advantages
mutation breeding involves exposing plant material, including seeds, to mutagens to produce variants that are then selected if they have desirable qualities	used since the 1920s and has produced over 2300 crop varieties has enhanced yields, improved nutritional quality and widened the adaptability of crops (e.g. wheat, rice, pulses, millets and oilseeds)	is a relatively easy process that is not subject to regulation
genetic engineering involves inserting genes from one organism into another (i.e. producing GMOs or recombinant DNA technology) current technology includes gene editing tools such as the use of CRISPR-Cas9 (see page 91) to add precision and efficiency to processes that disrupt gene function gene drives are used to magnify the impact of the deleted gene through a population	improves the sustainability and productivity of agriculture and protects plants, animals and humans from disease produces medicines and pharmaceuticals (e.g. human recombinant insulin, growth hormones, blood-clotting factors, hepatitis B vaccine and diagnosis of HIV infection) gene drives are currently being used in research to incorporate genes in order to prevent the ability of mosquitoes to carry the malaria parasite	new combinations of genes can dramatically improve the potential of agriculture, medicine, waste treatment and the pharmaceutical industry CRISPR-Cas9 has a target efficiency (i.e. it does its job in the correct place on the DNA) over 70% of the time; this compares very favourably with other methods of gene editing that range between 1% and 50%

Technology	Uses	Advantages
a knockout (KO) gene is an organism containing a gene that has been made inoperable	used to study gene function by comparing a knockout with normal organisms mice are frequently used as knockout animals because of similarities with humans	KO mice research has helped understand human diseases including cancer, obesity, heart disease, diabetes, arthritis, substance abuse, anxiety, aging and Parkinson's disease
RNA interference (RNAi) or gene silencing	RNAi technology can identify which genes are responsible for particular traits; plant breeders can then produce more productive but non-genetically modified plants researchers have used RNAi in chicken embryos, causing male to female sex reversal which assists understanding sex determination in animals in aquaculture the technology reduces stock losses due to diseases (e.g. developing an experimental anti-viral that prevents prawn mortality from Gill-associated virus)	a therapeutic agent used to control disease and prevent infection in plant and animal cells the CSIRO claims the process is fast, reliable, flexible, stable and controllable; the gene editing is temporary ability to produce plants with desirable properties without using GM so RNAi is more acceptable to many people in aquaculture, more sustainable feed formulations, enhanced growth, better survival and feed conversion rates and increased tolerance to viral diseases
marker-assisted breeding	use of genetic markers helps select organisms for selective breeding its use, combined with viral screening technology, has helped the tiger prawn aquaculture industry to develop breeds highly tolerant to endemic diseases	speeds up conventional selective breeding has increased pond yields of black tiger prawns by over 50%, easing pressure on wild stocks and making prawn aquaculture more sustainable
chromosome engineering, including manipulating chromosomes, results in rearrangements such as insertions, deletions and translocations also includes manipulation of meiosis, resulting in alterations in numbers of whole chromosomes or homologous chromosome sets	used to improve crop species such as corn, soybeans, rice, barley and potatoes	enables research into some diseases (e.g. myeloid tumours associated with chromosomal abnormalities)
microbial vectors involve microinjection using Ti plasmids and **microprojectile bombardment**	responsible for the majority of GE plants in commercial production although successful in fruits and vegetables, the Ti plasmid has generated limited success in grain crops though modification of the process has resulted in recent success in rice and sugar cane microprojectile bombardment involves delivering naked DNA to plant cells by 'shooting' them with microscopic pellets to which DNA has been adhered; this is effective in corn, rice and other cereal grains	overcame the difficulty of introducing cloned genes through cell walls and membranes into plant cells
genetic screening combined with **in-vitro fertilisation (IVF)**	embryos may be screened and selected to avoid inherited disease there has recently been the case of a human produced from three genetic parents where the maternal DNA (nucleus) was inserted into an enucleated egg from another female to avoid an inherited disease carried on the mother's mitochondria	harmful germ-line mutations can be avoided in offspring

microbial vectors: DNA segments from bacteria or viruses which are capable of carrying foreign genetic material into a cell

microprojectile bombardment: high-velocity particles that bombard plant cells, creating small wounds and enhancing entry of vectors or naked DNA

in-vitro fertilisation (IVF): a medical procedure to bring about fertilisation of an oocyte and a spermatozoon, traditionally occurring in a test tube or literally 'in glass'

➔ KEY QUESTIONS

1 **Identify five technologies that induce genetic change.**

2 **Identify the advantages of two named genetic technologies.**

3 **Identify a gene technology that produces fast, permanent genetic change.**

4 **Name three technologies that do not produce genetically modified organisms.**

Answers ➲ p. 114

2 Comparing the processes and outcomes of reproductive technologies

» Students compare the processes and outcomes of reproductive technologies, including but not limited to artificial insemination and artificial pollination.

- ➔ Reproductive technologies modify reproduction in order to result in more desirable outcomes in offspring.
- ➔ Cloning is a reproductive technology designed to produce genetically identical offspring.
- ➔ In Chapter 1 you reviewed reproductive technologies in aquaculture. In one example, pressure is used to create triploid fish which are sterile and grow faster. In other applications hormone treatments are used to promote sex reversal and fish of only one sex.
- ➔ Assisted reproductive technology refers to the treatment of infertility, usually in humans, by the manipulation of sperm, eggs or embryos.

Table 8.2 Process and outcomes of reproductive technologies

Reproductive technology	Processes	Outcomes
artificial insemination (AI)	collection, storage and transport of semen from male animals with desirable qualities used to fertilise a large number of females see page 20 of Chapter 1 for some variations in the use of AI in breeding cattle	larger numbers of more desirable animals less genetic diversity global benefits from storage and transport of semen
artificial pollination	involves the transfer of pollen to stigmas to enable pollination and fertilisation can be used to deliberately produce hybrids or for selective breeding natural pollination needs to be prevented by physically removing stamens or using other techniques to induce male sterility see page 18 of Chapter 1 for an explanation of the use of cytoplasmic male sterility (CMS) and environment-sensitive genic male sterility (EGMS) in rice	improved plant cultivars such as rice IR8 and many types of plant hybrids with hybrid vigour (i.e. increased productivity beyond that of either parent)
in-vitro fertilisation (IVF)	examples of the use of IVF in animals are outlined in Chapter 1 involves the collection and combination of oocytes and semen for fertilisation outside the body; embryos are then implanted into females hormones stimulate the development of oocytes and ensure the uterus of the female surrogate is ready for implantation	used in agriculture to improve the reproductive potential of desirable female animals and reduce intervals between generations used in humans to treat some infertility problems also used with embryo screening to reduce the chance of genetic diseases

artificial insemination (AI): the process of injecting semen into the vagina or uterus with the intention of achieving fertilisation

artificial pollination: the artificial transfer of pollen to control sexual reproduction in flowering plants

➔ KEY QUESTIONS

5 **Compare the processes of artificial pollination with artificial insemination.**

6 **Compare the outcomes of artificial pollination with artificial insemination.**

7 **What processes and outcomes are involved in in-vitro fertilisation?**

Answers ➲ p. 114

3 Investigating and assessing the effectiveness of cloning

›› Students investigate and assess the effectiveness of cloning, including but not limited to whole organism cloning and gene cloning.

➔ Three main types of cloning are distinguished as outlined below:
- Reproductive or whole organism cloning results in progeny genetically identical to the parent organism.
- **Gene (or DNA) cloning** results in many copies of a gene for use in gene technology.
- **Therapeutic cloning** uses a cloned embryo to produce **embryonic stem cells** for use in medical treatments or therapies. Tissues or organs derived from therapeutic cloning would theoretically not promote rejection and could be used in biomedical research (e.g. drug testing). Therapeutic cloning has been used to treat Parkinson's disease in mice.

➔ When you assess the effectiveness of the cloning procedure you will compare the achievements of the procedure against its purpose.

gene (or DNA) cloning: the production of exact copies of a gene or DNA sequence
therapeutic cloning: the transfer of a diploid nucleus from a somatic cell (SCNT) into an enucleated egg cell to harvest the resulting embryonic stem cells for replacing or repairing damaged tissues or organs
embryonic stem cells: cells taken from the inner cell mass of an embryo; capable of differentiating into any type of cell

Effectiveness of whole organism cloning

➔ Whole animal cloning produces multiple identical copies of organisms with selected characteristics.

➔ Plant cloning (e.g. tissue culture) is widely used to produce large numbers of the desired plant. See pages 16–17 for a discussion of plant cloning (asexual reproductive techniques in agriculture). Tissue culture is one method with a great capacity to produce identical progeny but it requires a sterile growth medium and aseptic techniques.

➔ Animal cloning occurs in two main procedures. In the first method, embryo splitting results in identical cells that develop into genetically identical individuals. This procedure is relatively easy but the qualities of the embryo are often unknown and the main aim of animal cloning is to produce copies of desirable or more productive animals. The second method, somatic cell nuclear transfer (SCNT), uses genetic material from mature (or even dead) animals to result in animals with identical genomes.

See pages 19–20 for an explanation of SCNT and Figure 1.21 for how Dolly was cloned. Since 1997, when Dolly was cloned, research and experimentation to clone useful livestock has continued and diversified. The attractiveness of cloning is the knowledge that the offspring will be genetically identical to the parent.

➔ The four steps for SCNT in animal cloning are described below:
- A differentiated somatic cell (one that has become specialised during development) has its diploid nucleus removed to provide the DNA source for the clone.
- An egg cell from a similar animal is recovered and the nucleus is removed, leaving only the cytoplasm and cytoplasm organelles.
- The nucleus is injected into the enucleated egg cell. Electric pulses assist the union to form a single diploid cell, which then begins normal cell division.
- The developing embryo is placed in a surrogate mother, who then undergoes a pregnancy.

➔ The effectiveness of SCNT is summarised in the points below:
- High pregnancy losses and high morbidity and mortality during the neonatal period reduce the effectiveness of SCNT, with a current success rate of 4%.
- The process may have benefits in preserving endangered species.
- Early reports of animal cloning suggest that premature aging might reduce the usefulness of the procedure. The length of the tips of chromosomes (telomeres) shortens with repeated cell divisions and age. If the telomeres of the clone were the same length as the already shortened telomeres from an aged donor nucleus then premature aging could result. Continued research now suggests that the telomere length can reset and that clones have a normal life expectancy and can breed normally.
- Cloning of animals is an expensive process and prone to failure, with a success rate between 0.1 and 2%. Cloning is generally restricted to produce breeding stock or research animals.
- Surrogate mothers suffer increased health risks.
- Using cloned animals for food remains a major concern to the public and is generally not considered acceptable.

Effectiveness of gene cloning

➔ Assessing the effectiveness of gene cloning needs to take into account time, cost and resources required as well as the number of steps in the process and the percentage of successfully cloned genes. Some researchers suggest that gene synthesis, which does not use a DNA or RNA template and requires knowledge of the nucleotide sequence, may soon replace gene cloning.

➔ Gene cloning using a vector plasmid results in multiple copies of genes, usually in a bacterium such as *E. coli*.

See Figure 8.1 for an outline of the steps. This process not only clones the gene but the polypeptide gene product can be produced within the host cell. This method is used extensively for the production of human hormones and proteins.

- Genes that have been expressed—mRNA has been transcribed from the gene DNA—can be cloned
- The polymerase chain reaction (PCR) is one method of copying DNA or genes. It is generally fast, efficient and allows high **throughput**. The PCR is useful when there are only small amounts of the required gene or DNA. This is the method used in DNA profiling.
- Gene cloning is rapidly being expanded and even replaced by an approach called **synthetic biology**. See www.ebrc.org/what-is-synbio. This relies on the well-established procedure of DNA sequencing and the relatively new process of DNA synthesis that produces specific DNA sequences 'on demand' in what is likened to a printing process using DNA nucleotides rather than letters. Bacterial and yeast cells are converted into miniature factories by the incorporation of a range of biological components. One milestone example of its use was the synthesis and assembly of a million base-pair bacterial genome which, when placed inside a bacterial cell host devoid of its own genetic material, was able to successfully generate an artificial living cell. In essence it was the process of gene cloning taken to such a level that it could be used to clone a bacterium. A video on synthetic biology can be seen at www.youtube.com/watch?v=iRO0-fMIW9I.
- New developments in microchip-based gene synthesis technologies (microarrays) hold the potential for significantly increasing the throughput and decreasing the cost of gene synthesis. Currently these procedures have a high error rate.

throughput: the amount of material passing through a system or process
synthetic biology: the use of an engineering approach (design and construction) that includes control mechanisms to solve biological problems using components such as cells, enzymes and genetic circuits

KEY QUESTIONS

8 Outline the effectiveness of plant cloning.

9 Outline the effectiveness of animal cloning.

10 How effective is gene cloning?

Answers ➲ p. 114

4 Describing the techniques and applications used in recombinant DNA technology

» Students describe techniques and applications used in recombinant DNA technology; for example, the development of transgenic organisms in agricultural and medical applications.

Recombinant DNA technology

- Producing **recombinant DNA (rDNA)** or chimeric DNA involves gene-shuffling. A specific segment of DNA is isolated and removed from one organism and attached to another piece of DNA, often from a different organism. This produces new characteristics in the host organism or enables the host organism to produce specific substances that scientists can use.

recombinant DNA (rDNA): DNA that has been artificially produced by combining DNA from two different species

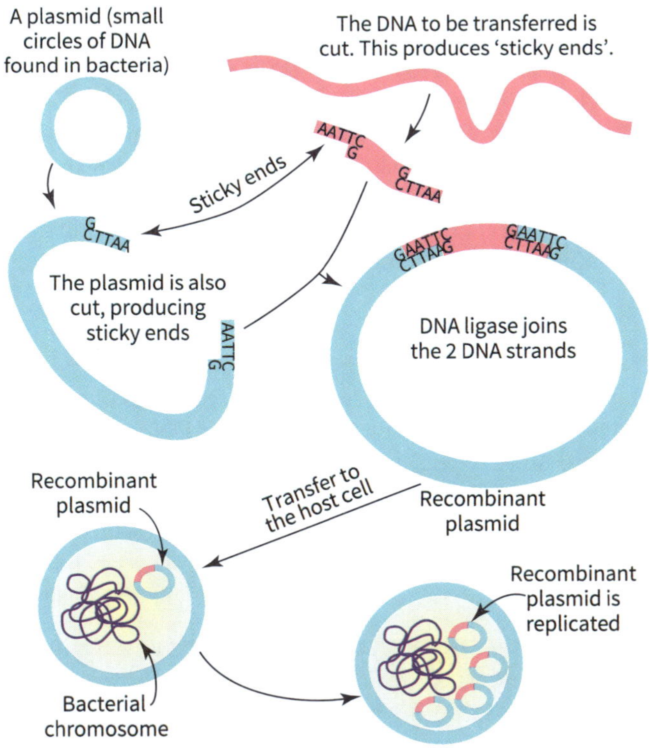

Figure 8.1 Gene cloning producing recombinant DNA

- Essential steps in the production and application of recombinant DNA are outlined below:
  - The gene of interest is identified and then the DNA containing the gene is extracted and purified.

- Cutting and joining the DNA or gene splicing (using a **restriction enzyme** and **ligases**) results in a single piece of DNA containing DNA from two different species joined together. An example is the human gene for insulin in a bacterial plasmid. Note the same restriction enzyme is used to cut the DNA of interest and the plasmid so they have the same base sequence (called 'sticky ends').
- Monitoring the cutting and joining—electrophoresis can be used to separate DNA fragments by their length. Often another gene is attached to the gene to be cloned. This might be a gene that confers antibiotic resistance and the resultant bacteria can be screened by applying the antibiotic. Only those bacteria with the recombinant DNA will survive.
- The recombinant DNA is amplified or modified.
- Transforming hosts, such as bacteria, with the recombinant DNA or the use of DNA **vectors** or a microinjection allows genes to be carried into nuclear DNA in multicellular organisms. A strong solution of calcium chloride can make the bacteria membrane more permeable so that the plasmid carrying the recombinant DNA can enter. Bacterial transformation can occur through opening pores in the bacteria. This is achieved by heating and passing an electric current through the bacterial culture.

➔ Applications of recombinant DNA are described in Table 8.1 under genetic engineering, knockout genes and microbial vectors. Table 7.1 includes past examples of applications of recombinant DNA, many of which are still in use in agriculture and medicine. Further agricultural and medical examples of current and potential future applications of recombinant DNA can be found in Chapter 7.

restriction enzymes: nucleases (enzymes that cleave nucleic acids) that can target specific sequences of bases and so be used to isolate and cut desired DNA fragments
ligase: an enzyme that joins two DNA fragments
vector: a small self-replicating DNA segment (usually a bacterial, viral or yeast plasmid) that carries the inserted DNA into the host (e.g. *E. coli*)

Example: the development of transgenic organisms in agricultural and medical applications

➔ The development of Golden Rice in agriculture has the potential to:
- save millions of children from vitamin A deficiency blindness
- prevent immune disorders resulting from vitamin A deficiency.

➔ Golden Rice is produced by adding two genes from daffodils and one from a bacterium, along with DNA promoters, into Ti plasmids that then infect rice embryos, transferring the genes for making beta carotene, the vitamin A precursor. Ti plasmids derive from a soil bacteria that causes crown gall disease. In the process of infecting its host the bacterial plasmid transmits DNA (called T DNA). This DNA integrates with the host's DNA and gets expressed. The resulting transgenic rice is then crossed with local varieties to ensure the rice is adapted to grow in local conditions.

➔ The development of transgenic mice (e.g. knockout mice or the OncoMouse) has major medical applications. These include the following:
- It is a tool for research into human disease, particularly cancer.
- It provides a key to understanding gene function, particularly in terms of susceptibility to diseases and their progression.
- It gives us a better understanding of responses to various therapies.

➔ The first genetically modified animal was a mouse created in 1974 by inserting simian vacuolating virus 40 (SV40) into early stage mouse embryos. SV40 is an oncovirus. Subsequent developments led to the OncoMouse, which was patented in 1988.

➔ The first animal containing recombinant DNA was a mouse created in 1980. It was the result of DNA being inserted into newly fertilised mouse eggs.

➔ Transgenic mice are produced by the microinjection of DNA into the pronuclei of fertilised oocyte; that is, in the interval between entry of the spermatozoon to the oocyte and fusion of the nuclei of both gametes. The egg is implanted into a surrogate mother for development.

➔ Knockout mice are created by using four-day-old embryonic stem cells. Modified viral vectors or linear fragments of bacterial DNA are carried into the cells. The cells are cultured for several days before being implanted into early stage mouse embryos that are then implanted into a surrogate. The inserted DNA may actually target a specific gene, which it inactivates, or be random and inactivate the polypeptide synthesis mechanism in such a way that the particular gene can be identified.

➔ KEY QUESTIONS

11 **What is recombinant DNA technology?**

12 **Outline the processes involved in recombinant DNA technology.**

13 **Compare the processes involved in recombinant DNA technology used for agriculture with those used in medical applications.**

Answers ➲ p. 114

5 Evaluating the benefits of using genetic technologies in agricultural, medical and industrial applications

» Students evaluate the benefits of using genetic technologies in agricultural, medical and industrial applications.

➔ In Chapter 1 you reviewed the impact of scientific knowledge on the manipulation of plant and animal reproduction in agriculture. In Chapter 7 you evaluated the potential benefits for society of research using genetic technologies. Many of those potential benefits derived from the use of gene technologies in agriculture, medicine and industry.

➔ Evaluating the benefits of using gene technologies versus the more traditional biotechnologies involves making judgements about the benefits compared with the costs or harms. Environmental, social, economic and cultural factors may need to be considered.

Agriculture

➔ Examples of some of the claimed benefits of using genetic technologies in agriculture include:
- the ability to more rapidly, precisely and cheaply breed organisms that are more suitable for growth in changing environments; for example, increased salinity, increased drought frequency or exposure to new diseases
- the ability to produce more nutritious foods
- the need for less or more effective use of pesticides
- less use of labour because of a reduced need for cultivation
- the ability to raise livestock more humanely
- the ability to control the gender of animals in order to increase productivity, along with the elimination of exotic pests such as carp.

Medicine

➔ Examples of some of the claimed benefits of using genetic technologies in medicine include the following:
- GM vaccines are safer as they are no longer made from denatured pathogens, which occasionally cause the disease.
- Genetic engineering has been used to mass-produce insulin, human growth hormones, follistim (for treating infertility), human albumin, monoclonal antibodies, blood clotting and thinning factors, vaccines, and many other drugs.
- Better analytical techniques can be used to detect substances such as specific proteins in cell and tissue samples; examples include ELISA and western blot technologies that can be used for research and diagnosis.
- Discovering the functions of certain genes has led to a better understanding of some diseases.
- Cancer-treatment peptides can stimulate the ability of the immune system to recognise the difference between self and non-self, resulting in more effective treatments and fewer side effects.

Industry

➔ Examples of some of the claimed benefits of using genetic technologies in industry are outlined in the points below:
- Mass quantities of proteins, amino acids and organic acids (e.g. lactic acid) can be produced by growing the transformed organisms (e.g. bacteria and yeast) in bioreactors or by using fermentation. The substances can then be purified.
- Gene technologies allow the development of highly sensitive analytical procedures.
- Processes and products that would be difficult to develop using traditional biotechnological techniques have been enabled, including some foods and food processing.
- The ability now exists to produce renewable and better fuels, along with products to clean up existing pollution and tools for recovering natural resources with a minimal production of pollutants.
- Biodegradable plastics and bioplastics reduce the demands on the petroleum industry and promote recycling.
- Bio-hydrometallurgy is an amalgamation of metallurgy and biotechnology. The extraction of metals from ores avoids pollution and energy-intensive approaches such as smelting. The process is environmentally friendly and can be used to extract metals from low-grade ores. New bacterial strains have increased resistance in arsenite and arsenate (which do not allow bacterial growth) and increased the recovery of gold from arsenopyrite-pyrite ores.
- Ethanol has found its use as an important chemical, solvent and fuel supplement. Gene technology is now being used to make ethanol production from biomass more efficient.

➔ KEY QUESTIONS

14 **Outline the benefits versus the harm of using gene technology in agriculture.**

15 **Outline the benefits versus the harm of using gene technology in medicine.**

16 **Outline the benefits versus the harm of using gene technology in industry.**

Answers ➲ p. 114

6 Evaluating the effect on biodiversity of using biotechnology in agriculture

» Students evaluate the effect on biodiversity of using biotechnology in agriculture.

➔ Without doubt the use of biotechnology in agriculture has reduced biodiversity on Earth. It has led to massive changes to ecosystems and reduced the number of species and varieties or breeds of plants and animals as more commercial ones have come to dominate.

➔ Some of the ways that this reduction in biodiversity has occurred are listed below:

- In Chapter 7 you evaluated the changes to the Earth's biodiversity due to genetic techniques. The sections in Table 7.3 relating to genetic pollution and genetic erosion apply to the effect of using biotechnology in agriculture. In addition the rise of monocultures has depleted biodiversity. Monoculture, irrigation and land clearing have all resulted in habitat destruction, rising salinity in soils and rivers, and eutrophication of waterways.
- GM plants containing genes that produce insecticides may reduce biodiversity because of declining populations of both target and non-target insects.
- Reduction in weeds as a result of herbicide-resistant crops has impacted populations of seed-eating birds. Insect-eating birds are potentially at risk from the use of GM crops with inbuilt insecticides.
- The natural selection and evolution of pests that are resistant to widely-used pesticides (either GM or not) also diminishes biodiversity as these proliferate at the expense of the varieties susceptible to pesticides.
- The ability to rapidly propagate or breed newly discovered or useful species has ensured less need to harvest from wild populations. In the case of aquaculture this is not without the risk of polluting waterways.
- Perhaps the most important reason why the use of biotechnology in agriculture has depleted biodiversity is that it has improved human nutrition, thereby increasing longevity and the size of the human population. If humans had remained hunters and gatherers the world population would be a fraction of the current population of 7.6 billion and the Earth's ecosystems would be very different and more diverse.

➔ KEY QUESTION

17 **Outline the effect that the use of biotechnology in agriculture has on biodiversity.**

Answer ➲ p. 114

7 Assessing the influence of social, economic and cultural contexts on a range of biotechnologies

SECONDARY-SOURCED INVESTIGATION

» Students interpret a range of secondary sources to assess the influence of social, economic and cultural contexts on a range of biotechnologies.

In determining the **social context** you would have considered which groups in society would benefit or suffer detriment, along with examining perceptions about the benefits and risks. One contributory factor to the lack of acceptance of GM foods was the perception that the benefit was for the farmer and the risk was for the consumer.

The social context puts pressure on governments and regulators. Studies investigated the relationship between level of understanding of genetic engineering and acceptance of the safety of foods derived from genetically modified sources. Such studies tend to indicate a lack of understanding of the processes involved in biotechnology.

There are inconsistencies across countries and continents in the regulations and labelling requirements for genetically modified foods. In 2002, when three million people faced severe food shortages or extreme hunger, the Zambian government rejected 32 000 tonnes of food aid because of the possibility that it could be GM.

A 2017 survey conducted on behalf of the Office of Gene Technology Regulation, which questioned over 1200 Australians, found support for GMOs varied. Support was particularly subject to conditions such as regulation and safety being ensured. Support also varied according to the type of modification and its purpose. There is a difference in support for GMOs in medical (63%), industrial (55%), environmental (54%) and food and crops (38%) situations.

The **economic context** relates to the potential of biotechnologies to create wealth for some groups or companies and the impact that competition and additional costs versus returns may have on producers and users.

social context: the immediate physical and social setting in which people live or in which an event happens or develops

economic context: the environment or situation relating to the production, development and management of material wealth

In 1980 the US Supreme Court ruled that General Electric could patent GM bacteria used to clean up an oil spill. This gave large companies the economic incentive to develop GMOs that were useful and profitable.

In 1992 FlavrSavr, a tomato designed for longer shelf life, was the first genetically modified food approved by the US Department of Agriculture following five years of extensive health and environment testing. The tomato was first sold in 1994 and taken off the market in 1997. Calgene, the company producing it, lost money and were taken over by Monsanto.

Between 1996 and 1999 1.8 million cans of tomato paste made from GM tomatoes, which were 20% cheaper than alternative products, were sold in labelled cans in two supermarket chains in the UK. Changing consumer perceptions of GM foods resulted in declining sales from 1998. Factors that were reported to influence the eventual demise of the product were perception of choice, media attention and campaigns, and labelling.

The **cultural context** encompasses the values and beliefs associated with the practices of biotechnology.

In Hawaii a new variety of papaya (Rainbow) has been genetically engineered to resist Papaya ringspot virus and now accounts for well over half of papaya production. Hawaiian taro (kalo) is regarded as part of the traditional culture so the acceptance of genetically engineered disease-resistant varieties was not automatic. There is a moratorium on research into such varieties at the University of Hawaii as a result of consultation with the local Hawaiian community. This decision is controversial as the number of kalo varieties has declined by almost 50% in the last 100 years due to invasive pests such as pocket rot, with up to half the kalo crop being lost on an annual basis as a result of disease. The Solomon Islands has also recently had taro crops devastated by viral attacks and there is concern that these diseases may be accidentally imported into Hawaii. Some argue for continued research into the non-Hawaiian varieties as it is better to have some taro, even if it is genetically engineered, rather than risk having none. The research could continue even if the GE plants are not released.

The European Union has banned GM foods whereas there has been general acceptance in the United States. The different approaches can be attributed to a combination of different economic, cultural and social influences. In the US corn-belt farmers are keen to adopt GM technology. Currently 90% of GM soybean and 70% of GM corn crops are herbicide tolerant, with 80% of GM corn containing the insect-resistant Bt gene. Some crops are GM for both insect resistance and herbicide tolerance.

cultural context: the environment or situation relevant to the beliefs, values, language, norms, shared attitudes and practices of the culture

Perceived benefits are reduced costs, less labour and simplified weeding regimes. Less soil cultivation reduces erosion and destruction of the physical structure of the soil. Between 1982 and 2007 the corn-belt farms nearly doubled in size and in a similar time frame the average age of farmers increased from 50 to 58. While farms are predominantly family-run businesses, there is a tendency for the young to move away from the farm. In addition increased size and mechanisation has added to some costs. Some concerns are the recent detection of weeds that, because of selection pressures, have evolved to be tolerant of herbicides (super weeds). This creates pressure for farmers to use more concentrated herbicides, increasing the likelihood of causing environmental damage in the run-off. GM crops have also been applied to small, intensive vegetable farms.

The social, economic and cultural contexts are closely related so if a product is culturally or socially unacceptable it will be unlikely to be economically viable.

Interpreting secondary sources

Your interpretation of secondary sources should consider their reliability and validity as well as the accuracy and relevance of the information you obtain. One example can be found at www.scientificamerican.com/article/first-genetically-engineered-salmon-sold-in-canada/. The source, *Scientific American*, is generally a reliable and current source of valid information. This article relates to the economic and social contexts of the introduction of GM salmon to the Canadian market.

Cross-referencing secondary sources is an excellent way to ensure the reliability of information. The viewpoint of the US Food and Drug Administration can be examined at www.fda.gov/AnimalVeterinary/DevelopmentApprovalProcess/GeneticEngineering/GeneticallyEngineeredAnimals/ucm473238.htm, while the opinion of a newspaper, *The Guardian*, can be found at www.theguardian.com/world/2017/aug/09/genetically-modified-salmon-sales-canada-aqua-bounty.

For a comprehensive discussion of GMOs and food from a reputable university go to sitn.hms.harvard.edu/signal-to-noise-special-edition-gmos-and-our-food/.

➔ KEY QUESTIONS

18 **How has social context influenced biotechnology in agriculture, medicine and industry?**

19 **How has economic context influenced biotechnology in agriculture, medicine and industry?**

20 **How have cultural contexts influenced biotechnology in agriculture, medicine and industry?**

Answers ➲ p. 115

CHAPTER SYLLABUS CHECKLIST

Are you able to answer these questions from the syllabus for this chapter? Tick each question as you go through the checklist if you are able to answer it. If you cannot answer a question, turn to the relevant page in the study guide to find the answer. For NESA key word meanings, go to www.educationstandards.nsw.edu.au and search 'key words'.

FOR A COMPLETE UNDERSTANDING OF THIS TOPIC:		PAGE NO.	✓
1	Can I investigate the uses and advantages of current genetic technologies that induce genetic change?	103–104	
2	Can I compare the processes and outcomes of reproductive technologies, including but not limited to artificial insemination and artificial pollination?	105	
3	Can I investigate and assess the effectiveness of cloning, including but not limited to whole organism cloning and gene cloning?	106–107	
4	Can I describe techniques and applications used in recombinant DNA technology; for example, the development of transgenic organisms in agricultural and medical applications?	107–108	
5	Can I evaluate the benefits of using genetic technologies in agricultural, medical and industrial applications?	109	
6	Can I evaluate the effect on biodiversity of using biotechnology in agriculture?	110	
7	Can I assess the influence of social, economic and cultural contexts on a range of biotechnologies?	110–111	

HSC EXAM-TYPE QUESTIONS

Objective-response questions

(1 mark each)

1 **Which genetic technology that produces genetic change has been used for nearly 100 years?**
   - A genetic engineering
   - B chromosome engineering
   - C mutation breeding
   - D artificial insemination

2 **What is the main difference between artificial pollination and artificial insemination?**
   - A One involves the transfer of male gametes, the other female gametes.
   - B One is used for crossbreeding flowering plants, the other for crossbreeding animals.
   - C One is used for breeding more productive organisms, the other for creating hybrids.
   - D One is used for sexual reproduction, the other for asexual reproduction.

3 **Which is the most widely used and effective form of cloning?**
   - A plant cloning
   - B animal cloning
   - C gene cloning
   - D therapeutic cloning

4 **Which of the following is a problem with using recombinant DNA technology in plants?**
   - A difficulty isolating the gene of interest
   - B difficulty cloning the gene of interest
   - C difficulty propagating the recombinant offspring
   - D difficulty inserting the recombinant gene into the plant cell

5 **What is the most significant long-term effect of the use of biotechnology in agriculture on natural biodiversity?**
   - A an increase because of the variety of crops and animals produced
   - B an increase because of the expansion of natural ecosystems
   - C a decrease because of the impact of recombinant organisms
   - D a decrease because of habitat destruction due to monoculture

Short-answer questions

6 **Read the following text which quotes Dr Keasling, Professor of Chemical Engineering and Bioengineering at the University of California, Berkeley, who used genetic technologies to produce a precursor to artemisinin, the main drug used to treat malaria, and answer the questions that follow.**

> Synthetic biology (gene technology) holds great promise for the production of everything from life-saving drugs to fuels, flavors, fragrances and chemicals currently produced from oil or scarce natural products—as well as a myriad of other applications. Dr. Keasling says, 'This is just the start of what will be possible through engineered biology, and the kind of work that Twist Bioscience is doing to make DNA (genes) inexpensive and widely available is key to this future.' Dr. Keasling sees his mission as 'making sure that we have all the science we need to do all that we need to do to make the planet better' with synthetic biology.

Source: twistbioscience.com/company/blog/dr-jay-keasling

   - a **Compare the comments of Dr Keasling with your own evaluation of the benefits of using genetic technologies in agriculture, medicine and industry.** (6 marks)
   - b **Briefly evaluate the statements of Dr Keasling as a secondary source of information on modern biotechnology.** (3 marks)

7 **Explain an example of how biotechnology might increase biodiversity and an example of how it might decrease biodiversity.** (4 marks)

8 **Outline the process of producing an organism containing recombinant DNA and its use.** (6 marks)

9 **Compare the processes and outcomes of artificial pollination with artificial insemination.** (6 marks)

10 **Read the following text and answer the question that follows. The first source comprises excerpts from the US Food and Drug Administration publication 'A Primer on Cloning and its Use in Livestock Operations'.**

> Large Offspring Syndrome (LOS) is seen in pregnancies of cattle and sheep that come from other ARTs (assisted reproductive technologies) and cloning. With LOS the fetus grows too large in the uterus, making problems for the animal and its surrogate dam.
>
> Most clones that are normal at birth become as strong and healthy as any other young animals.
>
> Clones are born just like other animals. They are similar to identical twins, only born at different times. Just as twins share the same DNA, clones have the same genes as the donor animal. A clone is not a mutant, nor is it a weaker version of the original animal.
>
> The main use of agricultural clones is to produce breeding stock, not food. Clones allow farmers to upgrade the overall quality of their herds by providing more copies of the best animals in the herd. These animals are then used for conventional breeding and the sexually reproduced offspring become the food-producing animals.

The second source comes from the *Canadian Journal of Veterinary Science* article: 'Benefits and problems of cloning animals'.

> Although cloning promises great advantages for commerce and research alike, its outcome is not always certain due to high pregnancy losses and high morbidity and mortality during the neonatal (newly born) period.

How do these accounts of the effectiveness of animal cloning reconcile with your investigation and assessment of this issue? (5 marks)

ANSWERS

KEY QUESTIONS

Key questions ➲ p. 105

1 Five technologies that cause genetic change include producing transgenic organisms (including using CRISPR-Cas9 and gene drives), mutation breeding, RNAi (temporary changes), chromosome engineering, knockout organisms and microbial injection.

2 Advantages of producing transgenic organisms are the speed and precision of the change. Advantages of using mutation breeding is that it requires relatively little technology (it has been done since the 1920s) and is acceptable by those opposed to transgenic organisms.

3 A gene technology that produces fast, permanent genetic change is microbial injection in plant crops.

4 Technologies that do not produce genetically modified organisms include embryo screening and IVF, RNAi and mutation breeding.

5 Artificial pollination is used on flowering plants whereas artificial insemination is used in sexually reproducing animals. Both reproductive technologies control reproduction by selecting the male gametes that are involved in fertilisation. In artificial pollination care has to be taken to ensure the correct pollen is transferred. This may involve the use of sterilised male flowers by physically removing stamens or cytoplasmic male sterility (CMS).

6 Artificial pollination can be used to produce hybrids whereas the main purpose of artificial insemination is the use of gametes from the more productive males in a species.

7 Processes involved in in-vitro fertilisation include the collection of oocytes and semen (which in the case of females may involve hormone treatment and surgical procedures to collect several oocytes), their combination under the appropriate conditions and the release of the embryo in a female that is receptive to implantation. If genetic screening of a cell from the embryo is carried out there may be selection of an embryo free of genetic disease. IVF is not always successful but in some cases of infertility there is an increased chance of conception and implantation of the embryo. In the case of agriculture it can be used to reduce the interval between generations and increase the reproductive output of genetically desirable female livestock.

Key questions ➲ p. 107

8 Plant cloning is extremely effective in producing large numbers of desirable plants by using a variety of techniques, many of which are only a simple manipulation of what plants do naturally. One example is tissue culture, which uses a sterile growth medium and very small samples of dividing tissue to propagate many new genetically identical plants.

9 Animal cloning is a lot more complex than plant cloning. It can be achieved by embryo splitting, which results in identical 'siblings' but these are of unknown quality. Somatic cell nuclear transfer (SCNT) uses the nucleus of a mature animal so the genetic qualities are known but it involves enucleating an ovum and transferring in to it the desired nucleus. This has a high failure rate. The other reason why animal cloning is less successful than plant cloning is premature aging and illness in the resultant animals. Apart from ethical concerns, human cloning is less likely to be successful because the spindle fibres are closely associated with the nucleus. This means that removing the nucleus of an ovum would probably damage the spindle to the extent that cell division would not occur.

10 Gene cloning is effective because of the use of technologies for choosing the appropriate restriction enzymes and ligases so the DNA is broken and rejoined correctly. Following the cloning, screening occurs so that only the cloned gene is extracted. New technologies such as CRISPR-Cas9 have improved the success rates associated with gene editing.

Key questions ➲ p. 108

11 Recombinant DNA technology is the process of producing transgenic or genetically modified organisms.

12 Recombinant technology involves gene cloning and inserting the cloned gene into another organism through a vector (e.g. bacterial plasmids) in order to produce a transgenic organism.

13 Recombinant technology is used in agriculture to produce transgenic crops and animals (e.g. fish). Modified genes must be incorporated into embryos or embryonic stem cells to ensure the organisms are GM. Microprojectile bombardment or Ti or Ri plasmids may be used as vectors. In medicine, recombinant technology is mostly used on microorganisms to result in bacteria such as *E. coli* producing pharmaceuticals. Bacteria naturally contain plasmids and can transfer plasmids from one bacteria to another so transformation of the bacteria with the recombined plasmid is relatively easy. One method is to heat and pass an electric current through the bacterial culture. Recombinant technology has recently been used to produce pharmaceuticals in goat milk. In both agriculture and medicine the vectors used to produce the transgenic organisms have to be appropriate for the different types of organisms.

Key questions ➲ p. 109

14 The benefits of gene technology in agriculture are increased production in a wider range of environments and more nutritious food. The negative effects include reduced biodiversity, damaged environments, increased initial expense which hampers equitable access for all farmers and the increased need for labelling to ensure consumer rights and choice.

15 The benefits versus the harm of using gene technology in medicine are better diagnoses and treatment or prevention versus the often increased costs.

16 The benefits of using gene technology in industry include a wider range of products that are often produced with less pollution. Harmful impacts include the costs of patents or intellectual property and the potential loss of business for smaller companies.

Key question ➲ p. 110

17 Biotechnology in agriculture, and in particular the practice of monoculture, has reduced biodiversity through habitat destruction, soil degradation and water pollution.

Key questions ➲ p. 111

18 Social context has influenced biotechnology in agriculture because of people having concerns about the content of their food. There is a perception that natural or organic food is better so there has been resistance by some groups, particularly to purchasing GM food. Medical treatments have generally been better accepted by society so GM vaccines, hormones and diagnostic practices are more freely used. Society generally supports biotechnology in industry because it is seen as resulting in more environmentally friendly processes and products such as biofuels and bioplastics.

19 Economic context has influenced biotechnology in agriculture because, although most of the benefits are savings for the producer or supplier, regulation has meant production has declined or stopped. The exceptions are GM corn, soybeans and canola for food and cotton for fibre. In medicine, economics has had an influence with GM human insulin being more expensive than insulin derived from animals. There is also some debate as to how much better human insulin is for diabetics when compared to animal insulin. Industry tends to operate for profit so economics plays a large role in determining whether biotechnology will be adopted. Biofuels (e.g. ethanol mixed with petrol) are widely accepted because they are cheaper than 100% petroleum alternatives.

20 Cultural contexts have influenced biotechnology in agriculture as some cultures do not accept GM food, though Amish farmers in Pennsylvania grow GM crops despite their rejection of many aspects of modern technology. Some religions do not accept medical interventions that may involve screening of embryonic stem cells and thus the possible disposal of defective embryos.

HSC EXAM-TYPE QUESTIONS

Objective-response questions

1 **C**. Mutagens have been used to produce variation in offspring since the 1920s. **A** and **B** are incorrect as they are modern genetic technologies. **D** is wrong because it does not induce genetic change.

2 **B**. Artificial insemination is used on animals and artificial pollination is used for breeding plants. **A** is wrong because they both transfer male gametes. **C** is wrong because although artificial pollination can be used to produce hybrids, the main purpose of both artificial insemination and artificial pollination is to produce more productive organisms. **D** is incorrect because they are both used in sexual reproduction.

3 **A**. Plant propagation can be achieved through a variety of cloning techniques. **B** is incorrect as animal cloning is still used mainly in research. **C** is wrong as gene cloning is restricted to laboratory procedures. **D** is incorrect because the use of embryonic stem cells for medical treatments is still limited.

4 **D**. Ti and Ri vectors have had to be developed to incorporate the recombinant DNA into the plant genome. **A** and **B** are incorrect as plants and animals are the same for isolating and cloning the gene of interest. **C** is wrong as once the gene has been inserted and incorporated into the cell it is relatively easy to clone recombinant plants from those cells.

5 **D**. Large-scale monoculture methods have damaged natural habitats for generations and caused gene erosion. **A** and **B** are wrong because the long-term effects are to reduce natural biodiversity. **C** is incorrect because any reduction of natural biodiversity due to recombinant organisms is through destruction of non-target organisms (e.g. insects), the use of herbicides or gene pollution, all of which are relatively minor compared to the impacts of monoculture.

Short-answer questions

6 **EM** Dr Keasling has expounded the potential benefits of the use of gene technologies (synthetic biology being an extreme version of gene technologies). You need to bring some critical analysis from your evaluation to add balance to the scenario.

a Dr Keasling cited examples of benefits that focused mainly around industrial products (e.g. fuels and chemicals) with a brief reference to life-saving drugs. Gene technology has the potential to enable the production of safer vaccines; for example, the Hepatitis B vaccine which has been approved and used to effectively reduce the incidence of this disease. (✓) Higher standards of purity, demanded by regulation, along with higher costs have restricted the further development of GM vaccines despite their potential in preventing a number of infectious human diseases. (✓)

Another major benefit of gene technology in medicine and agriculture is the use of gene editing to inactivate many pathogens, thereby reducing the need for pesticides and also increasing agricultural production by protecting crops from disease. (✓)

Gene technology (CRISPR-Cas9 and gene drives) could also potentially be used to prevent the mosquito vector of malaria being able to transmit the protozoan pathogen. This would benefit millions of people who live in the tropics by protecting them from the disease. This in turn would eliminate the need for ongoing vaccinations or drug treatments. (✓)

Gene technology has the potential benefit of increasing food production and creating more nutritious food. Golden Rice could greatly reduce the incidence of vitamin A deficiency blindness and immunity problems because it contains the precursor to vitamin A. (✓) An impediment to this benefit being realised is the resistance of social groups to genetically modified foods. Another limiting factor in agriculture is the possibility of gene erosion or gene pollution impacting on the biodiversity of native flora and fauna. (✓)

b You need to look for evidence of bias in the extract and outline its implications for the use of the secondary source.

In addition to his university work Dr Keasling appears to also be employed by Twist Bioscience (the URL suggests his comments were contained in a company blog) as he extols the benefits of the work of this organisation. (✓) The type of language used also suggests a bias in favour of synthetic biology. (✓) The information is quite general with little evidence given to support claims and no specific examples. (✓)

7 **EM** You need to state each example and then outline the process by which, as a consequence, biodiversity is either increased or decreased. The sample answer below has not used the obvious example of habitat destruction reducing biodiversity.

Selective or marker assisted breeding of new varieties of crops and livestock adds to the existing variety of living organisms and their genetic diversity. (✓) In a sense transgenic organisms have increased genetic diversity because they could not possess the introduced genes naturally. (✓)

In agriculture some species are considered pests and, in general, farmers attempt to reduce the amount of pests that either compete or otherwise negatively impact on their crops or livestock. (✓) Reduction of some weed species and insects has the potential to reduce the seed- or insect-eating birds living around farms, thereby reducing biodiversity. (✓)

8 **EM** Listing the steps involved in producing a named recombinant organism and stating how the organism is used should score full marks.

The steps required to produce recombinant DNA in *E. coli* so that human insulin can be made for the treatment of diabetes (✓) are:

- identify and extract the human gene for making the protein insulin (include an additional gene for antibiotic resistance) (✓)
- cut the gene(s) and open a bacterial plasmid using the same restriction nuclease so they have the same 'sticky ends' (✓)
- join the gene(s) into the plasmid using the enzyme ligase (✓)
- transfer the plasmid into *E. coli* using electroporation and let it divide (✓)
- screen the *E. coli* bacteria to ensure that they contain the human insulin gene (they will resist antibiotics) (✓)
- culture the bacteria and extract and purify the insulin that is produced. (✓)

9 **EM** You have to identify any similarities and differences between the processes involved in artificial insemination with artificial pollination and outline the outcomes of each process.

Both processes involve the artificial transfer of the male gamete (pollen in plants, semen containing spermatozoa in animals) to the female organ (stigma in plants, vagina or uterus in animals). Both processes are more productive than if natural breeding was allowed to occur. (✓) Semen can be stored by freezing while pollen would need to be isolated and kept at the correct temperature and humidity. (✓) You need to ensure that the female organism has not already been pollinated/fertilised, which involves removing the stamens (e.g. 'detasselling' in the case of rice), using cytoplasmic male sterility in corn or, in the case of animals, ensuring no other male animal has access to the females. (✓)

The outcome for plants and animals is that breeding is controlled in order to ensure the genetic origins of both parents. (✓) Hybrid plant varieties may be created that experience hybrid vigour. (✓) The outcome for animals is that many females can give birth to offspring that are more likely to have desirable characteristics. (✓)

10 **EM** You need to analyse the statements made in the two information sources and then compare them with the findings of your own investigation into the effectiveness of animal cloning.

Neither information extract refers directly to the cost of cloning animals and the lack of public acceptance of food from a cloned animal. (✓) Both extracts refer to problems with prenatal and neonatal development and health, along with the deleterious impacts on the surrogate female. (✓)

Neither source refers to early concerns about premature aging of cloned animals. These concerns have been largely overcome since the much publicised premature death of the first cloned sheep, Dolly. (✓)

Animal cloning is promoted as a means of preserving endangered species and even as a potential for bringing back recently extinct animals such as the woolly mammoth. (✓) Animal cloning is a way of amplifying the population of desirable animals in agriculture and producing genetically identical animals for research in drug testing and other medical applications. This raises ethical concerns about the use of animals. (✓)

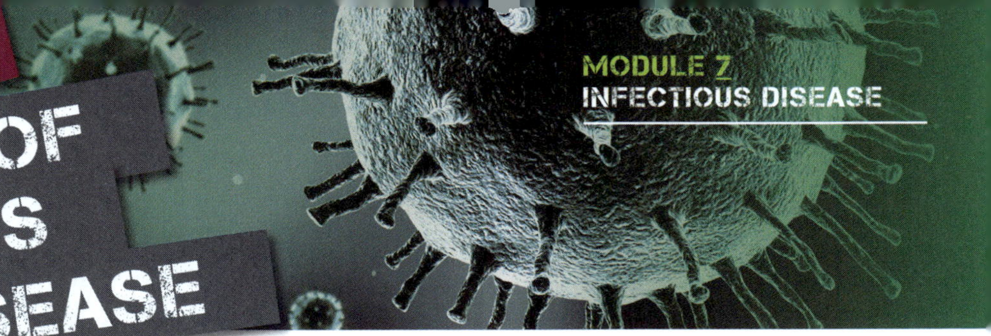

CHAPTER 9 CAUSES OF INFECTIOUS DISEASE

INQUIRY QUESTION:

How are diseases transmitted?

In a broad sense, disease is any condition that impairs normal living processes and is recognised by the presence of specific symptoms. In this chapter you review a variety of **infectious diseases** and how they are transmitted. Understanding how infectious diseases are transmitted is critical to preventing their spread.

1 Infectious diseases caused by pathogens

» Students describe a variety of infectious diseases caused by pathogens, including microorganisms, macroorganisms and non-cellular pathogens, and collect primary and secondary-sourced data and information relating to disease transmission, including classifying different pathogens that cause disease in plants and animals; investigating the transmission of a disease during an epidemic; designing and conducting a primary investigation relating to the microbial testing of water or food samples; investigating modes of transmission of infectious diseases, including direct contact, indirect contact and vector transmission.

Pathogens: microorganisms, macroorganisms and non-cellular pathogens

- An **infection** is the presence of a disease-causing organism in or on the body of a **host**. If the host is successful in controlling the disease-causing organism the disease does not occur.

infectious diseases: disorders caused by organisms such as bacteria, viruses, fungi or parasites; contagious or communicable diseases are those infectious diseases that can spread from one organism to another by direct or indirect transmission

infection: the entry of a pathogen into body tissues and cells followed by multiplication of the organism

host: a cell or tissue in which a microscopic pathogen multiplies; in the life cycle of parasites the host can be the primary or definitive host in which the parasite reaches maturity

- Infectious diseases differ from other diseases (e.g. genetic and lifestyle diseases) in that they are caused by invasion by a **pathogen** and can be transmitted from one host to another.
- It is important to note that even though a pathogen is the main cause of a particular infectious disease, environmental and hereditary factors influence:
  - whether the particular host becomes infected when exposed to the pathogen
  - the health of the host—people who are vulnerable after an accident or surgery, or whose immune system is weakened (those with AIDS for example), are often infected by the fungi *Candida albicans*
  - the severity of the disease
  - the development of the disease in a population and thus the extent and rate of transmission of the disease.
- In your study of infectious diseases and their transmission you have collected primary and secondary-sourced data. Collecting primary-sourced data may have involved the direct observation of infectious diseases in plants, for example. Collecting secondary-sourced data may have involved sourcing facts and data from health organisations such as NSW Public Health Units and the A to Z of Infectious Diseases at www.health.nsw.gov.au, the National Department of Health at www.health.gov.au, Health Direct at www.healthdirect.gov.au, www.sciencedaily.com/news/health_medicine or the University of NSW's Museum of Human Disease.

Classifying different pathogens that cause disease in plants and animals

- Pathogens can be prions, bacteria, viruses, protozoa or **protists**, fungi or macroscopic parasitic animals (see Chapter 1 for reproduction in protists).
- When you classified different pathogens you would have described their size and whether they were non-cellular or cellular. If the pathogen is cellular, you would have classified it as prokaryotic or eukaryotic. Your classification may also have described how they reproduce or replicate (see Table 9.1).

pathogen: any organism living in or on another organism that is capable of causing disease

protists: a group of eukaryotic, unicellular or colonial organisms that include protozoa such as *Amoeba*

- Microscopic pathogens include bacteria, viruses, fungi and protozoans.
- Macroscopic pathogens include parasites such as arthropods, flatworms and roundworms.

Table 9.1 Classifying different pathogens using distinguishing features such as size and cell structure

(Extremely small → Larger)

Type of pathogen	Some distinguishing features
prion	defective form of a protein molecule; does not contain DNA or RNA; mostly attacks brain or nerve cells
virus	non-cellular; contains DNA, RNA and protein coat; requires a living host cell to replicate
bacteria	prokaryotic cell; divides quickly and/or produces toxins
protozoan	eukaryotic cell (single-celled organism); may have a complex life cycle
fungi	eukaryotic cell with cell wall; spreads via spores or rapid division; some infect external skin and nails, while others enter the host's body
macroparasites	parasitic organisms visible without a microscope, e.g. tapeworms (endoparasites living inside the host) and ticks (ectoparasites living or feeding from hosts outside the surface)

- Pathogens cause disease in many different ways. The survival and success of all pathogens requires that they:
  - enter the host through a certain part of the body
  - reach their target site or niche in the body
  - attach to or invade the target site they are trying to infect so that they are not dislodged
  - multiply rapidly
  - obtain their nutrients from the host
  - avoid and survive attack by the host's natural defence mechanisms
  - have enough **virulence**; a virulent pathogen only needs to be present in small amounts to cause the disease
  - exit the infected host to spread to an uninfected host.

virulence: the measure of the pathogenicity of an organism; virulence is related to a number of variables, including the number of the particular pathogen needed to cause the disease

→ KEY QUESTIONS

1 What are pathogens?

2 What must happen for a pathogen to cause an infectious disease?

Answers ➲ p. 134

Prions

- Prions are infectious agents that consist only of protein. They have no nucleic acid and are therefore referred to as non-cellular pathogens.
- The term *prion* refers to transmissible abnormal, pathogenic agents. They are able to induce abnormal folding of specific normal cellular proteins known as prion proteins. These prion proteins are found most abundantly in the brain. The functions of normal prion proteins are still not completely understood. The abnormally folded protein spreads when prions in the blood interact with the normal version of the protein and pass on the deformation. The prions build up in the brain, leading to prion disease.
- Prion diseases are also called transmissible **spongiform** encephalopathies (TSEs). They are rare progressive neurodegenerative disorders that affect both humans and animals. The symptoms of the diseases include spongiform changes associated with loss of neurons (brain damage) and the body's inability to induce inflammatory response. Prion diseases are usually rapidly progressive and always fatal.
- Prion disease was discovered in 1982. The discovery of prion disease was a surprise to the scientific community because all previously known pathogens contain nucleic acid, which enables them to reproduce.
- Animal prion diseases include:
  - bovine spongiform **encephalopathy** (BSE); also known as Mad cow disease
  - chronic wasting disease (in deer)
  - scrapie (in sheep)
  - transmissible mink encephalopathy
  - feline spongiform encephalopathy.
- In humans, prion diseases are both infectious and hereditary (see page 188 for information on hereditary diseases). Human prion diseases include:
  - Gerstmann–Straussler-Scheinker syndrome (hereditary)
  - Creutzfeldt–Jakob disease (CJD) (hereditary)
  - Variant Creutzfeldt–Jakob disease; the human disease variant of Creutzfeldt–Jakob disease (vCJD) is believed to be a **zoonotic** disease caused by the BSE agent; the route of transmission of vCJD is not yet fully proven but it is generally accepted that it is transmitted through exposure to food contaminated by the bovine BSE prion.
- For more information see www.cdc.gov/prions/.

spongiform: the post-mortem appearance of the brain with large holes that show where cells have been destroyed

encephalopathy: diseases that damage the brain

zoonotic: diseases that can be transmitted from animals to humans

Viruses

- Viruses are microscopic pathogens. Their size ranges from 20 to 40 nanometres.
- A typical virus is a small piece of genetic material encased in a shell called a capsid. Viruses are also referred to as non-cellular pathogens. Even though they are not technically living, viruses have nucleic acid (DNA or RNA). On its own a virus can do little harm. However, if a virus enters a living cell, it starts copying itself (replicating) and causes disease. Viruses can replicate at an extraordinary rate and the fact that they contain nucleic acid means that they can mutate and evolve.
- Viruses can infect all types of living things from animals and plants to microrganisms such as bacteria. Viruses that infect bacteria are called bacteriophages.
- Rotavirus refers to a group of double-stranded RNA viruses.

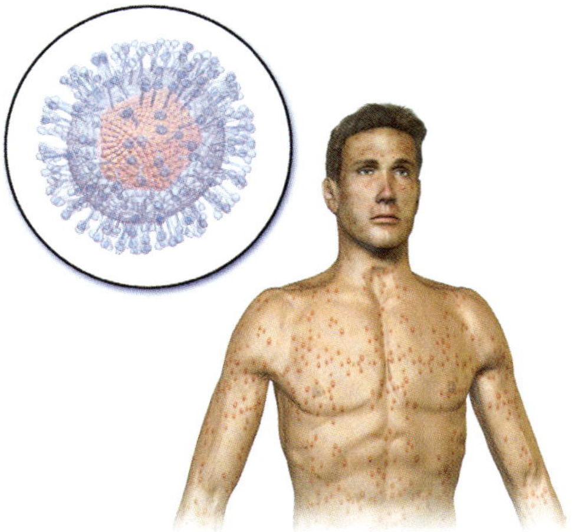

Figure 9.1 Diagram of a virus (*Varicella zoster,* chicken pox and human herpes virus 3) and its symptoms in the human body

Table 9.2 Examples of viruses, the diseases they cause, and modes of transmission

Virus	Disease caused	Transmission method
rotavirus	causes diarrhoea and gastroenteritis, mainly in children between six months and two years old	direct: transmitted by person-to-person contact, particularly with someone who has not washed their hands
human parvovirus B19	slapped cheek or fifth disease (mostly affects children); the virus lives in red blood cells	direct: spread through respiratory secretion when an infected person coughs or sneezes; can also be transmitted through blood and blood products
rhinovirus	colds	direct: can be transmitted by shaking the hand of a person who has a cold and who has just used their hand to wipe their dripping nose
hantavirus	found in mouse faeces, urine and saliva; causes serious and sometimes fatal respiratory infection in humans	indirect: dry mouse faeces, urine and saliva become fine airborne particles which can be inhaled
human immunodeficiency virus (HIV)	Acquired Immune Deficiency Syndrome (AIDS)	direct: can be transmitted through sexual intercourse or sharing used syringes contaminated with infected blood
strains of the Ebola virus	Ebola virus disease (EVD) is serious and often fatal; EVD was previously called Ebola haemorrhagic fever	direct: transmitted by direct contact with the blood, body fluids and tissues of infected animals or humans; transmission through sexual contact may occur up to seven weeks after clinical recovery
Zika virus	a virus that spreads through bites from tropical *Aedes* mosquitoes; babies of women infected with Zika virus during pregnancy have been born with microcephaly (smaller brains) and other birth defects	vector: transmitted by *Aedesa egypti* (a mosquito)
flavivirus	yellow fever; a skin-yellowing, often fatal disease	vector: transmitted by the mosquitoes *Aedesa egypti* and *Haemagogus* spp; kills tens of thousands of people a year
dengue viruses	dengue fever; ranked by the World Health Organization as one of the most critical mosquito-borne viral diseases in the world	vector: transmitted by the mosquitoes *Aedesa egypti* and *Aedes albopictus*; kills tens of thousands of people a year

Bacteria

- Bacteria are prokaryotic, unicellular microorganisms. DNA is contained in a single loop. Some bacteria have an additional circle of genetic material called a **plasmid**.
- Not all bacteria cause disease and many are beneficial: only about 1% of known bacteria are disease-causing **microbes**.
- Bacteria are about 1000 nanometres in size.
- Bacteria are classified according to their shape: spherical (*cocci*), rod (*bacilli*), spiral (*spirilla*), comma (*vibrios*) or corkscrew (*spirochaetes*). Bacteria are also aerobic or anaerobic. For example, the bacterium that causes tetanus is anaerobic and is found in deep wounds.
- See microbiologyonline.org/about-microbiology/introducing-microbes/bacteria
- Bacteria can cause disease by:
  - secreting toxins (chemical change)
  - invading cells (physical change)
  - forming bacterial colonies that disrupt normal cell function.

plasmid: a small DNA molecule found in cells (often bacterial cells) and separate from the chromosomal DNA; the plasmid may contain genes that give the bacterium some advantage over other bacteria (e.g. resistance to antibiotics)

- Figures 9.2 and 9.3 below show examples of disease-causing bacteria.

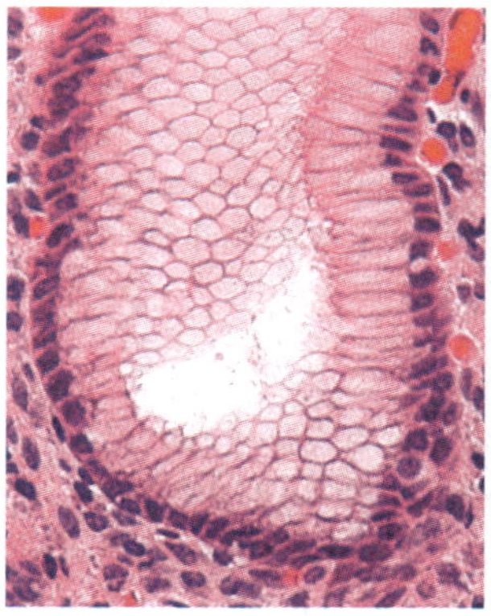

Figure 9.2 *Helicobacter pylori* (causes stomach ulcers)

Source: image by Nephron, (own work), [CC BY-SA 3.0 (https://creativecommons.org/licenses/by-sa/3.0)], via Wikimedia Commons

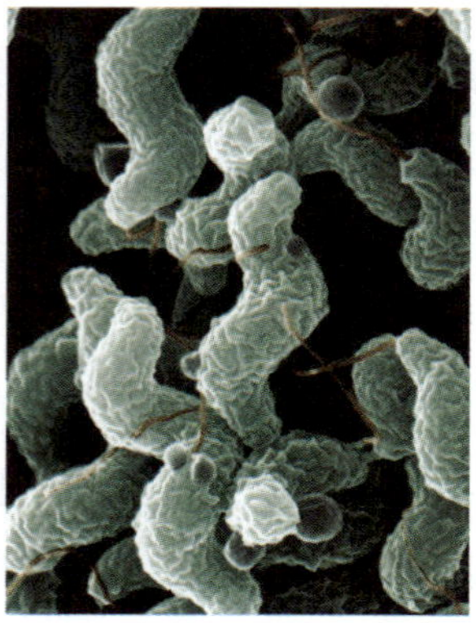

Figure 9.3 *Campylobacter jejuni* (causes bacterial food-related gastrointestinal illness)

Source: image by De Wood, Agricultural Research Service (ARS), the US Department of Agriculture, (public domain), via Wikimedia Commons

microbes: microorganisms; some definitions refer to microbes as 'microorganisms, especially bacteria that cause disease'

Table 9.3 Examples of bacteria, the diseases they cause, and modes of transmission

Bacteria	Disease caused	Transmission method
Salmonella typhi	typhoid fever or typhoid	indirect: transmitted through contaminated water
Vibrio cholerae	cholera	indirect: transmitted through contaminated water
Entero bacteria	gastroenteritis	direct: transmitted by person-to-person contact, particularly with someone who has not washed their hands
Campylobacter	food poisoning	indirect: can occur when the contaminated food is eaten; *Campylobacter* can be transmitted from raw meat (e.g. chicken) to hands or chopping boards
Chlamydia trachomatis	chlamydia	direct: via sexual contact
Bordatella pertussis	whooping cough	direct: spread through droplet transmission
Yersinia pestis	bubonic plaque	vector: transmitted by infected fleas
Borrelia burgdorferi	Lyme disease (the bacterium is carried in mouse blood)	vector: transmitted by ticks (*Ixodes* spp)
bacteria of the *Legionella* family	Legionnaires' disease (an infection of the lungs)	indirect: occurs when a person breaths in bacteria that are commonly found in the environment; *Legionella pneumophila* bacteria can contaminate air conditioning cooling towers, whirlpool spas, shower heads and other bodies of water; *Legionella longbeachae* can contaminate soil or potting mix
Rickettsia prowazekii	typhus fever, specifically epidemic typhus	vector: transmitted by mites

Protozoans

- Protozoans are eukaryotic microorganisms, larger than bacteria, with the DNA organised in a nucleus. Protozoans are protists (see Chapter 1). Many protozoans are free living and do not cause disease.
- The size of protozoans varies but most are less than 50 micrometres.

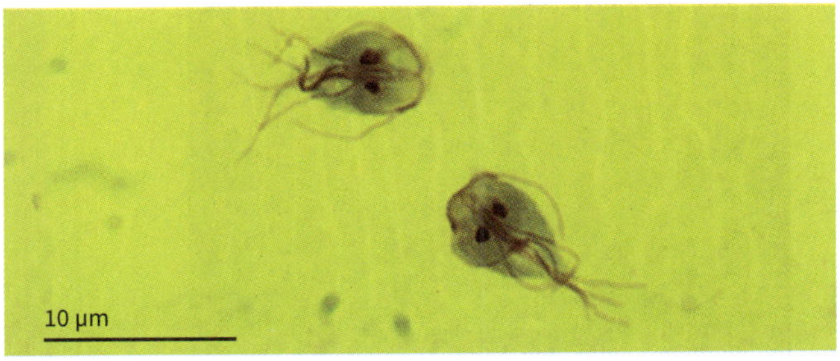

Figure 9.4 Diagram of the protozoan *Giardia intestinalis*

Source: image by Dr Josef Reischig (CSc), [CC BY-SA 3.0 (https://creativecommons.org/licenses/by-sa/3.0)], via Wikimedia Commons

- Many **parasitic** protozoans have complex life cycles involving several hosts. For example, malaria is transmitted by *Anopheles* mosquitoes. If a mosquito infected with the pathogen bites a human, the protozoan passes into the human bloodstream. If the human is already infected, the protozoan can pass from the human to the mosquito and is therefore spread to other humans. For more information see www.parasitesinhumans.org.

parasitic: living in or on another organism and obtaining nutrients from that organism

Fungi

- Fungi are eukaryotic organisms. They include yeasts, moulds and crop-destroying rusts and smuts. Like bacteria, many fungi are beneficial and play a key role as decomposers in ecosystems and in the fermentation process. Some fungi, however, cause disease.

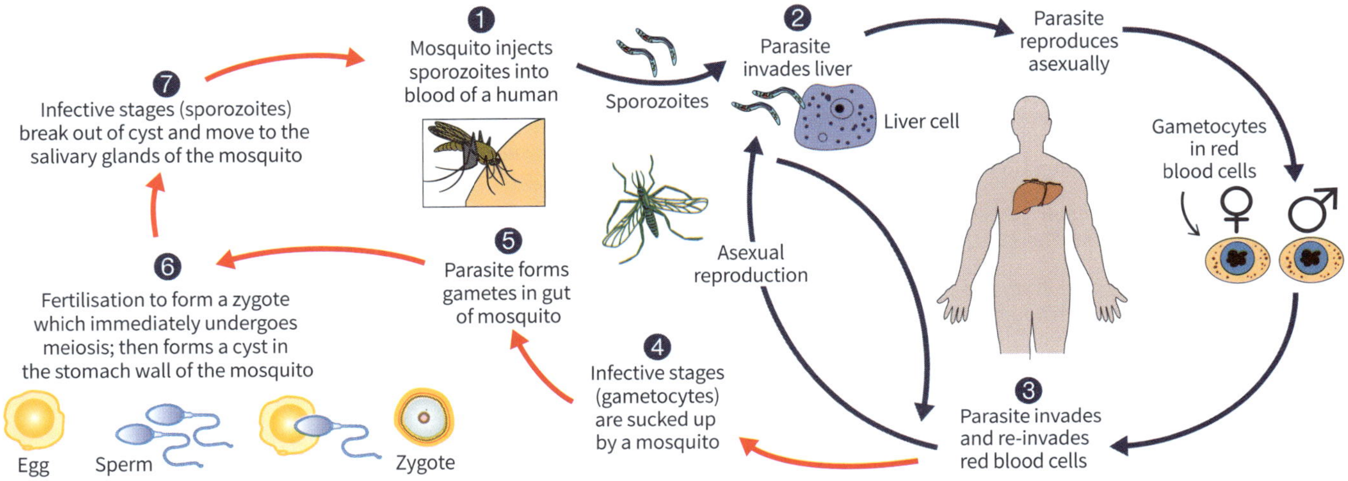

Figure 9.5 Life cycle of a parasitic protozoan

Table 9.4 Examples of protozoans, the diseases they cause and modes of transmission

Protozoans	Disease caused	Transmission method
entamoeba: an amoeba	amoebic dysentery	indirect: transmitted by ingestion of contaminated food or water during the cyst stage of the amoeba's life cycle
Naegleria fowleri: an amoeba	destroys the olfactory region and other parts of the human brain	direct: enters a swimmer's brain directly through the nose (nasal membranes)
giardia: a flagellated protozoan	giardiasis (a disease of the intestines)	direct: infections are usually person-to-person and associated with poor hygiene and poor sanitation; individuals can become infected by ingesting or coming into contact with food, soil or water contaminated with an infected carrier's faeces
Plasmodium falciparum: a sporozoan protozoan	malaria	vector: transmitted from person to person by the *Anopheles* species of female mosquito carrying the protozoan *Plasmodium falciparum;* malaria kills more than 400 000 people a year worldwide
Toxoplasma gondii: an intracellular parasitic alveolate	toxoplasmosis	indirect: found in cat faeces and can be transmitted to humans through contaminated soil; the most common transmission is through ingestion of the cysts in undercooked meat, particularly lamb and pork (this is the main way cats are infected)

Figure 9.6 A disease-causing fungus (*Rhizoctonia solani*, a fungal plant pathogen on cucumber)

Source: image by David B Langston, University of Georgia, Bugwood.org, [CC BY-SA 3.0 (https://creativecommons.org/licenses/by-sa/3.0)], via Wikimedia Commons

Table 9.5 Examples of fungi and the diseases they cause

Fungi	Disease caused
Candida albicans	vaginal thrush (vaginitis) and mucous membrane infection
Cryptococcus gattii	meningitis
Trichophyton mentagrophytes	athlete's foot or tinea
Trichophyton rubrum	ringworm
Batrachochytrium dendrobatidis: chytrid amphibian fungus	an introduced fungus causing chytridiomycosis (a disease threatening extinction of some native frogs)
Uredo rangelii: myrtle rust	an introduced fungus causing disease in members of the plant family *Myrtaceae*
Botrytis cineria	grey, furry growth on grapes, beetroots and lettuce
Phytophthora cinnamomi: root rot or dieback fungus	an introduced fungus causing root rot or dieback in some native plants

Macroparasites

- Macroparasites are macroorganisms that can cause disease. They include roundworms, flatworms and flukes, as well as arthropods such as ticks, mites, lice and fleas. Macroparasites can cause disease directly, or indirectly as vectors for disease.
- An example of direct cause is tick paralysis in dogs. When a certain species of female tick infests the skin of the dog and feeds on its blood, it releases a toxin through its saliva. This toxin causes paralysis.

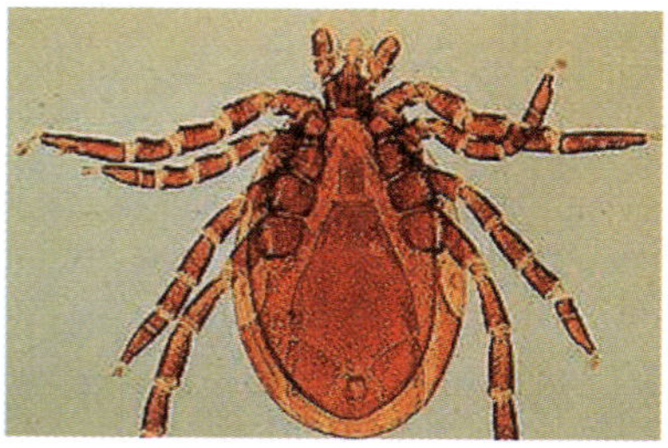

Figure 9.7 Deer tick (*Ixodes scapularis*)

Source: image from the US National Oceanic and Atmospheric Administration, (public domain), via Wikimedia Commons

- An example of macroparasites as a vector for disease is ticks of the *Ixodes* spp. These ticks transmit the bacterium *Borrelia burgdorferi* from mice to other animals, including humans. The bacterium *Borrelia burgdorferi* causes Lyme disease (see page 124 for vector transmission).

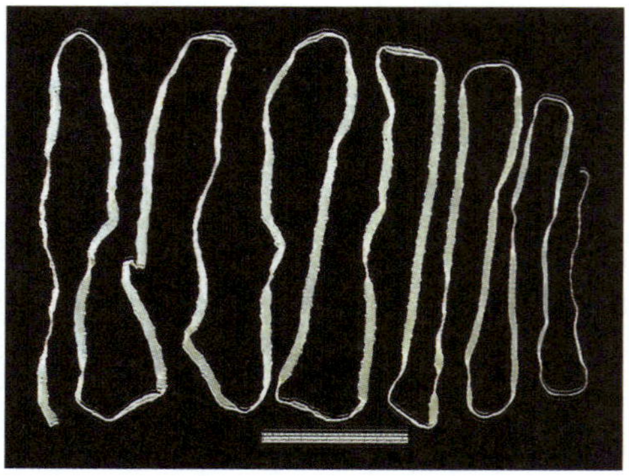

Figure 9.8 An adult tapeworm (*Taenia saginata*)

Source: image from the Centers for Disease Control and Prevention, US Department of Health and Human Services, (public domain), via Wikimedia Commons

Table 9.6 Examples of macroparasites and the diseases they cause

Macroparasites	Disease caused
trematodes (the blood fluke *Schistosoma mansoni*)	schistosoma causes a disease in humans called schistosomiasis; the disease causes the body to become weak and more susceptible to other diseases
trematodes (liver fluke)	liver fluke causes inflammatory disease of the bile ducts; sheep and cattle are most commonly infected with liver flukes but humans can also be infected
cestodes (flatworms), such as the tapeworm	tapeworms cause infection of the digestive tract; tapeworm infection is caused by ingesting undercooked meat, food or water contaminated with tapeworm eggs and larvae
nematodes, such as *Ascaris* (roundworm)	*Ascaris* causes infection in the small intestine in humans, pigs and horses
body lice (louse is singular)	head lice feed on human blood and cause an infection called pediculosis
sheep lice	sheep are prone to infestation by body, face and foot lice but only the body louse presents a serious problem
Varroa mites	a serious worldwide pest of honey bees; the mites infest the bee brood and cause the collapse of the bee colony
Golden potato cyst nematodes	the microscopic nematodes feed on the roots of potato plants; root development and tuber yield is reduced and plant growth is stunted

- Macroparasites have complex life cycles. Some parasites may also infect one or more intermediate hosts in which the parasite goes through several immature life cycle stages or reproduces asexually. For more information see www.parasiteecology.wordpress.com.

- For example, the life cycle of schistosomes includes two hosts: a definitive host and an intermediate host. The definitive host is a human, where the parasite undergoes sexual reproduction. The intermediate host, which is usually a freshwater snail, is where the parasite undergoes asexual reproduction, producing thousands of larvae in the process.

KEY QUESTIONS

3 **Describe a disease caused by a microscopic pathogen.**

4 **Describe a disease caused by a macroscopic pathogen.**

5 **Describe a disease caused by non-cellular pathogens.**

Answers p. 134

Microbial testing of food or water samples

- When designing a practical investigation you would have learnt about the importance of:
  - safe work practices
  - validity (whether the investigation is doing what it set out to do); this involves designing your investigations to use **independent**, **dependent** and **controlled variables** to ensure a valid procedure
  - reliability, which involves answering questions such as: What did students in your group find?; What did other groups find?; and Can the investigation be replicated?
- In the first-hand investigation following you need to manage the risks associated with inoculating nutrient agar plates and growing microbes along with **autoclaving** and the safe disposal of products. You also need to have employed safe work practices. These will have included:
  - the use of protective clothing (e.g. enclosed shoes, masks, lab coats, gloves and protective eyewear)
  - sterilising transfer loops, wires or needles with a flame immediately before and after the transfer of samples
  - correct labelling of all samples
  - thorough cleaning of work surfaces
  - washing hands thoroughly before leaving the laboratory
  - autoclaving agar plates to below 35 °C to ensure harmful microbes are destroyed.

independent variable: the one single factor that is changed; for example, the type of medium used in an agar plate
dependent variables: the factors that change as a result of the change to the independent variable; for example, the types of colonies that grow on the agar medium
controlled variables: the factors that are kept constant during an investigation; for example, the same amount of agar and the same temperature for each petri dish
autoclaving: a process used to sterilise equipment and samples

FIRSTHAND INVESTIGATION

Microbial testing of food samples

» Students design and conduct a practical investigation relating to the microbial testing of water or food samples.

Aim

to culture microbes growing on food such as bread and fruit, and identify them

Method

1 Collect an orange or another piece of fruit or old bread that has developed a furry growth of mould. Note: Many of the blue spots that grow on old oranges are penicillium-type moulds that are relatively safe.
2 Sterilise an inoculating loop or needle and use it to collect spores from the orange or bread. Spread the spores in a regular pattern over the agar plate(s).

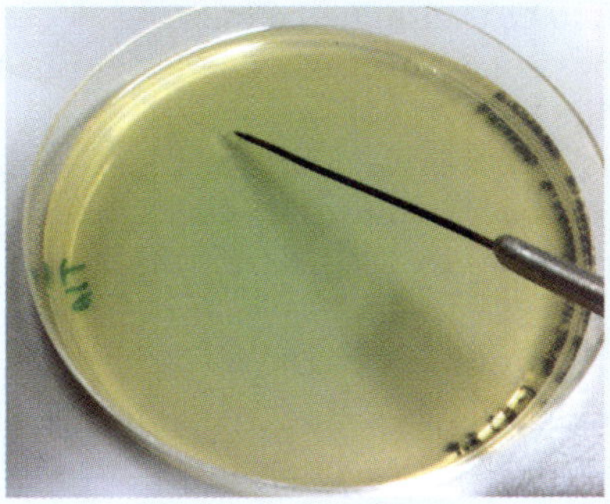

Figure 9.9 Inoculating an agar plate
Source: image by Budhavee SNA, [CC-BY-SA-4.0], via Wikimedia Commons

3 Place the lid on the plate, seal it, and invert and label the bottom plate. Note: Do not reopen the plate.
4 Incubate inoculated plates and a non-inoculated plate. Observe the growth of any microbes on the inoculated plates. Compare with the non-inoculated plate.
5 Dispose of the microbes on the plates by autoclaving.

Results

Typically you will have identified different colours and shapes of microbes on the inoculated plates. Generally the furry colonies that develop mycelium and spore-producing structures are fungi. Shiny smooth colonies are generally bacterial. However, different types of bacteria require different nutrients for growth and some do not grow in culture media at all. Characteristics of the individual bacterial colonies such as shape, size, colour, margin and surface are used to identify the bacterium.

If your techniques were sterile, the non-inoculated plate will have no microbial growth.

For more information see microbiologyonline.org/teachers/observing-microbes/observing-bacteria-in-a-petri-dish.

➔ KEY QUESTIONS

6 What general types of microbes did you investigate when you tested food or water?

7 Outline one safety precaution you used when conducting a practical investigation relating to the microbial testing of food and water.

Answers ➲ p. 134

Investigating modes of transmission of infectious diseases

- The spread of a specific disease involves a wide range of interrelated factors. These include the persistence of the pathogen within hosts, the transmission mechanism, the proportion of the population that is immune or has been immunised (see Chapter 12), and the mobility of individuals within the affected population.
- Transmission must occur for an infection to spread. If an infectious disease is to be transmitted it must:
  - escape from its host, or escape from where it normally lives and multiplies
  - survive transmission from one host to another—some pathogens can only be transmitted by direct contact because they cannot survive long outside the host
  - enter the new host
  - escape from the new host.
- A human who is a reservoir of a pathogen may be capable of transmitting the pathogen, depending on the stage of infection and the pathogen.
- An individual capable of transmitting a pathogen without displaying symptoms is referred to as a carrier. Carriers can be passive or active as outlined below.
  - A passive carrier is contaminated with the pathogen but not infected by the pathogen. A passive carrier can mechanically transmit the pathogen to another host.
  - An active carrier is an infected individual who can transmit the disease to others. An active carrier may or may not show signs or symptoms of infection. For example, pathogens such as the Hepatitis B virus, herpes simplex virus and HIV are frequently transmitted by carriers who do not show symptoms.
- Different pathogens have different modes of transmission. The main modes of transmission are direct contact, indirect contact and **vector** transmission.

Direct contact

- Direct transmission is through:
  - direct contact between people (through touch, saliva, mucus, contaminated blood or other bodily fluids)
  - **direct droplet transmission**, which occurs when an individual coughs or sneezes and small droplets of mucus, which may contain pathogens, are ejected as a fine spray.
- Pregnant women can transmit infectious diseases (e.g. sexually transmitted diseases) to unborn children via the placenta, during birth or through breastfeeding.
- Note: Transmission can be both direct and indirect. For example, SARS-CoV is usually spread from someone who has the disease via **airborne transmission**, or via direct contact with contaminated hands.
- For further information on direct transmission see Tables 9.2, 9.3 and 9.4.

Indirect contact

- Indirect transmission includes:
  - airborne transmission of droplets; respiratory pathogens are usually transmitted through the air
  - transmission from environmental reservoirs such as soil and water
  - transmission from **fomites**—infectious diseases can be transmitted from contaminated objects such as door handles, bedding, towels or contaminated blood products and medical supplies
  - transmission from contaminated food and drinking water—intestinal pathogens are usually spread through the presence of faeces in contaminated water or food
  - transmission from an infected animal—for example, from handling contaminated animal products or waste, or when an infected animal scratches or bites a person. Some animal diseases (e.g. rabies and anthrax) can be transferred from animals to humans. These animals are sometimes referred to as vectors of the disease they transmit.
- For further information on indirect transmission see Tables 9.2, 9.3 and 9.4.

vector: an organism that transmits a disease-causing pathogen; for example, mosquitoes are vectors of the malaria-causing parasitic protozoan; a vector often carries one part of the life cycle of the pathogen from one organism to the next

direct droplet transmission: droplet transmission of a pathogen to a new host over distances of one metre or less

airborne transmission: the transmission of droplets containing pathogens over distances greater than one metre

fomites: inanimate objects that become contaminated by pathogens from an infected individual or reservoir

Vector transmission

- Vector transmission can be mechanical or biological. Mechanical vector transmission involves an animal carrying a pathogen from one host to another without being infected itself. For example, a fly lands on faeces and later transmits bacteria from the faeces to food. Humans eat the food, resulting in diarrhoea or dysentery. Biological vectors include insects and other arthropods.
- For further information on vector transmission see Tables 9.2, 9.3 and 9.4.

➔ For more information see microbiologyonline.org/about-microbiology/microbes-and-the-human-body/routes-of-transmission and the *National Geographic* article 'This may be the deadliest creature on Earth', available at www.nationalgeographic.com/mosquitoes.

➔ KEY QUESTION

8 **Construct a table to describe three different modes of transmission of infectious diseases. For each mode of transmission provide an example of the disease.**

Answer ➲ p. 134

Investigating the transmission of a disease during an epidemic

➔ An **epidemic** is the rapid spread of (a usually) infectious disease to a large number of people in a particular population in a relatively short period of time. The transmission rate exceeds the extent of the disease normally found in the population. Epidemics often spread in leaps; that is, exponentially. Note: Epidemics do not only involve infectious diseases as there can also be epidemics of non-infectious diseases (e.g. obesity or drug addiction).

➔ Epidemics affect those in the population who do not have an acquired or inherent immunity (see Chapter 11 for information on acquired and inherent immunity).

➔ An epidemic that spreads to an entire nation or other countries is called a **pandemic**.

➔ Epidemics and pandemics are becoming more prevalent, with outbreaks of diseases such as Severe Acute Respiratory Syndrome (SARS) and avian flu at regional levels, and HIV/AIDS, diphtheria, malaria, measles and swine flu at a global level. Asia is one area that is increasingly susceptible to epidemics and pandemics due to increasing migration and global travel, high population density in urban areas and underdeveloped healthcare systems in some countries (see Chapter 12 for more details).

epidemic: an outbreak or unusually high occurrence of a disease in a population or region
pandemic: an outbreak of a disease over a whole country or on a global scale

➔ Transmission of a disease during an epidemic is usually rapid. There are many factors that increase exposure to pathogens and cause the spread of disease. These factors are often interrelated and include the factors outlined in the points below.

- A continual source of the disease is known as a reservoir. For example, an infected person during the AIDS epidemic, bacteria-carrying rats and fleas during the Black Death (bubonic plague), or contaminated water in cholera and typhus epidemics.
- In densely populated areas, such as in cities, and where there is overcrowding, such as in refugee camps and evacuation centres, pathogens can spread more easily and rapidly. This occurs when humans are in close proximity with each other and with other animals (e.g. farm animals). Densely packed livestock also allows for more rapid spread of diseases.
- Social and environmental factors such as contaminated water supply, limited or no water supply and poor sanitation facilities are key factors in poor, developing countries along with inadequate health systems and lack of access to affordable vaccines.
- Environmental factors include changes in climate. For example, Crimean–Congo haemorrhagic fever is found across Africa, Asia and southeast Europe. The virus is spread by ticks. In 2010 it appeared in western Europe and researchers believe that it is spreading as the tick's habitat expands in response to global warming.
- The loss of healthcare systems and shortages of medical supplies can result from events such as civil war.
- Natural disasters such as heavy rain, mudslides, cyclones, earthquakes and flooding can damage sewerage infrastructure, which causes contamination and pollution of water supplies.
- There can be an increase in the amount of the pathogen or new strains of the pathogen that can be spread from animals to humans. Viruses can evolve and spread faster than other pathogens (e.g. Avian Influenza (AI) or bird flu).
- Introduction of the pathogen into an area where it has not been before can occur when humans encroach on animals' natural habitats. This can occur when humans encroach on animals' natural habitats. An example is the Nipah bat virus, which caused human infections and deaths in Malaysia after pig farms were built in areas inhabited by the fruit bats. The transmission of the disease was through bats dropping half-eaten, contaminated fruit into pigsties.
- Enhanced modes of transmission of pathogens can result in more susceptible people being exposed. A disease can go from unknown to epidemic if a mutation of the pathogen makes it spread easier.
- Globalisation, including global trade, migration and global travel, can help the spread of disease. For example, Severe Acute Respiratory Syndrome (SARS), a disease that causes pneumonia, first appeared in southern China in late 2002. It then spread to several countries throughout the world between February and June 2003.
- The 2014 outbreak of the Ebola virus in western Africa infected 50 times more people than any previous outbreak of the disease. Part of the reason was the slow response by health agencies in diagnosing the disease and then, having done so, delays in isolating cases and quarantining those infected. Ebola also spread initially because of the cultural practice of preparing bodies for burial and therefore person-to-person contact. Researchers have also discovered that, as it spread, Ebola mutated to make transmission easier.

- A change in susceptibility in the population can come about as new strains of pathogens develop or, in the case of the host populations, because of starvation or increased levels of trauma and stress.

→ The World Health Organization website provides news on outbreaks, epidemics and pandemics (see www.who.int/csr/don).

→ Table 9.7 provides a summary of some epidemics and pandemics over time.

→ KEY QUESTION

9 Briefly outline, using an example, the transmission of one named disease during an epidemic.

Answer ➲ p. 134

2 Investigating the work of Robert Koch and Louis Pasteur

» Students investigate the work of Robert Koch and Louis Pasteur to explain the causes and transmission of infectious diseases, including Koch's postulates and Pasteur's experiments on microbial contamination.

Pasteur's experiments on microbial contamination

→ Before the discovery of lenses and microscopes it was believed that living matter, such as moulds and maggots, could **generate spontaneously** from non-living matter. Even after the invention of microscopes there was very little understanding of infectious diseases and the link between microorganisms, infection and disease transmission. In the 1850s the research of both Pasteur and Koch provided evidence that many diseases are caused by microorganisms.

→ Pasteur discovered that microbes such as bacteria and moulds can cause **contamination** and disease. With his famous swan-neck flask experiment Pasteur demonstrated that bacteria and mould cannot generate spontaneously.

spontaneous generation: an obsolete idea that certain life forms could arise from dust or dead tissue

contamination: in microbiology, the unintended or accidental introduction of infectious material or pathogens; commonly occurring through poor hygiene

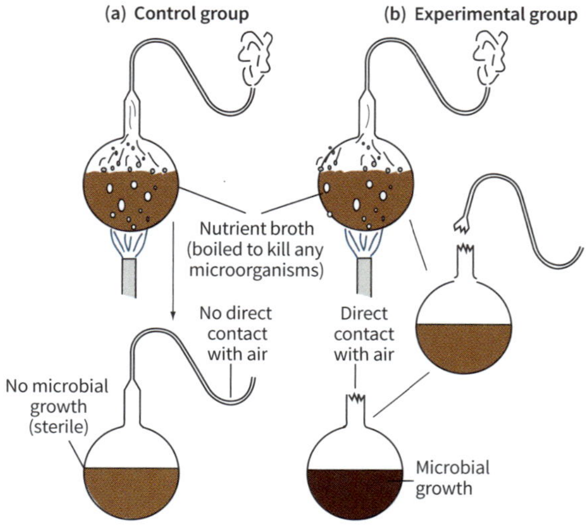

Figure 9.10 Pasteur's classic experiment

→ Pasteur discovered that microbes were responsible for the souring of alcohol and developed the process called pasteurisation.

→ Pasteur's work also led to the development of vaccinations for fowl cholera, anthrax and rabies.

→ One way to investigate Pasteur's work is to model his experiment in order to identify the role of microbes in contamination.

Table 9.7 A summary of some epidemics and pandemics over time

Year	Disease	Region or country/countries	Number of infections and deaths
541–42	bubonic plague	Europe and Asia	25–50 million
1346–53	Black Death (bubonic plague)	Europe, Asia, Africa	75–200 million
1545 and 1576	smallpox	Mexico	17 million
1665–66	great plague of London (bubonic plague)	Europe	100 000
1817–24	first cholera pandemic	Asia, Europe	> 100 000
1918–20	Spanish flu (influenza)	worldwide	20–50 million
1980s	HIV	worldwide	35 million
2003	SARS	China, Hong Kong, Taiwan, Hanoi, Toronto and Singapore	800 deaths
2013–16	Ebola	West Africa	>11 300 deaths
2017	plague: bubonic plague and pneumonic plague (a more virulent form of bubonic plague)	Madagascar	207

Source: MacKenzie, D, 'The coming plague', *New Scientist*, 25 February 2017, pp. 29–33

FIRSTHAND INVESTIGATION

Modelling Pasteur's experiment

» Students investigate Pasteur's experiment on microbial contamination.

Aim

to model Pasteur's experiment to identify the role of microbes in contamination

Method

1 Dissolve a meat extract cube in hot water.
2 Set up the equipment with a conical flask fitted with a one-holed stopper through which a glass tube bent into an S-shape is passed to prevent entry of air.
3 Set up an identical flask and stopper but containing a straight piece of glass tube.
4 Place equal amounts of broth in both flasks and gently boil for 15 minutes.
5 Observe for several weeks, looking for changes in broth colour and, if safe to do so, odour. Follow safe procedures when investigating the contaminated liquid.
6 Safely dispose of contaminated liquid by autoclaving.

Results

The liquid in the container opened directly to airborne microorganisms went cloudy, developed scum, bubbled and produced unpleasant odours; that is, became contaminated. The container with the S-shaped tubing closed to the air remained unchanged; that is, not contaminated.

Conclusion

Microbes cannot grow spontaneously. Contamination relies on the entry of microbes to the medium.

Koch's postulates

→ Koch showed that bacteria were the cause of a disease called anthrax in horses, sheep, cows and humans. He also demonstrated that bacteria caused tuberculosis in humans. During his work on diseases Koch designed four rules of procedure for showing that a particular microorganism is the cause of a particular disease. These are called **Koch's postulates** and are outlined below.

1 It must be shown that the microorganism believed to be the cause of the disease is always present in the diseased organism.
2 The microorganism must be isolated and grown in a pure culture; that is, a culture only containing that microorganism.
3 Microorganisms from the pure culture, when injected into a healthy organism without the disease, must produce the disease.
4 Microorganisms isolated from the experimental organisms, grown in pure culture, and compared with the microorganisms in the original culture, are shown to be identical.

Koch's postulates: the rules of procedure developed by Robert Koch

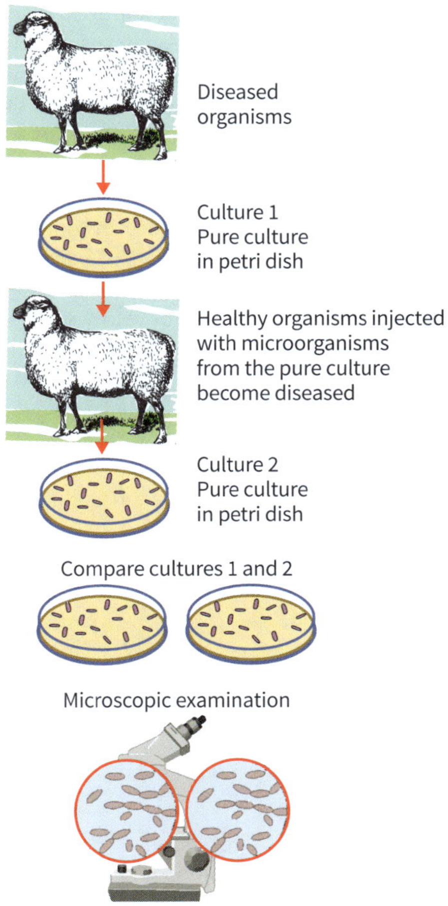

Figure 9.11 Koch's postulates

→ KEY QUESTIONS

10 **Briefly describe Pasteur's experiments on microbial contamination.**

11 **How did Pasteur's conclusions change our understanding of the causes and transmission of infectious diseases?**

12 **Describe Koch's experiments.**

13 **Explain the steps needed to demonstrate that a microorganism is responsible for a particular disease, as proposed by Koch.**

Answers ➲ p. 134

3 The causes and effects of diseases on agricultural production

» Students assess the causes and effects of diseases on agricultural production, including but not limited to plant diseases and animal diseases.

Animal diseases

- Australia is free of the world's worst animal diseases such as foot-and-mouth disease and Avian Influenza H5N1. Despite this, animal pests and diseases are a major threat to Australia's livestock and poultry industries. Outbreaks can be devastating to farmers and communities, and can impact on Australia's export industries.
- Table 9.8 provides examples of the causes and effects of some animal diseases on agricultural production.

Plant diseases

- Plant crops (e.g. cereals and potatoes) are often staple foods as well as being fodder for animals. This makes losses from diseases particularly damaging. Fortunately, Australia is free from many of the world's worst plant diseases. Table 9.9 provides examples of the causes and effects of some plant diseases on agricultural production.

notifiable: a disease that must be reported to agricultural authorities; listed as a major threat to Australian agricultural industries and biosecurity

Table 9.8 Examples of the cause and effects of some animal diseases on agricultural production

Disease	Cause	Effect
sheep lice	insect: *Bovicola ovis*	sheep lice costs producers in NSW over $100 million per year in lost production and treatment costs sheep with lice produce about 10% less wool and the wool has broken fibres, making the fleece more felted or matted and 10% less valuable; infected sheep are stressed through the discomfort of irritated skin, which causes rubbing, affects feed intake, and produces 'cockle' damage to hides; they are also more susceptible to flystrike
anthrax in humans, cattle, sheep, horses, goats and deer	bacterium: *Bacillus anthracis*	until the 20th century anthrax killed hundreds of thousands of people and animals each year anthrax is passed from infected animals to humans by inhaling the spores, eating infected meat or through skin lesions; the respiratory form of the disease has a mortality rate of 50 to 80% anthrax is a nationally **notifiable** disease and is subject to compulsory government controls, including quarantine, disposal of carcasses, vaccination and tracing of at-risk animals and their products spores have been known to survive 70 years in soil, skin, hair or hides
Newcastle disease in fowls, turkeys, pigeons, parrots and other domestic and wild birds	virus: *Avian pneumoencephalitis*	strains may vary in virulence; the disease is highly contagious and notifiable in Australia it is a global problem though vaccines are available
foot-and-mouth disease in cattle, water buffalo, sheep, goats, pigs, antelope, bison and deer	virus with seven serotypes: A, O, C, SAT1, SAT2, SAT3 and Asia1, and over 60 strains	the 2001 UK outbreak caused losses of $19 billion; highly contagious; the disease does not occur in Australia but it is a major threat and has prompted strong quarantine regulations, including support from neighbouring countries to reduce chances of entry into Australia

Table 9.9 Examples of the causes and effects of some plant diseases on agricultural production

Disease	Cause	Effect
potato blight	Fungi: *Phytophthora infestans*	caused one million deaths from starvation and over one million people to migrate from Ireland in the mid-1800s
Golden potato cyst nematode	*Globodera rostochiensis*	microscopic, soil-borne plant pests which feed on the roots of potato plants and are a serious threat to Australia's potato industry root development and tuber yield is reduced and plant growth is stunted; crop damage may appear as patches of poorly growing plants and can extend to complete crop failure

Disease	Cause	Effect
blast and bacterial blight in rice	blast fungus: *Magnaporthe grisea;* bacterial blight rice: *Xanthomonas oryzae* pv. *oryzae*	together these diseases cause half of the worldwide cereal food grain loss; high dependence on these cereals means that significant loss due to disease could result in starvation in local areas
stem rust and leaf blotch in wheat stem rust in barley	stem rust in wheat and barley: *Puccinia graminis*	
leaf blight and bacterial stalk rot in maize	leaf blotch in wheat: *Septoria* and *Stagnospora* species leaf blight in maize: *Heliminthorsporium turcicum* bacterial stalk rot: *Erwinia carotovora*	
grain smut in sorghum	grain smut in sorghum: *Sphacelotheca cruenta*	

➔ For more information see www.agriculture.gov.au/pests-diseases-weeds/plant#identify-pests-diseases and www.dpi.nsw.gov.au/biosecurity/plant/notifiable-list.

➔ KEY QUESTIONS

14 **Outline the cause of one plant disease that affects agricultural production.**

15 **Outline the cause of one animal disease that affects agricultural production.**

Answers ➲ pp. 134–135

4 The adaptations of different pathogens that facilitate their entry into and transmission between hosts

» Students compare the adaptations of different pathogens that facilitate their entry into and transmission between hosts.

Adaptations of different pathogens that facilitate their entry into hosts

➔ Pathogens have adaptations that facilitate their entry into cells and tissues and their transmission between hosts.

➔ A key to controlling diseases caused by pathogens is to identify these adaptations.

➔ Some pathogens have elaborate adaptations to exploit hosts' biology, behaviours and ecology in order to live in and move between hosts. An obvious example is vector-transmitted diseases. Many bacteria, viruses and protozoa have evolved the ability to survive in arthropods (e.g. insects and ticks). These pathogens:

- use the vector to gain entry into the host when the vector bites the host or sucks its blood; to do this the pathogens need adaptations to survive in the gut of the vector, and survive being pushed into the salivary glands before being injected into the next host
- use the life cycle of the vector to spread from one host to another; for example, the *Plasmodium* protozoan that causes malaria develops through several forms in its life cycle whereby some forms are adaptations for survival in humans and some are for survival in a mosquito; viruses such as dengue fever, yellow fever and viral encephalitis, which are spread by insects, have evolved adaptations to replicate in both insect and mammalian cells; these adaptations ensure that the virus is transmitted by a vector.

➔ In reviewing this section it is important to check your understanding of the term 'entry into a host'. Some pathogens can cause disease on the surface of the skin. For example, scabies is a disease caused by a burrowing mite. Most pathogens must enter a host to cause disease. Microbes enter into human hosts through the respiratory tract (mouth and nose), the gastrointestinal tract, the urogenital tract and skin surfaces. To do this they must get through protective barriers.

➔ In plants, protective barriers include physical barriers such as thick cell walls, thick cuticles, and the presence of a secondary cell wall or thick bark, while chemical barriers include antimicrobial compounds (see Chapter 10).

➔ In animals, protective barriers include physical barriers such as a thick, fairly tough covering of skin and other **epithelial surfaces** while chemical barriers include the pH of skin, along with body fluids (see Chapters 10 and 11).

epithelial surfaces: surfaces that line organs and blood vessels

➔ Table 9.10 summarises protective barriers in humans and how they function.

➔ Wounds in the epithelial layers, including the skin, provide direct entry into host tissue. For example:

- *Bacteroides* are anaerobic bacteria that live harmlessly in the large intestine. However, they can cause peritonitis (a life-threatening disease) if they enter the peritoneal cavity through a perforation in the intestine caused by trauma, surgery or infection.

Table 9.10 Protective barriers in humans and how they function to protect against pathogen entry

Protective barrier	Function against pathogen entry
skin	mechanical barrier: unbroken skin protects other tissues, and collects and holds pathogens if the skin is cut, the blood produces a blood clot that seals the wound and prevents pathogens from entering
surfaces of the body are covered by microbes (normal body flora)	these microbes act as a physical barrier; they also reduce the area available for pathogens to attach and become established
mucous membranes along the alimentary canal, respiratory surfaces and urogenital surfaces (e.g. bladder)	mucous traps pathogens
cilia in the respiratory tract	cilia beat in one direction to sweep mucous containing pathogens and small particles out of the lungs; coughing helps to remove materials and pathogens from the lungs
chemicals barriers found in body fluids and secretions, and the pH of fluids	low (acid) pH on the skin discourages growth of microbes; the stomach juices are highly acidic (pH ≈ 2) and this acidity, along with peristaltic washing and the thick mucus lining, make it difficult for microbes to colonise the stomach
other bodily secretions such as oil from oil glands and fatty acids	inhibit the growth of some bacteria and fungi; some of these secretions contain antibacterial factors

- Many healthy people carry the bacterium *Staphylococcus* on the skin and in the nose. The bacterium causes serious infections when it crosses the epithelial barrier through broken skin or skin with dermatitis.
- Fungi use direct penetration to enter the host plant through natural openings such as stomata, as well as through wounds caused by pruning, insect damage, hail or other diseases.
- In animals, fungal infections usually begin on the skin or in the lungs. Once a fungus colonises the skin or epithelial lining it can break the physical barrier with specialised hyphal penetration structures. The hyphal structure is a thickened, flattened tip of a branch of the hyphae.

➔ Many pathogens have adaptations for overcoming the protective barriers and the host's cleaning mechanisms, some of which are outlined below.

- Coughing, sneezing and saliva secretions act to wash microbes away from tissue surfaces. Some bacteria produce chemicals that bind them to the host tissue and resist the mechanical washing. For example, pathogens that infect the urinary tract have adaptations that resist the washing action of urine. They do this by adhering tightly to the epithelium with specific adhesins. Adhesins are proteins or protein complexes that recognise and bind to host-cell surface molecules. Adhesins can be hair-like structures that protrude from the surface of the bacterial cell or direct adhesins with no hair-like structure.
- *Helicobacter pylori* causes a bacterial infection of the stomach lining. It survives in the stomach by producing an enzyme (urease) which converts urea to ammonia and carbon dioxide. By doing so, the bacterium surrounds itself in a layer of ammonia that neutralises the stomach's natural acidity.
- The bacterium also produces adhesins to adhere to the stomach epithelium and **cytotoxins** that destroy the stomach's epithelial cells (creating stomach ulcers).
- Some plant pathogens penetrate the host by producing enzymes that degrade the waxiness of the cuticle. Other plant pathogens release enzymes that destroy the antimicrobial compounds produced by the host plant.

➔ Once they have gained access, many pathogens infect hosts without entering the host's cells. These are called **extracellular pathogens**. Most bacteria (e.g. *Streptococcus areus, Vibrio cholerae* and *Bordetella pertussis*) multiply in tissue fluids and not in host cells. Some bacteria invade cells but do not need them for growth.

➔ All viruses, some bacteria (e.g. *Rickettsia* spp) and some protozoa are **intracellular pathogens**. Intracellular pathogens have adaptations to facilitate their entry into the host cell, survival and replication within the cell, and exit from the cell. Pathogens are not accessible to antibodies once inside the host cell and they are not easily targeted by phagocytic cells (you will review antibodies and phagocytic cells in Chapter 11).

cytotoxins: substances that are toxic and destructive to cells
extracellular pathogens: pathogens that grow within the tissues of the body and do not need to be in a host cell for growth
intracellular pathogens: pathogens that grow and reproduce inside the cells of a host; all viruses are intracellular pathogens

➔ The first step for any intracellular pathogen is to bind to the surface of the host cell. Intracellular pathogens have acquired genes that encode proteins which interact specifically with particular molecules of the host cells.

→ Table 9.11 summarises and compares some of these adaptations and the ways pathogens change the biology of the host cells so they can survive in them. Note: Intracellular bacteria and parasites require the host cell for nutrients. Viruses, on the other hand, use the biochemistry of the host cell for their reproduction. Viruses depend on the host cell's ribosomes to produce proteins. Some also use the host cell's DNA and RNA for replication.

Adaptations of different pathogens that facilitate their transmission between hosts

→ All pathogens must have a mechanism of transfer from one host to another or they will die if the host dies.

→ Fungal spores facilitate transmission; for example, spores of *Aspergillus* are extremely hardy and can survive acidity, dehydration, freezing and high temperatures while spreading long distances on the slightest air currents.

→ Diseases that cause diarrhoea provide an example of adaptations that facilitate transmission. In humans, some bacterial infections produce massive outputs of diarrhoea and therefore massive outputs of the bacteria. This increases the density of bacteria in water and food, and therefore increases the transmission rate. This is one of the factors in transmission during an epidemic that you reviewed on page 125.

→ Some pathogens alter the behaviour of the host to facilitate their transmission between hosts as outlined in the examples below.

- Bacteria and virus infections cause mucus build-up which in turn causes coughing and sneezing, which help to clear the pathogens from the respiratory tract. They also help spread the disease.
- Tuberculosis is transmitted via airborne transmission when the pathogen *Mycobacterium tuberculosis* is released in small particles when people cough. Tuberculosis requires as few as ten microbes to cause a new infection.
- The tsetse fly transmits the protozoan parasite *Trypanosoma brucei* through its bite. This parasite causes sleeping sickness. Tsetse flies normally stop eating when satisfied but those that carry trypanosomes bite much more frequently and ingest more blood than uninfected flies. The presence of the trypanosomes impairs the receptors in the insect's stomach that measure blood flow and fullness so the fly thinks it is still hungry and continues to bite, therefore transmitting more of the disease.
- The bacterium *Yersinia pestis* causes bubonic plague. It is transmitted by fleas. *Yersinia pestis* has evolved a mechanism to ensure that a disease-carrying flea bites repeatedly. The bacterium multiplies in the flea's foregut, forming a mass that enlarges and blocks the digestive tract. As the flea starves and tries to feed, bacteria from the foregut are spread into the bite site and the disease is transmitted to a new host.

→ KEY QUESTIONS

16 **Construct a table to compare the adaptations of two different pathogens that facilitate their entry into hosts.**

17 **Briefly describe an adaptation of a specific pathogen that facilitates its transmission between hosts.**

Answers ⊃ p. 135

Table 9.11 Adaptations of intracellular pathogens to facilitate entry into host cells and to change the biology of the host cell

Viruses	after attaching to the host cell surface, the virus must enter to release its nucleic acid from the protective coat or lipid envelope viruses enter cells by either fusing with the membrane of the host cell or by forming a pore in the cell membrane of the host cell to deliver the genetic material into the cytoplasm
Bacteria	bacteria are larger than viruses and are therefore too large to enter a cell in the same manner some bacteria enter host cells by phagocytosis (a function of macrophages, which you will review in Chapter 11) some bacteria (e.g. *Mycobacterium tuberculosis*) have adapted to survive and replicate inside macrophages some bacteria invade cells by using adhesions, which bind them to deeper tissues (e.g. alveoli) bacterial toxins aid invasion by breaking down specific host proteins and disrupting signalling pathways
Protozoans	intracellular parasites are much larger than bacteria and must expend energy to invade cells through a variety of complex pathways; for example, when *Toxoplasma gondii* contacts a host cell, it protrudes a microscopic, ribbed tubule that connects with the surface of the host cell and allows the protozoan to push its way in as it moves into the cell, the protozoan secretes lipids that form a vacuole compartment in which the protozoan is protected from its compartment the protozoan takes up nutrients from the host cell
Fungi	some fungi can enter host cells; for example, *Candida albicans* can change form and enter the bloodstream, using camouflage and shedding tiny bits of its cell wall to avoid being detected *Cryptococcus neoformans* can cause meningitis by colonising white blood cells until the cell crosses the blood-brain barrier; *Cryptococcus* then forces the white blood cell to eject it, resulting in successfully entering the host's brain cells

CHAPTER SYLLABUS CHECKLIST

Are you able to answer these questions from the syllabus for this chapter? Tick each question as you go through the checklist if you are able to answer it. If you cannot answer a question, turn to the relevant page in the study guide to find the answer. For NESA key word meanings, go to www.educationstandards.nsw.edu.au and search 'key words'.

	FOR A COMPLETE UNDERSTANDING OF THIS TOPIC:	PAGE NO.	✓
1	Can I describe a variety of infectious diseases caused by pathogens, including microorganisms, macroorganisms and non-cellular pathogens?	117	
2	Can I collect primary and secondary-sourced data to classify different pathogens that cause disease in plants and animals?	117–123	
3	Can I design and conduct a practical investigation relating to the microbial testing of water or food samples?	123	
4	Can I collect primary and secondary-sourced data and information relating to modes of transmission of infectious diseases, including direct contact, indirect contact and vector transmission?	124–125	
5	Can I collect primary and secondary-sourced data and information relating to investigating the transmission of a disease during an epidemic?	125–126	
6	Can I investigate the work of Robert Koch and Louis Pasteur to explain the causes and transmission of infectious diseases, including Koch's postulates and Pasteur's experiments on microbial contamination?	126–127	
7	Can I assess the causes and effects of diseases on agricultural production, including but not limited to plant diseases and animal diseases?	128–129	
8	Can I compare the adaptations of different pathogens that facilitate their entry into and transmission between hosts?	129–131	

HSC EXAM-TYPE QUESTIONS

Objective-response questions (1 mark each)

1 **Which of the following is the best definition of a pathogen?**
- A any factor that can cause a disease
- B any organism that can bring about a disease
- C any organism that can be passed from one person to another
- D any microorganism, whether it be a virus, bacterium or protozoan

2 **What mode of disease transmission is illustrated in the following example: a person with the common cold sneezes, wipes their nose, and then transfers the mucus to a doorknob?**
- A indirect transmission
- B biological vector transmission
- C direct transmission
- D mechanical vector transmission

3 **Which of the following best summarises the impact of the work of Louis Pasteur?**
- A He proved the existence of bacteria.
- B He disproved the theory of spontaneous generation.
- C He provided evidence to question the theory of spontaneous generation.
- D He showed how to prove that a microorganism was the cause of a disease.

4 **Which of the following only contain possible pathogens?**
- A bacteria, environmental toxins, viruses
- B malaria, bacteria, viruses, macroparasites
- C protists, bacteria, viruses, macroparasites
- D bacteria, undercooked food, viruses

5 **Which of the following is most likely to be an adaptation of a pathogen that facilitates its entry into a host?**
- A The anthrax bacterium can survive in soil for many years in the form of spores.
- B The HIV virus can 'hide' in tissues such as the lymph glands and intestines.
- C Some viruses form a pore in the cell membrane of the host cell to deliver the genetic material into the cytoplasm.
- D Some bacteria have adaptations for extracting iron from host cell proteins.

Short-answer questions

6 **In an August 2016 article about mosquitoes in *National Geographic* the author stated, 'To this day, insects smaller than a child's thumbnail remain the most dangerous non-human animal on the planet' (p. 56).**
- a **What role do insects play in the transmission of some infectious diseases?** (1 mark)
- b **From your understanding of infectious diseases, why did the author make this statement about insects 'smaller than a child's thumbnail'?** (2 marks)

7 **The miasma theory of disease was the predominant theory up until the late 19th century. The theory held that diseases such as cholera, the Black Death and malaria were caused by miasma or bad air.**

Briefly explain how the work of Robert Koch and Louis Pasteur changed this predominant theory. (4 marks)

8 **Mary Mallon, also known as Typhoid Mary, is a famous historical example of an asymptomatic disease carrier. An Irish immigrant, Mallon worked as a cook for households in and around New York City between 1900 and 1915. In each household, the residents developed typhoid fever caused by *Salmonella typhi* a few weeks after Mallon started working. Later investigations provided evidence that Mallon was responsible for at least 122 cases of typhoid fever. In five cases the person died from the disease.**
- a **What do you think the term *asymptomatic disease carrier* means?** (1 mark)
- b **How do you think the typhoid fever was transmitted from Mary to household residents?** (2 marks)

9 **The yellow fever virus normally lives in forest monkeys. If the virus is transmitted to urban mosquitoes via humans, an explosive outbreak to a larger population could occur. World health authorities fear that the virus-carrying mosquitoes could spread to Asia where people have no immunity.**
- a **What is the term for 'an explosive outbreak' in a particular population?** (1 mark)
- b **Describe three factors in the transmission of disease during an explosive outbreak.** (3 marks)

10 **Construct a table to classify different types of pathogens, giving a specific example of the disease caused by each pathogen.** (7 marks)

ANSWERS

KEY QUESTIONS

Key questions ➲ p. 118

1 Pathogens are disease-causing organisms. Pathogens include prions, viruses, bacteria, fungi, protozoans and parasites.

2 To cause an infectious disease, pathogens must be able to enter the host organism in sufficient numbers without being destroyed by the host's defence system. They must then multiply, escape the host, survive transmission and enter another host.

Key questions ➲ p. 123

3 Hendra virus is caused by a microscopic pathogen. The virus mainly infects large fruit bats (flying foxes) and can be passed on to horses. The infection has occasionally been passed to people who have been in close contact with an infected horse. Hendra virus was discovered following an outbreak of illness in a large racing stable in the suburb of Hendra, Brisbane in 1994.

4 Macroscopic pathogens include macroparasites such as ticks, fleas and parasitic worms that obtain their nourishment from other organisms. Schistosomiasis is an example of a disease caused by the macroparasite blood fluke.

5 Some diseases are caused by a protein that has adopted an abnormal disease-causing form. Because prions are proteins they are considered to be non-cellular pathogens. The mis-folded form of the normal prion protein is known as PrP. PrP is found in the cytoplasm of cells and is normally in a stable state. Research shows that the protein flips into an abnormal disease-producing form. Prion diseases result in the degeneration of brain tissues, with the post-mortem characteristic of spongy cells.

Key questions ➲ p. 124

6 Microbes that can be detected and identified through preparing cultures in the school laboratory are bacteria and fungi. Bacteria can be identified as shiny colonies. Fungi can be identified as furry colonies.

7 The main safety precautions involve using sterilised equipment and sterilising all agar plates, solutions and samples after the investigation. Cleaning of work benches, hands and all equipment helps to minimise cross-contamination. Sterilisation by autoclaving destroys microbes that could be harmful to humans. The risk of infection from microbes is minimised by wearing protective clothing (e.g. masks, protective eyewear and gloves).

Key question ➲ p. 125

8

Table A9.1 Different modes of transmission of infectious diseases and examples of the disease transmitted

Mode of transmission of infectious disease	Example of the disease
direct	Ebola virus; the infection is transmitted by direct contact with the blood, body fluids and tissues of infected animals or humans
indirect	cholera, caused by a bacterium, is transmitted through contaminated water
vector	sleeping sickness (African trypanosomiasis) is caused by *Trypanosoma brucei*, a protozoan, spread by the bite of an infected tsetse fly

Key question ➲ p. 126

9 The Zika virus was discovered in the Zika forest in Uganda in 1947. For 50 years after discovery it caused small outbreaks of the disease in parts of Africa and southeast Asia. Since 2013 it has spread to Pacific Islands and then to Brazil and Mexico. In 2016 the disease spread into North America. The virus has adapted to thrive in the two species of mosquito it is transmitted by: *Aedesa egypti* (found in cities and feeds almost exclusively on human blood), and *Aedes albopictus* (an aggressive, adaptable species of mosquito that colonises habitats of other mosquitoes). The spread of the disease coincides with the mosquitoes' breeding season in warmer months.

Key questions ➲ p. 127

10 Pasteur is famous for experiments in which he demonstrated that microbes are the cause of contamination. Pasteur put broth into a flask with a swan-neck. Although the flask was open to the air, the shape of the neck restricted the entry of microorganisms and therefore contamination of the broth. Other experiments established the method of preservation through pasteurisation, and the process of vaccination by artificially weakening pathogens for anthrax, rabies and fowl cholera.

11 Pasteur's conclusions led to the wide acceptance that diseases are caused by microbes and can be transmitted from one host to another.

12 Koch discovered the microbial cause of tuberculosis and cholera. Koch succeeded in culturing the microbes and showed how they were transmitted. His precise experiments revealed the life cycle of the bacterium that causes anthrax disease. Koch developed a set of procedures that established a specific organism as the cause of a specific disease.

13 Step 1: establish the presence of the specific microorganism in all organisms with the disease. Step 2: grow the microorganism in a pure culture and inject it into a healthy organism. Step 3: once that organism has developed symptoms, isolate the microorganism and culture it. Step 4: the culture is now compared with the original culture: if they are the same, it indicates that the microorganism causes the disease.

Key questions ➲ p. 129

14 Wheat stem rust is a fungal disease. The disease is widespread in other countries and if it was accidently introduced into Australia could cause total devastation of Australia's cereal, wheat, barley, oats and rye crops. The disease is transmitted by windblown spores and through movement of infected plant material.

15 Equine influenza (EI) or horse flu is a highly contagious viral disease. The first outbreak in Australia occurred in 2007, and was later traced to horses imported from Japan. Horses in Australia were highly susceptible as they had no natural immunity and had not been vaccinated. In NSW 47 000 horses were infected. The outbreak impacted immediately on the horse-breeding industry and the future employment of industry workers. EI is transmitted directly through horse-to-horse contact, and humans can also transmit the disease to horses. Airborne transmission is possible over short distances.

Key questions ➲ p. 131

16

Table A9.2 Adaptations of two different pathogens that facilitate entry into hosts

Bacteria	Protozoan
E. coli is an important intestinal bacteria commonly found in the lower intestine; most strains are harmless	the amoeba *Naegleria fowleri* lives in water and enters a swimmer's brain directly through the nose
a strain of *E. coli—Enteropathogenic E.coli* (EPEC)—causes diarrhoea and is the second leading global cause of death in children younger than five	it attaches to the nasal mucus membranes, moves along the olfactory nerve, enters the cells through a porous part of the olfactory (smell) system and then to the olfactory bulbs
EPEC enters hosts by attaching tightly to the host cell membrane, causing disruption to the cell surfaces and destroying the micro-villi of the gut	it has food cups on its surface (like giant suckers) which allow it to attach to and eat the brain

Both the protozoan and the bacterium must first bind to the surface of the host cell or tissues before they can enter the host. The protozoan is larger than the bacteria and therefore must expend energy to enter into hosts through a more complex pathway.

17 One adaptation that facilitates transmission between hosts is for the pathogen to modify the behaviour of the host; for example, lyssavirus, which causes rabies. The pathogen modifies host behaviour to increase the spread of the disease. It does this by replicating in the nerve cells of the host, causing the infected host to become 'rabid'; that is, unusually aggressive with a strong desire to bite. When the infected host bites, the virus is shed in the saliva and transmitted through the bite wound into the bloodstream of the new host.

HSC EXAM-TYPE QUESTIONS

Objective-response questions

1 **B**. Pathogens must pass from host to host but ultimately the definition depends on the organism being the cause of an infectious disease. Note: Many microorganisms are beneficial so **D** is wrong. **A** is wrong because it includes non-infectious diseases that are not caused by pathogens. **C** is incorrect because some organisms can be passed from one person to another but are not always pathogens.

2 **A**. Transmission occurs indirectly when a host touches a fomite (contaminated object) such as a door knob, and potentially transmits the disease to another person. **B** is wrong because biological vector transmission involves a vector such as an arthropod. **C** is wrong because direct transmission is through direct contact. **D** is wrong because mechanical vector transmission is through another organism, such as a fly, not an inanimate object.

3 **C**. Even though it was widely accepted and challenged accepted thinking, Pasteur's work on spontaneous generation related to bacteria and moulds. His work did not provide categorical proof that spontaneous generation never occurred so **C** is a better answer than **B**. **D** is wrong because Koch's work showed how to prove that a microorganism was the cause of a disease. **A** is wrong because bacteria had already been observed (with the development of early microscopes) before Pasteur's time.

4 **C**. Pathogens can be protists, bacteria, viruses and macroparasites (as well as fungi and prions). **A** and **D** are wrong because environmental toxins and undercooked foods are not pathogens. **B** is wrong because malaria is a disease, not a pathogen.

5 **C**. Viruses must enter cells to replicate and reproduce. One way they do this is to create an opening or pore in the cell membrane of the host cell to release its nucleic acid. **A** and **B** are correct statements but they are adaptations to survive outside hosts, not adaptations to facilitate entry into hosts. **D** is also a correct statement but it is an adaptation for getting nutrients from the host cell, not an adaptation to facilitate entry into hosts.

Short-answer questions

6 **EM** Two steps are needed to answer this question. First, interpret the extract and conclude that the author is referring to mosquitoes. Second, recall the role of insects in vector transmission and some diseases transmitted by mosquitoes.

a Some mosquitoes are vectors of infectious diseases. (✓)

b A bite from a mosquito can transmit a range of deadly diseases (✓) such as malaria, Zika virus, yellow fever and dengue fever. (✓)

7 **EM** You need to be able to summarise the work of Koch and Pasteur, particularly in terms of the causes of infectious diseases. Make sure you provide an answer for both Koch and Pasteur.

Robert Koch developed a new experimental method called Koch's postulates (✓) to test whether a particular microorganism caused a disease. (✓) Louis Pasteur's experiments led to the wide acceptance that diseases are caused by microbes, not by bad air or miasma. (✓) Pasteur was able to develop vaccines through his understanding that microorganisms cause disease. (✓)

8 **EM** This question tests your understanding of how diseases are spread.

a An asymptomatic carrier carries a disease without showing any symptoms. (✓)

b Typhoid is transmitted through drinking water contaminated by human or animal faeces which contains disease-causing microbes. (✓) It is most likely that Mary transmitted the disease through contaminated food along with objects such as door handles and crockery. (✓)

9 **EM** This question asks for recall of the definition of an epidemic and factors in the transmission of disease during an epidemic. There are many factors and they are often interrelated. The ones you choose may not be the same as those in the sample answer below.

a An 'explosive outbreak' in a particular population is called an epidemic. (✓)

b Factors in the transmission of disease during an explosive outbreak include contaminated water supply, limited or no water supply, poor sanitation facilities, (✓) overcrowding (people in close proximity), (✓) limited or no medical response to the outbreak, and few medical supplies, including limited access to vaccinations. (✓)

10 **EM** The question asks you to classify different types of pathogens and use a table so it is important to include the column that lists the distinguishing features of each pathogen. You would receive marks for the points listed below.

Construction of the table (✓), the classification column (✓), prion (✓), virus (✓), bacteria (✓), protozoan (✓), fungi (✓) and macroparasites (✓).

Table A9.3 Classifying different types of pathogens

Type of pathogen	Example of disease	Some distinguishing features
prion	CJD: Creutzfeldt–Jakob Disease in humans; this is a neurodegenerative disease which also involves dementia; patients seldom survive a year	defective form of a protein molecule does not contain DNA or RNA mostly attacks brain or nerve cells
virus	hepatitis B: liver infection	non-cellular contains DNA, RNA and protein coat requires a living host cell to replicate
bacteria	enterobacteria: salmonella diseases and gastroenteritis	divides quickly and/or produces toxins prokaryotic cell
protozoan	amoebic dysentery	eukaryotic cell—single-celled organism may have a complex life cycle
fungi	tinea: ringworm and athlete's foot	eukaryotic cell with cell wall spreads via spores or rapid division some infect external skin and nails while others enter the host's body
macroparasites	flatworms (tapeworm)	eukaryotic cells—multicellular organism mostly arthropods or worms

CHAPTER 10 RESPONSES TO PATHOGENS

INQUIRY QUESTION:

How does a plant or animal respond to infection?

Pathogens must enter a host (and reproduce or replicate) to cause infection. In Chapter 9 you reviewed the adaptations of pathogens that facilitate their entry into hosts. Pathogens have developed highly specialised mechanisms for crossing cellular and biochemical barriers. Plants and animals have evolved innate defences to prevent entry of pathogens and highly specialised responses to pathogens when they do enter tissues and cells.

Together, these innate defences and responses are called the **natural or innate immune responses**. All plants and animals have innate immune systems. Note: Vertebrates have also evolved adaptive immune responses. You will review the innate and adaptive immune systems in the human body in detail in Chapter 11.

1 Responses of plants to pathogens

» Students investigate the response of a named Australian plant to a named pathogen through practical and/or secondary-sourced investigation; for example, fungal pathogens and viral pathogens.

Fungal and viral pathogens

- Australian plants are generally not impacted by viral pathogens although viral diseases are common in agricultural crops. Investigation into the causes of Mundulla Yellow disease of Eucalypts and other native plants shows that virus-like particles are present in some diseased plants (although abiotic factors such as iron deficiency in soils is also linked to the disease). Fungal diseases of Australian plants are common and include botrytis (grey mould), rusts, ink spot, powdery mildew and phytophthora. You will review the response of a named Australian plant to a fungal disease on page 139.

natural or innate immune responses: general, non-specific defence and response mechanisms, primed and ready to fight an infection at all times

- Plant and animal responses to infection can be summarised as a model as shown in Figure 10.1.

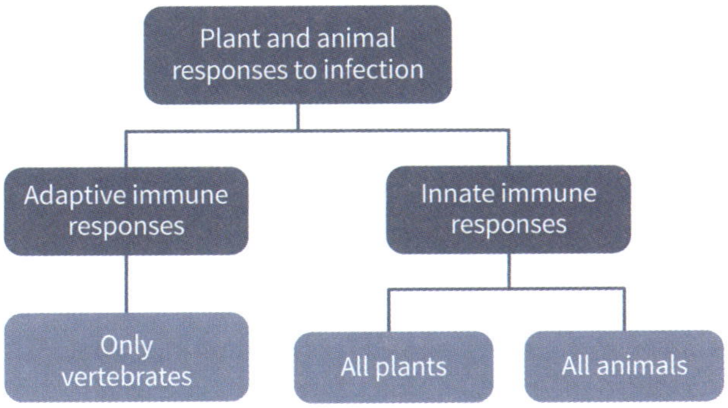

Figure 10.1 A simplified model of plant and animal responses to infection

- An infection is a battle between the invading pathogen and the host. The invasion of an organism's internal environment by pathogens challenges the effective functioning of cells, tissues and body systems. Pathogens do this in a number of ways:
  - The large numbers of pathogens present are too many for the host tissue to function normally.
  - The pathogens actually destroy cells or tissues.
  - The pathogens (e.g. bacteria) produce toxins.
  - Some pathogens (e.g. viruses) take over the polypeptide synthesis to produce more viruses and proteins.
  - The pathogen may not directly harm the host but an excessive immune response by the host may damage the host's tissue.
- To mobilise a response the immune system must be able to distinguish self from non-self. Non-self includes foreign particles and pathogens. This means that the host must be able to first recognise and then protect the body against non-self.
- Note: It is important when reviewing this section to make sure you understand the difference between the terms **responses to infection** and **symptoms of disease**.

responses to infection: the host's reactions to the invading pathogen, such as the inflammation response (see page 141)
symptoms of disease: the signs of possible infection (e.g. black marks on plant leaves or diarrhoea in animals)

- Plants must be able to protect themselves against a wide range of pathogens. Many diseases of plants are caused by macroscopic parasites (e.g. psyllids), microscopic parasites (e.g. nematodes) and fungal pathogens. You reviewed the growth and reproduction of fungi in Chapter 1.
- Plants can only rely on **preformed** defences and innate immune responses. An adaptive immune system does not exist in plants.
- In this section you will first review some general plant defences and responses to pathogens. You will then examine the response of a named Australian plant to a named pathogen.

preformed: meaning 'made beforehand', these are the defence mechanisms that are primed and 'ready to go', as compared to the responses that are induced (activated) by exposure to a pathogen

General defences and responses of plants to pathogens

- In your secondary-sourced investigation you would have gathered information, and constructed and used appropriate representations to analyse information. Figure 10.3 is an example of how to represent plant defences and their responses to pathogens.
- Reliable sources for your investigation would have included sites that provide a glossary. The following link is an example of a reliable source: bugs.bio.usyd.edu.au/learning/resources/PlantPathology/glossary.html.
- Plant defences and responses are complex and include:
  - passive, preformed defence mechanisms that are present before contact with a pathogen
  - recognition of the presence of a pathogen (the non-self); plants recognise a large number of signals from pathogens as non-self, including cell wall fragments, enzymes and other proteins from pathogens
  - early active defence mechanisms are activated only after pathogen recognition
  - later active responses, such as repair of damaged tissue to restrict the spread of infection, contain the damaged tissues, and prevent secondary infection; for example, some infected areas are sealed by wound cork and wounded tree trunks often secrete gums to seal the wound.
- Passive, preformed defence mechanisms are the physical and chemical defences that protect against entry of pathogens, including:
  - properties of the plant surface (e.g. waxy leaf cuticles and thickened, tough cell walls)
  - the presence of secondary cell walls in sclerenchyma, xylem or older plant tissue
  - secretory cells, glands and ducts which secrete defensive compounds
  - compounds that have antimicrobial properties; for example, the young fruit of mangoes and avocadoes have antifungal and antimicrobial compounds, making the unripe fruit less susceptible to disease
  - chemicals produced by some plants during normal growth that inhibit pathogen development; for example, marigolds secrete chemicals into the soil around their roots which are toxic to nematodes and provides protection for other plants growing in the same root zone
  - **phenolic compounds** secreted by some plants onto their leaves to discourage herbivores and inhibit some plant microorganisms, including potential pathogens
  - defence provided by some plant peptides against fungi, bacteria, insects and insect-transmitted viruses; the compounds are called plant **defensins** and interfere with pathogen nutrition, thereby retarding their development.
- The distinction between passive and active is not always clear because some preformed defences can be modified after infection. For example, many defensins accumulate as part of normal growth (preformed) while others are induced and increase in accumulation in response to an infection or wound.
- When a plant's protective barriers are breached, the invading pathogen encounters active defences. Because plants do not have a circulatory system and cell movement is restricted they rely on the innate immune responses of each cell. The resistance or susceptibility of a whole plant and plant community is, therefore, the sum of many individual cellular interactions.
- Active responses include:
  - the reinforcement of the cell wall and/or secondary cell wall, which occurs through intensification of cytoplasmic streaming and accumulation of the host's cytoplasm around the site of attempted invasion (see Figure 10.2 for an example of lignitubers)
  - **hypersensitive response**, which leads to rapid cell death at or near the invasion site; the death of the host cell stops the spread of the pathogen
  - the development of an **abscission layer** around the site of infection which causes the infected tissue to separate from the healthy tissue

phenolic compounds: compounds containing phenolic acid (as well as other chemicals); the compounds are used in treatment of cancers and are thought to have antioxidant and anti-inflammatory properties

defensins: antimicrobial or antifungal plant peptides

hypersensitive response: an immune response in plants to stop the spread of a pathogen by activating rapid cell death of the plant's cells in the area of the infection

abscission layers: layers of tissue formed by thin-walled cells that allow the tissue above and below the layer to separate; abscise means to cut off or away

- complex signalling cascades and hormone signalling
- production of specific receptors called resistance peptides (or R-protein receptors) that either kill pathogens or restrict their spread.

➔ A lignituber is a structure formed by the deposition of lignin around the tip of a fungal hypha (see Figure 10.2). The formation of a lignituber is an active defence mechanism and its function is to inhibit the growth of the hypha into other cells. Appressorium refers to the specialised cells from which an infecting tube grows.

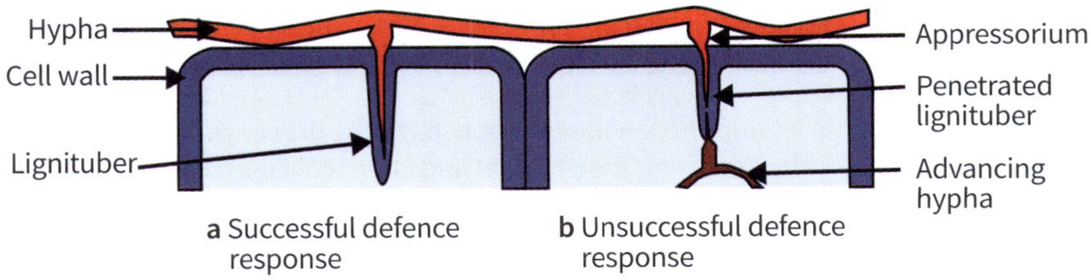

Figure 10.2 Lignituber

The response of a named Australian plant to a named pathogen

➔ In your study of infectious diseases you have used science inquiry skills to investigate a range of responses by plants to invasion by pathogens. In this investigation you studied an Australian plant. In the example below, Eucalyptus is used as the Australian plant.

➔ Your investigation may have involved:
- firsthand microscopic examination of Eucalyptus tissues, such as a thick leaf cuticle, the small number of stomata, the presence of a secondary cell wall, the presence of lignin or the presence of oil glands
- firsthand macroscopic observations of features of Eucalyptus trees, such as thick bark, the vertical orientation of the leaves and the presence of gum.

➔ The information below in Table 10.1 summarises some of the preformed defences and induced responses of Eucalyptus to pathogens.

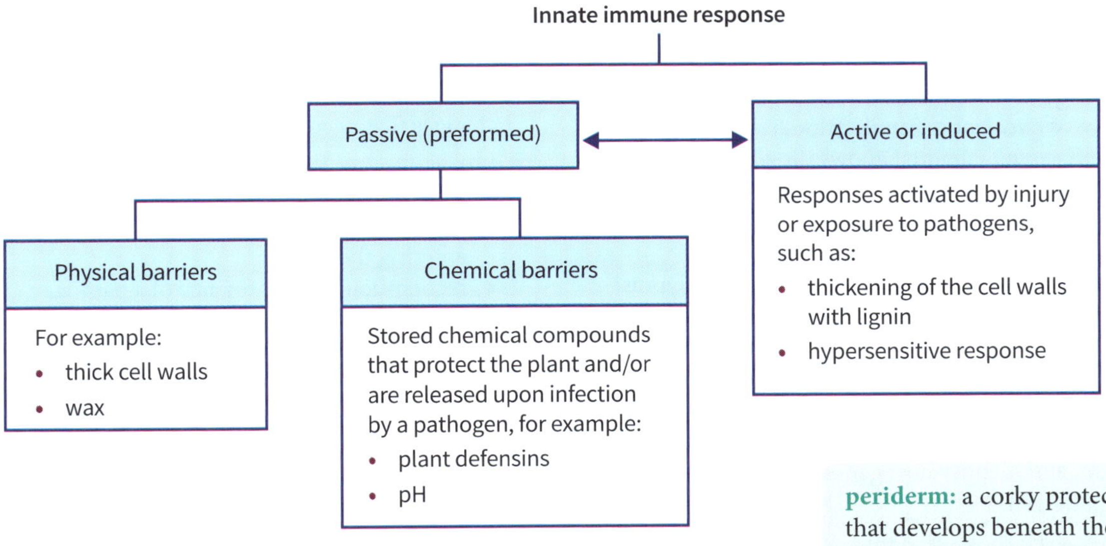

Figure 10.3 Passive and active defences and responses of plants to pathogens

periderm: a corky protective tissue that develops beneath the surface of damaged or infected tissue

Table 10.1 Responses of Eucalyptus to pathogens

Passive defences (preformed)		Active defences or induced responses
Physical barriers	**Chemical barriers**	
thick cuticle thick bark waxy leaves dry leaf surfaces vertically hanging leaves prevent the formation of moisture on the leaf which provides some protection from water moulds	stored oils function as preformed chemical defences, not only against defoliating animals but as antifungal and antibacterial agents or for priming of defences in both the host and neighbouring plants	the formation of barrier zones in the new tissue produced by the vascular cambium; these zones protect the healthy sapwood from damage by separating it from the adjacent infected or damaged tissue; in some Eucalypts these barrier zones contain gum the development of **periderm**, which separates damaged tissue from healthy tissue and prevents the spread of the disease(s) secretion of protective gums that seal the wound from further infection

- A number of pathogens pose a threat to a range of Eucalyptus in Australia. These include myrtle rust pathogen and the root rot pathogen *Phytopthera cinnamomi*. In the example below, *Phytopthera cinnamomi* (*P. cinnamomi*) is the named fungal pathogen.
- *P. cinnamomi* is a soil-borne water mould that produces a disease called root rot or die-back. This plant pathogen is one of the world's most invasive species, affecting native forests, agriculture and the forestry industry in over 70 countries. In Australia *P. cinnamomi* has caused extensive damage to native ecosystems including *Eucalyptus marginata* (jarrah) forests in Western Australia.
- The development of gene technologies such as genome and RNA-sequencing (see Chapter 8) has enabled researchers to better understand the host–pathogen relationship between Eucalyptus and *P. cinnamomi*, as well as the immune responses of Eucalyptus to infection by *P. cinnamomi* and why some Eucalyptus are more susceptible to infection.
- A better understanding of the defence response of plants to *P. cinnamomi* will help in the development of resistant plants.
- When you investigate the study of responses of Eucalyptus to infection by *P. cinnamomi* you will find that the research often involves:
  - the use of genome sequences, such as the genome sequence of *Eucalyptus grandis* (flooded gum)
  - dual-RNA sequencing that allows the simultaneous study of RNA transcripts of both the host and the pathogen; this technique allows for the detection of minute amounts of pathogen RNA in the host
  - **stem inoculation** of particular species of Eucalyptus; for example, *Eucalyptus nitens* (shining gum) is considered to be resistant to *P. cinnamomi* in its native environment but is susceptible to the pathogen when grown in plantations. By inoculating *Eucalyptus nitens* with *P. cinnamomi* researchers are able to study defence responses and identify host genes that are suppressed by the pathogen.
- Ongoing research shows that Eucalyptus responses to the pathogen *P. cinnamomi* are multi-layered and come about through different mechanisms, including:
  - mass production of defence proteins such as defensins and **phytoalexins** which are synthesised in cells and accumulate rapidly at the site of the infection
  - activation of defence pathways that involve the production of a number of defence hormones (e.g. salicylic and abscisic acids)
  - hypersensitive response, or rapid cell death, at or near the invasion site
  - lignin synthesis to reinforce physical barriers such as cell walls.

Source: www.ncbi.nlm.nih.gov/pmc/articles/PMC4773608/

→ KEY QUESTIONS

1 **Using the information in Table 10.1 and this section construct a flow diagram to model some preformed defences and induced responses of Eucalyptus to *P. cinnamomi.***

2 **Briefly outline how you investigated the response of a named Australian plant to a named pathogen.**

Answers ➲ p. 145

2 Responses of animals to pathogens

» Students analyse responses to the presence of pathogens by assessing the physical and chemical changes that occur in the host animals' cells and tissues.

Innate immune defences and responses of animals to pathogens

- In this section you will analyse responses of animals to infection by pathogens by examining and relating these responses to some of the specific physical and chemical changes that occur.
- Innate immune defences and responses of animals to pathogens include:
  - preformed defences that help prevent pathogens from entering the host
  - recognition of the presence of a pathogen (the non-self); the innate immune system relies on recognition of particular types of molecules that are common to pathogens but absent in the host (pathogen-associated **immunostimulants**)
  - immune responses that are rapidly activated when the pathogen crosses protective barriers and begins to replicate; these mechanisms pre-exist in all animals and act within minutes of infection (see Figure 10.4).
- In Chapter 9 you reviewed some of the physical and chemical barriers that help prevent pathogens from entering the host. These barriers are key elements of the innate immune system. Barriers occur at the tissue and cell levels.

stem inoculation: the introduction of the pathogen into the host, in this case through the stem
phytoalexins: antimicrobial proteins or plant antibiotics that accumulate rapidly in response to infection
immunostimulants: pathogen associated molecules that stimulate innate immune responses

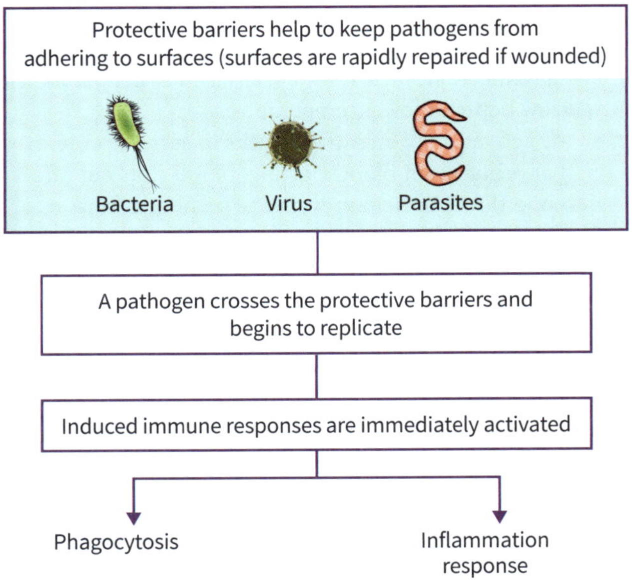

Figure 10.4 A simplified model of the innate immune defences and responses of animals to pathogens

- Physical barriers at the tissue level include external and internal **epithelial** surfaces; for example, thick, dry and/or scaly skin; the closed surface of unbroken skin and mucous membranes of the internal epithelia. The slimy mucous coating helps prevent pathogens from adhering to surfaces. Many amphibians and fish have a mucous layer covering the skin.
- Chemical barriers at the tissue level include antimicrobial substances produced by epithelia; for example, an antibacterial enzyme is secreted in saliva and tears. Other chemical barriers include the acid pH of the stomach and digestive enzymes, and antibacterial and antifungal peptides (defensins) produced in the lower intestinal tract. Antimicrobial peptides play a key role in the immune defences of animals, including insects.
- Most epithelial surfaces also contain normal body flora that compete with pathogens for nutrients and attachment sites on cells.
- At the cellular level physical barriers include the cell membrane, the nuclear envelope and compartmental borders. Chemical barriers include antimicrobial peptides in the plasma membrane of cells. **Lysosomes** contain enzymes that have a number of functions (e.g. immune defence).

epithelial: the skin (external covering) and to the lining of internal tissues such as the gastrointestinal, respiratory and urogenital tracts

lysosomes: membrane-bound cell organelles containing enzymes which break down extracellular and intracellular materials, pathogens engulfed by macrophages, virus particles and bacteria; lysosomes also play a role in the processing and secretion of inflammation signals

- If a pathogen crosses the host's barriers and begins to replicate it is immediately recognised by the host as non-self. In animals, the recognition of a pathogen rapidly activates two types of innate immune responses: the inflammation, or **inflammatory response**, and **phagocytosis** by cells such as **macrophages** and **neutrophils**.
- Both responses bring about physical and chemical changes in cells and tissues. You will review the inflammation response in detail in this section and phagocytosis, neutrophils and macrophages in more detail in Chapter 11.

inflammatory response: inflammation of the tissue in response to an infection; designed to isolate and destroy the foreign particles, it also prepares the tissue for healing

phagocytosis: the active process in which phagocytic cells enclose a pathogen and digest it in their interior

macrophages: white blood cells found in tissues and in especially large numbers in the gastrointestinal tract, lungs, liver and spleen; long-lived cells that circulate in the blood

neutrophils: short-lived, white blood cells that are abundant in blood but not in normal, healthy tissue

Physical and chemical changes in the presence of pathogens

- A pathogen that has crossed protective barriers and begins to replicate is immediately recognised by **phagocytes** called macrophages (see Chapter 11). Macrophages and neutrophils recognise, ingest and destroy pathogens.
- Once they have ingested the pathogen, macrophages and neutrophils produce a number of toxic chemicals that help kill the engulfed microorganism. These include hydrogen peroxide, superoxide and nitric oxide.
- When pathogens interact with macrophages, the macrophages are activated to release **cytokines**. Cytokines along with other chemicals such as **histamine** set up the inflammatory response.

phagocytes: white blood cells that engulf and destroy microbes

cytokines: small proteins secreted by a broad range of cells, they are signalling molecules that help cell to cell communication in infection, immune responses and inflammation response (see Chapter 11)

histamine: a chemical compound released by special types of white blood cells (called mast and basophil cells) in response to injury or infection by pathogens; histamines regulate both the innate and adaptive immune responses

- The inflammatory response is activated within minutes of infection or wounding. The inflammatory response has three key roles in fighting infection. These are:
  - to bring plasma proteins and phagocytes to the site of infection to help kill the pathogen
  - to provide a physical barrier to prevent further spread of the infection and make the host aware of the infection
  - to promote the repair of damaged tissue.
- Physical and chemical changes in the cells and tissues as a result of the inflammatory response include:
  - an increase in the diameter of blood vessels and a reduction in the velocity of blood flow, which allows blood flow to the area of infection to increase; this means that more red and white blood cells, as well as defence compounds, can travel to the site; heat and redness characterise this response
  - the activation of the cells lining the blood vessels to secrete adhesion chemicals which help the circulating white blood cells bind to the site of infection
  - the increased permeability of blood vessels which allows for the white blood cells and defence proteins to move out of the bloodstream and accumulate in the tissues at the site of infection; swelling and pain characterise this response
  - fever and an increase in temperature of the tissues and body temperature; fever helps the immune system to fight infection, since most bacterial and viral pathogens grow better at lower temperatures
  - the activation of endothelial cells to secrete proteins that bring about blood clotting; blood clots cut off the blood flow and help prevent the pathogen from entering the bloodstream and spreading to other parts of the body.
- Note: The same inflammatory response that can effectively control infections by pathogens can also create problems. For example, septic shock is a result of over signalling of the inflammatory response in the blood; that is, too much dilation of blood vessels, loss of plasma volume and widespread clotting. An extreme inflammatory response can also cause other problems such as asthma.

→ KEY QUESTION

3 **What chemical and physical changes occur in the cells and tissues in the presence of a pathogen?**

Answer ➲ p. 145

CHAPTER SYLLABUS CHECKLIST

Are you able to answer these questions from the syllabus for this chapter? Tick each question as you go through the checklist if you are able to answer it. If you cannot answer a question, turn to the relevant page in the study guide to find the answer. For NESA key word meanings, go to www.educationstandards.nsw.edu.au and search 'key words'.

FOR A COMPLETE UNDERSTANDING OF THIS TOPIC:		PAGE NO.	✓
1	Can I investigate the response of a named Australian plant to a named pathogen through practical and/or secondary-sourced investigation; for example, fungal pathogens and viral pathogens?	137–140	
2	Can I analyse responses to the presence of pathogens by assessing the physical and chemical changes that occur in the host animals' cells and tissues?	140–142	

HSC EXAM-TYPE QUESTIONS

Objective-response questions (1 mark each)

1 Rapid deposition of lignin following infection reinforces the cell walls of the host and increases resistance to pathogens. Lignin can also bind to hyphal tips and bacteria, stopping them from entering the host and restricting the diffusion of their enzymes and toxins into the cell wall.

What does this provide an example of?

A an immune response in animals
B an immune response in plants
C acquired immunity in animals
D acquired immunity in plants

Use the following information to answer Questions 2 to 5. The information summarises experimental methodology where the stems of *Eucalyptus nitens (E. nitens)* are inoculated with a fungal pathogen *Phytopthera cinnamomi (P. cinnamomi)* to study the defence responses of the plant to the pathogen.

E. nitens seedlings were obtained from a third-generation breeding plantation program. They were all selected to have the same wood density. All seedlings were grown in pine bark seedling mix until they had a stem thickness of more than 0.5 cm.

One group of seedlings (Group A) was inoculated with *P. cinnamomi* at two sites 10 cm apart. 200 ml/L of mycelia plug was put into each of the inoculation sites. The mycelia plug is a plug of agar that contains the fungal *P. cinnamomi* mycelia. Mycelia are the mass of branching thread-like hyphae.

The other group of seedlings (Group B) was mock-inoculated. That is, the seedlings were treated identically to the inoculated Group A, but instead of the mycelia plug they had a sterile agar plug put into each 'inoculation' site.

Inoculation sites were covered with a damp sterile cheesecloth, tinfoil and wax film. Five days later 3 cm of stem tissue was taken from each plant 1.5 cm above and 1.5 cm below the inoculation site.

Examination of the tissue samples showed that hyphae of *P. cinnamomi* were present in stem tissues in Group A.

2 In this methodology the experimental group is:

A the third-generation breeding stock.
B the seedlings grown in pine bark.
C the plants inoculated with *P. cinnamomi*.
D the mock inoculated plants.

3 Which one of the following is the dependent variable in this experiment?

A the thickness of the stem
B the inoculation of Group A seedlings with mycelia
C the covering of the inoculation site with damp, sterile cheesecloth, tinfoil and wax film
D the examination of tissue samples for hyphae

4 Plants that had not been inoculated (Group B plants) would be expected to:

A have hyphae growing only in the 1.5 cm above the site.
B have hyphae growing only in the 1.5 cm below the site.
C have no hyphae detected.
D have hyphae growing 1.5 cm above and below the site.

5 Six weeks later the inoculated seedlings (Group A) were re-examined. Based on your understanding of plant responses to pathogens, a likely observable response in the seedlings could be:

A evidence of extra thickening of cell walls.
B evidence of extra immune white blood cells in the stem tissue.
C evidence of bacteria in the stem tissue.
D evidence of immune cells circulating in the stem.

Short-answer questions

6 Use the information below to answer Questions a and b.

Research into the immune system of insects shows that some recognise the chemicals in the cell wall of a bacterial pathogen and this triggers a rapid immune response. Once the bacteria have been killed, molecules in the insect digest chemicals from the bacterial cell wall and this dampens down the immune response.

a What term is used for the rapid immune response in animals? (1 mark)
b Why is it important to 'dampen' the immune response once the bacterial pathogen has been killed? (1 mark)

7 *Phytopthera infestans*, a plant fungal pathogen, manipulates one of its hosts (potatoes) by suppressing the hypersensitive response. Other plant pathogens of some Australia plants are able to inhibit the synthesis of compounds that reinforce the cell wall. From your understanding of plant responses to pathogens, explain how each of these characteristics could provide an adaptive advantage to the pathogen. (4 marks)

8 In your studies you have analysed the physical or chemical changes that occur in the cells and tissues of animals in the presence of pathogens. Summarise chemical changes that occur in the presence of pathogens as well as the physical changes that occur during the inflammatory response. Use a table to show your answer. (6 marks)

9 Use the information below to answer Question 9.

Under the same environmental conditions and amount of infection, a northern NSW variation of the species *Eucalyptus nitens (E. nitens)* is more resistant to *Mycosphaerella* leaf disease than a southern NSW variant. *Mycosphaerella* is a fungal pathogen. Comparisons were made of the two variations after infection by the disease. The results showed that leaves of the northern NSW *E.nitens* were thinner, had a higher proportion of dense leaf tissue, and reduced intracellular spaces. After infection there was an increased level of cell division in resistant leaves and infected leaves were rapidly partitioned off with abscission layers.

Analyse this information to write a possible explanation for the resistance of the northern NSW variation of the species *Eucalyptus nitens* to *Mycosphaerella*. (4 marks)

10 Use the following information to construct a Venn diagram to represent the differences and similarities between plant and animal immune systems.

Plant and animals have evolved similar defences against invasion by pathogens. Both have innate immune systems and both attack pathogens with bursts of superoxide, hydrogen peroxide, nitric oxide and toxic antimicrobial peptides.

Plants lack mobile immune cells whereas animals have mobile immune cells. Plants also lack a circulatory immune system. This means that each plant cell is largely responsible for its own defence and must be able to respond independently.

Animals have an immune circulatory system. This means that responses to pathogens can be exerted throughout the whole body. (6 marks)

ANSWERS

KEY QUESTIONS

Key questions ➲ p. 140

1

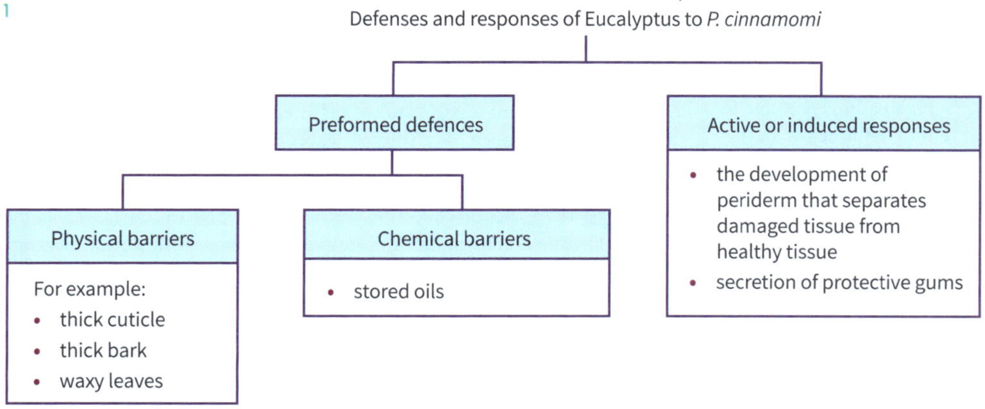

Figure A10.1 Defences and responses of Eucalyptus to *P. cinnamomi*

2 The investigation had a number of parts to it, as outlined below.

a Secondary-sourced investigations to gather information about plant responses to infections in general. Understanding of general responses was used to investigate responses of an Australian plant to a pathogen.

b Secondary-sourced investigations to find out about Australian plant pathogens and choose a plant and a pathogen for study.

c A practical investigation to make firsthand observations of plant defences in the chosen Australian plant.

d Secondary-sourced investigations to analyse research about the Australian plant's response to the named pathogen.

Key question ➲ p. 142

3 Specific white blood cells (macrophages) produce a number of chemicals (e.g. cytokines) in the presence of a pathogen. Other types of white blood cells produce histamine. These chemicals cause physical changes to cells and tissues such as dilation of blood vessels around the site of infection to allow increased blood circulation (the tissues become red and warmer). Other responses include a change in the permeability of blood vessels and development of gaps in the cell membranes surrounding the infected area to allow the larger immune cells of the blood to pass out of the bloodstream into the infected tissue. Different types of cells of the immune system move to the site of inflammation, along with a large supply of proteins to fuel the immune response. The tissues become swollen, due to the increased amount of blood and proteins. The tissues in the area are red and warm, as a result of the large amount of blood reaching the site, and the area is painful due to the expansion of tissues putting pressure on nerve cells.

HSC EXAM-TYPE QUESTIONS

Objective-response questions

1 **B.** Deposition of lignin is an example of an induced or active innate immune response in plants. **A**, **C** and **D** are incorrect. Lignin deposition does not occur in animals and is not an example of acquired immunity, which is associated with adaptive immune response.

2 **C.** This is the experimental group; that is, the group that receives the fungal mycelia so that scientists can observe what happens. **A** and **B** are controlled variables (the constants) as they are same generation plants using the same medium for growth. **D** is the control group where everything is kept the same except this group does not receive the fungal mycelia.

3 **D.** This is the dependent variable. A dependent variable in an experiment is the response that is measured or observed. **B** is incorrect because it is the independent variable; that is, the one thing different between the experimental group and the control group. **A** and **C** are controlled variables (the constants), with the same thickness of stem and same covering of the inoculation site.

4 **C.** The mock inoculated plants (Group B) did not receive the fungal mycelia and therefore should not have hyphae, making **A**, **C** and **D** incorrect.

5 **A.** Thickening of cell walls is one possible observable response in plants in the presence of a fungal pathogen. **B** and **D** are incorrect because plants do not have special immune cells such as white blood cells and they do not have circulating immune cells. **C** is incorrect as unless there is secondary infection of the stem you are not likely to observe bacteria.

Short-answer questions

6 **EM** This question is mostly straight recall.

a the innate immune response (✓)

b Regulating the immune system is important because if left unchecked it can harm the insect by attacking its own body. (✓)

7 EM This question is more than straight recall. You need to use your understanding of some plant responses to pathogens and recall that plants sometimes use the hypersensitive response or reinforce cell walls to fight pathogens. You also need to analyse the examples of how pathogens can avoid being controlled by the plant's natural or innate immune responses in order to answer this question.

Innate immune responses of plants to pathogens can include reinforcement of the cell wall and hypersensitive response, which leads to rapid cell death at or near the invasion site. (✓) The death of the host cell stops the spread of the pathogen. (✓) If a pathogen has evolved to suppress the hypersensitive response, it will be more successful in invading the plant tissues. (✓) If a pathogen is able to inhibit the synthesis of compounds that reinforce the cell wall, the plant will have less protection against the pathogen and the pathogen will be more successful in invading the plant cells. (✓)

8 EM Make sure you read all parts of the question. The question asks for chemical changes that occur in response to pathogens and physical changes that occur during the inflammatory response. Remember to include a heading for your table.

You would receive marks for the heading; (✓) for table construction; (✓) two marks for four chemical changes (✓)(✓) and one mark for two, (✓) plus two marks for four physical changes (✓)(✓) and one mark for two. (✓)

Table A10.1 Summary of physical and chemical changes that occur in the cells and tissues of animals in the presence of pathogens

Chemical changes	Physical changes
toxic hydrogen peroxide, superoxide and nitric oxide produced by phagocytes help kill the pathogen production of cytokines (that set up the inflammatory response) secretion of adhesions during the inflammation response secretion of chemicals that bring about blood clotting	an increase in the diameter of blood vessels an increase in permeability of blood vessels an increase in temperature of the tissues and body temperature development of blood clots

9 EM To analyse something you must examine it methodically and in detail. You must also identify components and the relationships between them. There are two key parts to your answer for this question: the features of the northern variant compared to the southern variant and the comparisons of the response of each variant after infection.

The leaves of the northern NSW *E. nitens* have a higher proportion of dense leaf tissue and reduced intracellular spaces. (✓) These characteristics provide a better physical barrier to the entry of pathogens. (✓)

An abscission layer after infection in the resistant variants means that the plant was able to cut off infected leaves and stop the spread of the pathogen. (✓) Thinner leaves mean they can be partitioned off more easily. (✓)

10 EM To answer this question you need to be able to construct a Venn diagram and organise the information to show both similarities and differences.

You would receive marks for construction of the Venn diagram; (✓) for showing similarity; (✓) two marks for showing differences in plant immune systems (✓) (✓) and two marks for showing differences in animal immune systems. (✓) (✓)

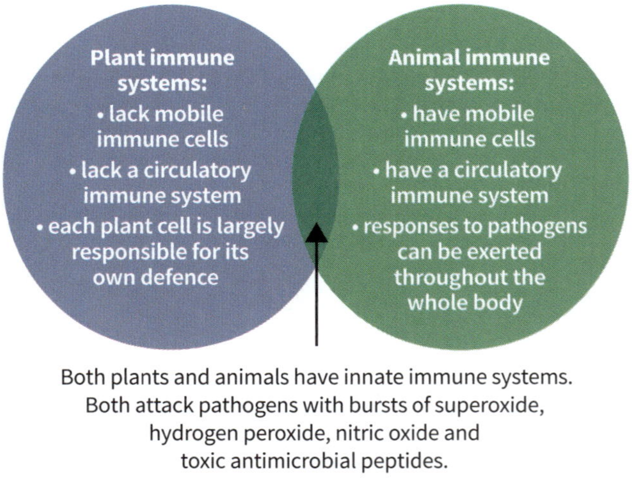

Figure A10.2 Venn diagram

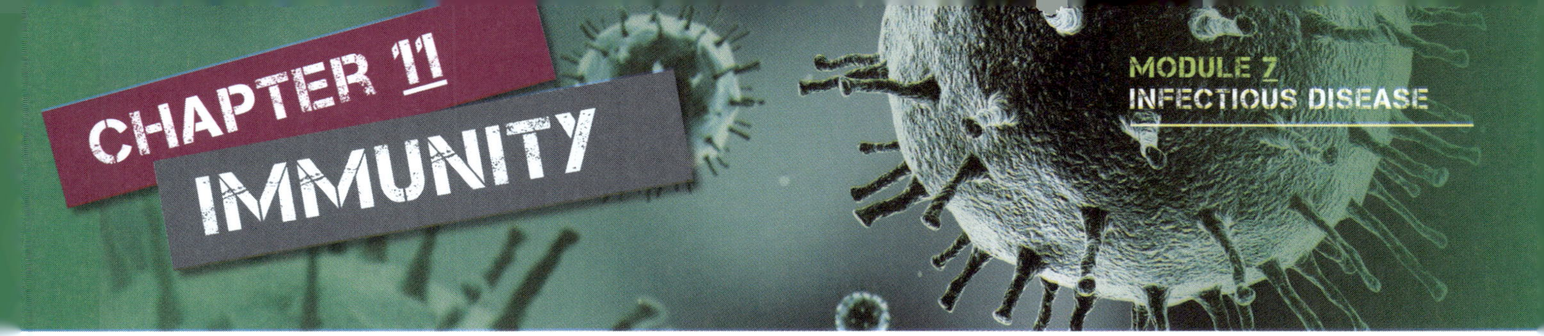

CHAPTER 11 IMMUNITY

MODULE 7 INFECTIOUS DISEASE

INQUIRY QUESTION:

How does the human immune system respond to exposure to a pathogen?

Humans are exposed to millions of potential pathogens each day. Immunity is the ability of the immune system to control or eradicate pathogens (or other potentially damaging foreign bodies).

Human history is intertwined with infectious diseases and their impacts. From the first early understandings about cells and cell theory, scientific research has not only focused on the causes of infectious diseases in humans and how the body responds, but on how to prevent, control and treat them. You will review the prevention, control and treatment of infectious diseases in Chapter 12.

1 The innate and adaptive immune systems in the human body

» Students investigate and model the innate and adaptive immune systems in the human body.

Human immune systems

- The human immune system responds to the presence of non-self in two ways: through the innate immune system, which provides a fast-acting, non-specific and broadly effective general defence, and through the specialised and specific **adaptive** immune system.
- Both systems have special defense cells that are directly involved in eliminating pathogens. In addition, both systems have several soluble substances found in blood and other body fluids (e.g. proteins such as enzymes and antibodies).
- The two systems work closely together to create a complex response.
- In your study of immunity and immune system responses you used models to represent this complexity.

adaptive: the ability to adapt; that is, to modify behaviours, physiology and/or structures in order to survive

- Before you review each of the systems in detail, use the diagram below to check your understanding. The diagram represents one way to model the innate and adaptive immune systems in the human body.

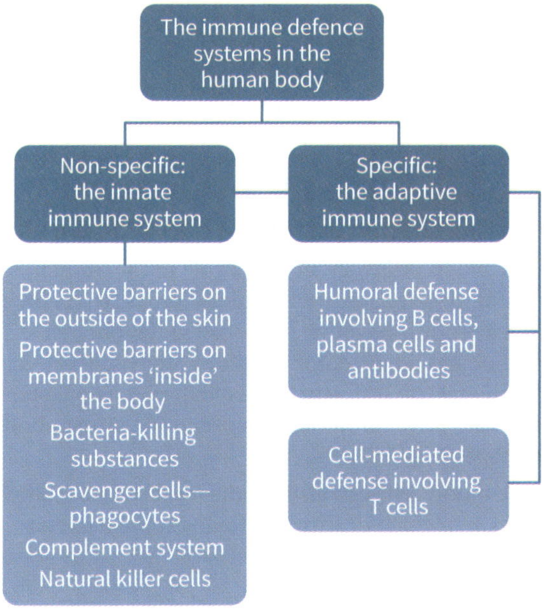

Figure 11.1 Model representing the innate and adaptive immune systems in the human body

The innate immune system in the human body

- In Chapter 10 you reviewed the innate immune system in plants and animals. The innate immune system is non-specific. Therefore its main role is to prevent pathogens from entering and spreading in the body. The system provides a front line of defence by engaging pathogens directly and acting immediately. It is unaltered in its ability to fight the same or different pathogens time and time again.
- The innate immune system in the human body is made up of:
  - the skin and all mucous membranes (epithelial surfaces), which form external and internal barriers (see Chapters 9 and 10)
  - defence cells formed from white blood cells (leukocytes) such as phagocytes.
  - various substances circulating in the blood and in body fluids, such as cytokines and antimicrobial peptides (defensins)

- complement system
- natural killer (NK) cells.

➔ If natural killer cells find cells with a changed surface, they dissolve them using cell poisons called cytotoxins. NK cells destroy virus-infected cells by inducing the infected cell to kill itself before the virus has had a chance to replicate.

complement system: a system of soluble proteins that activate one and other in a chain reaction to support (complement) defence mechanisms

natural killer (NK) cells: cells that specialise in identifying cells that are infected by a virus or have become cancerous by looking for changes in cell surfaces

➔ KEY QUESTION

1 **What is the innate immune system in the human body?**

Answer ➲ p. 156

Defence cells formed from white blood cells

➔ White blood cells play a critical role in the human immune system. Some white blood cells are phagocytes. Some are non-phagocytic cells called lymphocytes. These lymphocytes are receptor cells for antigens. T cells and B cells are the two different types of lymphocytes responsible for the adaptive immune response in the human body (see pages 151–152).

➔ White blood cells can also produce chemicals (called antitoxins) which destroy the toxins some bacteria produce when they invade the body.

➔ All blood cells originate in the bone marrow from a single unspecialised cell called a stem cell. The stem cell differentiates into white blood cells, red blood cells and platelets. White blood cells differentiate into five types of cells, including lymphocytes. Those lymphocytes that continue their development in the bone marrow are called B cells (B for bone marrow). Those that migrate in an immature state from the bone marrow and continue their development in the thymus are called T cells (T for thymus). The shape and structure of white blood cells and lymphocytes helps identify their core function (see Figure 11.2).

➔ The bone marrow produces and releases more white blood cells in response to infection.

➔ If a pathogen crosses the protective barriers and begins to replicate it can be eliminated directly by phagocytes. This is a localised response in the tissues. Phagocytes patrol the blood and tissues for foreign substances, including pathogens, and migrate to the site of the infection. The phagocytes are activated and begin their function of digesting and destroying the pathogen through a process called phagocytosis.

➔ Macrophages are the first phagocytes to encounter pathogens in the tissues but they are soon reinforced by neutrophils. Macrophages and neutrophils can recognise pathogens by means of cell-surface receptors that distinguish self from non-self.

➔ Phagocytes move by amoeboid action. They send out a protrusion that allows them to surround the invading pathogen(s) and engulf them. Phagocytes release digestive enzymes which break down the trapped pathogen(s) (see Figure 11.3).

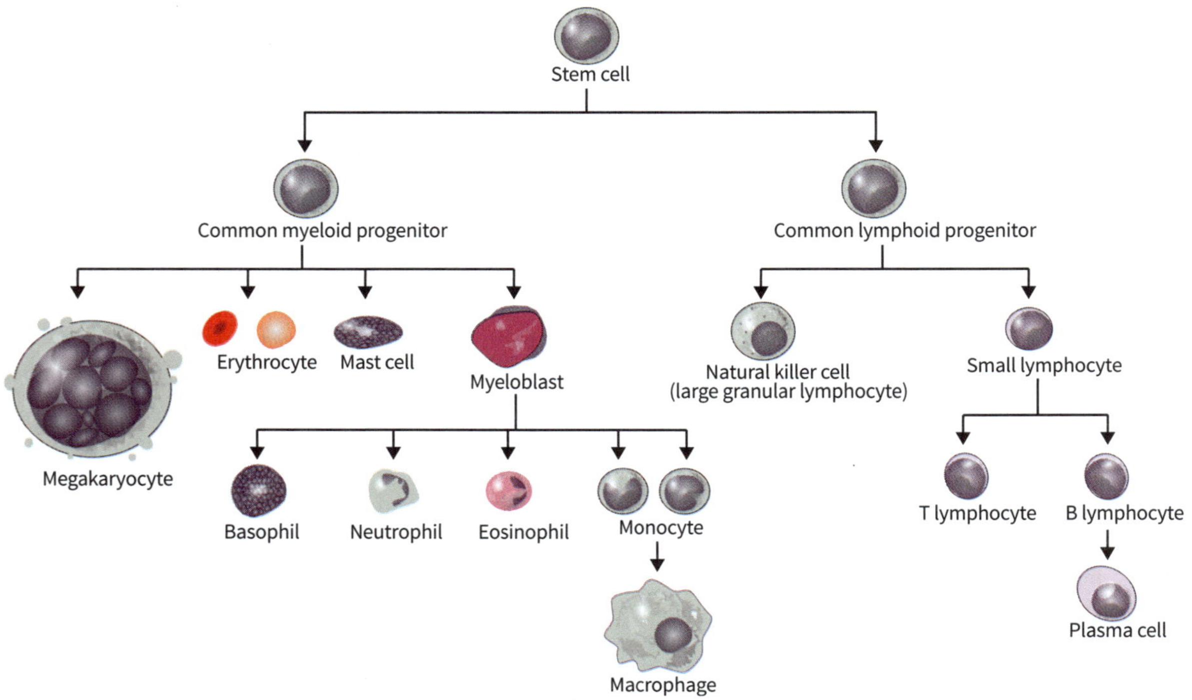

Figure 11.2 White blood cells and lymphocytes play key roles in the immune response to pathogens

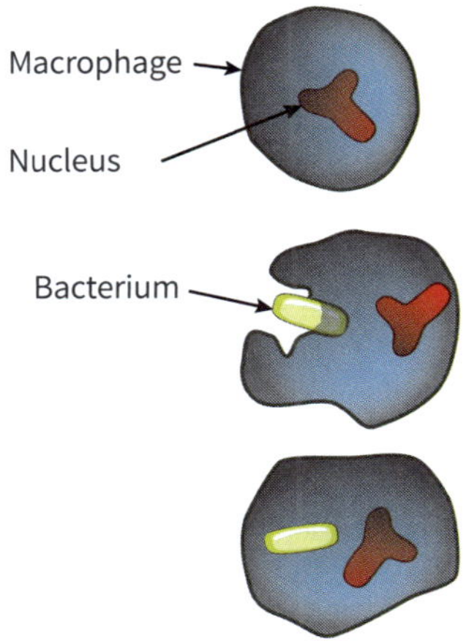

Figure 11.3 Diagram modeling the process of phagocytosis by a macrophage

- Neutrophils are short-lived cells that die soon after phagocytosis. Dead and dying neutrophils are a major component of the pus that forms in some infections.
- In Chapter 10 you reviewed the fact that the interaction between macrophages and pathogens activates the macrophages to release cytokines. The following section examines cytokines, along with defensins, in more detail.

Cytokines and defensins

- Cytokines are a group of proteins secreted by specific cells of the immune system (see Table 11.1). Cytokines include **interleukins** and **interferons**. Cytokines play a role in the innate immune system either by direct action, such as inhibiting viral replication, or indirectly by activating many other cells or stimulating them to grow and to mature. For example, cells that have been infected by a virus release interferons in order to protect neighbouring cells from the virus. The neighbouring cells bind interferons to their surface, which makes them less susceptible to the virus. Note: Cytokines also play a role in the adaptive immune system as hormone-like messenger substances.

interleukins: cytokines that can alter the behaviour of cells; produced by white blood cells

interferons: produced by cells infected with viruses, interferons trigger a series of cell responses to inhibit viral replication and activate natural killer cells and cytotoxic T lymphocytes

- Defensins are proteins; that is, defence peptides that act against bacteria, fungi and some viruses by disrupting the membranes of the pathogen. Defensins are produced by phagocytic cells and lymphocytes as well as the epithelial lining of the gastrointestinal and urogenital tracts. When defensins are produced by cells they signal the body to initiate, mobilise and amplify the adaptive immune response.
- If a pathogen is too large to be successfully phagocytosed, a group of macrophages and neutrophils will gather around the invader. They then secrete their defensins and toxins to destroy the pathogen.

Complement system

- The complement system is also rapidly activated in response to the presence of pathogens. The complement system enhances (complements) the immune defence mechanisms and provides circulating innate immunity.
- The system consists of about 20 interacting soluble proteins (enzymes) that are made by the liver and circulate in the blood and extracellular fluid. The different enzymes activate one another in a process similar to a chain reaction: one enzyme of the first stage alerts several enzymes of the second stage, each of which again activates several enzymes of the third stage, and so on. This is sometimes referred to as the complement cascade. The products of the complement reaction or cascade have a number of functions, including that they:

Table 11.1 Examples of cytokines involved in the immune response

Cells in the immune system that secrete the specific cytokine	The name of the cytokine	What does it do?
macrophages	interleukin-1	stimulates cells such as T cells to initiate inflammation; sends a message to the hypothalamus to increase body temperature
T cells	interleukin-2	causes proliferation of activated T and B cells; sends message for antibody synthesis
	interleukin-3	sends message for growth and differentiation of immune cells in bone marrow
	interleukin-4	promotes B cell growth and differentiation
	interleukin-5	sends message for differentiation of B cells
	interleukin-6 (also produced by macrophages)	stimulates T cells; sends message for growth in B cells
	Gamma-interferon (also produced by NK cells)	activates macrophages

- clear microbes and damaged cells
- disrupt membranes of microorganisms
- mark pathogens, making them more attractive for phagocytes
- attract macrophages and neutrophils from the blood
- activate local phagocytes and local mast cells that release granules containing histamine
- help trigger an inflammatory response
- dissolve the cell walls of bacteria so that they lose fluids and minerals, resulting in their death
- fight viruses directly by destroying the virus envelopes, or indirectly by destroying cells infected by viruses.

➔ It is important to note that the initial response of the immune system to exposure to a pathogen varies depending on the site which has been infected and on the type of pathogen that has invaded, as outlined below.
- Some bacteria (e.g. *Staphylococcus aureus*) produce chemicals when they enter the body to try to camouflage themselves against the immune system. The chemicals produced by the bacteria activate phagocytes that consume and destroy the invading bacteria.
- Some bacteria are recognised by the complement system, which produces the chemical messengers that warn other cells of the invasion.
- If the infection is in an area of the body mainly defended by macrophages, such as the lungs and intestines, then macrophages will be 'first on the scene'. The macrophages begin to digest the pathogen and, in doing so, stimulate other cells of the immune system into action.
- The pathogen may be first recognised by the acquired immune system (the lymphocytes). These cells will either directly fight the infection themselves or activate other immune system cells to do so (see page 153 for acquired immunity).

The adaptive immune system in the human body

➔ The innate immune response often succeeds in stopping the infection from being established.This can take a number of days. Sometimes the innate host defences are bypassed, evaded or overwhelmed. For example, the pathogen may spread through the lymph system or bloodstream. Some bacteria have ways to avoid being ingested by phagcoytes while others have evolved ways to grow inside macrophages (such as *mycobacteria*). In some cases the pathogen reproduces so rapidly that it may overwhelm the innate immune system.

➔ If the innate response fails to clear the infection, then a different response is needed and the adaptive immune response is required. Macrophages and other cells activated in the immediate innate response help initiate the adaptive immune response.

➔ Even though the adaptive response takes longer it has advantages over the innate system, including that:
- it targets the pathogen more accurately
- it can remember the pathogen and acts specifically against certain antigens (see page 153 for information on acquired immunity).

➔ The adaptive immune system is made up of antibodies (as soluble proteins in the blood), B cells and T cells.

Antibodies

➔ When an **antigen** (the pathogen) enters the body, the adaptive immune system produces **antibodies** against it. It is like a battle. The army (the body or antibody) tries to fight off the invader (the antigen). Each antibody has a unique binding site shape which locks onto the specific shape of the antigen. The antibodies destroy the antigen, which is then engulfed and digested by macrophages.

➔ You reviewed the antigen-antibody complex in Chapter 3 when you studied the structure and function of proteins (see Figure 3.5).

antigen: any substance, microbe or pathogen that is usually foreign to an organism's own body and thus triggers an immune response
antibody: a protein capable of interacting with a specific antigen

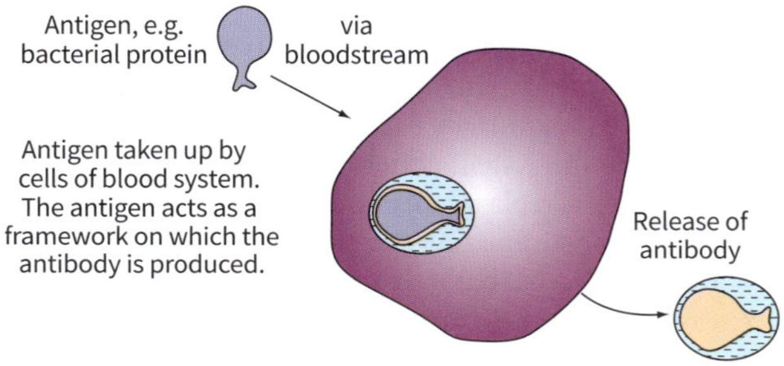

Figure 11.4 How antibodies are made

➔ Antibodies work in three ways by:
- neutralising the pathogen by either attaching directly to the surface through the antigen–antibody complex or by attaching to the toxins produced by bacteria
- activating other defense cells (e.g. phagocytes in the tissue and mast cells), which are responsible for many allergic reactions; phagocytes can fight pathogens better if they are packed with antibodies around them
- activating the complement system.

➔ There are different forms of antibodies in the human body, some of which are outlined below.
- ImmunoglobulinG (IgG)—antibodies produced mostly towards the end of a new infection; responsible for fighting already known pathogens; they activate the complement system and mark pathogens by attaching to them, making them palatable to the scavenger cells.
- Immunoglobulin M (IgM)—very big molecules capable of making infected cells clump together quickly; the first antibodies to be produced when a pathogen is identified in order to swiftly stop it from spreading.

- Immunoglobulin A (IgA)—specialists found in all body fluids (e.g. saliva, mucus in the bronchi or the bowel).
- Immunoglobulin E (IgE)—plays an important role in defending against infections by macroparasites; also responsible for many allergic reactions; attach to mast cells, which are responsible for typical allergic symptoms (e.g. itchy hives on the skin or asthma-like breathing difficulties).

➔ For more information see www.informedhealth.org/the-defense-mechanisms-of-the-adaptive-immune.2256.en.html.

➔ KEY QUESTIONS

2 **What are antigens?**

3 **What are antibodies?**

4 **What does the following model show?**

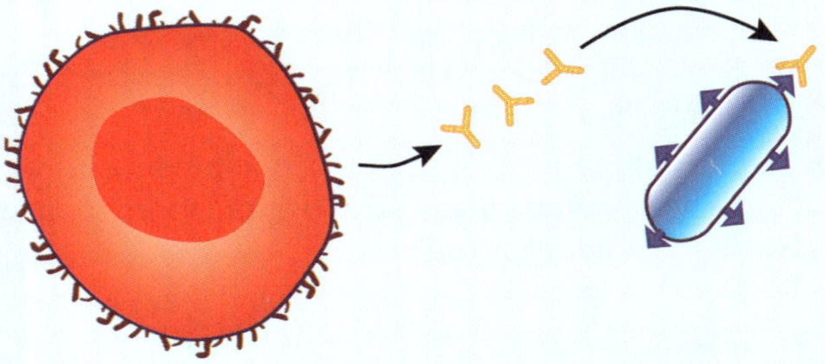

Figure 11.5 Sample model

Answers ➲ p. 156

B cells—B lymphocytes

➔ B cells develop in the bone marrow. They accumulate in the spleen and lymph nodes and circulate in the blood and lymphatic systems.

➔ The presence of an antigen stimulates the B cells to begin to divide to produce more B cells and specialist cells, called plasma cells. Plasma cells make large amounts of the specialised antibody and release them into the blood.

➔ There are no free antibody proteins circulating in the blood, only antibodies that are attached to the surface of B cells where they wait until needed. From the beginning of an infection until enough of the right antibodies are produced takes a week on average. This is the time when most people feel very unwell.

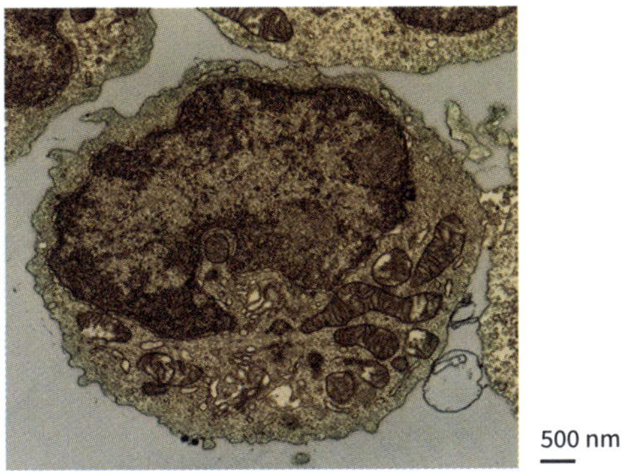

Figure 11.6 Human B lymphocyte

Source: image by NIAID via Wikimedia Commons

T cells—T lymphocytes

➔ T cells are lymphocytes that destroy antigens directly. T cells do not generate antibody-producing plasma cells.

➔ Immature T cells migrate from the bone marrow and develop in the thymus. They enter the bloodstream and are carried to lymph nodes and the spleen.

lymphokines: a subset of cytokines, they are protein molecules that signal between cells to stimulate an adaptive immune response

Table 11.2 Some examples of specialised T cells

T cells	Functions
helper T cells (also called T4 cells)	release certain messenger substances called **lymphokines** stimulate the production of antibodies by B cells and their dispersal through the bloodstream and therefore the entire body, meaning that the infection can be fought on many fronts
regulatory T cells (also referred to as suppressor T cells)	slow down the workings of the body's defense so that the immune system 'calms down' again after the infectious agent has been destroyed the regulatory T cells have the receptor molecule CD8 on their surface; together with the killer T cells they are called the T8 cells
killer T cells (also referred to as cytotoxic T cells) comes from the ancient Greek words *zyto*, meaning 'cell', and *toxikon*, meaning 'poison'	can directly destroy bacteria or cells that have been infected by a virus or have become cancerous identify pathogens only when they are 'presented' with the help of other defense cells in the body when killer cells are activated they release a substance that gets into the wall of the bacterium or of the infected cell and makes a hole in the wall; this results in the bacterium or infected cell, together with the pathogen, dying from the loss of fluids and electrolytes
memory T cells	remember information about the pathogen and have the ability to recognise the original invading antigen for a longer time; memory T cells can pass on this information immediately if there is another infection, resulting in a quick response (see page 153 for acquired immunity)

➔ T cells have characteristic structures on their surfaces that pathogens can bind to. A pathogen that exactly fits a T cell stimulates that T cell to multiply quickly and develop into specialised T cells (see Table 11.2). T cells need help from other defense cells because they can only identify a pathogen when it is bound to the surface of another defense cell.

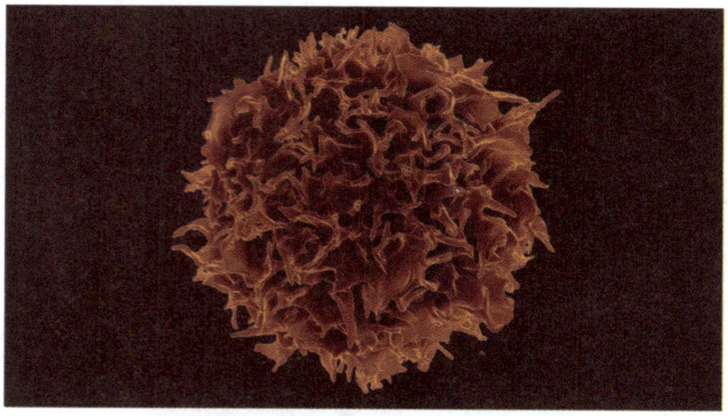

Figure 11.7 Human T lymphocyte
Source: image by NIAID via Wikimedia Commons

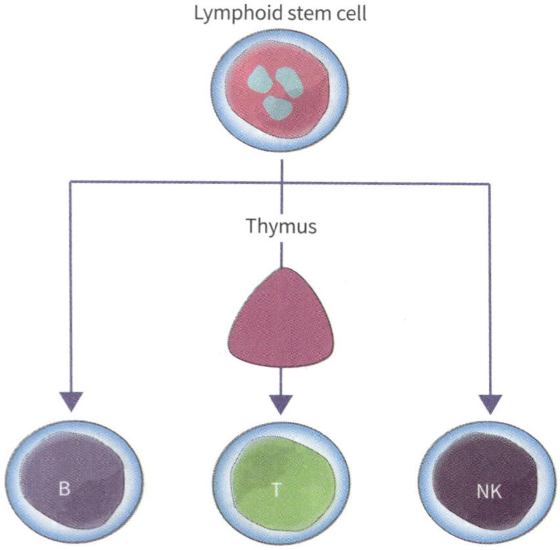

Figure 11.8 Development of B cells, T cells and NK cells

➔ KEY QUESTIONS

5 **What are B cells and what is their role in the human immune system's response to a pathogen?**

6 **What are T cells and what is their role in the human immune system's response to a pathogen?**

Answers ➲ p. 156

Humoral and cell-mediated immune responses

➔ The adaptive immune system has several parts that react in different ways, depending on the place in the body where the pathogen is and the type of foreign substance or cell.

➔ Antibodies are made available for a pathogen outside the cells in the blood and in body fluids (an extracellular infection). This is called the **humoral** immune response. To eliminate pathogens that are inside tissues (an intracellular infection) a **cell-mediated** immune response is needed.

➔ The humoral immune response involves B cells that recognise antigens and pathogens. The steps in the response are:
- the antigen binds to B cells
- interleukins or helper T cells activate B cells
- B cells **proliferate** and produce plasma cells, which make antibodies; the antibodies are released and circulate through the body to antigens
- B cells produce memory cells.

➔ The cell-mediated immune response involves mostly T cells. Cell-mediated immunity does not involve antibodies. Rather, it involves the activation of T lymphocytes and the release of various cytokines in response to an antigen. The steps in the response can go one of two ways. The first is that the:
- antigens bind to T cells
- interleukins activate T cell function
- T cells proliferate, producing cytotoxic T cells which destroy the antigens.

➔ The second possibility is that the T cells produce helper T cells which release interleukins, which in turn stimulate B cells to produce antibodies.

humoral: substances such as antibodies found in the body fluids; comes from the Latin *humor* meaning 'fluid'
cell-mediated: concerning T lymphocytes (T cells) rather than antibodies
proliferate: to increase rapidly in number

➔ KEY QUESTIONS

7 **The model below is a representation showing how T cells and B cells can interact to produce antibodies. Interpret the model to explain, in your own words, Mechanism 1 and Mechanism 2.**

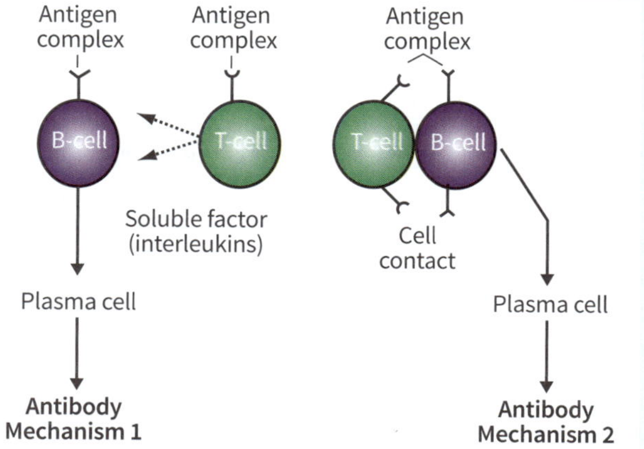

Figure 11.9 Model showing how T cells and B cells can interact to produce antibodies

→ KEY QUESTIONS (continued)

8 **What are the adaptive immune responses in the human body?**

9 **Explain one way you could model the innate and adaptive immune systems in the human body.**

10 **How does the human immune system respond to exposure to a pathogen?**

Answers ➲ p. 156

2 How the immune system responds after primary exposure to a pathogen

» Students explain how the immune system responds after primary exposure to a pathogen, including innate and acquired immunity.

→ **Innate immunity** is naturally present. This type of immunity depends on variations in a species. That is, some people are naturally more resistant to specific diseases. It does not involve a response to an antigen and does not confer long-term immunity.

→ In Section 1 you reviewed B cells, antibodies and the fact that antibodies are disease (pathogen) specific. **Acquired immunity** to a disease is achieved through the adaptive immune system and antibodies, B cells and T cells. It can confer long-term immunity. The adaptive immune system can remember the antigen because it produces memory cells. This is also the reason why there are some illnesses a person can only get once in their life, because afterwards their body becomes immune. After first (primary) contact with the pathogen (antigen), it takes several days for the adaptive immune system to respond. After a second infection with the same pathogen the adaptive immune system will recognise it and immediately release the antibodies needed to fight it. Usually there are no symptoms or, if there are, they will be weaker.

→ Acquired immunity is active or passive and natural or artificial (see Figure 11.10).

→ Active immunity results when exposure to a disease triggers the immune system to produce antibodies. Exposure can occur through:
- infection with the actual disease (natural immunity)
- **vaccination**, referred to as vaccine-induced or artificial immunity (you will review artificial immunity through vaccination in Chapter 12).

→ Active immunity is long lasting and sometimes lifelong.

→ Passive immunity is provided when a person is 'given' antibodies. Examples include:
- IgG antibodies can be passed on from the mother to the unborn child; after birth this protects the child from many pathogens until its own immune system has matured
- if a baby is breastfed, IgA antibodies pass from the mother's milk to the infant; they are not ingested by the baby but line internal surfaces, such as the intestine, and help block the entry of pathogens
- a person may have been given immunoglobulin for protection from a specific disease; immunoglobulin replacement therapy or antibody serum injections are for people with primary immunodeficiency disease who have inadequate antibodies to fight infection; protection is immediate but only lasts for a few weeks.

innate immunity: immunity that is naturally present rather than due to prior exposure to an antigen

acquired immunity: immunity through exposure to a specific antigen (e.g. through an infection or vaccination)

vaccination: the ingesting or injection into the body of antigens from living, dead, weakened or non-virulent strains of microorganisms

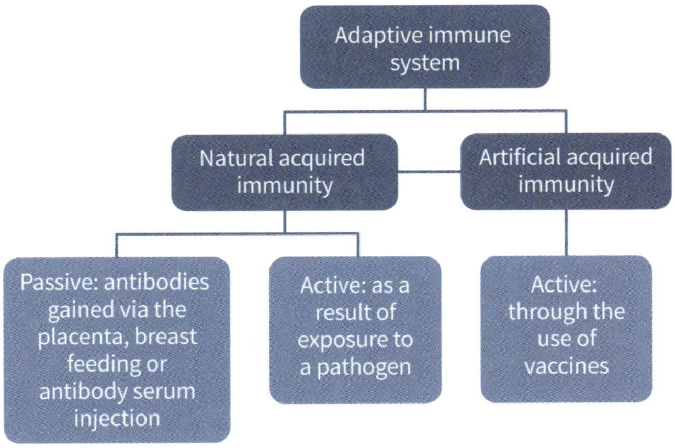

Figure 11.10 Acquired immunity represented as a model

→ KEY QUESTIONS

11 **Explain how the human immune system responds after primary exposure to a pathogen.**

12 **Explain how the human immune system responds after secondary exposure to the same pathogen.**

Answers ➲ p. 156

CHAPTER SYLLABUS CHECKLIST

Are you able to answer these questions from the syllabus for this chapter? Tick each question as you go through the checklist if you are able to answer it. If you cannot answer a question, turn to the relevant page in the study guide to find the answer. For NESA key word meanings, go to www.educationstandards.nsw.edu.au and search 'key words'.

	FOR A COMPLETE UNDERSTANDING OF THIS TOPIC:	PAGE NO.	✓
1	Can I investigate and model the innate and adaptive immune systems in the human body?	147–153	
2	Can I explain how the immune system responds after primary exposure to a pathogen, including innate and acquired immunity?	153	

HSC EXAM-TYPE QUESTIONS

Objective-response questions (1 mark each)

1 **Which one of the following is a correct statement about the adaptive immune system in humans?**

A The adaptive immune system provides non-specific defence against foreign bodies and includes defence cells such as B lymphocytes.

B The adaptive immune system provides general defence and includes scavenger cells.

C The adaptive immune system provides specific defence and includes defence cells such as T lymphocytes.

D The adaptive immune system provides specific defence and includes the washing action of cilia.

2

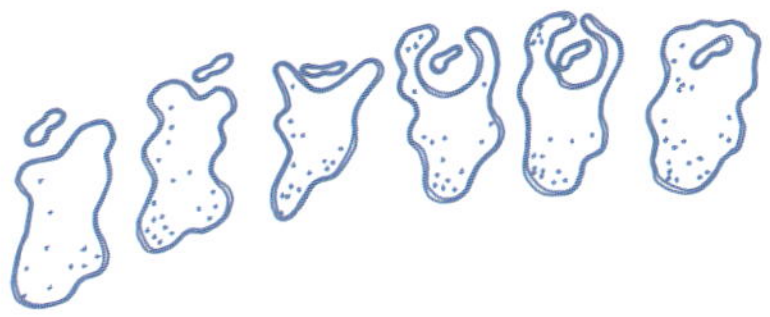

Figure 11.11 Model of part of the defence adaptations in the human immune system

Figure 11.11 above shows part of the defence adaptations in the human immune system. What process is represented by the diagram?

A lymphocytosis

B phagocytosis

C pathogenicity

D cell toxicity

3 **An antigen can be defined as:**

A a part of the lymph system involved in fighting disease.

B proteins produced by the body that help to destroy pathogens.

C the product of mucous membranes which help to trap pathogens.

D a protein that the body recognises as non-self and therefore triggers an immune response.

4 **A key role of the B cells in the immune response is the production of which of the following?**

A antigens

B antidotes

C antibiotics

D antibodies

5 **To vaccinate against diphtheria an inactive form of the toxin produced by the bacteria that causes the disease is used. This provides the basis for the formation of antibodies along with memory B cells and T cells. What does this provide an example of?**

A active acquired immunity

B passive acquired immunity

C innate immunity

D natural immunity

Short-answer questions

6 **The term phagocyte comes from the Greek *phagein*, meaning 'to eat'. From your understanding of the role of phagocytes in the human immune system, explain why this is an apt term.** (2 marks)

7 **When the immune system is weakened by the HIV virus, it is mostly a specific group of T cells that is reduced in number (the so-called T helper cells). The lack of T cells means the immune system has difficulty fighting pathogens. The body is especially susceptible to infections by what are usually harmless pathogens. For example, people who have immune deficiency AIDS often develop a streptococcal bacterial inflammation of the lungs or get fungal *Candida* infections.**

From your understanding of diseases and the role of T cells as part of the human body's response to disease, explain why people with HIV/AIDS are prone to infections by what are normally harmless pathogens. (2 marks)

8 **Briefly explain how the acquired immune system in humans responds after primary and secondary exposure to pathogens.** (3 marks)

9 **Construct a flow chart to summarise the difference between cell-mediated and humoral responses. Include the role of B cells and T cells in your answer.** (6 marks)

10 **Use the following analogy to answer Question 10.**

In a YouTube clip from vcebiology.edublogs.org the immune system response is modelled using the analogy of invaders on the Great Wall of China. The first line of defence is like towers and a moat preventing invasion. If the invaders breach the first line of defence, there is a heated battle where invaders are indiscriminately killed. Next are the trained killers that recognise their targets and actively seek them out and destroy them.

a **Use the analogy to summarise how the human immune system responds to exposure to a pathogen.** (2 marks)

b **Construct a table to summarise how the human immune system responds to exposure to a pathogen.** (2 marks)

ANSWERS

KEY QUESTIONS

Key question ⊃ p. 148

1 The innate immune system in the human body is the general defense system. It is not specialised for specific pathogens and therefore does not need a long start-up phase. It is primed and ready to go. It has a broad effect that aims at stopping pathogens from first entering the body and then, if they do enter, from spreading. During the first critical hours or days of infection humans rely on the innate immune system.

Key questions ⊃ p. 151

2 Antigens are protein molecules usually found on the surface of pathogens. Antigens are recognised by the host as non-self. The presence of antigens triggers an immune response through the production of antigen-specific antibodies.

3 Antibodies are proteins produced by the body in response to the presence of a specific antigen. By joining with the antigens and clumping them together the antigens can be more easily destroyed by macrophages. Some antibodies can neutralise toxins produced by antigens.

4 The red colour is a protein antibody. The Y-shaped end of the antibody (yellow colour) recognises and binds to the antigen (blue colour) on the surface of the specific pathogen.

Key questions ⊃ p. 152

5 In the adaptive immune system, B lymphocytes (B cells) are a type of white blood cell that develop in the bone marrow but accumulate in the spleen and lymph nodes and, in the presence of antigens, divide to produce plasma cells that make antibodies. Some B cells are stored as memory cells in the lymph nodes.

6 In the adaptive immune system, T lymphocytes (T cells) are a type of white blood cell capable of recognising self from non-self cells. T cells have been sensitised by the thymus. T cells destroy an antigen directly or attract phagocytes and macrophages to destroy it. T cells interact with B cells to stimulate the production of antibodies.

Key questions ⊃ p. 152–153

7 Mechanism 1: The T cells produce a soluble factor (a cytokine) after interaction with an antigen. The B cells react with the soluble factor and the specific antigen to produce plasma cells that make an antibody for that antigen.

Mechanism 2: Close contact between the T cells and B cells comes about because of interaction with the antigen. This contact allows the T cells to signal the B cells to produce plasma cells that make an antibody for that antigen.

8 Adaptive immune responses to specific antigens include the humoral immune response through the production of antibodies by B lymphocytes and the provision of cell-mediated immunity by T lymphocytes. In both cases memory cells are produced that confer long-term immunity to the specific antigen.

9 Modelling the innate and adaptive immune systems in the human body could involve flow charts and dichotomous diagrams to summarise the two systems and/or diagrams such as Figure 11.1 to show how specific parts of the systems work.

10 The human immune system has a diverse set of mechanisms to respond to exposure to a pathogen. The first is the innate immune system. These mechanisms are preformed and ready to protect against and respond to invasion by a pathogen at any time. This immediate, non-specific response includes activation of phagocytes such as macrophages and the triggering of an inflammatory response. If the pathogen is not destroyed by the innate system then the adaptive immune system takes action. This system targets the pathogen more accurately. The adaptive immune system remembers the pathogen and acts specifically against certain antigens. If there is new contact with an antigen that is already known, the defense response can then be quicker. The adaptive immune system is made up of antibodies, B cells and T cells.

Key questions ⊃ p. 153

11 Before primary exposure to a pathogen there is a negligible amount of antibody in the immune system and no memory cells for the pathogen. Once the body is exposed to antigens, through disease or vaccination, the B cells are activated. They divide to produce plasma cells and memory cells. Plasma cells produce antibodies. As the number of plasma cells increase, the concentration of antibodies increases. Once the pathogen has been destroyed, antibody production decreases as fewer or no antigens are present. The memory cells are still present.

12 If the body is exposed to the same pathogen again (called secondary exposure), memory T cells activate the memory B cells, which quickly differentiate into antibody secreting plasma cells which speedily produce large amounts of antibody. There is a higher peak of antibody concentration due to more plasma cells secreting antibodies. Once the infection has been dealt with, antibody production slows. Eventually the antibody decreases as fewer antigens are present. A higher level of immunity is gained as more memory B and T cells are formed.

HSC EXAM-TYPE QUESTIONS

Objective-response questions

1 **C**. The adaptive immune system provides specific defence and includes defence cells such as T lymphocytes. **A** and **B** are incorrect because the adaptive immune system is specific, not non-specific or general defence. **D** is incorrect because even though the adaptive immune system provides specific defence, it is the innate immune system that includes the washing action of cilia.

2 **B**. Phagocytosis occurs when the phagocytic cell encloses the pathogen and digests it in its interior. **A** is wrong because lymphocytosis refers to an increase in lymphocytes. **C** is wrong because pathogenicity is the ability of an organism (pathogen) to cause disease. **D** is incorrect because cell toxicity means being toxic to cells.

3 **D**. An antigen is a protein found on the surface of a pathogen. It is recognised as non-self and triggers the body to produce antibodies. **A** is incorrect because an antigen is part of the pathogen. **B** is wrong because this refers to antibodies. **C** is wrong because again this refers to how the body responds to pathogens.

4 **D**. B cells produce plasma cells, which produce antibodies. **A** is incorrect because antigens refer to the pathogens. **B** is wrong because antidotes are medicines to counteract the effects of poisons (e.g. snakebite antidotes). **C** is wrong because antibiotics are substances that can kill microorganisms.

5 **A**. Vaccination is active acquired immunity. **B** (passive acquired immunity) is an incorrect alternative. **C** is incorrect because vaccines bring about acquired, not innate, immunity. **D** is wrong because vaccination is artificial, not natural, immunity.

Short-answer questions

6 **EM** It is important when answering this question to relate the meaning 'to eat' to the process of phagocytosis.

'To eat' means to take in and ingest. (✓) The word is apt because phagocytes engulf, digest and destroy a pathogen through a process called phagocytosis. (✓)

7 **EM** This question asks you to bring together some of your understandings from both this chapter and Chapter 9. It is important to make the link between reduced numbers of T cells, reduced ability to produce antibodies, and the development of infections from normally harmless pathogenic microorganisms.

The HIV virus infects helper T cells. This is why in AIDS, the helper T cells are affected and reduced in number. (✓) Helper T cells enhance antibody production by B cells. Due to the lack of T cells, the immune system has difficulty fighting against the pathogens, including usually harmless microorganisms. (✓) For example, streptococci can live naturally in the nose and throat of healthy people (✓) and *Candida albicans* is a natural microorganism of the human body. (✓)

8 **EM** This is a recall question. Make sure your answer includes responses to primary and secondary exposure.

Acquired immunity means that once a person has had a disease they do not normally catch it again because during primary exposure to a pathogen (antigen) the body produces memory cells specific to that antigen. (✓) The memory cells remember the pathogen which caused the disease and rapidly make the correct antibody if the body is exposed to infection again (secondary exposure). (✓) The pathogen is quickly destroyed, thereby preventing symptoms of the disease. (✓)

9 **EM** This is a high-level question that asks you to make a model of the adaptive immune system using a flow chart. Marks are given for the overall flow diagram and the correct information under the cell-mediated and humoral sections.

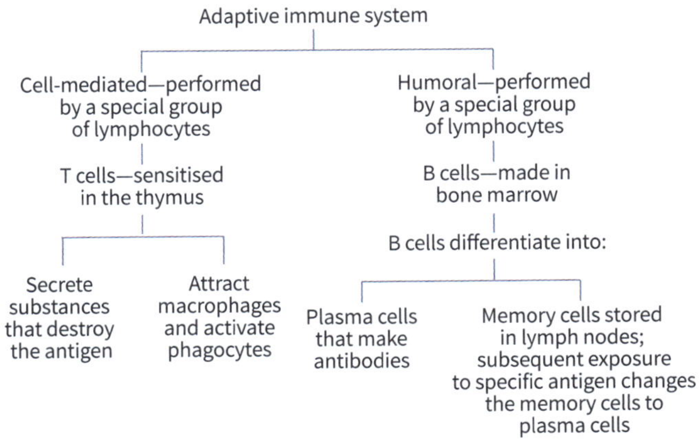

Figure A11.1 Adaptive immune system

10 **EM** The answer to this question has a number of parts. First you must explain the analogy (or model) and relate it to the innate and adaptive immune systems. Then you need to use a table to summarise the innate and adaptive immune systems.

The first line of defence is the innate immune system. This includes physical barriers to prevent entry of the pathogen (the moat and towers that surround the Great Wall represent these physical barriers). (✓)

The second line of defense is the heated battle. This is also the innate immune system's response. If the pathogen crosses the protective barriers and enters the host a local inflammatory response is triggered. (✓) The inflammatory response causes a number of physical changes, including increased temperature of the tissues around the site of the invasion (infection). (✓) **EM** The clues here are the words *heated* in 'heated battle' and *indiscriminately* in 'indiscriminately killed'.

The third line of defense in the analogy represents the adaptive immune system response, which is specific and targeted with defence cells such as killer T cells. (✓) **EM** The clue here is the word *specific* in 'specific target'.

The table is worth four marks. One mark is awarded for headings (✓) while supplying the correct information will result in 3 marks.

Table A11.1 How the human immune system responds to exposure to a pathogen

Innate immune system—non-specific		Adaptive immune system—specific and targeted
Physical and chemical barriers	**'Internal' defences**	
skin and mucous membranes defensins produced by mucous membranes	phagocytes inflammatory response complementary system natural killer cells	B cells, antibodies and the humoral response T cells and the cell-mediated response

CHAPTER 12 PREVENTION, TREATMENT AND CONTROL

INQUIRY QUESTION:

How can the spread of infectious diseases be controlled?

In Chapter 9 you reviewed the causes of infectious diseases and how they are transmitted. An infectious disease occurs when a pathogen succeeds in evading or overwhelming the innate immune defences to establish a site of infection where it then reproduces or replicates, allowing for further transmission.

Prevention of infectious diseases refers to the range of measures that aim to reduce the risks of pathogens or stop infections in the first place. Treatment of a disease depends on a range of factors that include the type of pathogen, its virulence, the incubation period, the severity of the disease and the drugs available. Control usually refers to procedures employed to contain and stop the spread of the disease.

1 Investigating and analysing the range of interrelated factors involved in limiting local, regional and global spread of a named infectious disease

» Students investigate and analyse the wide range of interrelated factors involved in limiting local, regional and global spread of a named infectious disease.

- Prevention, treatment and control of infectious diseases are interrelated. Using influenza as an example, Figure 12.1 represents a model that shows the interrelationship between prevention, treatment and control.
- In Figure 12.1 the major control measure for prevention of influenza is vaccination. **Antiviral** therapy is used as a treatment strategy as well as a prevention strategy. **Quarantine** is a control measure to prevent the spread of influenza.

antivirals: manufactured drugs that either inhibit a virus's ability to replicate or strengthen the host body's immune response to the viral infection
quarantine: from the Italian word *quarantina* meaning 'forty days', this refers to a range of measures (e.g. strict isolation) that aim to control the spread of disease

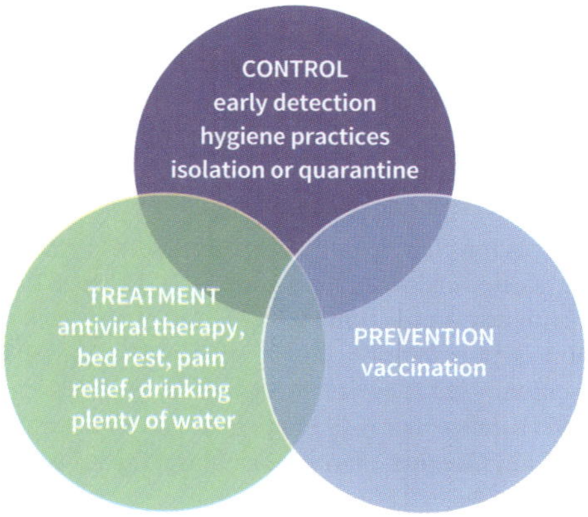

Figure 12.1 Model representing the interrelationship between prevention, treatment and control measures of a disease

- Some reliable and authoritative sources for investigations are listed below:
  - the NSW Health guidelines on disease control, treatment and prevention at www.health.nsw.gov.au
  - the Office of Health Protection, Department of Health at www.health.gov.au/internet/main/publishing.nsf/content/ohp-about.htm
  - a National Framework for Communicable Disease Control at www.health.gov.au/internet/main/publishing.nsf/Content/ohp-nat-frame-communic-disease-control.htm
  - Australian guidelines for prevention and control of infection in health care at www.nhmrc.gov.au/guidelines
  - the World Health Organization at www.who.int.
- In Chapter 9 you reviewed modes of transmission and interrelated factors involved in the spread of infectious diseases. Analysis of these factors enables prediction of the potential for an outbreak as well as evaluation of strategies to control the spread.
- The named infectious disease in this section is human seasonal epidemic and pandemic influenza, a highly contagious viral disease mainly affecting the respiratory system. Note that you may have studied a different example.
- Influenza is a global disease that does not respect borders. An estimated 10–20% of the world's population is affected each year by seasonal epidemic influenza. Influenza pandemics come about through changes in the influenza antigen, resulting in a new virus that is not related to previous human seasonal influenza viruses.

➔ This means there is a lack of **immunity** in the population to the pandemic virus. Influenza pandemics and epidemics can have substantial health and socio-economic impacts. Globalisation and increased travel and trade within and between countries have resulted in the rapid spread of influenza. For example, Severe Acute Respiratory Syndrome (SARS) spread rapidly across three continents within weeks in 2003 and then globally as A/H1N1 influenza in 2009. Between 2006 and 2013 (excluding the 2009 pandemic) more than 80 deaths and approximately 4800 hospitalisations occurred annually in Australia alone.

➔ To understand how to limit the spread of a disease such as influenza it is important to understand a number of factors as outlined below:

- How it is transmitted—the main mode of transmission for influenza is direct spread through respiratory droplets although it can also be spread indirectly through contact with respiratory mucous or contaminated **fomites** such as hands or towels. Droplets only travel a short distance through the air so close contact is needed for direct transmission.
- Where the virus replicates—upper respiratory tract infections seem to be the main site of infection for seasonal and pandemic influenza. This means that the virus is more likely to being expelled and circulated through coughing or sneezing.
- The type of setting where spread is most likely to occur—settings with poor hygiene practices or closed environments with close contact; for example, nursing homes and healthcare facilities, schools, refugee camps, prisons, aeroplanes and cruise ships are places where a disease like influenza can easily and rapidly spread.
- Individual susceptibility to the disease—influenza is most likely to affect children, especially those less than two years old; people aged 65 years and older; individuals with pre-existing medical conditions such as **chronic** lung disease; and, with some strains, pregnant women. Pre-existing immunity in the population is also a factor.
- Transmission at different stages of infection—influenza is most infectious during the first three days of illness.
- Seasonality of infection—influenza epidemics exhibit clear seasonal patterns. The disease occurs seasonally during winter months in temperate regions such as Australia but continually circulates in the tropics, especially in Southeast Asia.

➔ Preparedness, early detection, containment at its source and disease outbreak responses are some of the factors involved in limiting the spread of seasonal and pandemic influenza at the local, regional and global level (see Table 12.1).

Table 12.1 Interrelated factors involved in limiting local, regional and global spread of seasonal influenza

Local	Regional	Global
limiting person-to-person transmission through personal hygiene measures such as covering the mouth and nose when coughing or sneezing, and washing hands with soap and water; using personal protective equipment such as masks, gloves and gowns; isolating individuals and social distancing	limiting person-to-person transmission through border protection and entry screening although this may have limited effectiveness, is difficult to implement and requires extensive resources	limiting person-to-person transmission through travel bans or restrictions and screening of travellers exiting areas with the disease
public health and community-level measures such as early detection; public health policies for tracing close contacts and infection control guidelines; closing schools and workplaces; public education; antiviral medication and vaccination to protect healthcare workers; emergency preparedness for emerging (new) strains	public health and community-level measures such as antiviral medication; **surveillance**; area quarantining; regional networking and collection of data; promoting prevention and control practices within communities; infection control guidelines; emergency preparedness for emerging (new) strains	public health and community-level measures such as surveillance; international networking; redistribution of antiviral stockpiles to countries lacking antivirals; infection control guidelines; emergency preparedness for emerging (new) strains

Source: World Health Organization, *Limiting the spread of pandemic, zoonotic and seasonal epidemic influenza*, 2010

immunity: the state of protection that occurs when a person has been vaccinated or has had an infection and recovered (see Section 2 for more details)

fomites: objects or materials that carry infection

chronic: a wide group of conditions, illnesses and diseases, generally characterised by their long-lasting and persistent effects as distinguished from short-term, intense acute conditions

surveillance: early detection and the collection, analysis and dissemination of data; surveillance data is used in planning interventions and public health campaigns (see Section 6 for more details)

➔ KEY QUESTIONS

1. **Use the model in Figure 12.1 to summarise the prevention, treatment and control of an infectious disease you have studied.**
2. **List some of the worldwide challenges and issues related to the prevention, treatment and control of infectious diseases.**

Answers ➲ p. 174

2 Investigating procedures that can be employed to prevent the spread of disease

» Students investigate procedures that can be employed to prevent the spread of disease, including but not limited to hygiene practices; quarantine; vaccination, including passive and active immunity; public health campaigns; use of pesticides; and genetic engineering.

Hygiene practices

- Effective hygiene practices prevent and control the spread of disease by reducing the chance of fomites or food becoming contaminated and infectious diseases being contracted. Practices include personal hygiene, such as washing hands with soap and water frequently; staying home if sick; when sneezing or coughing turn the head away and cough into the arm rather than the hands; using single-use tissues and disposing of the tissue immediately; washing hands after coughing, sneezing or using tissues; not touching eyes, nose or mouth and not sharing cups, glasses, dishes or cutlery.
- Hygienic cleaning practices include using separate cleaning cloths for wiping surfaces, food preparation and drying hands. Proper handling, preparation and cooking of food and proper disposal of contaminated materials also help prevent the spread of disease.
- Clinical settings such as hospitals, aged-care facilities and prisons often have an increased incidence of infectious diseases. Higher rates of transmission are usually caused by characteristics of the environment itself and the high concentration of individuals more susceptible to disease. Extra special efforts must be taken in these settings to limit the risks of infection, including training of healthcare workers and strict hygiene practices.

Quarantine

- As an island Australia has some advantage over many other countries because its borders are easier to protect against disease-carrying vectors, materials, humans and other organisms. However, as global trade and air travel become more prevalent it is increasingly important for Australia to protect its population, agricultural industry and unique natural environment through continued quarantine measures.
- At regional and global levels, limiting or controlling human mobility through travel restrictions and quarantine is one of the first measures applied by national and international public health institutions in order to prevent the spread of infectious diseases, especially in the case of a pandemic or an outbreak caused by a new pathogen.
- Quarantine includes surveillance, monitoring, screening and clearance procedures at airports and shipping ports. Quarantine policy is determined through bilateral and multinational negotiations and, in Australia, involves protection of Australia's animal and plant health status, Australia's international obligations, the trade impact of quarantine policies and environmental protection.
- Australian **biosecurity** operates under legislation called the *Biosecurity Act 2015*, which replaced the *Quarantine Act 1908*. Listed human diseases under the Act include human influenza with pandemic potential, plague, SARS, Middle East respiratory syndrome, smallpox, viral haemorrhagic fevers and yellow fever. See the Australian Government Department of Agriculture and Water Resources website at www.agriculture.gov.au/biosecurity. Note: The glossary on this site is a valid and reliable source for your investigations.
- Each state or territory has their own regulations in place to prevent the introduction of infectious diseases; for example, New South Wales is currently free of potato cyst nematodes and it is illegal to move potatoes, soil and associated packaging or machinery into NSW unless certified as having met specific import conditions.

KEY QUESTIONS

3 **How do hygiene practices help prevent the spread of disease? Provide one example to illustrate your answer.**

4 **How does quarantine help prevent the spread of disease? Provide one example to illustrate your answer.**

Answers p. 174

Vaccination, including passive and active immunity

- In Chapter 11 you reviewed passive and active acquired immunity. Active immunity results when exposure to a disease triggers the immune system to produce antibodies. Active immunity is long lasting and sometimes lifelong. Artificial active acquired immunity comes about through vaccination. Passive immunity is provided when a person is 'given' antibodies (see Figure 11.10).
- Vaccination is a procedure that uses **vaccines** to prevent diseases in populations. This is the key purpose of **immunisation** programs.

biosecurity: measures designed to protect populations, the economy and the environment from harmful biological and chemical substances

vaccine: a substance that is introduced into the body to stimulate the body's acquired immune response; it is given to prevent an infectious disease from developing and the person becoming ill

immunisation: the process of using vaccines to protect against illnesses caused by infection

- An effective vaccine protects an individual against a specific infectious disease and its various complications. Some vaccines only require a single vaccination to provide lifelong protection, such as the hepatitis B vaccine. Other vaccines require booster doses to maintain protection. These include the vaccines for whooping cough (pertussis), tetanus and polio along with the pneumococcal and meningococcal vaccines.
- Figure 12.2 below shows the effect of giving booster doses of vaccines. After the first immunisation of a non-immune person a small and brief response occurs. When booster doses are given B cells created during the initial response are switched on to generate a much more rapid and longer-lasting protective response.

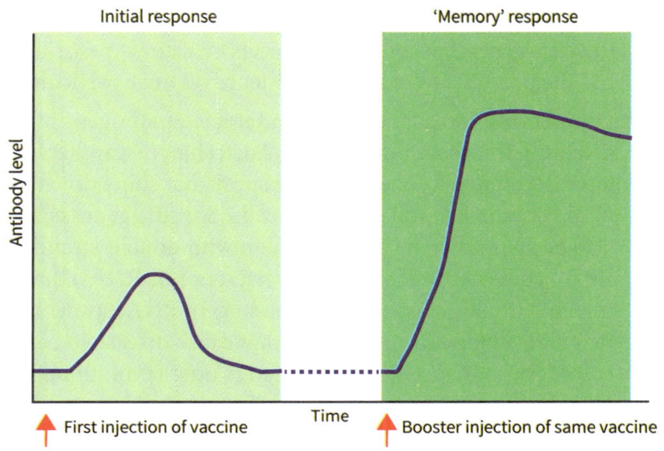

Figure 12.2 Effect of giving booster doses of vaccines; the figure shows the levels of antibodies from B cells after first and booster vaccinations.

Source: www.science.org.au/learning/general-audience/science-booklets/science-immunisation/3-who-benefits-vaccines

- Vaccines are very effective in stopping infections and this reduces the need for antibiotics or antivirals which, in turn, help limit the spread of antibiotic resistance.
- All vaccines currently in use in Australia confer high levels of protection that are sufficient to prevent disease in the great majority of vaccinated individuals and in the wider community (see herd immunity below).
- The effectiveness of vaccines in preventing infectious diseases is illustrated by the disappearance, or near disappearance, in Australia of deaths from diphtheria, whooping cough, tetanus, polio and measles, and more recently from cases of Haemophilus influenza type B and meningococcal type C infections. The influenza vaccine reduces the risk of infection by 50–60% in adults and children.
- Vaccines are also responsible for the prevention, control and even eradication of some infectious diseases that were once globally common.
- Vaccines generally have two major types of ingredients: antigens and **adjuvants**. Antigens are designed to cause the immune system to produce antibodies and/or T cells against a specific pathogen or its toxin. Adjuvants amplify immune responses more generally by targeting the defence cells and chemicals of the innate defence system. In doing so they may cause mild local reactions (soreness, redness and swelling) at the injection site. These reactions are a healthy indicator of the strength of the immune response. An example of a vaccine that contains an adjuvant is the childhood diphtheria-tetanus-pertussis vaccine.
- Vaccines have the potential to treat rather than prevent disease. For example, most people with HIV need to take antiretroviral drugs for the rest of their lives to stop the virus from replicating and causing damage to the immune system. These drugs are expensive and can cause nasty side effects. Researchers are developing vaccines that they hope will stop the virus replicating without the use of antiviral drugs.
- Vaccines not only protect the individual but can also provide **herd immunity**. Herd immunity comes about when a high proportion of the population has been immunised and has acquired immunity. If this proportion is high enough then the small number of people who have no immunity will be protected as there are not enough susceptible people to allow ready transmission of the pathogen. The percentage of a population that needs to be immune for herd immunity to take effect is high. For example, for polio it is 80–86%, for diphtheria 85% and for measles more than 95%.
- References for investigations include the following:
  - the Australian Academy of Science website at www.science.org.au/learning/general-audience/science-booklets/science-immunisation/
  - the National Centre for Immunisation Research and Surveillance (NCIRS) for fact sheets on vaccine preventable diseases at www.ncirs.edu.au.

adjuvants: chemicals such as alum that are added to vaccines to boost the immune response and stimulate stronger, longer-lasting immune responses
herd immunity: when a significant proportion of individuals within a population are protected against a disease through immunisation

KEY QUESTIONS

5 **What is vaccination and how does it help prevent the spread of disease?**

6 **What is the difference between active and passive immunity?**

7 **How do vaccines work?**

8 **Define the term *immunisation* and outline its purpose.**

Answers ⊃ p. 174

Public health campaigns

- **Public health campaigns** are focused on prevention and control through strategies such as:
  - health promotion information focusing on behavioural risk factors relating to sexually transmitted diseases such as HIV/AIDS, intravenous drug use and measures to reduce risk of infection.
  - immunisation campaigns such as the New South Wales 'Save the date to vaccinate' campaign that includes television advertisements and brochures available from public health centres and www.health.nsw.gov.au in multiple languages.
  - campaigns to reduce antibiotic resistance by providing prescription guidelines to practitioners and the public, in particular parents with young children.
  - disease prevention programs run through public organisations such as hospitals, health centres and schools.

public health campaigns: procedures aimed at persuading the public to engage in or stop specific behaviours and to communicate information about symptoms, treatment and public health support

Use of pesticides

- In Chapter 9 you reviewed vector-transmitted diseases such as mosquito-borne malaria, Zika virus, dengue fever and yellow fever as well as diseases spread by ticks, mites and rodents.
- Pesticides are used to prevent and control these disease vectors. Pesticides are also used in agriculture to control insect infestations and plant diseases. Pesticides include insecticides, fumigants, fungicides, herbicides and antimicrobial pesticide products such as sanitisers and disinfectants.
- In the past insecticides effectively prevented and controlled many diseases including vector-transmitted diseases. For example, in the late 1940s the World Health Organization (WHO) initiated programs to eradicate malaria around the world with the use of the now-banned insecticide DDT.
- In addition to some pesticides having negative impacts on human health and the environment, resistance of insects to pesticides is now one of the greatest problems facing the control of vector-borne diseases around the world, particularly in developing countries.
- Alternative approaches to pesticide use include integrated crop protection and integrated pest management, such as understanding the life cycles of pathogens and having knowledge of relevant disease prediction models. Other approaches include better farm sanitation, using resistant or tolerant plant varieties, using clean (pathogen-free) transplants and seeds, and testing soil for soil-borne diseases. Biological controls (e.g. native fish feeding on mosquito larvae), improved hygiene practices and sanitation, environmental management (e.g. getting rid of vector breeding sites and using window screens and netting) and genetic engineering are also alternatives to the use of pesticides.

Genetic engineering

- In Chapter 8 you reviewed the current genetic technologies that induce genetic change. Genetic engineering makes it possible to improve human medicines (see Chapters 7 and 14) and engineer pathogenic bacteria and viruses to make vaccines such as hepatitis B vaccine and human papillomavirus vaccine, which protects against cervical cancer.
- Genetic engineering can create disease-resistant plants; for example, plant biologists are working with crop plants to delete genes that code for chemicals that attract pests. This has the potential to reduce dependence on toxic pesticides.
- Engineering can silence genes in targeted pathogens by disabling the gene; for example, Australian researchers have developed a gene-silencing spray that shuts down viral genes inside tobacco plant cells. Specific genes can also be shut down by spraying them with double-stranded RNA sequences (called RNA interference or RNAi) that match with target genes of the pathogen. RNAi sprays are being developed to spray mites that cause disease in bee colonies (Source: 'Gene-silencing farming revolution', *New Scientist*, 14 January 2017).
- Engineering the genes of vectors such as mosquitoes can control the spread of diseases including malaria, dengue and Zika. For example, with CRISPR-Cas9 technology it is possible to alter an *Anopheline* or *Aedes* mosquito's genes so it cannot spread the malaria parasitic protozoan or Zika virus, to genetically modify mosquitoes so that they bear sterile offspring, or to alter the mosquito's genes to prevent the birth of females so that the disease stops spreading and the insect dies out.
- Genetic engineering can improve processes for pathogen discovery and diagnosis; for example, screening for and identification of pathogens using genome sequencing in samples collected at hotspots for emerging infectious diseases. Genome-wide sequencing and microarray technology are also used for disease surveillance and the study of population host factors that determine disease susceptibility.
- You reviewed CRISPR-Cas9 briefly in Chapter 7. CRISPR-Cas 9 has two components: the first is an enzyme (Cas9) that functions as a cellular scalpel to cut the DNA. In nature bacteria use it to sever and disarm the genetic code of invading viruses. The other component consists of an RNA guide that leads the scalpel to the precise nucleotides it needs to cut. To patch the break, the cell inserts the chain of nucleotides that has been delivered by the CRISPR-9 package. CRISPR-Cas 9 has been used to rid pigs of viruses that would otherwise prevent their organs from being transplanted into humans.

- CRISPR-Cas9 is also experimentally combined with a gene drive. As you reviewed in Chapter 7, any sexually reproducing animal receives one copy of a gene from each parent. Some genes, however, are 'selfish'. That is, evolution has bestowed on them a better than 50% chance of being inherited. Scientists combine CRISPR with a gene drive to alter the genetic code of a species by attaching the desired DNA sequence onto a 'selfish' gene before releasing the animal to mate naturally. This would force a genetic trait through a population (Source: M Specter, 'DNA Revolution', *National Geographic*, 2016).
- The use of genetic engineering to control a disease vector has opponents as well as advocates because it raises a number of ethical questions. Opponents of the idea are concerned that the modified gene could escape the species boundary of, for example, the mosquito and cause problems in other species, leading to unforeseen ecological consequences. Opponents are also wary of the programs that are controlled by for-profit corporations as this creates the potential for conflicts of interest and possibly concentration of the control of technologies in the hands of a few companies. Note: You reviewed the social and ethical implications of genetic engineering in Chapter 7.

→ KEY QUESTIONS

9 **What does a public health campaign do?**

10 **Provide an example of a health campaign that has been employed to prevent the spread of a disease.**

11 **Provide an example of how pesticide use has been employed to prevent the spread of a disease.**

12 **Provide an example of how genetic engineering can be employed to prevent the spread of a disease.**

Answers ➲ p. 174

3 Investigating and assessing the effectiveness of pharmaceuticals as treatment strategies for the control of infectious disease

» Students investigate and assess the effectiveness of pharmaceuticals as treatment strategies for the control of infectious disease; for example, antivirals and antibiotics.

- Substantial reductions in the incidence of infectious disease **morbidity** and **mortality** have been achieved through the use of **pharmaceuticals**, immunisations, access to better health care and other preventive practices such as improved sanitation and personal hygiene.
- The 20th century saw the beginning of the use of pharmaceuticals as treatment strategies in the control of infectious diseases.
- The first experimental antivirals were developed in the 1960s. Antivirals are drugs used to treat viral infections. Antivirals are difficult to develop because they can damage the host cells that contain the virus. Antivirals treat a narrow range of viruses and include drugs for the treatment of influenza, HIV (antiretroviral), herpes, hepatitis B and hepatitis C.
- The **antibiotic** penicillin was discovered in the late 1920s but it was not until the 1940s that manufacture of the drug began. Antibiotics are drugs used to treat bacterial infections. Antibiotics are made to target both specific bacterial pathogens as well as a broad spectrum of pathogens. Antibiotics do not work against viral infections such as colds or influenza. When used properly antivirals and antibiotics are highly effective and save lives.
- Factors that challenge and limit the effectiveness of antivirals and antibiotics include the following:
  - They can cause an increase in antibiotic-resistant bacteria (sometimes called 'super bugs'). Examples include erythromycin-resistant *Streptococcus pyogenes* and *Streptococcus pneumonia*, vancomycin-resistant enterococci and methicillin-resistant *Staphylococcus areus*. You studied antibiotic resistant strains of bacteria as an example of evolutionary change in Year 11 Biology (see page 137 of the *Excel Year 11 Biology* study guide).
  - There can be side effects to taking antibiotics. When we take antibiotics we wipe out friendly bacteria along with harmful ones and this change to the body's ecosystem can allow once-friendly bacteria from elsewhere in the body to colonise and become pathogenic.
  - They can lead to the emergence and evolution of new infectious diseases, particularly zoonotic viral diseases.
  - There may be unregulated access to and inappropriate use of pharmaceuticals in the global market.
  - There are a number of ethical issues; for example, other than in well-resourced settings the drugs may not be available in sufficient quantities. The high cost of drugs and vaccines and community mistrust of pharmaceuticals presents a particular challenge for developing countries. Pharmaceutical companies that

morbidity: the rate of disease in a population
mortality: the number of deaths in a population
pharmaceuticals: manufactured compounds used as medicinal drugs
antibiotics: manufactured drugs that either kill bacteria or stop them from growing so that the immune system can more easily destroy them

hold a drug **patent** control the cost and availability of the drug. For example, estimates suggest that 80 million people in the world have hepatitis C but only 5.4 million people have access to the antiviral Sofosbuvir because the pharmaceutical company that manufactures the drug controls the patent and therefore has a monopoly over the drug in some regions (Source: *New Scientist*, 1 April 2017, p. 6).

- There is a time and cost factor involved: the production of antiviral drugs and vaccines can take years. Deploying them is also critical in the control and prevention of viral diseases. Products must be affordable, which means that the cost of the final product needs to be de-linked from the cost of developing them.

patent: a form of legal protection that confers a sole right to the ownership and use of a product, invention or design

➔ KEY QUESTION

13 **Describe two factors that can limit the effectiveness of pharmaceuticals such as antibiotics and antivirals in controlling diseases.**

Answer ➲ p. 175

4 Investigating and evaluating environmental management and quarantine methods used to control an epidemic or pandemic

» Students investigate and evaluate environmental management and quarantine methods used to control an epidemic or pandemic.

➔ Environmental management and quarantine methods aim to control the spread of a disease by reducing exposure to the pathogen through regulation of the environment and modification of behaviours.

➔ In evaluating methods to control an epidemic or a pandemic you would have made a judgement to determine the effectiveness of the management and control based on some sort of criteria or measures. The rate and extent of spread and severity of a disease determines the urgency and scope of the response. A rapid-spreading severe epidemic or pandemic with high morbidity and mortality requires more extreme measures. As a first priority the disease must be identified, contained and controlled at its source as quickly as possible. Factors that could help the spread, such as geopolitical, social and cultural contexts, need to be rapidly identified and managed to ensure successful control. For example, many potentially exposed people in Asia during the avian flu (H5N1) epidemic believed that antiviral drugs would give them the flu rather than protect them.

➔ Key measures for environmental management and quarantine methods used to control epidemics or pandemics in human populations include:
- pre-preparedness
- alerts
- management of healthcare facilities with strict procedures for treatment, including personal protective equipment (PPE) for all health workers and special methods for handling infected patients
- the use of chemical spray and disinfectants upon entry and exit from facilities
- the biosecure disposal of all contaminated medical materials
- quarantine and isolation of all suspected individuals
- exit screening at airports for individuals departing from infected countries.

➔ KEY QUESTION

14 **For a named epidemic or pandemic summarise some of the measures you would use to evaluate the environmental management and quarantine methods used to control the spread of the disease. Construct a table to answer the question.**

Answer ➲ p. 175

5 Interpreting data relating to the incidence and prevalence of infectious disease in populations

» Students interpret data relating to the incidence and prevalence of infectious disease in populations; for example: mobility of individuals and the proportion that are immune or immunised; and malaria or dengue fever in Southeast Asia.

Mobility of individuals and the proportion that are immune or immunised

➔ The **incidence** and **prevalence** of infectious diseases involves a wide range of interrelated factors that includes the **mobility** of individuals and the proportion of the population that are **immune** or immunised.

incidence: a measurement of how many people (number of cases) are affected by a disease for the first time over a period of time
prevalence: a measurement of the number or proportion of cases of disease in a population at a given time
mobility: the movement of individuals; for example, commuting, travel, trade, migration and refugee movements
immune: the state of protection or immunity that occurs when a person has had an infection and recovered or has been successfully vaccinated (immunised)

- Analysis of these factors can enable prediction of the potential for an outbreak of an epidemic or pandemic as well as evaluation of strategies to control the spread of disease.
- The mobility of individuals in a population influences the incidence and prevalence of disease by changing the frequency of contacts between infected and susceptible individuals. More susceptible individuals include those with no immunity to the disease or those who have not been immunised.
- For most of human history regional and global populations have been isolated from one another. New infectious diseases could only spread as fast or as far as people could walk. Wars, exploration and trade brought greater population movement and mobility of individuals and therefore the spread of diseases into populations that had evolved with relatively small gene pools along with no exposure to the disease and therefore no immunity. An example is the decimation of Aboriginal populations due to exposure to diseases such as smallpox, tuberculosis, scarlet fever and venereal disease to which they had no prior exposure and no immunity.
- The speed and volume of mobility of individuals and human population movement is unprecedented in human history. Mobility of individuals in a population has been shown to play a key role in the spread of influenza, cholera, malaria, rubella, measles, dengue fever, polio, Ebola and HIV/AIDS. Given the speed of travel it is possible that an infected individual can circle the globe within the incubation period of a disease, causing an exponential increase in contacts; for example, HIV has a long incubation period and therefore mobility of infected individuals can be a high risk factor in the spread of the disease.
- Mobility of individuals also contributes to the incidence and prevalence of infectious diseases and the re-introduction of diseases for the following reasons:
  - Mobility increases the contact between humans and vectors; for example, tourism and migration bring humans who are more susceptible or have no immunity to vector-borne diseases such as malaria to areas where vectors thrive.
  - New arrivals, refugees, displaced populations and legal and illegal workers can bring disease vectors and drug-resistant strains, such as anti-malarial drug resistance, into areas where they are not present. Australia has seen an increase in the incidence of tuberculosis with 80–90% of cases coming from overseas arrivals. The increase is a result of the spread from neighbouring countries of multidrug resistant strains of TB.
  - Changes in global land and animal use have significantly increased the risk of zoonotic diseases in humans.
  - Humans encroaching into vector habitats increases the prevalence of diseases for which there is no previous exposure or immunity.
- The proportion of the population immunised significantly reduces the incidence and prevalence of infectious diseases. In 1980 the World Health Organization announced that the smallpox disease had been eradicated worldwide. Global vaccination programs have reduced poliomyelitis by 99.9% and in the six years following the introduction of the polio vaccine it was calculated that more than 150 000 cases of paralytic polio and 12 500 deaths were prevented worldwide. India eliminated polio in 2014 and it is now only found in a few countries including Afghanistan, Pakistan and Nigeria. WHO predicts that the virus will be eradicated by 2019. See www.polioeradication.org/polio-today/polio-now/this-week/ for weekly updates on reports of polio cases.

KEY QUESTIONS

15 **Provide an example that shows the spread of a disease due to mobility of individuals in a population.**

16 **Provide an example of data that shows how the incidence or prevalence of disease is related to the proportion of the population that are immune or immunised.**

Answers ➲ p. 175

Malaria and dengue fever in Southeast Asia

- Each year millions of people die of mosquito-borne diseases. Malaria and dengue fever are the most common mosquito-borne diseases in the Southeast Asia region and scientists are concerned that the morbidity and mortality from these diseases will increase with the rise in global temperatures, drug resistance and international travel.
- Data of dengue fever and malaria cases can be obtained from the Health Information Platform at the World Health Organization website. The website is a reliable source of data relating to the incidence, prevalence and death rates in populations due to disease.

Malaria

- Approximately 40% of the world's population, mostly those living in the world's poorest countries, are at risk of malaria. Every year more than 500 million people become severely ill with the disease. Post-eradication epidemics surfaced in the mid-1970s and in 2015 there were 296 million cases of malaria worldwide, resulting in an estimated 731 000 deaths.
- More than 80% of the population is at risk of contracting malaria in Asia and the Pacific. Malaria incidence is between 2.5 to 2.8 million cases and high-risk groups include infants, young children, pregnant women and mobile population groups.
- Factors relating to increases in the incidence and prevalence of malaria include:

- the resistance to chloroquine and other commonly available anti-malarial drugs
- the increasing proliferation in some regions of low-quality and counterfeit drugs
- changes in malaria spread as a result of population movement and rapid urbanisation, both of which are largely caused by unemployment and environmental deterioration
- the high proportions of young people in some populations causing low community immunity.

→ See also www.adb.org/features/malaria-asia-and-pacific-numbers and www.wpro.who.int/southpacific/programmes/communicable_diseases/malaria/page/en/index2.html.

Table 12.2 Data relating to malaria prevalence rate per 10 000 people and death rate per 10 000 people (all ages) for four Southeast Asian countries compared to data for ten Asian countries in total, 2006–11

Country	Indicator(s)	2006	2007	2008	2009	2010	2011
Bangladesh	prevalence death rate	2.308 0.120	4.159 0.155	5.821 0.086	4.344 0.035	3.758 0.055	3.440 0.055
India	prevalence death rate	15.428 0.068	12.853 0.068	12.869 0.072	12.946 0.082	13.065 0.071	10.555 0.069
Myanmar	prevalence death rate	43.573 0.531	79.708 0.420	94.618 0.425	91.608 0.385	87.736 0.335	96.261 0.288
Republic of Korea	prevalence death rate	2.684 0.000	0.894 0.000	3.394 0.000	2.975 0.000	2.711 0.000	3.140 0.000
ten Asian countries	prevalence death rate	486.728 3.642	635.043 2.900	613.898 2.263	570.589 2.044	550.123 1.974	298.970 1.500

Source: www.hindawi.com/journals/mrt/2015/620598/

Dengue fever

→ The global prevalence of dengue and severe dengue (dengue haemorrhagic fever) has grown dramatically in recent decades. Before 1970 only nine countries had experienced dengue fever epidemics but the disease is now endemic in more than 120 countries, with almost half of the global population at risk. Three-quarters of those most at risk live in the Asia-Pacific region where vector populations are increasing. Dengue is found in tropical and subtropical climates worldwide, mostly in urban and semi-urban areas. Severe dengue is a leading cause of serious illness and death among children in some Asian and Latin American countries. Not only is the number of cases increasing as the disease is spreading to new areas but explosive outbreaks are also occurring. In 2015 Delhi in India recorded its worst outbreak since 2006 with over 15 000 cases. The WHO estimates suggest there may be at least 50 million cases of dengue infection worldwide each year.

→ The year 2016 was characterised by large worldwide dengue outbreaks. The Asia-Pacific region reported more than 375 000 suspected cases of dengue in 2016. Of these, the Philippines reported 176 411 and Malaysia 100 028 cases.

→ Factors relating to increases in the incidence and prevalence of dengue fever and severe dengue include the following:
- population growth and unprecedented rapid urbanisation resulting in an increased number of people living in poor housing with lack of proper waste disposal and inadequate drainage which creates favourable breeding conditions for the mosquito vectors.
- underfunding for prevention and control in many countries.
- partial or temporary immunity: there are four distinct but closely related **serotypes** of the virus that cause dengue. Recovery from infection by one serotype provides lifelong immunity against that particular serotype but post-recovery immunity to the other serotypes is only partial and temporary. Subsequent infections by other serotypes increase the risk of developing severe dengue.
- with increased global temperatures comes expanding geographic distribution of the four dengue serotypes and their mosquito vectors, the most important of which is the predominantly urban species *Aedes aegypti*.
- the increased spread of *Aedes albopictus*, a secondary dengue vector in Asia; *Aedes albopictus* is highly adaptive and has spread due to its tolerance to temperatures below freezing along with its ability to hibernate and shelter in microhabitats.

serotypes: distinct variations within a species of pathogen

→ For more information see www.searo.who.int/entity/vector_borne_tropical_diseases/data/data_factsheet/en/index2.html and the World Health Organization's 2012 *Global Strategy for Dengue Prevention and Control 2012–2020* at www.who.int/denguecontrol. See also *Changing epidemiology of dengue in South-East Asia*, Department of Communicable Diseases, World Health Organization, South-East Asia Regional Office.

→ KEY QUESTION

17 **Interpret the following graph that shows ten populations in Southeast Asia that are at risk of malaria or malaria free. For example, which countries are most at risk and which countries are least at risk?**

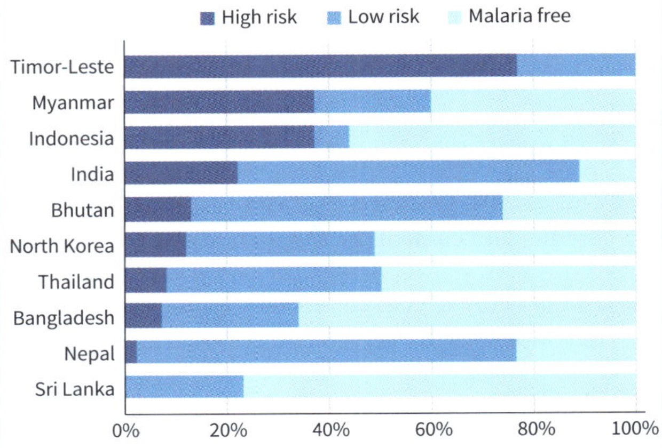

Figure 12.3 Population at risk of malaria in Southeast Asia
Source: www.searo.who.int/entity/malaria/data/sear12.pdf

Answer ➲ p. 175

6 Evaluating historical, culturally diverse and current strategies to predict and control the spread of disease

» Students evaluate historical, culturally diverse and current strategies to predict and control the spread of disease.

Historical strategies to predict and control the spread of disease

- Predictions need some kind of mathematical model. In evaluating strategies to predict and control the spread of disease you would have made judgements about models and systems and how well they help predict outbreaks of disease and the extent to which a range of strategies might control the disease. You would have also evaluated current strategies, including new technologies.
- Historically there was very little understanding of the causes of diseases and therefore minimal knowledge of how to prevent, treat or control them. For example, most people believed that diseases such as cholera and bubonic plague were caused and spread by miasma (bad air) as a punishment from God. Even so, the ancient Greeks knew that people who had recovered from the bubonic plague 'acquired immunity'. Based on this observation, authorities in Athens used survivors from previous epidemics to nurse sufferers when the same infection re-emerged.
- The earliest account of mathematical modelling of the spread of disease was carried out in 1766 by Daniel Bernoulli. Bernoulli created a model to predict increased life expectancy if a population was inoculated against smallpox. Bernoulli's work preceded the understanding of germ theory and therefore was limited in its application.
- The 20th century saw pioneering work in infectious disease modelling, including Kermack's and McKendrick's epidemic model in 1972. The model describes the relationship between susceptible, infected and immune individuals in a population and the spread and possible means of control of a disease or epidemic.

Culturally diverse strategies to predict and control the spread of disease

- Different cultures have different understandings about diseases and therefore different strategies to predict and control their spread. These include cultural and religious beliefs about health, health care, disease, pharmaceuticals and vaccinations. Cultural beliefs must be understood in order to predict and control the spread of disease; for example, cultural practices and beliefs played a key role in the spread of the African Ebola epidemic from 2013–16. The African culture prides itself on strong family ties and community relationships where hugs and handshakes are common ways of greeting. In addition, individuals are cared for by the immediate and extended families. Many individuals who fell ill with Ebola did not go to medical centres for treatment because there was a belief that they would die isolated from loved ones and their community. Cultural practices also include burial and mourning rituals that involve members of the family having very close contact with the corpse of the deceased.

Current strategies to predict and control the spread of disease

- In Section 5 you reviewed factors relating to the incidence and prevalence of infectious disease in populations, such as mobility of individuals. In our globalised world pathogens can spread quickly both regionally and internationally; for example, among travellers returning to Europe from low- and middle-income countries dengue is the second most diagnosed disease after malaria. *Aedes albopictus* has spread from Asia to North America and more than 25 countries in the European region largely due to the international trade in used tyres (a breeding habitat) and other goods such as 'lucky' bamboo.
- Surveillance, prediction, preparedness and continued public health interventions are critical for maintaining disease control (see Table 12.1).
- Modelling is used to:
  - predict future occurrences of disease
  - project and simulate how a disease will progress and spread
  - simulate the effects of possible interventions

- inform public health interventions and distribution of resources such as mass vaccination programs
- identify hotspots of emerging (novel) infectious diseases and provide early warnings
- increase understanding of human mobility patterns and the impact of mobility restriction measures such as travel restrictions
- establish the degree to which the past is an accurate guide to the future.

➔ Traditionally surveillance relied on routine reporting by healthcare facilities of structured predefined information regarding events and diseases. This could mean a delay of weeks between finding out about the disease and the implementation of control measures, in which time the disease can rapidly spread. With advances in technology, event-based surveillance has been developed to complement traditional surveillance. For example, in the 2010 Haitian cholera outbreak epidemic mapping was enabled by using information from social and news media, leading to a more rapid and targeted response.

➔ Computer technologies provide fast processing capacity and enormous levels of data storage. This means that modelling has increased in its complexity, with new variables examined and new relationships found such as the relationships between epidemic frequency and location and factors such as population size, environmental change and antibiotic resistance.

➔ For more information see www.ncbi.nlm.nih.gov/pmc/articles/PMC4720248/.

➔ KEY QUESTIONS

18 **Outline two examples of current strategies used to predict and control the spread of disease.**

19 **Evaluate the effectiveness of technological developments in the prediction or control of diseases.**

Answers ➲ p. 175

Table 12.3 Evaluation of some global epidemic intelligence, surveillance, prediction and reporting systems

Organisation/Network	What it does and examples of its effectiveness
the World Health Organization (WHO)	uses both traditional surveillance methods and informal event-based data and coordinates international outbreak responses using data from GOARN data sets that combine existing data on climate, vaccination and population immunity are used to predict outbreaks while networks are established to improve communications among people during an epidemic
the Global Public Health Intelligence Network (GPHIN) was developed by Health Canada in collaboration with WHO and is an internet-based, multilingual early-warning tool	event-based surveillance system that systematically scans a multitude of informal sources, including news reports, online newspapers, social media and internet-based searches in 2002 and 2003 GPHIN issued the first alert of unusual respiratory illness in Guangdong province, China to WHO and GOARN members, triggering an international response
the Global Outbreak Alert and Response Network (GOARN) www.who.int/csr/outbreaknetwork/en/	established in 2000 by WHO; gathers infectious disease intelligence in order to promptly detect and verify outbreaks, issue real-time alerts and rapidly respond to global or national public health threats during the H1N1 influenza pandemic GOARN was activated at the WHO organisational level in order to monitor disease activity, offer technical guidance, and implement screening strategies and containment measures in order to halt the pandemic
Global Epidemic and Mobility (GLEaM) model www.gleamviz.org	integrates global demographic data, air travel and short-range mobility data for the simulation of worldwide spread of influenza-like illnesses
Google Flu Trends www.google.org/flutrends/	one of the first web-based tools developed for near real-time monitoring of influenza activity; based on the fact that the number of people searching the web using influenza related keywords is correlated with the number of people actually having influenza symptoms and hence an increase in searches may signal a rise in disease activity
Google Trends www.google.com/trends/	used to track other communicable diseases like West Nile virus (WNV) and dengue, and has been applied in mathematical models for the prediction of peak influenza activity
One Health network www.onehealthinitiative.com	launched in 2009 on the basis that human and animal health are essentially interconnected and the majority of emerging infectious diseases are zoonotic the One Health network brings together medical, veterinary, public health and other environmentally related professionals to integrate surveillance systems, diagnostic tools and preventive interventions

7 Investigating the contemporary application of Aboriginal protocols in the development of particular medicines and biological materials in Australia

» Students investigate the contemporary application of Aboriginal protocols in the development of particular medicines and biological materials in Australia and how recognition and protection of Indigenous cultural and intellectual property is important; for example, bush medicine and smoke bush in Western Australia.

Aboriginal protocols, medicines and biological materials

- Over the millennia indigenous peoples have developed a close and unique connection with the lands and environments in which they live. They have established distinct systems of knowledge, innovation and practices relating to the uses and management of biological diversity on these lands and environments. Much of this knowledge forms an important contribution to research and development, particularly in areas such as pharmaceuticals, agriculture and cosmetic products. Indigenous peoples demand not only recognition and protection of this **traditional knowledge** but also the right to share equitably in any benefits derived from the uses of this knowledge (Source: *Native Title Report*, 2008).
- Research into and development of traditional medicines and biological materials in Australia usually requires establishing partnerships between Aboriginal communities and industries or professions. In doing so it is important that **Aboriginal protocols** are acknowledged, respected and applied.
- The *Protocol for Aboriginal Knowledge and Intellectual Property* was developed by the Desert Knowledge Cooperative Research Centre (Desert Knowledge CRC). The protocol serves as a guide towards best practice in ethics, confidentiality, equitable benefit sharing and managing research information. The protocol respects and acknowledges that Aboriginal communities and groups will have their own protocols that must also be observed, understood and engaged with as an essential ongoing part of any process with indigenous people (Source: www.desertknowledgecrc.com.au).

traditional knowledge: the knowledge, innovations and practices of indigenous and local communities around the world

Aboriginal protocols: appreciation of the knowledge, standing and status of people within a local Aboriginal community; protocols are ways of behaving, communicating and showing respect

- The *Nagoya Protocol* is a supplementary agreement to the 1992 Convention on Biological Diversity and was adopted as a global agreement in Nagoya, Japan in 2010. Australia signed the agreement in 2012. The protocol provides a legally binding framework that respects the value of indigenous knowledge, focusing on how resources are accessed and the fair and equitable sharing of monetary benefits from their use.
- For more information see www.environment.gov.au/indigenous/publications/pubs/recommendations.pdf, link.springer.com/chapter/10.1007/978-3-319-26021-1_7 and www.ipaustralia.gov.au/sites/g/files/net856/f/uts__recognising_and_protecting_aboriginal_knowledge.pdf.

Recognition and protection of Indigenous cultural and intellectual property

- A major concern of Aboriginal people is that their traditional knowledge is being exploited and that their **Indigenous cultural and intellectual property** rights are not recognised. These concerns include that fact that cultural knowledge of plants, animals and the environment is being used by scientists, medical researchers, nutritionists and pharmaceutical companies for commercial gain, often without Aboriginal peoples' informed consent and without any benefits flowing back to indigenous communities (see the smoke bush case study below). In addition, unique resources and the associated traditional knowledge are often unlawfully collected.
- Practices that do not acknowledge traditional knowledge are not only offensive to many Australian indigenous people but represent the continuing dispossession of their communities.
- In Australia there are a number of national and state government arrangements that attempt to address the lack of protection domestically, including cultural heritage legislation. Some policies provide for the recognition of indigenous traditional knowledge through environmental protection regulations, particularly concerning knowledge held by indigenous people about biological resources. The *Native Title Act 1993* established principles for the recognition of customary property rights, including rights in knowledge, based on the traditional laws and customs observed and practiced by the native title holders. However, existing intellectual property laws, such as the *Copyright Act 1968*, offer limited scope for the recognition of indigenous peoples' rights in biodiversity-related knowledge and practices.

Indigenous cultural and intellectual property: objects, sites, cultural knowledge, cultural expression and the arts that have been transmitted or continue to be transmitted through generations as belonging to a particular Indigenous group or Indigenous people as a whole or their territory

While native title, cultural heritage and environmental laws provide some recognition and protection they are considered to be insufficient. For more information see www.aph.gov.au/About_Parliament/Parliamentary_Departments/Parliamentary_Library/pubs/rp/RP9697/97rp20.

Bush medicine

- It is estimated that traditional medicines are relied on by up to 80% of the world's population for primary health care and there is evidence to suggest that approximately three-quarters of the plants used in pharmaceuticals were originally used in traditional medicine (Source: UTS Indigenous Forum, 2014).
- Native plants have been used as bush medicine by Aboriginal people for generations. Many plants contain antibacterial and anti-inflammatory compounds that are known to Western medicine. For example, in the 1920s tea tree oil (*Melaleuca alternifolia*) was found to have antiseptic potency far stronger than the commonly used antiseptics of the time. Since then the oil has been used to treat everything from fungal infections of the toenails to acne. Eucalyptus oil is used in a wide range of pharmaceuticals such as throat lozenges.
- As another example, concoctions of emu bush leaves were used by Northern Territory Aboriginal tribes to wash cuts and sores. In the last decades chemicals from the leaves have been found to have the same strength as some established antibiotics.

  Source: www.australiangeographic.com.au/topics/history-culture/2011/02/top-10-aboriginal-bush-medicines/.

Smoke bush in Western Australia

- The smoke bush of Western Australia has been traditionally used by Aboriginal Australians for healing. In the 1960s the WA government granted the US National Cancer Institute a licence to collect plants. In 1981 specimens of smoke bush were sent to the National Cancer Institute to test for the presence of cancer-fighting properties. The specimens were found to be ineffective but held in storage until the late 1980s when they were tested again in the search for a cure for HIV/AIDs. Out of 7000 plants screened from around the world the WA smoke bush was one of only four species found to contain a substance that when tested, even in low concentrations, destroys the HIV virus. The discovery was then patented. The National Cancer Institute has awarded Amrad, a Victorian pharmaceutical company, an exclusive world license to develop the patent. In the early 1990s the WA government also awarded Amrad the rights to the smoke bush species to develop an anti-AIDs drug.
- Indigenous people are concerned that they have not received any acknowledgment, financial or otherwise, for their role in discovering the healing properties of smoke bush. The WA legislation disregards any potential intellectual property rights that Indigenous peoples have on their lands. Multinational drug companies could be sold exclusive rights to entire species of flora, preventing anyone from using these species for any other purposes without the consent of those companies.

→ KEY QUESTION

20 **What are some key concerns that Aboriginal people have about recognition and protection of their Indigenous cultural and intellectual property?**

Answer ➲ p. 175

CHAPTER SYLLABUS CHECKLIST

Are you able to answer these questions from the syllabus for this chapter? Tick each question as you go through the checklist if you are able to answer it. If you cannot answer a question, turn to the relevant page in the study guide to find the answer. For NESA key word meanings, go to www.educationstandards.nsw.edu.au and search 'key words'.

	FOR A COMPLETE UNDERSTANDING OF THIS TOPIC:	PAGE NO.	✓
1	Can I investigate and analyse the wide range of interrelated factors involved in limiting local, regional and global spread of a named infectious disease?	158–159	
2	Can I investigate procedures that can be employed to prevent the spread of disease, including but not limited to hygiene practices; quarantine; vaccination, including passive and active immunity; public health campaigns; use of pesticides and genetic engineering?	160–163	
3	Can I investigate and assess the effectiveness of pharmaceuticals as treatment strategies for the control of infectious disease; for example, antivirals and antibiotics?	163–164	
4	Can I investigate and evaluate environmental management and quarantine methods used to control an epidemic or pandemic?	164	
5	Can I interpret data relating to the incidence and prevalence of infectious disease in populations; for example, mobility of individuals and the portion that are immune or immunised and malaria or dengue fever in Southeast Asia?	164–167	
6	Can I evaluate historical, culturally diverse and current strategies to predict and control the spread of disease?	167–169	
7	Can I investigate the contemporary application of Aboriginal protocols in the development of particular medicines and biological materials in Australia and how recognition and protection of Indigenous cultural and intellectual property is important; for example, bush medicine and smoke bush in Western Australia?	169–170	

HSC EXAM-TYPE QUESTIONS

Objective-response questions (1 mark each)

1 A study in Pakistan found that providing soap along with education about hand washing reduced the incidence of the pneumonia disease in children below the age of five by 50%. This study is an example of:
- A a public health campaign employed to treat pneumonia.
- B a public health campaign employed to prevent the spread of pneumonia.
- C hygiene practices and a public health campaign employed to treat pneumonia.
- D hygiene practices and a public health campaign employed to prevent the spread of pneumonia.

2 A bird flu epidemic caused by the H5N8 virus struck poultry farms across Europe and the Middle East in 2016. As a result 800 000 free-range geese and ducks exposed to the virus were slaughtered in southwest France (*New Scientist*, 14 January 2017). Which of the following best explains why the geese and ducks were slaughtered?
- A The birds were slaughtered as a quarantine method to control the epidemic.
- B The birds were slaughtered to stop the H5N8 virus spreading.
- C The birds were slaughtered as an environmental control method to stop the H5N8 virus spreading and contain the epidemic.
- D The birds were slaughtered just in case they had the disease.

3 Which of the following is a current strategy to predict and control the spread of disease?
- A bed rest
- B genetic engineering
- C global surveillance of emerging diseases
- D vaccination

4 Scientists have introduced a new form of the pneumococcal vaccine which protects against disease caused by the bacterium *Streptococcus pneumoniae*. In addition to protecting susceptible infants and young children from the disease, the vaccine also helps to reduce the circulation of the bacteria in the community. Consequently older people are also protected even though they have not been vaccinated against this organism. What does this example illustrate?
- A pathogen resistance
- B disease modelling
- C herd immunity
- D incidence and prevalence data

5 A study of a population's pre-existing immunity to the 2009 influenza pandemic (H1N1) has been used by researchers to explain in part the different rates of infection observed across age groups. For example, up to a third of adults 65 years or older in a population were found to have pre-existing partial immunity to the virus. This contrasts with less than 10% of younger adults in the population and almost no children.

Below are some possible conclusions from the study. Which statement is most likely to be a (possibly) correct conclusion?
- A Partial immunity may be responsible for the low incidence of cases in children and younger adults.
- B Partial immunity may be responsible for the relative fewer cases of H1N1 in older adults.
- C Adults 65 years or older have pre-existing partial immunity.
- D Ten percent of younger adults in the population had partial immunity.

Short-answer questions

6 Pathogens transmitted indirectly via fomites are a major cause of healthcare-associated infections. Briefly describe three hygiene procedures that can be employed to prevent the spread of disease from fomites. (3 marks)

7 Briefly describe the contemporary application of a named Aboriginal protocol in the development of particular medicines and biological materials in Australia. (3 marks)

8 Read the following extract and then answer Question 8.

> For all our high-tech modernity, and in many ways because of it, the risk that new infectious diseases will evolve is actually getting worse. Pathogens began circulating regularly among humans only after we started farming and settled in towns. One reason was that we caught infections from our livestock … Now we are crowding into cities and travelling more, especially within the tropics where pathogen diversity is highest. That, plus globalised trade, migration and climate change, reshuffle wildlife, people and pathogens.
>
> Source: D MacKenzie, 'The coming plague', *New Scientist*, 25 February 2017, pp. 29–33

From your studies of mobility of individuals in relation to the spread of disease outline an example of how people and pathogens can be 'reshuffled'. (4 marks)

9 Interpret the data in Table 12.4 below to answer Question 9.

a Which two regions had the highest incidence of new cases of tuberculosis and measles in 2004? (1 mark)

b Southeast Asia and Western Pacific together had the highest episodes of which illness in 2004? (1 mark)

c What percentage of the incidence of worldwide dengue was found in Southeast Asia in 2004? (1 mark)

d In 2004 the lowest number of new cases of HIV infection was found in which regions? (1 mark)

10 From your study of infectious diseases answer parts a and b below.

a Evaluate the environmental management and quarantine methods used to control a named epidemic against the following criteria: preparedness, treatment, containment and control. (4 marks)

b For the disease named in part a, construct a table to summarise some factors that could be involved in limiting the spread of the disease locally, regionally and globally. (4 marks)

Table 12.4 Incidence (in millions) of selected diseases by six WHO regions data in 2004

Disease	World	Africa	The Americas	Eastern Mediterranean	Europe	Southeast Asia	Western Pacific
Tuberculosis[a]	7.8	1.4	0.4	0.6	0.6	2.8	2.1
HIV infection[a]	2.8	1.9	0.2	0.1	0.2	0.2	0.1
Diarrhoeal disease[b]	4620.4	912.9	543.1	424.9	207.1	1276.5	1255.9
Pertussis[b]	18.4	5.2	1.2	1.6	0.7	7.5	2.1
Measles[a]	27.1	5.3	0.0[c]	1.0	0.2	17.4	3.3
Tetanus[a]	0.3	0.1	0.0	0.1	0.0	0.1	0.0
Meningitis[b]	0.7	0.3	0.1	0.1	0.0	0.2	0.1
Malaria[b]	241.3	203.9	2.9	8.6	0.0	23.3	2.7
Dengue[b]	9.0	0.1	1.4	0.5	0.0	4.6	2.3

a = new cases
b = episodes of illness
Source: Disease incidence, prevalence and disability from www.who.int/global_burden_disease

ANSWERS

KEY QUESTIONS

Key questions ➲ p. 159

1 Dengue fever (or simply dengue) is used as the example in this answer. There is no specific treatment for dengue or severe dengue but early detection and access to proper medical care lowers fatality rates below 1%. Dengue prevention and control depends on effective vector control measures. In late 2015 and early 2016 the first dengue vaccine was registered in several countries for use in individuals nine to 45 years of age who were living in endemic areas. Other vaccines are under development.

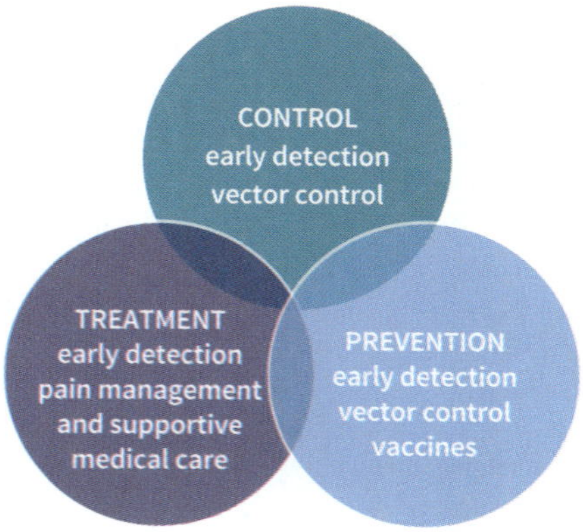

Figure A12.1 Model to summarise the prevention, treatment and control of dengue

2 Worldwide challenges and issues in relation to the prevention, treatment and control of infectious diseases include antibiotic-resistant bacterial infections due to the overuse or misuse of antibiotics. Other factors include the re-emergence of once common diseases such as dengue, measles, mumps, whooping cough, chlamydia and hand, foot and mouth disease. There has also been the emergence of new (novel) diseases due to changing interactions between humans, the environment and other organisms such as avian flu (H7N9) and Ebola. Emergence of novel infections is linked to human factors such as population density, travel and trade, and ecological factors such as climate change and agricultural practices. The speed and scale of national and international travel and trade is another challenging issue along with the continued risk of food-borne diseases spreading through global food supply chains and the increased risk of zoonotic diseases from contaminated animal feeds.

Key questions ➲ p.160

3 Good hygiene practices can prevent and reduce the spread of infections by reducing harmful levels of disease-causing pathogens. Examples include personal hygiene (e.g. washing hands with soap and water), workplace procedures (e.g. keeping work areas clean and sanitised), and rules and guidelines for the handling and preparation of food.

4 Quarantine practices can prevent and reduce the spread of infections by preventing exposure to people, objects or animals that have (or potentially have) infectious diseases or contain vectors of a disease. Quarantine practices include separation and isolation of people and animals, restrictions of movements, and surveillance and screening. Two examples are during the international outbreak of mad cow disease (BSE) Australia put a complete ban on imported cattle from all countries with reported cases. During a 2007 foot-and-mouth epidemic in the UK all passengers entering Australia from the UK were screened for soil, clothing and footwear contamination.

Key questions ➲ p. 161

5 Vaccination means being given a vaccine; that is, the process of getting an injection, nasal spray or oral vaccine. Vaccination confers immunity by switching on the host's own immune system to make the antibodies and memory cells needed to provide long-term protection against infection. Vaccinations have greatly reduced the prevalence of diseases and continue to be important for global health. The World Health Organization estimates that vaccination prevents an estimated two to three million deaths every year.

6 Active immunity comes about through having an infection and recovering from it or by getting vaccinated. Active immune response takes seven to 21 days to fully develop. Sometimes if an infection is overwhelming the immune system antibodies obtained from healthy blood donors are given to the patient. These can act much more quickly and provide immediate protection. This is called passive immunity.

7 A vaccine stimulates an individual's immune system to develop immunity to pathogens such as viruses and bacteria. The antigens used in a vaccine are designed to trigger a specific protective response by the immune system towards a particular pathogen and each vaccine therefore contains a different set of antigens.

8 Immunisation means both receiving a vaccine and becoming immune to a disease post-vaccination. The purpose of immunisation is to prevent people from acquiring infections and to protect them against the short- and longer-term complications of those infections. Longer-term complications can include illnesses such as cancer.

Key questions ➲ p. 163

9 Effective public health campaigns convey key messages to large sections of a population, provide useful tools to promote health, and are increasingly important in the prevention and control of diseases. The success of a campaign is usually measured by the decrease in morbidity and mortality.

10 An example of a health campaign is the mass media, advertising and educational campaigns created by the Australian and state governments in 1987 to control the spread of HIV/AIDS. The governments used a series of graphic and imaginative television and radio commercials. The initial campaign, built around the image of the Grim Reaper, was intentionally controversial and designed to have the maximum impact on public opinion. The commercials were complemented by the publishing of a variety of printed materials about HIV/AIDS and steps to be taken to prevent transmission of the virus. HIV/AIDS community groups were also part of the campaigns, delivering condoms, clean needles and syringes to venues where unsafe practices were most likely to occur.

11 Pesticides such as insect repellents are used to control vector-transmitted diseases such as malaria and the Zika virus. Antimicrobial pesticides are used to kill avian influenza viruses and are used by the poultry industry; for example, to disinfect breeding facilities.

12 To fight dengue researchers have designed a CRISPR-9 package that could delete a natural gene from the wild mosquito parent (the vector) and replace it with a gene that would confer sterility in the offspring. If enough of the mosquitoes are released and mate then in a few generations—about two to three weeks—the entire species would carry the genetically engineered version.

Key questions ➲ p. 164

13 The evolution of pathogens can limit the effectiveness of pharmaceuticals. Bacteria and viruses replicate very quickly, evolve rapidly and spread fast. Bacteria and viruses can become resistant to antibiotics or antivirals. Emerging zoonotic pathogens are becoming increasingly resistant to antimicrobial agents due to the widespread use of antibiotics in animal reservoirs.

14 The named epidemic is foot-and-mouth disease. Foot-and-mouth is a highly contagious viral disease that affects cloven-hoofed animals such as cattle, sheep and pigs.

Table A12.1

Environmental management measures	Quarantine methods
the establishment of environmental exclusion or control (quarantine) zones slaughter (or culling) and the biosecure disposal of all animals infected and exposed to the pathogen or those showing symptoms control of the movement of people and vehicles decontamination of infected premises, vehicles and equipment destruction and biosecure disposal of all animal products and produce restrictions on farm activity restrictions on movement of animals and bans on animal sales	border protection trade bans on imports: trade bans are backed up by the international regulations and procedures for control and eradication set by the World Organisation for Animal Health (OIE) screening of products and passengers from regions with reported cases of the disease

Key questions ➲ p. 165

15 HIV, first identified in 1981, is an example of the impact that the incidence and prevalence of infectious disease can have globally. AIDS has claimed more than 25 million lives, more than 60 million people have been infected with HIV, and more than 90% of the cases are in developing countries. HIV spread from the first zoonotic transmission from chimpanzees to people around 70 years ago in central Africa. Data suggests that the pandemic started in the mid-1970s and by 1980 had spread through air and sea travel and human migration. Studies have shown that certain mobile and sexually active individuals are important in seeding local epidemics.

16 Avian influenza (AH7N9) has been detected in birds in the past but was reported in humans in China in 2013. An increase in sporadic cases in human infections has been reported since 2016. Population surveys in 2015 reported zero or very low human population immunity against AH7N9.

Source: www.who.int/influenza/human_animal_interface/en

Key question ➲ p. 167

17 The following is a sample answer. There is a lot of data that you can interpret from the graph. Ten countries make up the Southeast Asia region. In 2011 more than 75% of the population in Timor-Leste were at risk of malaria: the highest risk population in the region, followed by Myanmar. According to the 2011 data no country in the region was free of malaria although Sri Lanka had the lowest risk.

Key questions ➲ p. 168

18 By using satellite imaging to collect information on sea temperature, surface height and chlorophyll A levels researchers were able to construct an environmental model that could accurately predict the actual incidence rate of a cholera outbreak in Bangladesh. Satellite imaging is also used to trace vector activity such as the distribution of the *Anopheles* species that drives malaria transmission in Africa.

Global surveillance for emerging infectious diseases aims to rapidly detect changes in the incidence rates of endemic diseases. Surveillance also helps to promptly recognise and characterise disease syndromes caused by previously unknown pathogens that have epidemic potential.

19 New technologies are becoming increasingly available for the rapid molecular identification of pathogens but also for the more accurate monitoring of infectious disease activity. Web-based surveillance tools and epidemic intelligence methods, used by all major public health institutions, help with risk assessment and timely outbreak detection. For example, mobile phone data is increasingly being used to track mobility of individuals in an attempt to study the transmission dynamics and routes of importation of infectious diseases from region to region.

Key question ➲ p. 170

20 Aboriginal people claim that their rights as traditional holders and custodians of knowledge are not adequately recognised or protected and they do not share equitably in any benefits from the use of their knowledge or practices.

HSC EXAM-TYPE QUESTIONS

Objective-response questions

1 **D**. The example provides two procedures: hygiene practices and a public health campaign. **A** and **C** are incorrect because the procedures do not treat the disease. **B** is incorrect because it is more than a public health campaign.

2 **C.** This option provides the complete answer: it is an environment control method designed to stop the spread. **A** is incorrect because it does not describe a quarantine method. **B** provides part of the answer but is not the best explanation. **D** is also partly correct but again it is not the best explanation.

3 **C**. The question asks for a current strategy and for a strategy that predicts and controls spread. **A** is incorrect because it is a treatment strategy. **B** is incorrect because it does not provide a prediction strategy. **D** is wrong because vaccination is a treatment strategy to help control spread.

4 **C**. This example illustrates herd immunity**. A** is incorrect because the example does not illustrate pathogen resistance. **B** is wrong because the example does not provide data about modelling**. D** is incorrect because the example does not provide data about incidence and prevalence.

5 **B**. This option is the best example of a conclusion. **A** is incorrect because the study showed that there was a high incidence of cases in children and younger adults. **C** and **D** are incorrect because they are results of the study and not possible conclusions.

Short-response questions

6 **EM** This is a straight recall question. It also tests your understanding of the terms *indirect transmission* and *fomites*.

Three hygiene procedures are the thorough washing of hands with soap and water, (✓) the correct disposal of all contaminated materials, equipment and objects, (✓) and the disinfection of all areas. (✓)

7 **EM** This is a straight recall question. Make sure you provide a specific example of a protocol and how it can be applied.

The *Community Guide* adapted from the Desert Knowledge Cooperative Research Centre protocol (✓) is an example of a protocol that enables Aboriginal people to share knowledge for research and development without its loss, misuse or abuse. (✓) The guide states that Aboriginal people own their knowledge, all research must respect Aboriginal culture and knowledge, everyone should be equal in the research and any knowledge passed to a researcher should not be used in any way without the final agreement of the traditional owners. (✓)

8 **EM** This question requires you to read and comprehend the extract and then provide an example to show how humans are moving into new environments and therefore changing the incidence of and risks associated with disease transmission. It asks you to provide an example. Two examples are provided in the sample answer but note that they may not be the same as the ones you have studied.

Humans are moving into vector habitats, increasing the prevalence of diseases for which there is no previous exposure or immunity. (✓) For example, the development of previously uncultured land in tropical areas such as Southeast Asia has intensified the problem of insect resistance, especially in *Anopheline* mosquitoes. (✓) Humans living in the tropics had rarely farmed these areas due to the prevalence of illness and death from malaria.

Malaria eradication efforts enabled farmers to reach previously uncultivated areas. (✓) The new vegetation covering these areas provides a habitat for mosquitoes while drainage and irrigation ditches generate suitable breeding sites for the mosquito larvae. These factors have led to an increase in the incidence of malaria. (✓)

9 **EM** When we interpret we draw meaning from something: in this case data presented as a graph. You are asked to interpret the data to answer parts **a**, **b** and **d**. For part **c** you will need to do a calculation.

a Southeast Asia and Western Pacific (✓)

b diarrhoeal disease (✓)

c 51% (✓)

d Eastern Mediterranean and Western Pacific at 0.1. (✓)

10 **EM** This is a higher-order question that asks you to describe a case study you have learnt about, to evaluate it against specific criteria and to construct a table to summarise factors that can limit the spread of the same named disease. Make sure you answer both parts of the question. The example used in the sample answer is the Ebola haemorrhagic fever (Ebola) epidemic in Africa which may not be the same as the one you have studied. There has been a lot written about the epidemic and many lessons learned from the poor first response and the initial rapid spread of the disease in the region.

The source for the sample answer below can be found at www.who.int/csr/disease/ebola.

a Ebola haemorrhagic fever (Ebola) outbreaks constitute a major public health issue in Sub-Saharan Africa. The disease was discovered in 1976. Ebola is a highly contagious disease caused by the filovirus. It is spread through direct contact with infected blood, organs and bodily fluids and indirectly through contaminated objects. In Africa fruit bats are the natural hosts of filovirus. The infection of humans has occurred through the handling of infected animals such as chimpanzees, monkeys, bats and antelopes.

The 2013–16 Ebola epidemic was the most widespread outbreak and it caused a conservative estimate of more than 11 000 deaths and over 28 000 suspected cases. The outbreak peaked in October 2014 and then declined gradually due to significant international intervention. (✓)

Table A12.2 Evaluation of environmental management and quarantine methods used to control the Ebola epidemic 2013–16

Criteria for evaluation	Evaluation
preparedness	there was a lack of pre-epidemic preparedness because up until the outbreak the international community did not think that Ebola was endemic to the region (✓)
treatment	a vaccine for Ebola is being trialled but there is no specific treatment as yet; treatment focused on looking after individual patients; in the early stages, health workers treated patients without using basic precautions and isolation techniques and became infected with the disease (✓)
containment and control	it is not always easy to identify patients with Ebola as initial symptoms are non-specific and therefore early cases were misdiagnosed, giving the disease several months to spread before it was confirmed as an Ebola outbreak many people in the community did not accept medical team explanations of the epidemic and continued to both treat family members at home and follow 'hands-on' burial rituals when they died, causing the disease to spread rapidly between family members (✓)

b See Table A12.3 below. One mark is given for the construction of the table and one mark each for correct content for local, regional and global.

Table A12.3 Factors limiting the spread of Ebola locally, regionally and globally

Local	Regional	Global
rapid alert and early detection; strict infection control in all treatment facilities; specially trained burial teams; PPE that includes the use of impermeable and fluid-resistant gowns and coveralls, respiratory protection and boot covers; training of medical staff; disinfecting of all contaminated areas, including homes; public health campaigns and communication (✓)	construction of regional treatment centres, special wards and treatment beds to isolate the sick; setting up of surveillance to interrupt transmission routes; stockpiling of PPE and other equipment; collaboration with wildlife teams to stop infection from wildlife; public health campaigns and communication (✓)	coordination by WHO of aid from international teams such as the Red Cross and UNICEF; travel warnings against all non-essential travel; quarantine and monitoring of anyone coming from countries with the epidemic; screening at international airports (✓)

CHAPTER 13 HOMEOSTASIS

MODULE 8
NON-INFECTIOUS DISEASE AND DISORDERS

INQUIRY QUESTION:

How is an organism's internal environment maintained in response to a changing external environment?

In order to survive, organisms must be able to maintain system structure and function in the face of changes in their external and internal environments. Changes in temperature and water availability, the incidence and spread of infectious disease and the effects of non-infectious diseases present significant challenges for organisms and require coordinated system responses. **Homeostasis** involves mechanisms that control an organism's responses to environmental change as long as the changes are within the organism's tolerance limits.

1 Constructing and interpreting negative feedback loops that show homeostasis

» Students construct and interpret negative feedback loops that show homeostasis by using a range of sources, including but not limited to temperature and glucose.

→ Living organisms need to control:
- body temperature and metabolic rate
- concentrations of dissolved salts and minerals
- concentrations of nutrients such as glucose
- input and output of water
- quantities of wastes (e.g. nitrogenous wastes)
- oxygen and carbon dioxide concentrations
- removal of malfunctioning cells, pathogens and foreign substances
- repair of damaged cells.

→ Homeostasis can be explained using a stimulus-response model in which a change in the external or internal environmental conditions is detected and appropriate responses occur via a **negative feedback mechanism**. It involves:
- a stimulus which is a change in the environment; that is, a disturbance to homeostasis
- a receptor that detects the change in the environment
- a control centre that controls the response
- an effector that receives the message and carries out the response
- a response that mitigates the original stimulus, hence the negative feedback.

homeostasis: the process by which organisms maintain a relatively stable internal environment despite changes to the external environment

negative feedback mechanism: a self-regulating mechanism that maintains balance or homeostasis as a coordinated response or series of responses to a stimulus or stimuli

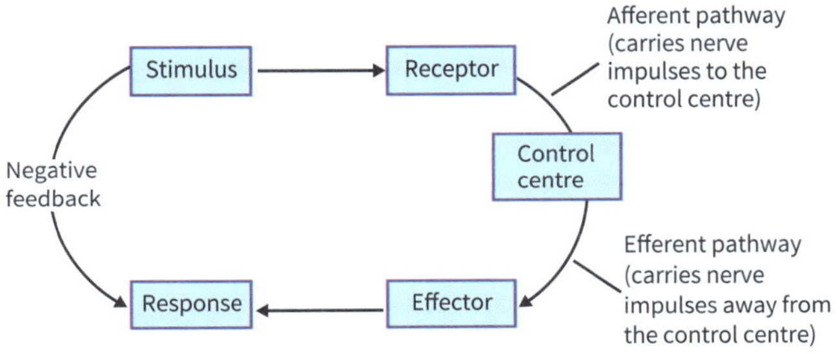

Figure 13.1 Simplified model to represent a stimulus-response model

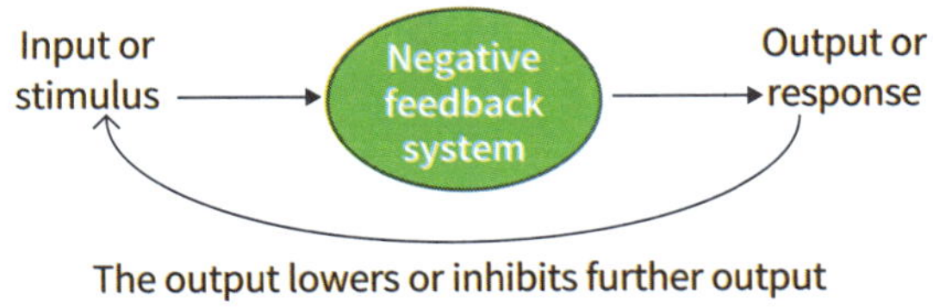

Figure 13.2 Simplified diagram to represent a negative feedback loop

Negative feedback loops that maintain homeostasis for temperature

→ Negative feedback loops act to keep a certain variable in check by opposing the stimulus. When you constructed negative feedback loops you would have put the parts of the stimulus-response pathway together. Most likely you would have used flow diagrams and circular diagrams such as shown in Figure 13.3.

→ All organisms have a range of temperatures in which they can survive; for example, humans can generally only survive external temperatures of about 0 °C to 45 °C and can only be normally active in a range of internal body temperatures between 30 °C to 37.5 °C.

- It is important for living organisms to maintain a constant temperature in order to maintain metabolic processes.
- The temperature in mammals is detected by receptors called thermoreceptors. External temperature changes are detected by receptors in the skin. Internal changes are detected by receptors in the **hypothalamus** in the brain. These inputs are then sent via the **nervous system** to the appropriate effectors such as muscles and glands. The subsequent response brings about changes depending on whether it is hot or cold (see Table 13.1).
- In humans the following negative feedback mechanisms maintain homeostasis for temperature:
  - Detecting changes from the stable state: if our body temperature rises, the temperature rise in the blood stimulates the brain's anterior hypothalamus (and inhibits the posterior hypothalamus). Alternatively when we experience cold, skin receptors increase activity and send nerve impulses to the posterior hypothalamus.
  - Counteracting changes from the stable state: after detecting the rise in body temperature the hypothalamus then stimulates heat loss by increasing blood circulation through the skin, increasing sweating and decreasing metabolic and muscular activity. This lowers the body's temperature. After detecting a drop in temperature, activity in the posterior hypothalamus stimulates the sympathetic nervous system to activate mechanisms that conserve heat.

hypothalamus: a part of the brain that controls certain metabolic processes; it secretes certain hormones and links the nervous system to the endocrine system via the pituitary gland
nervous system: a complex system of nerves and cells that carry messages to and from the brain and spinal cord

Negative feedback loops that maintain homeostasis for glucose

- Glucose is the main source of fuel for cells so it is important that the amount of glucose in the blood is carefully controlled.
- The pancreas is responsible for producing two hormones—**insulin** and **glucagon**—that regulate the amount of glucose in the blood.
- Insulin lowers blood glucose (blood sugar) levels by facilitating glucose transport from the blood into cells. After you have eaten a meal the blood glucose levels rise because the carbohydrates in the food are digested and absorbed.

insulin: a hormone made by the pancreas that helps maintain homeostasis of blood glucose levels
glucagon: a hormone made by the pancreas that prevents blood glucose levels from dropping too low

Table 13.1 Simplified model of thermoregulation in humans

Stimulus ↓	increase in body temperature when exercising or when external temperature is hot	decrease in body temperature when external temperature is cold
Receptors ↓	thermoreceptors in the skin detect change in temperature; receptors in the hypothalamus detect rise in blood temperature	thermoreceptors in the skin detect change in temperature; receptors in the hypothalamus detect decrease in blood temperature
Afferent pathway to control centre ↓	heat loss centre in the hypothalamus is activated (heat promoting centre shuts off)	heat promoting centre in the hypothalamus is activated (heat loss centre shuts off)
Efferent pathway from the control centre to the receptors ↓	message sent from the hypothalamus to blood vessels in the skin and sweat glands	message sent from the hypothalamus to blood vessels in the skin and skeletal muscles
Receptors such as glands, blood vessels and skeletal muscles ↓	blood vessels dilate; capillaries become flushed with blood and sweat glands are activated	blood vessels constrict; blood is diverted from skin capillaries to deeper tissues; skeletal muscles are activated
Response ↓	heat radiates from the skin's surface so that the body cools down evaporative heat loss due to perspiration	heat loss from skin surface is minimised; shivering begins and heat is produced
Feedback	blood temperature drops	blood temperature rises

- This rise is detected by beta cells in the pancreas, which produce more insulin. This insulin binds to receptor proteins in cell membranes, particularly in the liver. This causes **protein channels** to open so that more glucose can enter the cell. Insulin also inhibits glucose production from amino acids and fatty acids, and triggers enzymes to convert glucose to glycogen for storage. As a result of these processes blood glucose levels fall.
- Glucagon raises blood glucose levels by metabolising glycogen into glucose. If blood glucose levels fall due, for example, to strenuous exercise, alpha cells in the pancreas produce glucagon. Glucagon triggers the release of an enzyme that breaks down glycogen to glucose and stimulates the release of glucose from the liver and glucose production from amino acids and fatty acids. As a result of these processes blood glucose levels rise.
- Glucose levels in the blood are usually measured in terms of milligrams per decilitre (mg/dl), with a normal range of 70 to 110 mg/dl. If glucose levels move out of this range the amounts of insulin and glucagon produced by the pancreas will be adjusted to bring glucose levels back to homeostasis. When the system is functioning properly there is always some insulin and some glucagon being produced, creating a balance between glucose release into the blood and glucose uptake into cells.

protein channels: these allow the transport of specific substances across the cell membrane

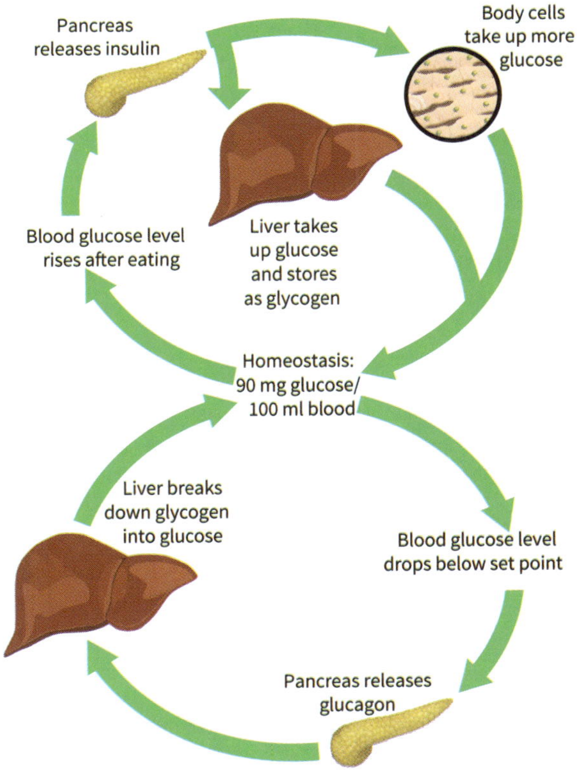

Figure 13.3 Model of negative feedback loops that show homeostasis for glucose

→ KEY QUESTIONS

1. What is homeostasis?
2. Explain the importance of homeostasis.
3. What is a negative feedback mechanism?
4. What is the difference between a positive feedback mechanism and a negative feedback mechanism?
5. Construct a simple model of a negative feedback loop to show how temperature is maintained. Briefly explain the mechanism you have modelled.

Answers ⮌ p. 185

2 Investigating the mechanisms used by organisms to maintain their internal environment within tolerance limits

» Students investigate the various mechanisms used by organisms to maintain their internal environment within tolerance limits, including:

- trends and patterns in behavioural, structural and physiological adaptations in endotherms that assist in maintaining homeostasis
- internal coordination systems that allow homeostasis to be maintained, including hormones and neural pathways
- mechanisms in plants that allow water balance to be maintained.

Trends and patterns in behavioural, structural and physiological adaptations in endotherms that assist in maintaining homeostasis

- Endotherms such as birds and mammals generate their own body heat internally and therefore generally maintain a constant body temperature. Body temperatures are controlled largely by altering metabolic processes and by adaptive mechanisms that control the rate of heat exchange with the environment.
- The concentration of water in cells must also be maintained within a narrow range for optimal function. Water is the major solvent for chemical reactions and the amount of water will affect the concentration of materials in cells. Temperature tends to stay relatively constant in cells that contain adequate water.
- In Year 11 you learnt about behavioural, structural and physiological adaptations (see Chapter 7). Structural features, changes in an organism's metabolic activity, and changes in physiological processes and behaviour enable an organism to maintain its internal environment within tolerance limits.

- Trends and patterns in behavioural, structural and physiological adaptations in endotherms that assist in **thermoregulation** include changing behaviour; for example, curling up into a ball to keep warm. Other adaptations include:
  - increasing or decreasing metabolic heat production; for example, muscle contraction through behaviours such as shivering and rubbing hands increases muscle activity and boosts heat production. Fat tissue (called **adipose tissue**) contains many mitochondria that release energy directly as heat.
  - controlling the exchange of heat with the environment: this includes structural and physiological circulatory mechanisms such as altering blood-flow patterns. For example, **vasoconstriction** of the blood vessels that supply the skin reduces blood flow and helps to retain heat while **vasodilation** increases blood flow and increases heat loss to the environment.
- Table 13.2 summarises some behavioural, structural and physiological adaptations in endotherms that assist in maintaining homeostasis.

thermoregulation: processes designed to maintain core internal temperature and return the body to homeostasis
adipose tissue: connective tissue in which fat is stored
vasoconstriction: the narrowing of the diameter of blood vessels to slow the flow of blood, which increases blood pressure
vasodilation: the widening of the diameter of blood vessels to increase the flow of blood

Internal coordination systems that allow homeostasis to be maintained: hormones and neural pathways

- Receptors and effectors in vertebrates are linked via a control centre by nervous or neural and hormonal pathways.
- The nervous system provides rapid and short-term coordination of internal organ systems. The hormonal system provides slower and longer-lasting response coordination.
- The nervous system consists of the central nervous system—the brain and spinal cord—and **peripheral** nerves. The basic unit of the nervous system is the neuron or nerve cell. Nerve cells transmit signals by electrochemical changes in their membranes. The signals travel like a wave from dendrites to the axon and then have to cross synapses to other neurons (synaptic transmission, see Figure 13.6). The long axon is covered in a myelin sheath that insulates the fibre. In humans axon fibres in motor neurons (neurons that activate muscles) can be one metre long and in sensory neurons (neurons that convey messages to the central nervous system) they can be 1.5 metres in length.
- Special endings on the sensory nerves, such as heat sensors or thermoreceptors in the skin, detect stimuli such as changes in temperature. These receptors relay messages that are processed in the central nervous system and then messages are conveyed to effector organs or muscles that bring about the response.

peripheral: on the outside, in this case the nerves and cells outside the brain and spinal cord

Table 13.2 Behavioural, structural and physiological adaptations in endotherms that assist in maintaining homeostasis

	Behavioural adaptations	Structural adaptations	Physiological adaptations
Temperature regulation for cooling down or losing heat	burrowing nocturnal activity seeking shade cooling down in water stretching out to increase the surface area for heat loss	large ears increase the surface area for evaporative heat loss a larger surface area to volume ratio helps to maximise heat loss	sweating to increase evaporative heat loss increased blood flow to the extremities: more blood enters skin capillaries and heat is lost
Temperature regulation for staying warm	hibernation migration curling up in a ball or huddling together to decrease surface area	insulation such as fat layers, blubber, hair, fur and feathers a smaller surface area to volume ratio helps to minimise heat loss	reduced blood flow to extremities which keeps the core of the body warm shivering causes rapid contraction and relaxation of skeletal muscles resulting in more heat being generated increases in metabolic activity

KEY QUESTIONS

6 **What are endotherms?**

7 **List two trends and patterns in behavioural adaptations in endotherms that assist in maintaining homeostasis.**

8 **List two trends and patterns in structural and physiological adaptations in endotherms that assist in maintaining homeostasis.**

Answers ➲ p. 185

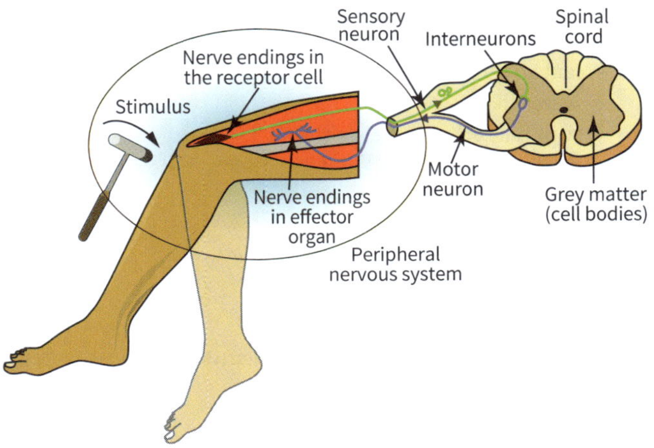

Figure 13.4 Neural pathways consist of cells that transport nerve impulses from sensory receptors to neurons and on to effectors.

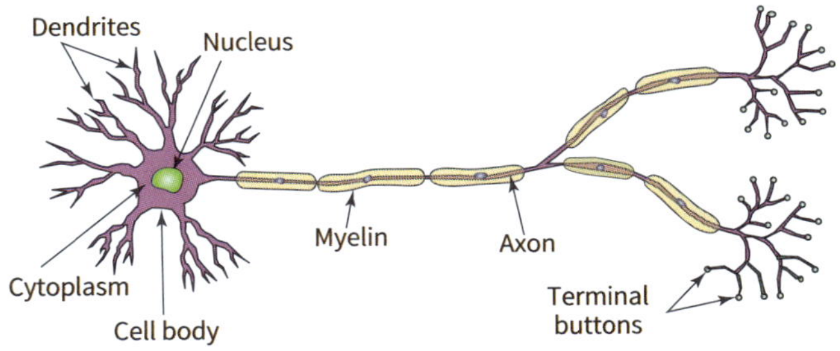

Figure 13.5 A neuron

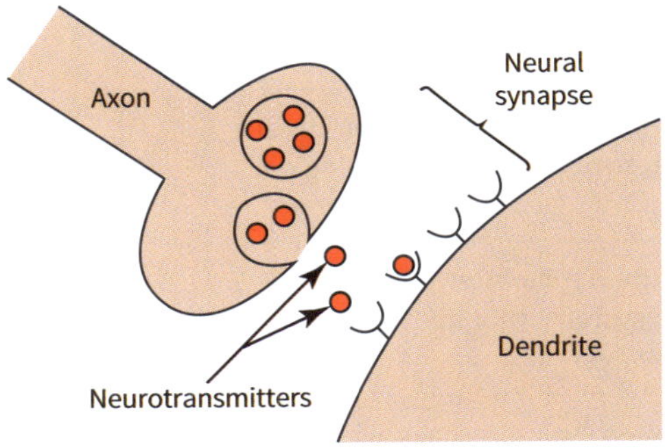

Figure 13.6 Synaptic transmission involves neurotransmission and signal transduction.

- Hormones are produced by the **endocrine system**. Hormones alter the metabolism of target cells, tissues or organs by increasing or decreasing their activity. A specific hormone will fit a specific receptor protein and this brings about a change in that cell's activity. Most hormones in animals are produced in endocrine glands as a result of nervous or chemical stimulation and are released by the glands into the bloodstream. They then travel via the circulatory system to the target cells, tissues or organs.
- Examples of endocrine glands include the hypothalamus, pituitary gland, thyroid, adrenals, pancreas, ovaries and testes.
- In Section 1 you reviewed hormones responsible for regulating blood glucose. Other examples of hormones include:
  - adrenalin, which increases heart rate, blood pressure and changes the body's metabolism
  - cortisol: a **steroid** hormone that regulates processes such as metabolism and immune response
  - antidiuretic hormone (ADH) and aldosterone—together these hormones help maintain water balance.
- Aldosterone brings about retention of salts within the body. ADH brings about reabsorption of water within the body.
- If a person begins to dehydrate, their blood volume drops. This change is detected by the hypothalamus, which stimulates the pituitary to secrete ADH. ADH stimulates the **nephrons** in the kidney to reabsorb more water. This decreases urine production, increases urine concentration and maintains the concentration of water in the blood.

endocrine system: a collection of glands that secrete hormones directly into the nervous system
steroid: a group of hormones derived from cholesterol
nephrons: the functional unit of the kidneys (see Chapter 17 for details)

KEY QUESTIONS

9 **Describe the two internal coordination systems that allow homeostasis to be maintained in endotherms.**

10 **What is a key difference between the nervous system and the hormonal system?**

11 **Provide an example of how neural pathways allow homeostasis to be maintained.**

12 **Provide an example of how hormones allow homeostasis to be maintained.**

Answers ⮌ p. 185

Mechanisms in plants that allow water balance to be maintained

- In Year 11 you studied leaf structure and gas exchange (see pages 55 and 84 of *Excel Year 11 Biology*) and the transpiration stream in plants (see pages 60 and 80). Water is taken up by the roots and is transported to the leaves and lost to the air through transpiration. Water enters root hair cells by osmosis.
- Mechanisms in plants that allow water balance to be maintained are complex and include specific adaptations for obtaining water and minimising water loss as well as metabolic processes.
- Different plants have different structures, different capacities for water uptake or minimising water loss, and different responses in times of water stress. Australian sclerophyll plants have a range of adaptations to minimise

water loss (and at the same time allow for gas exchange) such as thin leaves with a waxy cuticle, sunken stomata and stomata only on the underside of the leaf. Leaves hang vertically to reduce the surface area exposed to the sun. Plants often grow sparsely to minimise competition for limited water and have shallow, extensive root systems to enable uptake of water as soon as it rains. Other plants have deep root systems to access water from deeper soil layers.

→ Other mechanisms in plants that allow water balance to be maintained include:
- decreasing photosynthesis in response to water stress: this occurs through a number of mechanisms such as closing stomata, which avoids excessive water loss by reducing leaf transpiration
- dropping leaves in response to water stress through the production of abscisic acid
- leaf shape, such as small, thick leaves to minimise water loss
- root hairs that increase the surface area, ensure close contact with the soil and greatly increase the rate of water uptake
- fleshy leaves that store water.

→ KEY QUESTION

13 **Describe two mechanisms in plants that allow water balance to be maintained.**

Answer ⮌ p. 185

CHAPTER SYLLABUS CHECKLIST

Are you able to answer these questions from the syllabus for this chapter? Tick each question as you go through the checklist if you are able to answer it. If you cannot answer a question, turn to the relevant page in the study guide to find the answer. For NESA key word meanings, go to www.educationstandards.nsw.edu.au and search 'key words'.

FOR A COMPLETE UNDERSTANDING OF THIS TOPIC:		PAGE NO.	✓
1	Can I construct and interpret negative feedback loops that show homeostasis by using a range of sources, including but not limited to temperature and glucose?	177–179	
2	Can I investigate the various mechanisms used by organisms to maintain their internal environment within tolerance limits, including trends and patterns in behavioural, structural and physiological adaptations in endotherms that assist in maintaining homeostasis?	179–180	
3	Can I investigate the various mechanisms used by organisms to maintain their internal environment within tolerance limits, including internal coordination systems that allow homeostasis to be maintained, including hormones and neural pathways?	180–181	
4	Can I investigate the various mechanisms used by organisms to maintain their internal environment within tolerance limits, including mechanisms in plants that allow water balance to be maintained?	181–182	

HSC EXAM-TYPE QUESTIONS

Objective-response questions (1 mark each)

1 **Which of the following is a model of a feedback mechanism?**
- A Marine fish have the ability to excrete salt.
- B Animals that live in the desert are often inactive during the day.
- C When an oven reaches the set temperature this is detected by the thermostat which then turns off the heating element.
- D Haemoglobin assists blood to transport more oxygen when needed.

2 **The photo below shows a bilby. The bilby is an endotherm that lives in dry, hot habitats. It has large ears and a small body size. Which of the statements below best explains the adaptations that assist the bilby in maintaining homeostasis?**

Figure 13.7 Bilby

- A The large ears reduce the surface area for evaporative heat loss.
- B The large ears and small body size are both structural adaptations to help maintain a constant body temperature.
- C The large ears and small body size are both physiological adaptations to help maintain a constant body temperature.
- D The small body size decreases the surface area to volume ratio and allows for heat loss.

3 **The diagram below shows a negative feedback loop. What is the most accurate description for the boxes labelled X, Y and Z?**

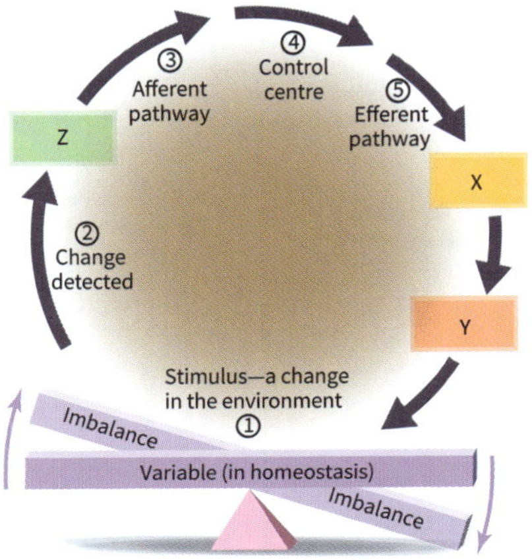

Figure 13.8 Negative feedback loop

	X	Y	Z
A	receptor	response	effector
B	effector	response	receptor
C	response	effector	receptor
D	receptor	effector	response

4 **Which statement below represents a trend or pattern in behavioural adaptations in endotherms that assists in maintaining homeostasis?**
- A licking the skin to stay warm
- B activating the heat loss and heat promoting centre of the brain
- C metabolic rate going up and down
- D cooling down by stretching out and keeping warm by curling up in a ball

5 **Under what conditions is homeostasis important?**
- A in all organisms all of the time
- B when feedback mechanisms stop working
- C when organisms have a stable external environment
- D when organisms have a stable internal environment

Short-answer questions

6 The following diagram is a representation of a negative feedback mechanism involving the antidiuretic hormone (ADH). Antidiuretic hormone is also called vasopressin. The hormone helps maintain the balance of water and salt levels in the blood. Interpret the diagram to complete the labels for X and Y. (2 marks)

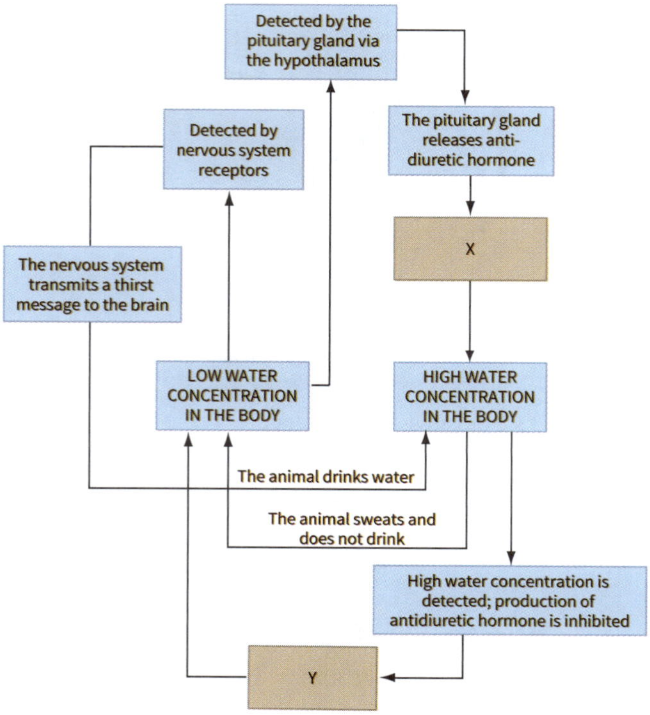

Figure 13.9 Feedback mechanism involving antidiuretic hormone (ADH)

7 Explain the similarities and differences between the nervous system and the endocrine system. Use dot points to summarise your answer. (4 marks)

8 Complete the following table to summarise trends and patterns in structural and physiological adaptations in endotherms that assist in maintaining homeostasis. (4 marks)

Table 13.3 Table template

Maintaining homeostasis if the environment is hot	Maintaining homeostasis if the environment is cold

9 Read the following description of a negative feedback loop involving the hormones CRH, ACTH and glucocorticoids and then answer the questions which follow.

The hypothalamus secretes a hormone called CRH. CRH stimulates the pituitary gland to secrete a hormone called ACTH. In turn, ACTH directs the adrenal cortex to secrete glucocorticoids. Glucocorticoids then prevent the release of further secretions from both the hypothalamus and the pituitary gland.

a Use the information to construct a negative feedback loop. (3 marks)

b Explain how hormone secretion is regulated by a negative feedback loop. (3 marks)

10 Researchers have conducted water stress experiments on plants under field conditions in order to investigate the mechanisms that allow water balance to be maintained. They have observed that when water stress is imposed slowly on plants, one of the first events to take place is stomatal closure in response to the migration of chemicals synthesised in dehydrating roots. The decline in intercellular CO_2 following stomatal closure brings about a down-regulation in photosynthesis.

a Explain why stomatal closure is used by plants as a mechanism that allows water balance to be maintained. (2 marks)

b Construct a simple flow diagram as a stimulus-response model to represent the mechanisms observed in this study. (3 marks)

ANSWERS

KEY QUESTIONS

Key questions ⊃ p. 179

1 Homeostasis is the process by which organisms maintain their internal environment within tolerance limits despite changes in behaviour and the external environment.

2 Living organisms are exposed to frequent changes in their external environment. The internal environment is also constantly changing. Despite these changes organisms must be able to maintain a constant internal environment for optimum metabolic activity; for example, for a high level of efficiency for running cells, enzymes must have the optimum temperature and pH conditions.

3 Negative feedback is the mechanism by which the body maintains conditions within particular tolerance limits. The body will do this by opposing a change that deviates from the normal; for example, if glucose levels in the blood go up insulin is released by the pancreas into the bloodstream, facilitating the uptake of glucose by cells, and the blood glucose levels go down.

4 A positive feedback mechanism is the exact opposite of a negative feedback mechanism. With negative feedback the output reduces or inhibits the original effect of the stimulus. In a positive feedback system the output enhances the original stimulus. An example of a positive feedback mechanism is blood clotting. Once a vessel is damaged platelets start to cling to the injured site and release chemicals that attract more platelets. The platelets continue to pile up and release chemicals until a clot is formed.

5 The stimulus is a change in temperature. Thermoreceptors are located in the skin for sensing external temperature and the hypothalamus for internal temperatures of the blood. The control centre is the hypothalamus. The response takes place in effectors such as the smooth muscles of the blood vessels, sweat glands, muscles on the skin (making goosebumps), skeletal muscles (causing shivering) and hormonal glands to increase metabolic rate. All of these processes cause the body temperature to drop.

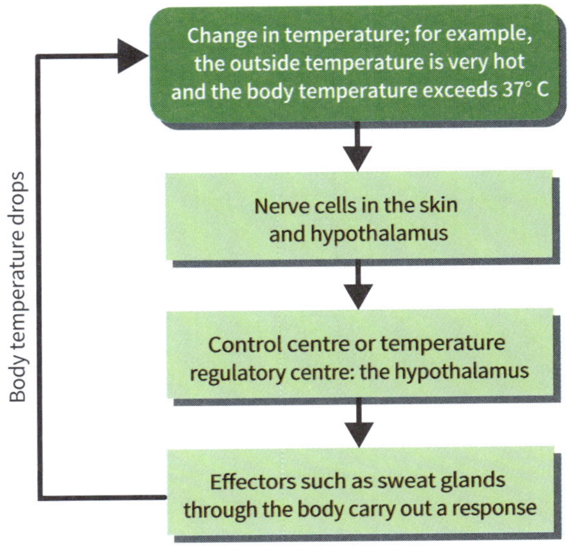

Figure A13.1 Simplified model of a negative feedback loop to show how temperature is maintained

Key questions ⊃ p. 180

6 Endotherms maintain a constant body temperature. Endothermic animals have varying thermoregulatory mechanisms that involve structural features, behavioural responses, and physiological and homeostatic mechanisms to control heat exchange and metabolic activity.

7 Increased temperature causes small blood vessels in the skin to dilate, increasing the blood flow to the skin and therefore the amount of heat lost, while the activity of the sweat glands increases.

8 Responses to cold involve a decrease in blood circulation through the skin. The activity of the sweat glands decreases metabolic activity and muscular activity increases, causing hairs on the skin to become erect in order to trap an insulating layer of air (seen as goosebumps).

Key questions ⊃ p. 181

9 Internal coordination of homeostasis is controlled by two systems: the nervous system and the endocrine system (the system of hormones).

10 Signals are sent throughout the body in two main ways. The first is through nerves of the nervous system where they are sent as nerve impulses that travel through nerve cells called neurons. These impulses are sent to other neurons or specific target cells at a location in the body that the neuron extends to. The second way is through the circulatory system. These signals are transmitted by specific chemicals (hormones) which travel through the circulatory system.

11 Most of the signals that the human body uses to regulate body temperature are sent through the nervous system.

12 Blood glucose levels are maintained through hormone signalling.

Key question ⊃ p. 182

13 One mechanism is metabolic processes such as decreasing photosynthesis in response to water stress through stomatal closure. Another is structural adaptations such as deep root systems that can access water from deeper in the soil.

HSC EXAM-TYPE QUESTIONS

Objective-response questions

1 **C**. This option demonstrates the process of detecting a change and switching off further action so that the stable state can be achieved. **A** is a statement about adaptations of fish, not a feedback mechanism. **B** is a statement about responses of desert animals to minimise heat loss, not a feedback mechanism. **D** is a statement about the adaptive advantage of haemoglobin but does not relate to feedback.

2 **B**. Large ears and small body size are both structural adaptations to help maintain a constant body temperature. **A** is incorrect because large ears increase surface area. **C** is wrong because these are not physiological adaptations. **D** is incorrect because small body size increases the surface area to volume ratio.

3 **B**. This is the only correct alternative. In the model X is the effector, Y is the response and Z is the receptor.

4 **D**. Cooling down by stretching out increases the surface area for heat loss and curling up in a ball decreases the surface area, thus minimising heat loss. **A** is incorrect as though licking skin is behavioural it is used to increase evaporative heat loss (physiological) to cool down. **B** is incorrect because activating the heat loss and heat promoting centre of the brain describes what happens in the control centre: the hypothalamus. **C** is incorrect because metabolic rate going up and down is a physiological, not behavioural, response.

5 **A**. Very rarely is the internal or external environment unchanging so homeostasis is always required. **C** and **D** are incorrect because homeostasis is the process that responds to changes in the environment. **B** is incorrect because feedback mechanisms are the processes that allow homeostasis to occur and if feedback stopped working homeostasis could not occur.

Short-response questions

6 **EM** The question asks you to interpret a model of a negative feedback mechanism. Even though you may not have studied this example you need to be able to interpret—draw meaning from—negative feedback loops.

X = ADH hormone stimulates the kidney to reabsorb more water into the blood. (✓)

Y = the kidney responds by decreasing the reabsorption of water into the bloodstream and produces dilute urine. (✓)

7 **EM** In this question you are asked to explain similarities and differences; that is, to compare and contrast. Make sure you do both in your answer.

Both systems are involved in internal coordination that allows homeostasis to be maintained. (✓)

The nervous system provides rapid and short-term coordination of internal organ systems whereas the hormonal system provides slower longer-lasting response coordination. (✓)

The nervous system consists of the central nervous system (the brain and spinal cord) and peripheral nerves. Nerve cells transmit signals by electrochemical changes in their membranes. (✓) whereas the endocrine system consists of glands that secrete specific hormones. Hormones alter the metabolism of target cells, tissues or organs by increasing or decreasing their activity. Signals (hormones) travel via the circulatory system. (✓)

8 **EM** Make sure your answer explains a pattern and that your examples are for both structural and physiological adaptations.

Table A13.1 Structural and physiological adaptations in endotherms

Maintaining homeostasis if the environment is hot	Maintaining homeostasis if the environment is cold
large surface areas are structural adaptations that help increase the rate of evaporative heat loss (physiological) (✓)	small surface areas are structural adaptations that help minimise the rate of evaporative heat loss (physiological) (✓)
increased blood flow to the extremities means more blood enters skin capillaries and heat is lost (✓)	reduced blood flow to extremities keeps the core of the body warm (✓)

9 **EM** There are two parts to this higher-order question: constructing a negative feedback loop from information given and writing a general explanation about how hormone secretion is regulated. The most important part of the diagram is the return loop showing the negative feedback.

a

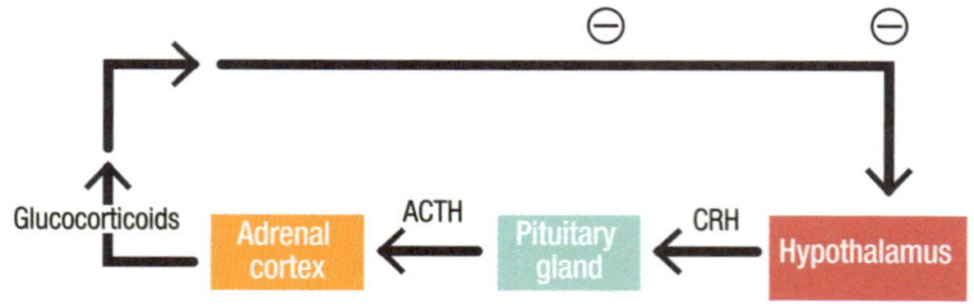

Figure A13.2 A negative feedback loop involving the hormones CRH, ACTH and glucocorticoids

b For hormone secretion regulated by the negative feedback loop: when gland X releases hormone X this stimulates target cells to release hormone Y. (✓) When there is an excess of hormone Y (✓) gland X senses this and inhibits its release of hormone X. (✓)

10 **EM** The question has two parts. Part **a** is straight recall and part **b** asks you to interpret the information provided to construct a flow diagram. Make sure you answer both parts. One mark is given for each part of the diagram.

a Stomatal closure is used by plants as a mechanism that allows water balance to be maintained. (✓) Closing stomata avoids excessive water loss by reducing leaf transpiration. (✓)

b

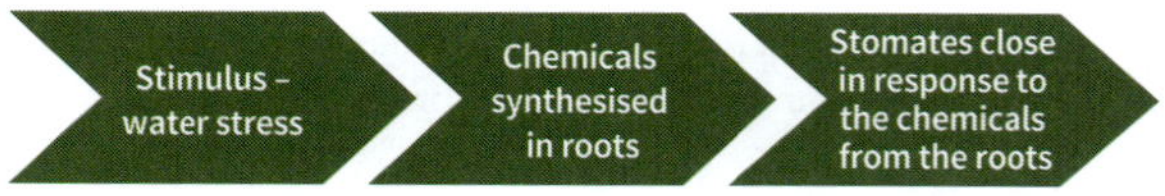

Figure A13.3 Simple flow diagram of the stimulus-response model to represent the mechanisms observed in this study

MODULE 8
NON-INFECTIOUS DISEASE AND DISORDERS

CHAPTER 14 CAUSES AND EFFECTS

INQUIRY QUESTION:

Do non-infectious diseases cause more deaths than infectious diseases?

In Chapters 14, 15 and 16 you will review **non-infectious diseases** in humans, including their causes and effects on human health, the epidemiology of non-infectious diseases in populations and the technologies used in treating diseases.

In Chapter 13 you reviewed homeostasis. Many diseases are best understood as a disturbance of homeostasis. In this chapter you review non-infectious diseases and their causes and effects on human health.

1 Investigating the causes and effects of non-infectious diseases in humans

» Students investigate the causes and effects of non-infectious diseases in humans, including but not limited to genetic diseases, diseases caused by environmental exposure, nutritional diseases and cancer.

- Non-infectious diseases have now exceeded infectious diseases as the leading cause of death and disability worldwide. Disease rates from these conditions are accelerating globally and advancing across every region and pervading all socio-economic classes. The World Health Organization's *World Health Report 2002* indicates that the mortality, morbidity and disability attributed to the major chronic diseases accounted for almost 60% of all deaths and 43% of the global **burden of disease**. By 2020 their contribution is expected to rise to 73% of all deaths and 60% of the global burden of disease. It should be noted that 79% of the deaths attributed to these diseases occur in developing countries.
- Figure 14.1 below provides data showing global patterns in diseases from 2000 to 2015. Of the 56.4 million deaths worldwide in 2015 more than half were due to the top ten

non-infectious diseases: diseases and disorders caused by something other than a pathogenic organism; also referred to as non-communicable diseases (NCDs)
burden of disease: quantitative measures of the impact of a disease on an individual or population

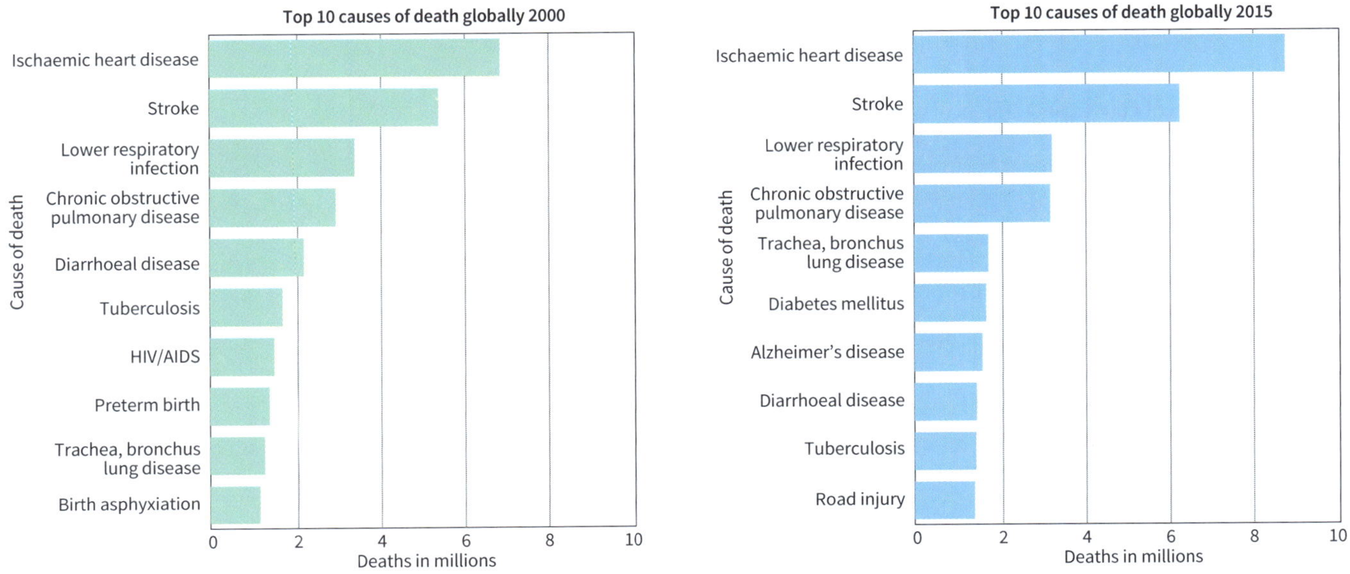

Figure 14.1 Data showing the top ten causes of death worldwide in 2000 compared to 2015
Source: www.who.int/mediacentre/factsheets/fs310/en/

causes. **Ischaemic heart disease** and stroke have the highest mortality rates, accounting for a combined 15 million deaths in 2015. These diseases have remained the leading causes of death globally in the last 15 years. Chronic obstructive pulmonary disease claimed 3.2 million lives in 2015 while lung cancer (along with trachea and bronchus cancers) caused 1.7 million deaths. Diabetes killed 1.6 million people in 2015, up from less than one million in 2000. Deaths due to dementias more than doubled between 2000 and 2015, making it the seventh leading cause of global deaths in 2015.

- Note: Infectious diseases remain a major cause of preventable illness and death in many countries in the world, particularly in developing countries. According to the World Health Organization (WHO) infections still account for about 40% of recorded deaths worldwide and each year more than 1.5 million children die from diseases that could be prevented by vaccination (Source: data.unicef.org/topic/child-health/immunization/).
- Non-infectious diseases are caused by factors such as genetic disorders, environmental exposure, nutrition and cancer. Some non-infectious diseases can be caused by a combination of factors. For example, **mitochondrial diseases** can be caused by mutations in mitochondrial DNA or in nuclear genes that code for mitochondrial components. They may also be acquired due to environmental causes such as the adverse effects of drugs or injections.

ischaemic heart disease: ischaemia refers to reduced or blocked blood supply; ischaemic heart disease includes heart attacks and chest pain; also called coronary heart disease

mitochondrial diseases: a group of disorders such as mitochondrial myopathy caused by malfunctioning mitochondria

- Complex interactions between genetics, lifestyle (such as nutrition) and environment increase the risk factors in diseases such as diabetes (see page 190 below).

The causes and effects of genetic diseases in humans

- **Genetic diseases** are inherited disorders caused by genetic mutations passed down from parent(s) to offspring (see Table 14.1). In Chapters 4 and 6 you reviewed mutations, which are permanent changes in DNA or RNA. Mutations in germ-line cells result in a large proportion of inherited disorders.

The causes and effects of diseases caused by environmental exposure

- Environmental diseases are caused by **environmental exposure** to factors such as high levels of stress, noise, overcrowding, drugs and pollutants (see Table 14.2). In Chapter 6 you reviewed the role that mutagens play in increasing the risk of mutations. Mutagens can be classed as physical, such as ultraviolet radiation, or chemical. Mutagenic chemicals include nicotine, pesticides, **asbestos**, mercury, lead and food additives. Physical and chemical mutagens in the environment have been linked to diseases such as cancer.

genetic disease: an inherited disorder caused by genetic mutations passed down from parent(s) to offspring

environmental exposure: exposure to a range of potential mutagens in the environment often created by humans; includes hydrocarbons, pesticides, certain foods, pollution and cigarette smoke

asbestos: a fibrous, silicate mineral that was once used in a wide range of materials including fibro, brake linings, fabrics and insulation

Table 14.1 Examples of causes and effects of genetic diseases in humans

Examples	Cause of the disease	Effects
sickle-cell anaemia	caused by a mutation in a gene that tells the body to make haemoglobin; blood cells become misshapen and break down; it is an autosomal recessive disorder; that is, both parents must carry the gene (see Chapter 4)	there are not enough healthy red blood cells to carry adequate oxygen and this leads to fatigue, painful swelling in hands and feet, and frequent infections
Down syndrome (trisomy)	caused by an additional chromosome 21	mild to moderate intellectual disability and poor muscle tone; higher risk of disorders such as congenital heart defects
haemophilia	a sex-linked genetic disorder; the gene that controls blood clotting proteins occurs on the sex chromosomes	males with haemophilia do not have the genetic information for blood clotting
Huntington's disease (see also Chapters 5 and 6)	a faulty gene makes an abnormal, toxic version of the Huntington protein; the toxic protein collects in the brain, causing damage to brain and nerve cells; it is an autosomal dominant disorder	progressive brain disorder causes uncontrolled movements, emotional problems and loss of cognitive function

Table 14.2 Examples of the causes and effects of diseases caused by environmental exposure

Examples	Cause of the disease	Effects
mesothelioma	a rare type of cancer affecting the membrane that covers and protects the lungs (pleura) it is almost always caused by exposure to asbestos though it can take many years after exposure for mesothelioma to develop see pages 190–192 for more details on cancer	if tiny fibres of asbestos are breathed in or swallowed they can cause inflammation in the protective membranes called mesothelium and form a tumour most cases occur in the pleural mesothelium of the chest cavity where the malignant cells can invade and damage nearby tissue effects include shortness of breath, chest pain or pain in the shoulder and upper arm, weight loss and eventually respiratory failure and pneumonia; it is a fatal disease for more information see www.asbestos.com
skin cancer (e.g. malignant melanoma)	excessive exposure to ultraviolet radiation melanoma is a type of skin cancer that usually develops when a mole becomes cancerous see Chapter 15 and pages 190–192 for more information on cancer	melanoma is the most lethal form of skin cancer and tumours start in the pigment-producing melanocytes in the skin new or unusual growths on the skin or changes to moles are often the first signs of the effect of melanoma malignant cells can spread to other parts of the body
asthma	an inflammatory disease of the airways though the causes are not fully understood people with asthma have sensitive airways in their lungs that swell, become narrow and produce more mucous, usually in response to environmental triggers such as exposure to cigarette smoke (especially as a baby or young child), pollen and other **allergens**, air pollution and some chemicals	effects can vary in severity from person to person and include breathlessness, a wheezing, tight feeling in the chest, and a continuing cough asthma attacks can be fatal see www.asthmaaustralia.org.au for more information

The causes and effects of nutritional diseases in humans

→ **Nutritional diseases** are caused by the nutritional deficiency of an important compound, such as a protein, vitamin or mineral, and/or undernutrition. Other causes include excesses in diet along with other behavioural risk factors such as unhealthy eating, alcohol consumption, poor food choices, tobacco smoking and lack of physical exercise. Eating disorders such as bulimia and anorexia nervosa are also classed as nutritional diseases although the causes are complex.

allergens: substances that cause an allergic response; that is, an over response of the immune system

nutritional disease: nutrient-related diseases that include deficiencies or excesses in diet

Table 14.3 Examples of nutritional diseases caused by nutritional deficiency

Examples	Cause of the disease	Effects
vitamin K deficiency	lack of green leafy vegetables in the diet and therefore low levels of vitamin K	vitamin K is critical for the formation of coagulation (clotting) factors and a deficiency may prevent blood from clotting, leading to serious blood loss
anaemia	iron deficiency	iron is necessary for the production of healthy red blood cells so the blood is unable to carry enough oxygen to the tissues
night blindness	vitamin A deficiency	vision problems as vitamin A is critical for the formation of photoreceptive (light-receptive) pigment in the retina (also see Chapter 17)
kwashiorkor	a severe form of malnutrition caused by a protein deficient diet a disease of poverty, it is usually found in countries where rice and starchy food are the basic diet	protein is the basic building block for many compounds in the body and without protein, growth and normal body functions slow down people with kwashiorkor have an emaciated body except for the ankles, feet and belly, which swell with fluids

- Examples of diseases caused primarily by excesses in diet and other behavioural **risk factors** such as unhealthy eating and lack of physical exercise include **obesity** and **diabetes mellitus**, in particular type 2 diabetes. Note: Researchers have evidence that genetic factors make some people more vulnerable than others to type 2 diabetes. While it makes sense that obesity is a nutritional disease, researchers also consider other risk factors such as social, cultural and family contexts and food choices. Food choices depend upon the balance of environmental factors, including access to affordable and nutritious foods, the increase in fast-food outlets, widespread access to highly processed foods, trends in portion sizes of food and drinks, marketing and advertising of unhealthy foods, and access to food labelling information on ingredients.
- Overweight is defined for the purpose of population studies as a **body mass index** of 25 or over. Obesity is classified as moderate through to very severe depending on how much over 30 a person's body mass index is. Obesity is caused by an energy imbalance and it is generally agreed that these imbalances are caused by large-scale changes in the modern environment. Research shows that obese adults are nearly five times more likely than normal weight or underweight adults to have high triglycerides (fats in the blood) and more than twice as likely to have lower than normal levels of good cholesterol. These are significant risk factors for cardiovascular disease. Research also shows that obesity is a major risk factor for other chronic health conditions such as depression, type 2 diabetes and liver disease. Sources: Department of Health at www.health.gov.au and the Australian Institute of Health and Welfare at www.aihw.gov.au/reports/australias-health/australias-health-2016/contents/ill-health.
- Diabetes mellitus (usually referred to as simply diabetes) is a chronic condition in which the body cannot properly use glucose, its main energy source (see Chapter 13). It is caused either by the inability to produce insulin or by the body not being able to use insulin effectively, or both.

  Diabetes is marked by an abnormal build-up of glucose in the blood and it can have serious short- and long-term effects. The three main types of diabetes are outlined below:
  - Type 1 diabetes, also known as juvenile-onset or insulin-dependent diabetes, is believed to be caused by an interaction of genetic and environmental factors. It is a disease where the immune system suddenly attacks and irreversibly destroys the insulin-producing cells of the pancreas.
  - Type 2 diabetes is a progressive condition in which the body becomes resistant to the normal effects of insulin and/or gradually loses the ability to produce enough insulin in the pancreas. Impaired glucose tolerance is a condition in which blood glucose levels are higher than normal but less than required for a diagnosis of diabetes. This, which signals an increased risk of developing type 2 diabetes, is often called pre-diabetes. Modifiable risk factors that can lead to type 2 diabetes include insufficient physical activity, saturated fat intake, obesity, tobacco smoking and alcohol consumption.
  - Gestational diabetes is a form of diabetes that is first diagnosed during pregnancy (gestation). It may disappear after pregnancy but signals a high risk of diabetes occurring later on.
- Diabetes may result in a range of other health complications including heart disease, kidney disease, blindness and lower-limb amputation (Source: www.diabetesaustralia.com.au).

risk factor: any factor associated with an increased likelihood of a person developing a disease or disorder; there are different types of risk factors, some of which can be modified and some cannot

obesity: a condition in which excess body fat has accumulated

diabetes mellitus: a group of diseases where people have trouble regulating their blood glucose

body mass index: a commonly used method of assessing whether a person is normal, under or overweight, or obese; it is calculated by dividing the person's weight (kg) by their height squared (m^2)

The causes and effects of cancer in humans

- **Cancer** is a disease of the body's cells. Normally cells grow and multiply in a controlled way but if something causes a mutation to occur in the cells' DNA (or mitochondria) this control can be lost (see also Chapter 6).
- Cancer or **tumour** is the term used to describe the collections of these cells growing and potentially spreading within the body. Cancerous cells can arise from almost any type of tissue.
- Tumours are either benign (do not spread beyond the immediate area in which they arise) or malignant (cells spread into surrounding areas or to different parts of the body).
- Mutagens that cause cancer are referred to as carcinogens.
- Causes of cancer are usually described as risk factors. Each form of cancer has different risk factors. About one third of deaths worldwide from cancer are caused by five behavioural and nutritional risks: being overweight or obese, having a low fruit and vegetable intake, lack of physical exercise, tobacco use and alcohol use (Source: www.who.int/mediacentre/factsheets/fs349/en/). Aging is another key factor for the development of cancer.

cancer: a disease caused by uncontrolled division of the body's cells

tumour: the abnormal growth of tissue

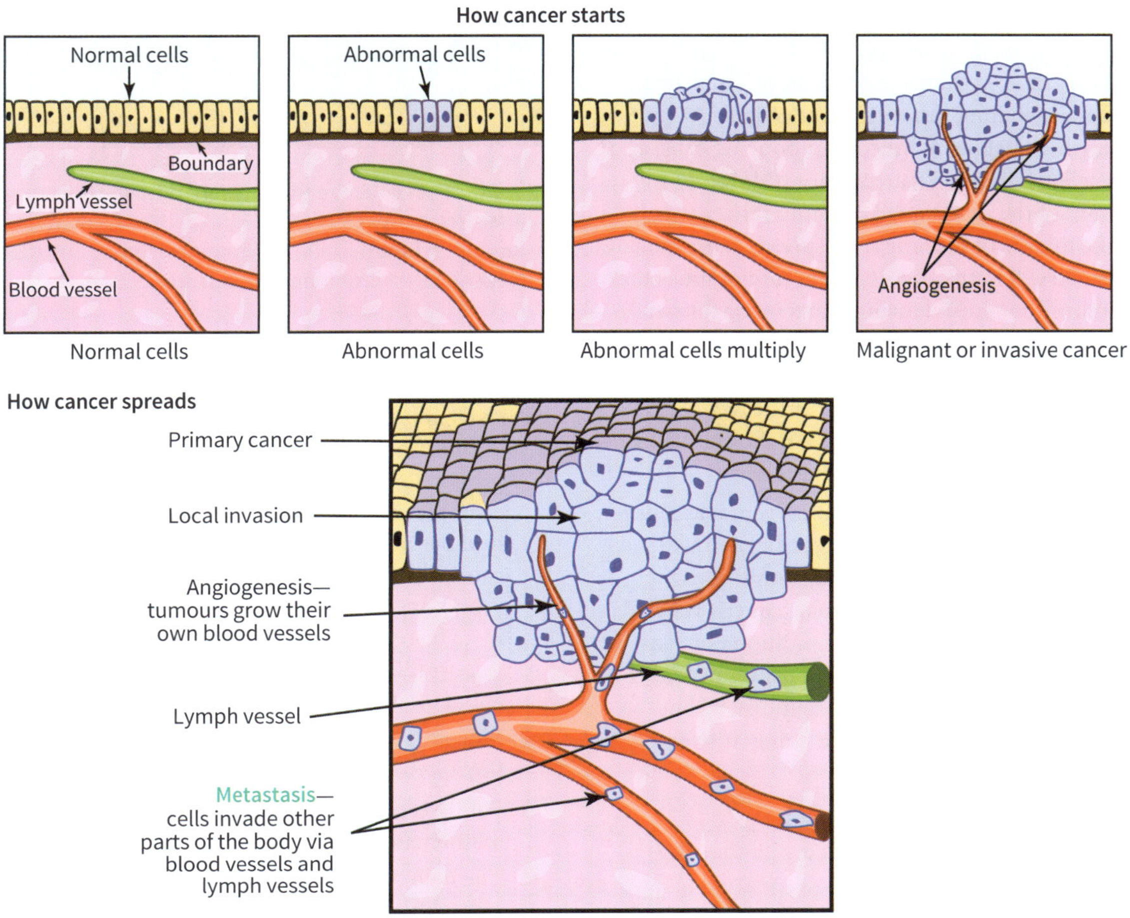

Figure 14.2 How cancer starts and spreads
Source: www.cancer.org.au/about-cancer/what-is-cancer/

- In Australia about 3% of all cancers can be attributed to infectious agents. These cancers include cervical cancer and adult T-cell leukaemia.

metastasis: the spread of cancer in the body away from the primary tumour

- The cause and effects of three types of cancer—breast cancer, blood cancers and lung cancer—are summarised below:
- Breast cancer is the abnormal growth of the cells lining the breast lobules or ducts. Cancer cells can move beyond the primary tumour and be transported via the lymph or blood vessels to other parts of the body. Breast cancer develops when there are mutations in genes due to carcinogen exposure leading to DNA damage. It can also develop when there are errors in DNA replication but these mutations can occur quite randomly. Over 75% of patients with advanced stage breast cancer will have developed metastases into their bones, causing significant pain and side effects such as fractures. Once the cancer reaches this stage chances of successful treatment are much slimmer. The main risk factors for breast cancer are a strong family history of breast cancer, the presence of an inherited faulty gene and early onset (before the age of 12) of menstrual periods (Source: www.abcr.com.au/about/about-breast-cancer/).
- Blood cancers include leukaemia, lymphoma and myeloma. Leukaemia is a cancer that affects the blood and bone marrow. Malignant cells crowd the marrow and affect the body's ability to make normal blood cells. Lymphomas are cancers that affect the lymphatic system. Lymphomas arise when developing lymphocytes undergo a malignant change and multiply in an uncontrolled way. Abnormal lymphocytes are called lymphomas. Over time lymphoma cells replace normal lymphocytes, weakening the immune system's ability to fight infection. Myeloma is a cancer of plasma cells which help fight infection. Risk factors for blood cancers include smoking, exposure to petrochemicals, chemotherapy, and a family history of blood cancers and viruses such as the human T-lymphotropic virus and HIV (Source: www.leukaemia.org.au/).
- Tobacco smoking causes the majority of lung cancers both in smokers and in people who are exposed to cigarette smoke. When inhaled the carcinogens in cigarette smoke damage the cells that line the lungs. The risk of

lung cancer increases with the length of time and the number of cigarettes that someone smokes. There are two common types of lung cancer: non-small cell and small cell. Non-small cell lung cancer includes adeno carcinoma, which begins in mucous-producing cells, and squamous cell carcinoma, that commonly develops in the larger airways of the lung. Small cell lung cancer usually begins in the middle of the lungs and spreads more quickly than non-small cell lung cancer. Adeno carcinoma is the most commonly diagnosed form in current or former smokers as well as passive non-smokers and makes up about 40% of lung cancers. Lung cancer can cause complications such as **emphysema**, shortness of breath, bleeding in the airways, fluid build-up in the chest and severe pain. Lung cancer can also metastasise.

- Tobacco smoking is one of the largest preventable causes of death and disease in Australia. It is associated with an increased risk of a wide range of health conditions including heart disease, diabetes, stroke, cancer, renal disease, eye disease and respiratory conditions such as asthma and bronchitis. Other potential causes of lung cancer include **radon gas**, exposure to industrial and chemical carcinogens, air pollution, a family history of lung cancer and previous lung diseases.

emphysema: a disease in which the alveoli of the lungs are damaged and enlarged

radon gas: the second leading risk factor for lung cancer around the world, it is a naturally occurring radioactive gas released from the normal decay of uranium in rocks and soil

KEY QUESTIONS

1 **Briefly explain the causes and effects of one genetic disease in humans.**

2 **Briefly explain the causes and effects of one disease caused by environmental exposure.**

3 **Briefly explain the causes and effects of one nutritional disease in humans caused by nutrient deficiency and one nutritional disease in humans caused by diet.**

4 **What is cancer?**

5 **Briefly explain the causes and effects of one type of cancer in humans.**

Answers ➲ p. 197

2 Data that shows the incidence, prevalence and mortality rates of non-infectious disease

» Students collect and represent data to show the incidence, prevalence and mortality rates of non-infectious diseases; for example, nutritional diseases and diseases caused by environmental exposure.

Non-infectious diseases

- In Chapter 12 you reviewed the incidence and prevalence of infectious diseases. Prevalence is the number or proportion of cases in a population at a given time. For example, in relation to cancer, prevalence refers to the number of people alive who had been diagnosed with cancer in a prescribed period (usually one, five, ten or 26 years). Incidence is the number of new cases diagnosed in a given period of time. For example, Australia's Health Report (2016) estimated that in 2016 there would be about 130 000 new cases of cancer diagnosed in Australia.
- Mortality refers to the number of deaths in a specified time period.
- Future estimations for incidence and mortality are a mathematical extrapolation of past trends.
- The data below shows the incidence, prevalence and mortality rates for some non-infectious diseases in Australia. These are diabetes (a nutritional disease), lung cancer and melanoma skin cancer (diseases caused by environmental exposure), breast cancer and blood cancer (Source: Australian Institute of Health and Welfare (AIHW) at www.aihw.gov.au).

Nutritional diseases

- Data for diabetes in Australia is outlined in the points below:
  - In 2015 there were 16 400 deaths due to diabetes with roughly half of these (55%) due to type 2 diabetes.
  - In 2016 diabetes was Australia's seventh leading cause of death accounting for 3% of all deaths.
  - According to the Australian Bureau of Statistics 2014–15 National Health Survey an estimated 1.2 million (5.1%) people had diabetes based on self-reported data (ABS 2015). The majority of these (85%) had type 2 diabetes and 15% had type 1 diabetes.
- For more information see www. diabetesaustralia.com. au/type-2-diabetes.
- Figure 14.3 shows the proportion of persons with diabetes 2001 to 2014-15.

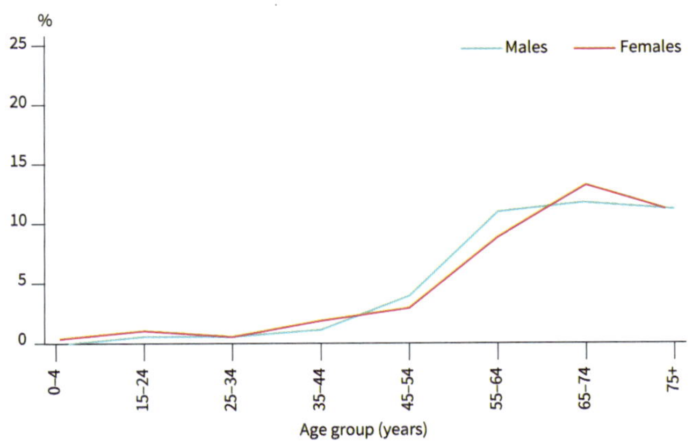

Figure 14.3 Proportion of persons in Australia with diabetes 2001 to 2014–15
Source: Australian Bureau of Statistics

Diseases caused by environmental exposure

Lung cancer

➔ Data for lung cancer in Australia is outlined in the points below:
- Lung cancer is the leading cause of cancer death and in 2013 was the fifth most common cancer diagnosed in Australia. It was estimated that it would remain the most common cause of death from cancer in 2017.
- In 2014 there were 8251 deaths from lung cancer in Australia (4947 males and 3304 females). In 2017 it was estimated that this would increase to 9021 deaths (5179 males and 3842 females).
- In 2014, 11 556 new cases of lung cancer (including small cell and non-small cell lung cancers) were diagnosed in Australia. This accounted for close to 9% of all cancers diagnosed. The risk of being diagnosed with lung cancer by age 85 was one in 13 for men and one in 21 for women.
- In 2015 there were 8466 deaths caused by lung cancer.

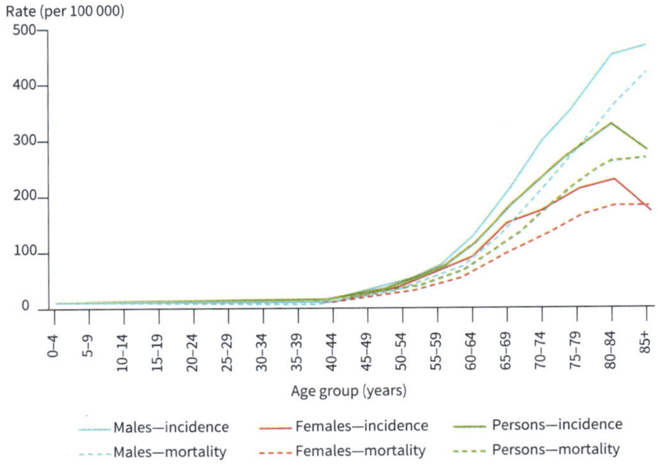

Figure 14.4 Estimated incidence and mortality rates for lung cancer, by sex and for different age groups for 2017
Source: AIHW data from lung-cancer.canceraustralia.gov.au/statistics

- The AIHW estimated that in 2017 12 434 new cases would be diagnosed—7094 males and 5340 females—and that the **age-standardised incidence rate** would be 42 cases per 100 000 people (52 for males and 35 for females).
- The incidence rate is expected to increase with age. It was estimated that in 2017 the risk of an individual being diagnosed with lung cancer before their 85th birthday would be one in 17 (one in 14 for males and one in 21 for females).

Melanoma skin cancer

➔ Data for melanoma skin cancer in Australia is outlined in the points below:
- Melanoma skin cancer was the fourth most commonly diagnosed cancer in Australia in 2013 and it is estimated that it will remain the fourth most commonly diagnosed cancer. The AIHW estimated that in 2017, 13 941 new cases of melanoma would be diagnosed (8392 males and 5549 females).
- In 2014 melanoma skin cancer was the tenth leading cause of cancer deaths in Australia. It was estimated that it would become the eighth most common cause of death from cancer in 2017 and the number of deaths from melanoma would be 1839 (1280 males and 559 females).
- In 2013 the age-standardised incidence rate was 50 cases per 100 000 people (62 for males and 40 for females). In 2017 it was estimated that the age-standardised incidence rate would be 50 cases per 100 000 persons (62 for males and 39 for females). The incidence rate of melanoma skin cancer will generally increase with age.
- It was estimated that in 2017 the risk of an individual being diagnosed with melanoma skin cancer by their 85th birthday will be one in 17 (one in 13 males and one in 23 females). See also melanoma.canceraustralia.gov.au/statistics.

age-standardised incidence rate: a method of removing the influence of age when comparing populations with different age structures; this is usually necessary because the rates of many diseases vary strongly (usually increasing) with age

Other non-infectious diseases

Breast cancer

➔ Data for breast cancer in Australia is outlined in the points below:
- Breast cancer is one of the most commonly diagnosed cancers in Australia. AIHW 2017 data showed that each day 48 women were diagnosed with breast cancer and one in eight women were diagnosed with breast cancer by the age of 85.

- The AIHW estimated that in 2017, 17 586 new cases of breast cancer would be diagnosed and the mortality rate would be 3087 women. For 2018 it is estimated that 18 235 new cases of breast cancer will be diagnosed and that breast cancer will make up 13.2% of all new cases of cancer.
- Breast cancer also affects Australian men: the AIHW estimated that in 2017, 144 cases of breast cancer in men would be diagnosed.
- In 2014 breast cancer was the fourth leading cause of cancer death. Between 1982 and 2014 breast cancer mortality dropped from around 30 deaths per 100 000 women to 20 deaths per 100 000 women. In 2018 it is estimated that the number of deaths from breast cancer will be 3128 for females and 28 for males.
- The incidence of breast cancer increases with age. The AIHW estimates that around 21% of new cases will be diagnosed in women younger than 50 years and 42% in those aged 65 plus. Given the ageing population, the number of women diagnosed with invasive breast cancer is expected to increase.

➔ For more information see breast-cancer.canceraustralia.gov.au/statistics.

Blood cancers

➔ Data for blood cancers in Australia is outlined in the points below:

- It was estimated that leukaemia would remain the eighth most commonly diagnosed cancer in 2017 and the risk of an individual being diagnosed with leukaemia by their 85th birthday would be one in 59 (one in 46 males and one in 79 females). The risk of an individual dying from leukaemia by their 85th birthday would be one in 115 (one in 86 males and one in 164 females).
- It was estimated that lymphoma would remain the sixth most commonly diagnosed cancer in 2017. The risk of an individual being diagnosed with lymphoma by their 85th birthday would be one in 36 (one in 30 males and one in 45 females). The risk of an individual dying from lymphoma by their 85th birthday would be one in 146 (one in 114 males and one in 196 females).
- In 2013 leukaemia and lymphoma were among the most common types of cancer diagnosed in children aged 0 to14 years. In 2016 it was estimated that the leading cause of death from cancer in children in 2017 would be leukaemia.

➔ For more information see the Australian Institute of Health and Welfare publication *Cancer Compendium: information and trends by cancer type* (2017), retrieved from www.aihw.gov.au/reports/cancer/cancer-compendium-information-and-trends-by-cancer-type/report-contents/leukaemia-in-australia.

➔ KEY QUESTIONS

6 **What is meant by the term *incidence rates*?**

7 **What is meant by the term *prevalence rates*?**

8 **Interpret the following graph to briefly outline the trends in the proportion of people in a specific population who were diagnosed as obese in 1995 compared to 2011–12.**

Figure 14.5 Proportion of persons in Australia aged 18 years and over who were obese, 1995 to 2011–12; based on body mass index for persons whose height and weight was measured

Source: *Australian Health Survey 2011–12*

Answers ➲ p. 197

CHAPTER SYLLABUS CHECKLIST

Are you able to answer these questions from the syllabus for this chapter? Tick each question as you go through the checklist if you are able to answer it. If you cannot answer a question, turn to the relevant page in the study guide to find the answer. For NESA key word meanings, go to www.educationstandards.nsw.edu.au and search 'key words'.

	FOR A COMPLETE UNDERSTANDING OF THIS TOPIC:	PAGE NO.	✓
1	Can I investigate the causes and effects of non-infectious diseases in humans, including genetic diseases?	187–188	
2	Can I investigate the causes and effects of non-infectious diseases in humans, including diseases caused by environmental exposure?	188–189	
3	Can I investigate the causes and effects of non-infectious diseases in humans, including nutritional diseases?	189–190	
4	Can I investigate the causes and effects of non-infectious diseases in humans, including cancer?	190–192	
5	Can I collect data that shows the incidence, prevalence and mortality rates of non-infectious disease; for example, nutritional diseases?	192–193	
6	Can I collect data that shows the incidence, prevalence and mortality rates of non-infectious disease; for example, diseases caused by environmental exposure?	193	

HSC EXAM-TYPE QUESTIONS

Objective-response questions (1 mark each)

1 Which of the following groups all contain possible causes of non-infectious disease?

A pathogen, aging, repetitive strain
B poor hygiene, bacteria, air pollution
C viruses, malnutrition, poorly preserved food
D environmental toxins, poor diet, deficient chromosomes

2 Which of the following is an example of a non-infectious disease?

A polio B malaria
C diphtheria D nutritional deficiency

3 Melanoma is a disease linked to ultraviolet radiation. To what category of disease does melanoma belong?

A viral B infectious
C nutritional D environmental

4 The table below shows projected data for some of the most common cancers diagnosed in Australia. What does the data represent?

A the burden of disease B incidence rates
C prevalence rates D mortality rates

Table 14.5 Projected data for 2017 for some of the most common cancers diagnosed in Australia.

Cancer type	New cases 2017
breast (among females)	17 586
colorectal (bowel)	16 682
melanoma	13 941
lung	12 434

5 What do scientists mean when they say that a disease is genetic?

A It is inherited.
B You acquire it through your lifetime.
C It is congenital.
D It is communicable.

Short-answer questions

6 What is the difference between infectious and non-infectious disease? (2 marks)

7 From your investigations do non-infectious diseases cause more deaths than infectious diseases? (2 marks)

8 In your studies you investigated the causes and effects of non-infectious diseases in humans and collected data on their incidence, prevalence and mortality rates. Outline some of the most important aspects to keep in mind when sourcing data for investigations. (3 marks)

9 Study the table below which shows leading causes of death in 2013 in Australia for males and females and then answer the questions which follow.

Table 14.6 Leading causes of death in 2013 in Australia for males and females

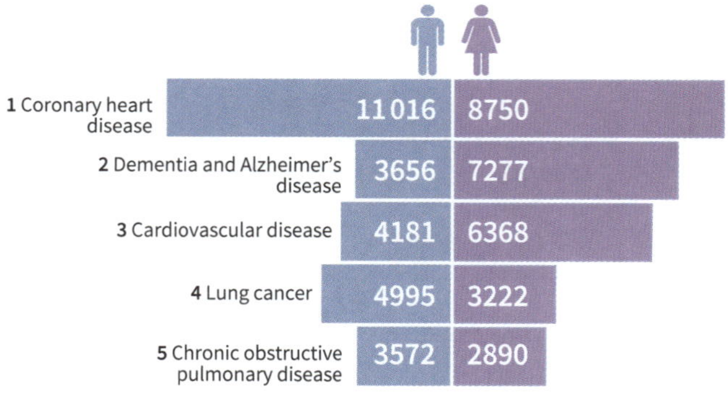

Source: AIHW National Mortality Database in *Australia's Health Report 2016*, page 11.

a What were the leading causes of death in Australia in 2013? (2 marks)

b Compare the leading causes of death for males with those of females. (2 marks)

10 In your studies you have collected data that shows the incidence, prevalence and mortality rates of some non-infectious diseases. The data below is a collection of incidence and prevalence data for some of the most commonly diagnosed cancers in Australia. The data is taken from page 83 of the Australian Institute of Health and Welfare's *Australia's Health Report 2016*. Read the data to construct and complete four tables to summarise the information. (8 marks)

In 2016 it was estimated that the most commonly diagnosed cancers in males would be prostate cancer (18 100 cases), colorectal cancer (9810), melanoma of the skin (7850), lung cancer (7130) and head and neck cancers (3430). The most commonly diagnosed cancers in females would be breast cancer (15 900 cases), colorectal cancer (7700), melanoma of the skin (5440), lung cancer (5070) and uterine cancer (2650).

In Australia at the end of 2010 around 385 000 people were alive who had been diagnosed with cancer within the previous five years: this represented 1.7% of the Australian population. Five-year survival for males was highest for prostate cancer (42% of the total male 5-year prevalence), followed by melanoma of the skin (13%) and colorectal cancer (12%). For females, 5-year survival was highest for breast cancer (36% of the total female 5-year prevalence) followed by colorectal cancer (13%) and melanoma of the skin (12%).

ANSWERS

KEY QUESTIONS

Key questions ➲ p. 192

1 Cystic fibrosis is a serious genetic disease caused by a defect in the CFTR gene. The gene normally makes a protein that controls movement of salt and water into and out of the body. In people with cystic fibrosis the protein does not work very well. Mucous and secretions from glands are too thick and sticky, affecting the lungs, sweat glands and other organs. The effects include frequent chest infections and a much-shortened life expectancy.

2 Occupational lung diseases result from breathing in harmful dusts or fumes such as silica, asbestos and coal dust. This exposure typically occurs in the workplace. Scarring of the lung tissue caused by inhaled dust is one of the most common forms of occupational lung disease.

3 Anaemia is caused by insufficient iron in the diet. The effects are drowsiness and fatigue that can extend to impacts on growth and learning if long term. Blood pressure is the force exerted by blood on the wall of the arteries. High blood pressure or hypertension is caused by poor diet (especially high salt intake), being overweight or obese, excessive alcohol consumption and physical inactivity. Hypertension is a risk factor for chronic diseases including stroke, coronary heart disease, heart failure and chronic kidney disease.

4 In normal circumstances cells grow and divide as our body requires new cells. When cells get damaged or grow old they are replaced by new cells. Sometimes this natural process goes wrong and new cells appear without our body needing them while old or damaged cells may build up rather than die. These cells grow uncontrollably and have the potential to spread to other parts of the body. Sometimes these cells may attach to other tissues in the body and cause more tumours or growths.

5 Bowel cancer, also called colorectal cancer, develops in the lining of the bowel and is usually preceded by growths in the bowel called polyps. Causes of bowel cancer include genetic risk factors, inflammatory bowel disease, nutritional factors such as a diet high in red meat and environmental factors such as smoking. Symptoms include blood in stools, abdominal pain and weight loss.

Key questions ➲ p. 194

6 Incidence is the number of new cases of diseases and injuries diagnosed in a given period of time.

7 The prevalence of an illness or condition is the number of individuals who have the condition or the number or proportion of cases diagnosed in a given period of time.

8 The prevalence of obesity has increased in Australia over time. Obesity varies with age. The proportion of people who are obese has increased across all age groups over time.

HSC EXAM-TYPE QUESTIONS

Objective-response questions

1 D. Environmental toxins, poor diet and deficient chromosomes are all causes of non-infectious diseases. **A** is wrong because pathogens cause infectious disease. **B** is incorrect because poor hygiene and bacteria can cause infectious disease. **C** is wrong because viruses and poorly preserved food can cause infectious disease.

2 D. **A**, **B** and **C** are all incorrect because they are examples of infectious diseases.

3 D. The direct cause is exposure to ultraviolet radiation, which is environmental. **A** and **B** are wrong because melanoma is not caused by pathogens. **C** is incorrect because it is not linked to diet.

4 B. Incidence is the number of new cases diagnosed in a given period of time. **A** is wrong because burden of disease is a measure of the impact of a disease on an individual or population. **C** is wrong because prevalence is the number or proportion of cases in a population at a given time. **D** is incorrect because mortality refers to the number of deaths in a specified time period.

5 A. Genetic diseases are inherited. **B** is incorrect because genetic disease is not acquired through a lifetime. **C** is wrong as this is a congenital disease; that is, one that occurs at birth. **D** is wrong because a genetic disease is an example of a non-infectious or non-communicable disease.

Short-answer questions

6 **EM** This is a recall question that asks you to bring together your knowledge from two modules: infectious and non-infectious diseases.

An infectious disease is caused by pathogens and can be transmitted from one host to another. (✓) Non-infectious diseases such as cancer are not passed directly from one person to another through transmission but are caused by factors such as aging, cell malfunction, environmental factors and diet. In some cases several of these factors may interact to cause the disease. (✓)

7 **EM** It is important to be able to answer the inquiry questions in the syllabus. This question asks you to answer the inquiry question for the unit of study on non-infectious diseases and disorders called Causes and Effects.

Non-infectious diseases such as cardiovascular diseases, cancer, diabetes and chronic respiratory diseases are the leading global cause of death and are responsible for 70% of deaths worldwide. (✓) These diseases share key modifiable behavioural risk factors like tobacco use, unhealthy diet, lack of physical activity and the harmful use of alcohol, which in turn lead to being overweight or obese along with having raised blood pressure and cholesterol levels. These diseases are an important public health challenge globally as well as in Australia. (✓)

8 **EM** This question is asking you to 'outline' and it therefore requires you to provide a general answer in relation to secondary sources that you have used and why you have used them.

One of the most important things to keep in mind when sourcing data for investigations is that the sources are valid, authoritative and reliable and that they are as up-to-date as possible. (✓)

For global statistics the World Health Authority is an authoritative source for data. (✓) In Australia, the Australian Bureau of Statistics and the Australian Institute of Health and Welfare provide health and welfare statistics and data reports such as the National Health Measures survey and National Health Reports. (✓)

9 EM The question asks you to interpret data in a graph. Remember to answer both parts of the question and make sure you compare for part **b**; that is, show how the data is similar or different.

a In 2013 coronary heart disease was the leading cause of death in Australia (✓) followed by dementia and Alzheimer's disease and cerebrovascular disease. (✓)

b Coronary heart disease was the leading cause of death for both males and females. (✓) More females died from dementia and Alzheimer's disease and cerebrovascular disease compared to males. More males died from lung cancer and chronic obstructive pulmonary disease than females. (✓)

10 EM This is a higher-order question that has a number of parts to it. The stem gives a clue in the wording 'The data below is a collection of incidence and prevalence data for some of the most commonly diagnosed cancers in Australia' and assumes that you will be able to work out which data is incidence data and which is prevalence data. You are then asked to organise the data into tables. Your answer would receive one mark for each table and correct content and one mark for each table heading.

Table A14.1 Cancer incidence projections in Australia for most commonly diagnosed cancers for males 2016

Type of cancer	Projected new cases
prostate	18 100
colorectal	9810
melanoma of the skin	7850
lung	7130
head and neck	3430

Table A14.2 Cancer incidence projections in Australia for most commonly diagnosed cancers for females in 2016

Type of cancer	Projected new cases
breast	15 900
colorectal	7700
melanoma of the skin	5440
lung	5070
uterine	2650

Table A14.3 Cancer survival and 5-year prevalence in Australia for common cancers in males, 2010

Type of cancer	5-year prevalence (as a % of the total male 5-year prevalence)
prostate	42
melanoma of the skin	13
colorectal	12

Table A14.4 Cancer survival and 5-year prevalence in Australia for common cancers in females, 2010

Type of cancer	5-year prevalence (as a % of the total female 5-year prevalence)
breast	36
colorectal	13
melanoma of the skin	12

CHAPTER 15 EPIDEMIOLOGY

MODULE 8
NON-INFECTIOUS DISEASE AND DISORDERS

INQUIRY QUESTION:

Why are epidemiological studies used?

Epidemiology is the study of the patterns and causes of disorders and diseases in defined populations. Epidemiological studies involve the systematic and ongoing collection, statistical analysis and interpretation of large quantities of data.

In this chapter you will review the epidemiology of non-infectious diseases in populations. You will also review the technologies and methods used to treat and manage non-infectious diseases.

1 Analysing patterns of non-infectious diseases in populations

›› Students analyse patterns of non-infectious diseases in populations, including their incidence and prevalence, including but not limited to nutritional diseases and diseases caused by environmental exposure.

Patterns, incidence and prevalence of non-infectious diseases

- In Chapter 14 you reviewed the collection of data that shows the incidence, prevalence and mortality rates of non-infectious diseases. In this chapter you will review the analysis of patterns of non-infectious diseases in populations. This requires identifying trends in data such as changes over time; patterns in specific populations, such as geographical remoteness, socio-economic status, rural, urban or Aboriginal communities; incidences and prevalence in males compared to females in a population; **comorbidity**; age distributions; and survivorship population.
- The survivorship population is measured using prevalence data; that is, the number of people alive who have previously been diagnosed with the disease in a given period of time.
- The incidence and prevalence of diseases and causes of death are commonly used as indicators to measure a population's health over time.
- The World Health Organization (WHO) reports on global health; for example, data from 2005 showed that the total number of **cardiovascular disease** deaths (mainly coronary heart disease, stroke and rheumatic heart disease) had increased globally to 17.5 million from 14.4 million in 1990. Of these, 7.6 million were attributed to **coronary heart disease** and 5.7 million to stroke. More than 80% of the deaths occurred in low- and middle-income countries (see Chapter 14).
- Every two years the Australian Institute of Health and Welfare (AIHW) reports on the health of the Australian population. The reports include data on the leading types of ill health, the burden of disease, and specific risks and health inequalities faced by different groups within our population. The snapshot provided in the 2016 report included the information outlined below:
  - Chronic diseases are the leading cause of ill health and death in Australia and have been for some decades, largely replacing the infectious diseases of 50–100 years ago such as pneumonia and tuberculosis
  - In 2014–15 50% of Australians had at least one of eight selected chronic conditions, these being arthritis, asthma, back problems, cancer, chronic obstructive pulmonary disease, cardiovascular disease, diabetes mellitus, or a mental or behavioural condition.
  - Australia's population has one of the world's highest life expectancies.
  - Smoking rates are declining (see Figure 15.1).
  - People living in rural and remote areas, people living in the lowest socio-economic areas, Aboriginal and Torres Strait Islander Australians, and people living with disability fare worse than others in terms of their health and longevity.

epidemiology: the study of the patterns and causes of disorders and diseases in defined populations; comes from Greek *epi* meaning 'upon' and *demos* meaning 'populace'

comorbidity: the occurrence of two or more diseases, sometimes unrelated, in a person at one time; for chronic diseases they are often related; for example, a common comorbidity is arthritis with cardiovascular disease

cardiovascular disease: disease of the circulatory system, namely the heart (cardio) or blood vessels (vascular)

coronary heart disease: disease due to blockages in the heart's coronary arteries, expressed as angina or a heart attack

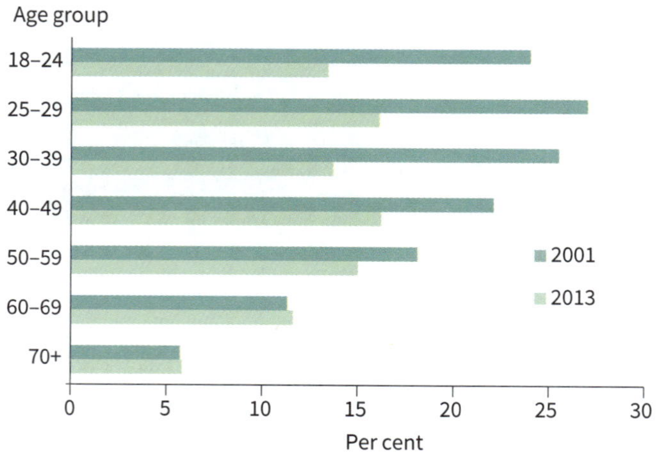

Figure 15.1 Daily smokers by age, 2001 and 2013

Source: *Australia's Health 2016*, page 173

➔ In this section of your study you will analyse patterns of non-infectious diseases in populations, including their incidence and prevalence, along with nutritional diseases and diseases caused by environmental exposure. Examples of data for type 2 diabetes and overweight or obesity (nutritional diseases) and lung cancer and melanoma cancer (diseases caused by environmental exposure) are provided below.

Nutritional diseases

Examples of data for type 2 diabetes in the Australian population

➔ One million people (4.4%) had type 2 diabetes in 2014–15, up from 840 000 people (3.8%) in 2011–12. Type 2 diabetes accounts for 85% of all diabetes and is increasing. See www.diabetesaustralia.com.au/diabetes-in-australia.

➔ Compared with non-Indigenous Australians, Indigenous Australians are 3.5 times as likely to have diabetes, 4.0 times more likely to be hospitalised for diabetes and 4.0 times as likely to die from the disease.

➔ Compared with those living in major cities, people in combined remote and very remote areas are 1.8 times as likely to be hospitalised for diabetes and 1.9 times as likely to die from diabetes.

➔ Compared with those living in the highest socio-economic areas, people living in the lowest socio-economic areas are 3.6 times as likely to have diabetes, 1.8 times as likely to be hospitalised for diabetes and 2.0 times as likely to die from diabetes.

Examples of data for overweight or obesity in the Australian population

➔ The proportion of people who are obese or overweight is increasing. This increase is reflected in the general increase in weight and body mass index (BMI) over time. Figure 15.2 compares the distribution of BMI scores for people aged 18 years and over between 1995 and 2011–12. This graph shows that not only is there a greater proportion of people who are overweight and obese since 1995, but also that there has been an increase in the proportion of people with much higher BMI scores. For example, in 1995 5% of people aged 18 years and over had a BMI of 35 and over whereas in 2011–12 the corresponding proportion was 9.6%. Between 1995 and 2011–12 the average adult man's weight increased by 3.6 kg while the average adult woman's weight increased by 4.0 kg.

➔ In 2011–12 males living in inner regional, outer regional and remote areas of Australia were more likely to be overweight or obese (74.4%) compared with males living in major cities (67.7%). This pattern was also consistent for females, with women living in inner regional, outer regional and remote areas more likely to be overweight or obese (63.2%) than women living in major cities (52.5%).

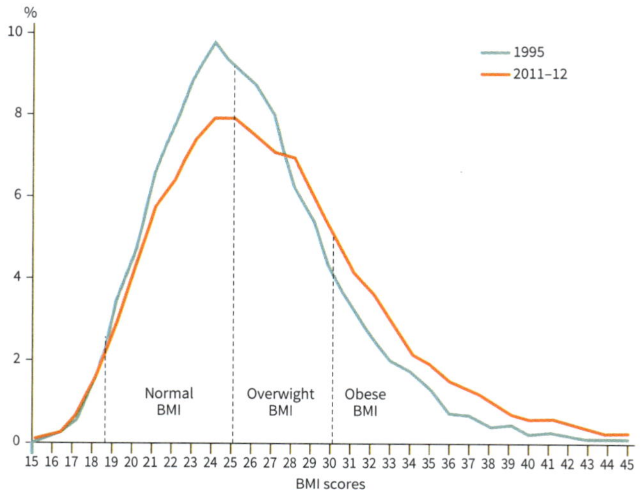

Figure 15.2 Body mass index scores, based on BMI for persons whose height and weight was measured, for persons aged 18 years and over, 1995 and 2011–12

Source: *Australian Health Survey: updated results, 2011–12*

Diseases caused by environmental exposure

Examples of data for lung cancer in the Australian population

➔ The 5-year relative survival from lung cancer improved from 9% between 1984 and 1988 to 16% between 2009 and 2013.

➔ The number of new cases of lung cancer diagnosed increased from 5951 (4691 males and 1260 females) in 1982 to 11 174 in 2013. Over the same period the age-standardised incidence rate decreased from 47 cases per 100 000 people (85 for males, 18 for females) in 1982 to 43 cases per 100 000 people in 2013 (see lung-cancer.canceraustralia.gov.au/statistics).

Examples of data for melanoma cancer in the Australian population

➔ In 2009–13 individuals diagnosed with melanoma skin cancer had a 90% chance (88% for males and 93% for females) of surviving for five years compared to their counterparts in the general Australian population.

- From 1984–88 and 2009–13, 5-year relative survival from melanoma skin cancer improved from 86% to 90%.
- The number of new cases of melanoma skin cancer diagnosed increased from 3527 (1733 males and 1794 females) in 1982 to 12 744 in 2013. Over the same period the age-standardised incidence rate increased from 27 cases per 100 000 persons (28 for males and 26 for females) in 1982 to 50 cases per 100 000 people in 2013 (see melanoma.canceraustralia.gov.au/statistics).
- For more data visit the Australian Bureau of Statistics National Health Survey website. For more detailed data on nutrition and physical activity visit the Australian Health Survey website. For more statistics on diabetes, cancer, burden of disease and a range of other health conditions visit the Australian Institute of Health and Welfare website. For more information on global health statistics visit the Organisation for Economic Co-operation and Development (OECD) health statistics website at www.oecd.org/els/health-systems.

KEY QUESTIONS

1 **What is epidemiology?**

2 **What information is gained in analysing patterns of non-infectious diseases in populations, including their incidence and prevalence, and how is this information used?**

3 **When you analysed patterns of non-infectious diseases in populations what did you look for?**

Answers ➲ p. 209

2 Investigating the treatment/management and possible future directions for further research of non-infectious diseases

» Students investigate the treatment/management, and possible future directions for further research, of a non-infectious disease using an example from one of the non-infectious diseases categories listed in Section 1.

Treatment and management of non-infectious diseases

- Treatment and management of non-infectious diseases includes primary health care, specialised treatment services and targeted services for populations most at risk.
- A fundamental aim of any healthcare system is to prevent disease so that people remain as healthy as possible for as long as possible. When people become ill they need to have access to timely and appropriate treatment and health care. In this section you will review treatment, management and future directions for research. In Chapter 16 you will review prevention.
- Survival rates from non-infectious diseases have improved significantly in Australia since the 1980s with medical research contributing to improvements in the detection, diagnoses, treatment and management of disease. For example, children diagnosed with leukaemia in Australia in the late 1990s had about a 65% chance of survival but by 2018 survival rates had increased to about 90%.
- The treatment and management of melanoma of the skin and type 2 diabetes is summarised in the section below.

Treatment and management of melanoma of the skin

- Treatment for melanoma depends on the stage of the disease, the location and the severity of symptoms. Surgery is the main initial treatment. Thin, early-stage melanomas can be removed by minor surgery. A wide margin of healthy skin around the edge of the tumour is removed to ensure that no cancer cells are left behind. The thicker the tumour, the wider the margin required in terms of both area and depth. If the melanoma has spread to nearby lymph nodes these lymph nodes are usually removed by surgery. Surgery is not necessarily a cure but it might prolong survival and avoid the pain resulting from the cancer growing in the lymph nodes. Additional treatments are outlined in the points below:
  - **Chemotherapy**, which is less effective for melanoma than for some other types of cancer but may be used to prolong survival or relieve symptoms.
  - **Radiotherapy**, which is not usually used to treat the original melanoma in the skin but may be used to treat melanoma that has recurred, to kill any cancer cells remaining after surgery or to relieve symptoms caused by spread of the melanoma.
  - **Targeted therapy**: BRAF is a human gene that encodes a protein called B-Raf. About half of all melanomas have mutations in the BRAF gene. Melanoma cells with these changes make an altered BRAF protein that helps them grow and divide more rapidly. The medicines used for treatment of melanoma target cells that have a damaged BRAF gene, shrinking or slowing the growth of the tumour.

chemotherapy: the use of cytotoxic chemicals called chemotherapy drugs to try to destroy cancer cells
radiotherapy: the use of X-rays to kill, damage or shrink cancer cells
targeted therapy: the use of medicines designed to specifically attack cancer cells without harming normal cells; these types of medicines affect the way that cancer cells grow, divide, repair themselves or interact with other cells

- **Immunotherapy** uses medicines to block proteins on immune system cells (T cells) that stop these cells from attacking other cells in the body (see PD-1 inhibitors below). This boosts the immune system's response against melanoma cells but may have side effects, including causing the immune system to attack other healthy parts of the body (see melanoma.canceraustralia.gov.au/treatment).

➔ Melanoma cannot be inherited but a small number of cases in Australia (less than 2%) involve an inherited predisposition to develop melanoma cancer. These individuals have inherited a faulty copy of a melanoma 'cancer protection' gene. The two 'cancer protection' genes that have been identified are CDKN2A and CDK4. They are inherited in an autosomal dominant pattern. Future directions in treatments include the further development and trialling of drugs and vaccines that target these mutations (see below). For more information see www.genetics.edu.au/publications-and-resources/facts-sheets/fact-sheet-34-genetics-and-melanoma/view.

immunotherapy: the use of medicines to stimulate the immune system to attack cancer cells

Treatment and management of type 2 diabetes

➔ Type 2 diabetes is a largely preventable disease often caused by modifiable risk factors (see Chapter 14). It can therefore be managed initially through changes in lifestyle, including a healthy diet (eating well helps manage blood glucose levels and body weight), regular exercise (exercise helps insulin work more effectively, lowers blood pressure and reduces the risk of heart disease) and monitoring of blood glucose levels which tests whether lifestyle changes are working.

➔ The aim of diabetes management is to keep blood glucose levels close to the target range to prevent both short- and long-term complications.

➔ Type 2 diabetes is also a progressive condition and sometimes healthy eating and exercise is not enough to keep the blood glucose levels down. If the body is still producing enough insulin, anti-diabetic medication, usually in tablet form, is given to control the amount of glucose in the blood. If the body is no longer able to produce enough insulin of its own, insulin is taken or insulin injections are given to manage blood glucose levels. The tablets or injections are intended to be used together with healthy eating and regular physical activity, not as a substitute for these. For more information see www.diabetesaustralia.com.au/managing-type-2.

Possible future directions for further research of non-infectious diseases

➔ In Chapter 7 you reviewed the medical uses and applications of genetic techniques. Some of the most recent and intensive medical research involves application of genetic techniques to better understand the factors that influence cancer risk and develop anti-cancer treatments. Future directions for further research will come from an increasingly better understanding of the metabolic pathways of cells, genetic sequencing and single nucleotide polymorphisms (SNPs) and genomes (see Chapter 4). The hope is that the continued discovery of gene alterations (mutations) and how they are linked to cancer will provide novel targets for the development of anti-cancer drugs. Note: You will review genetic techniques used in the prevention of disease in more detail in Chapter 16.

➔ Most conventional cancer treatments target tumour cells with drugs or radiation. These treatments are non-specific and damage both healthy cells and cancer cells. They are also unable to keep up with the mutations cancer cells undergo in response to treatments. Examples of future directions for further research for the treatment of cancer in general, and melanoma of the skin in particular, are summarised in the points below:
- Focused radiotherapy concentrates the radiation on a tumour which lessens the impact on healthy tissue. The technology uses CT scans and X-rays to deliver the precise dose of radiation and a machine called a Cyberknife, with a moveable robotic arm, to focus the beam at different angles.
- Proton therapy is another type of focused radiation treatment and uses high-energy proton beams. The protons deliver the specific radiation dose into the tumour, damaging the DNA of the tumour cells.
- Cancer vaccines can be tailored to an individual's cancer mutations. The research began by sequencing genes that encode proteins in individual patient's tumours. The basis of the vaccine comes from selected mutant proteins that are more likely to generate an immune response. The vaccines are injected into patients and are often combined with immunotherapy treatment.
- Because of developments in immunotherapy researchers can now use an ever-increasing understanding of the human immune system and of how tumours metabolise and spread in order to treat cancers. For example, one of the reasons cancer cells are able to spread in the body is that they 'quieten' the immune system. Researchers have discovered a protein found on the surface of T cells (see Chapter 11) they call PD-1 or programmed cell death 1. PD-1 normally helps prevent the immune system running out of control. Researchers have found a molecule on cancer cells that they call programmed death ligand-1 (PD-L 1) that links with the PD-1 on the surface of T-receptor cells. Once the link is made the T cell remains inactive, allowing the tumour to grow. Researchers have developed and are trialling immunotherapy drugs called PD-1 inhibitors or checkpoint inhibitors that block the PD-1 protein.

Checkpoint inhibitors focus the immune attack on cancer cells rather than healthy cells and boost the immune system's ability to fight cancer. These immunotherapy drugs are being used to treat melanoma and lung cancer. For more information see 'Blocking Cancer', *New Scientist*, 14 January 2017, pp. 36–39.

- A monoclonal antibody (see Chapter 7) that targets the CDK4 antigen has been developed and is being used to treat advanced melanoma.
- A BRAF inhibitor has been used in patients whose melanoma tumours contain a specific mutation in the BRAF gene.
- Novel technologies used for early detection have been developed. These include gene expression platforms using adhesive patch or tape strips that can distinguish between benign and malignant tumours along with technologies that allow individuals to image their moles and have them classified by a computer algorithm.

→ KEY QUESTIONS

4 **Outline the treatment/management for one nutritional disease or one disease caused by environmental exposure.**

5 **Briefly outline possible future directions for research of one non-infectious disease you have studied.**

Answers ➲ p. 209

3 Evaluating the method used in an epidemiological study

» Students evaluate the method used in an example of an epidemiological study.

Evaluating methods in epidemiological studies

→ Various methods are used to carry out epidemiological investigations, including surveillance and descriptive studies to study distribution of a disease and analytical studies to investigate possible causes.

→ The steps used in scientific methodology can be used to evaluate an epidemiological study. These steps are outlined in the points below:

- Establish that a problem exists by confirming that all occurrences of the disease are true cases of the disease.
- Evaluate the importance of the research.
- Collect as much information as possible such as incidence and prevalence rates.
- Categorise the events in terms of their epidemiological factors; for example, non-environmental factors that increase the likelihood of getting a disease such as genetic history, age and gender. Other categories include factors relating to the environment that either increase or decrease the likelihood of disease, such as exercise and good diet, along with possible sources of exposure, such as a toxin or pathogen. Other factors that compound the likelihood of getting a disease include repeated exposure or excessive environmental stresses.
- Look for patterns and trends.
- Formulate a hypothesis as to the nature of the relationship between the potential disease-causing agent and the disease.
- Test the hypothesis: methods that have evolved to assess the strength of the relationship between the disease-causing agent and the disease include Koch's postulates, the **Bradford-Hill criteria** and the **five evidence grading criteria** of the World Cancer Research Fund.

An example of an epidemiological study

→ In Chapter 5 you reviewed two large-scale collaborative projects to identify trends, patterns and relationships using population genetics studies that determined the inheritance of a disease or disorder. Your review included examples of genome-wide association studies into diseases such as stroke and type 2 diabetes (see page 73).

→ When you evaluated an epidemiological study you would have made judgements about the study using criteria such as the methods outlined above, the value of the study, the scale of the study, the length of time, the specificity of the population(s), the specificity of the site(s), the assessment of evidence and limitations in the study. For example, the linking of cause and effect may require many studies with large populations over long periods of time and rigorous scientific method to eliminate variables.

→ The following example of an epidemiological study comes from a global **meta-study** on the epidemiology of cardiovascular disease. The source is *Promoting Cardiovascular Health in the Developing World: A Critical Challenge to Achieve Global Health* from www.ncbi.nlm.nih.gov/books/NBK45688/.

→ Cardiovascular disease (CVD) is a dominant contributor to global mortalities. Global trends in cardiovascular disease are based on models that use country-specific data from a diverse range of developed and developing countries. By gathering and comparing data from different countries incidence and prevalence patterns, contributions

Bradford-Hill criteria: a set of nine criteria used to provide epidemiological evidence of the relationship between a possible cause and an observed effect

five evidence grading criteria: criteria for grading evidence to support a judgement of a relationship with cancer

meta-study: a method for systematically combining data from several selected studies for analysis of risk factors and levels of risk can be used to explain the increase of CVD in some countries and the low levels and decline of CVD in others. The studies also help to establish lessons that can be applied to countries where the incidence is increasing; that is, developing countries.

- Some of the limitations to the validity of the global study include limitations in available data as outlined in the points below:
  - Many countries have established health surveillance systems but the quality of the data collected varies substantially across countries.
  - In many countries health statistics are often based on surveillance that does not cover all areas of the country, is incomplete in the areas it does cover or is collected by undertrained staff.
  - Similarly analysis of long-term trends using data from the early 20th century is difficult because the data is often limited, particularly in developing countries.
- Despite these limitations health statistics are often the only complete or comparable data available and thus remain a key tool for evaluating the status of a CVD epidemic within and between countries.
- The following are specific examples of some of the epidemiological studies used in the meta-study:
  - A study in the United Kingdom followed 20 040 people over 11 years to determine the risk of stroke incidence. Four measures of health behaviours were used to determine risk of stroke. These four factors were smoking, low physical activity, low plasma vitamin C levels (used as a proxy for fruit and vegetable intake) and not drinking alcohol in moderation (consuming more than 14 drinks per week). The four factors were found to contribute to more than a two-fold increase in stroke incidence.
  - A study on the rise of coronary heart disease mortality in Beijing from 1984 to 1990 found that **blood lipid** increases were the largest contributor and responsible for 77% of increased coronary heart disease mortality.
  - The Global Burden of Disease study estimated that in 2000, 880 000 deaths from coronary heart disease and 412 000 deaths from stroke were attributable to tobacco smoking.
  - Studies in Finland found that dietary changes instituted in the 1970s explained nearly all the reduction in cholesterol observed in the Finnish population. This is significant because the reduction in cholesterol was the most important factor in the overall reduction of coronary heart disease mortality in Finland between the 1970s to the early 1990s. Since the 1970s both stroke and coronary heart disease mortality have declined 75 to 80% and the average life expectancy has increased by five to six years.
- A snapshot of the results from the epidemiology of cardiovascular disease studies can be seen in the points below:
  - The same set of major risk factors consistently plays a large role in explaining trends in CVD incidence and mortality across the world. Taken together the data indicates that poor diet, tobacco use, physical inactivity, excess alcohol use and psychosocial factors are the major contributors to CVD increases.
  - Mortality rates from non-infectious diseases generally appear to be most closely linked to a country's stage of **epidemiological transition**. In underdeveloped countries infectious diseases predominate but as the economy, development status and health systems of these countries improve the population moves to a later stage of epidemiological transition, and chronic non-infectious diseases become the predominant causes of death and disease.
  - An emerging body of evidence suggests that rapid dietary changes associated with nutritional transition along with a decrease in levels of physical activity in many rapidly urbanising societies play a particularly important role in the rise of CVD in developing countries. Especially significant is the increasing consumption of energy-dense diets high in unhealthy fats, oils, sodium and sugars along with an exponential growth in the use of tropical palm oil. Palm oil has a saturated fatty acid content of 45%.
  - There is a rise in tobacco smoking globally, particularly in low- and middle-income countries. By 2020, if current smoking and population growth trends continue, the global annual cigarette consumption could rise to between 6.7 and 6.8 trillion cigarettes.
  - A strong and robust base of evidence shows that excessive sodium intake significantly increases CVD risk. Numerous studies have found that there is a continuous and graded relationship between increased salt intake and increased blood pressure.
  - Increased body mass index (BMI) is associated with a greater risk of stroke in both Asian and Western populations. The association between obesity and CVD is partly, but not completely, related to hypertension, high cholesterol and diabetes.
  - Diabetes is growing increasingly common and is a significant contributor to CVD risk. Diabetes is emerging as a particular concern in Asia where more than 110 million individuals were living with diabetes in 2007. Asians tend to develop diabetes at a relatively young age with a low BMI and by 2025 the number of individuals with diabetes in the region is expected to rise to almost 180 million, of which approximately 70 million will be in India and about 60 million in China.

blood lipids: fatty acids and cholesterol in the blood
epidemiological transition: changes in the predominant forms of disease and mortality burdening a population that occur as its economy and health system develops

- Numerous epidemiological studies in both developed and developing regions of the world have found that both short-term (several hours to a few days) and long-term exposure to particulate matter air pollution significantly increases cardiovascular events and CVD deaths.
- Psychosocial factors such as depression, anxiety, anger, hostility, acute and chronic life stressors and lack of social support have been consistently associated with both the onset and the progression of CVD in epidemiologic studies in multiple populations and regions. Despite this these conditions remain under-recognised when compared with more traditional CVD risk factors.
- Key contributors to the decline of CVD mortality rates in some countries include declines in tobacco use and exposure, healthful dietary shifts, population-wide prevention efforts and treatment interventions.

➔ KEY QUESTION

6 **Outline two examples of epidemiological studies that you evaluated.**

Answer ➲ p. 209

4 Evaluating the benefits of engaging in an epidemiological study

» Students evaluate, using examples, the benefits of engaging in an epidemiological study.

➔ The key benefit of engaging in an epidemiological study is that it potentially saves lives.

➔ Analyses of changes in prevalence and incidence data over time and/or differences in prevalence among populations help to identify risk factors, determine the resources needed for health care, education and research, and develop targeted public health interventions that are cost and resource effective.

➔ The examples below illustrate some of the benefits of engaging in an epidemiological study. Each example includes a brief summary of the reason why the study was conducted.

- Studies have uncovered the cause of a high incidence of an unpredictable disease; for example, in the late 1950s more than 10 000 cases of birth defects were reported in over 46 countries. Babies were born with missing or abnormal legs, arms, feet and hands; spinal cord defects; cleft lip or palate; absent or abnormal external ears; heart, kidney and genital abnormalities and abnormal formation of the digestive system. By the early 1960s epidemiological studies had found the link between these severe birth defects and the drug thalidomide. Thalidomide was first marketed in 1957 for morning sickness and soon became the drug of choice for pregnant women.
- Studies have identified associations between exposure to environmental substances (environmental risk factors) and adverse health effects such as the link between exposure to asbestos and mesothelioma (as well as other asbestos-related diseases). Mesothelioma is a rare form of cancer and was misdiagnosed for many years. Even so, many studies showed that mesothelioma mortality rates were much higher than would be statistically expected for a diverse range of people working in and living near asbestos mines and people working and living with asbestos products. Recently the number of new cases has increased globally and the disease is now more common. Epidemiological studies established the link between exposure to asbestos and the disease and it is now understood that reducing exposure to asbestos is the most likely way to lower the risk of mesothelioma. In response to these findings Australia has implemented a code of practice for managing and disposal of asbestos. For more information see www.safeworkaustralia.gov.au.
- Studies have identified behavioural or lifestyle risk factors. The knowledge gained in identifying risk factors for non-infectious diseases is used to direct further research and to investigate and determine treatment, management and prevention strategies. For example, the burden of lung cancer in Australia has increased significantly over time, leading to the most comprehensive study of cancer causation in Australia. The study estimated that 81% of lung cancers in 2010 were caused by tobacco smoking, making control of tobacco essential for lung cancer prevention.
- Studies have led to breakthroughs in understanding genetic factors involved in the predisposition to and development of some cancers. This has helped understanding of why not everyone exposed to a particular cancer-causing chemical will develop cancer and why cancer rates vary across populations, among some groups and within families. For example, the risk of a woman being diagnosed with breast cancer in Australia by their 85th birthday is one in eight. Extensive studies into breast cancer are being conducted, including studies to understand more about genetic links and breast cancer risk and what can be done to reduce the risks (see below and Chapter 16).
- Studies have also led to breakthroughs in the understanding of the causes of disease; for example, the link between cervical cancer and the human papillomavirus (HPV), a sexually transmitted virus that causes genital warts (see below).

➔ Because family history is a predictor of a woman's chance of developing breast cancer, epidemiological studies have involved cancer-prone families—those with a high incidence of cancer in several generations—to find specific inherited gene alterations that are passed on from one generation to the next. In one study researchers tested the DNA in blood to see which of the more than 5000 volunteers had an alteration. Then, using the family cancer histories reported by the volunteers, the scientists estimated the cancer risk by comparing the histories of cancer in the relatives of the volunteers with the alteration to the histories of cancer in the relatives of the volunteers without the alteration. After a long search, two genes were found that are altered in many families with hereditary breast cancer. The first—BRCA1 (for BReast CAncer gene)—was discovered in 1994 and the second—BRCA2—in 1995. BRCA1 and BRCA2 are genes that normally prevent a woman developing breast or ovarian cancer. Once the genes were isolated it was possible to analyse the specific alterations inherited in each cancer-prone family. Studies show that the risks of breast cancer are strongest if two or more relatives have the mutated genes and other characteristics such as being diagnosed before the age of 50 and/or being of Ashkenazi Jewish descent. The search for other genes continues. Source: www.genome.gov/10000940/brca1brca2-study-faq/.

→ Cervical cancer is the fourth most common cancer in women worldwide and disproportionally affects women in lower- and middle-income countries. In the last two decades epidemiological studies have shown that cervical cancer is a rare outcome of a common, sexually transmitted infection whose **etiologic** association is restricted to a few human papillomavirus (HPV) types. Using optimal testing systems HPV DNA can be identified in nearly all specimens of invasive cervical cancer. Epidemiological studies have clearly and consistently shown that the key determinants of infection in women are the number of sexual partners, the age at which sexual intercourse was initiated, and the likelihood that at least one sexual partner was an HPV carrier. The evidence is consistent worldwide. Studies also identify co-factors that increase the risk of cervical cancer for HPV DNA-positive women that include the use of oral contraceptives for five or more years, smoking, five or more full-term pregnancies and previous exposure to other sexually transmitted diseases such as *Chlamydia* and herpes simplex virus type 2. Women exposed to the human immunodeficiency virus (HIV) are also at high risk for HPV infection, HPV DNA persistence, and the progression of HPV lesions to cervical cancer. Findings from the studies have resulted in the use of **pap smears** and more recently screening and vaccination strategies for the prevention of cervical cancer (see Chapter 16). Source: 'The epidemiology of human papillomavirus infection and its association with cervical cancer', *International Journal of Gynaecology and Obstetrics*, 94 (Supplement 1), S8-S21, 2006.

etiologic: *etiology* is a term used in research to mean cause or origin of disease

pap smear: a cervical screening test to detect pre-cancer and cancer in the cervix

→ KEY QUESTIONS

7 **Why are epidemiological studies used? Provide an example.**

8 **Evaluate the benefits of epidemiological studies using one example you have investigated.**

Answers ➲ p. 209

CHAPTER SYLLABUS CHECKLIST

Are you able to answer these questions from the syllabus for this chapter? Tick each question as you go through the checklist if you are able to answer it. If you cannot answer a question, turn to the relevant page in the study guide to find the answer. For NESA key word meanings, go to www.educationstandards.nsw.edu.au and search 'key words'.

FOR A COMPLETE UNDERSTANDING OF THIS TOPIC:		PAGE NO.	✓
1	Can I analyse patterns of non-infectious diseases in populations, including their incidence and prevalence, including but not limited to nutritional diseases and diseases caused by environmental exposure?	199–201	
2	Can I investigate the treatment/management and possible future directions for further research for nutritional diseases and diseases caused by environmental exposure?	201–203	
3	Can I evaluate the method used in an example of an epidemiological study?	203–205	
4	Can I evaluate, using examples, the benefits of engaging in an epidemiological study?	205–206	

HSC EXAM-TYPE QUESTIONS

Objective-response questions

(1 mark each)

1 **Below is a graph showing causes of death in Australia and the percentages of all deaths for particular diseases for males and females.**

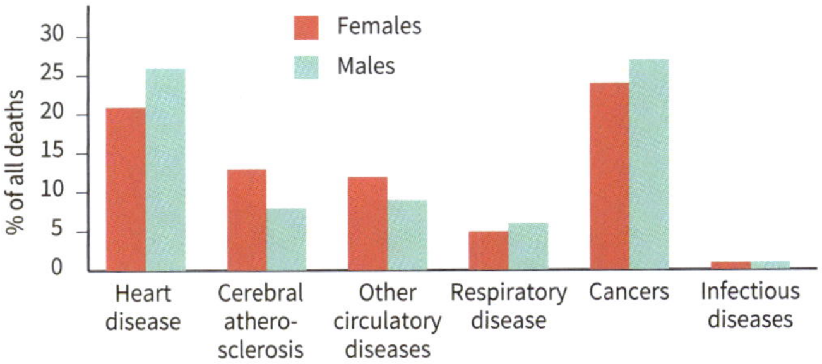

Figure 15.3 Percentages of all deaths in Australia for particular diseases for males and females

What can we conclude from the graph?

- A The mortality for lung cancer is a greater for males than females.
- B More males die of circulatory disease than females.
- C Breast cancer is the major form of cancer affecting females.
- D More Australians die from non-infectious diseases than from infectious diseases.

2 **What is the following data an example of?**

Projected data in Australia from 1982 to 2016 showed that the rate for all cancers combined would increase by 22% from 383 new cases per 100 000 of the population to 467 per 100 000.

- A prevalence rates
- B proportions in population rates over time
- C incidence rates
- D mortality rates

3 **A study called INTERHEART enrolled approximately 15 000 cases and 15 000 controls from 52 countries to examine the impact of risk factors on incidence of cardiovascular disease. The study identified three major dietary patterns: oriental (high intake of tofu and soy); western (high in fried foods, salty snacks, eggs and meat); and prudent (high in fruits and vegetables).**

What is the study an example of?

- A a nutritional study
- B an epidemiological study
- C a clinical trial
- D a disease study

4 **Which of the following is an example of future directions in the treatment of non-infectious disease?**

- A taking stem cells from people with haemophilia, correcting the genetic mutation and them reinjecting them into the patient
- B using CRSPR technology to eliminate HIV from the DNA of cells
- C gene alteration that makes a mosquito resistant to malaria
- D supplementing diets for people with vitamin A deficiency

5 **Which one of the following patterns can be identified in the data below?**

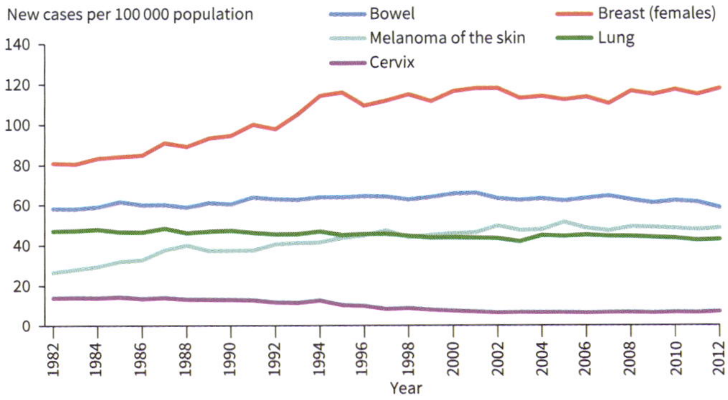

Figure 15.4 Age-standardised incidence rates for selected cancers in Australia, 1982–2012

Source: *Australia's Health Report 2016*, page 389

- A The incidence rate of melanoma fell between 1982 and 2002.
- B For cancer of the cervix the incidence rate rose between 1982 and 2001 and then remained steady.
- C Between 1982 and 2012 incidence rates for lung cancer and bowel cancer remained relatively steady overall.
- D The age-standardised rate for breast cancer in 2012 was 130 new cases per 100 000 females.

Short-answer questions

6 The following data provides examples of variations among populations in Australia for coronary heart disease (CHD).

Compared with non-Indigenous Australians, Indigenous Australians are two times as likely to have CHD; 2.4 times as likely to be hospitalised for CHD; 1.6 times as likely to die from CHD. Compared with those living in major cities, people in combined remote and very remote areas are 1.6 times as likely to be hospitalised for CHD; and 1.3 times as times as likely to die from CHD. Compared with those living in the highest socio-economic areas, people living in the lowest socio-economic areas are 2.2 times as likely to have CHD; 1.5 times as likely to be hospitalised for CHD; and 1.4 times as likely to die from CHD.

Source: *Australia's Health 2016*

Write ONE conclusion from your analysis of a pattern in this data. (2 marks)

7 From the following information construct a table to summarise the occurrence, cause, effects, symptoms and treatment/management of the disease phenylketonuria. (5 marks)

Most babies born in Australia have a blood test in the first few days of their life to test for the recessive genetic disease phenylketonuria (PKU). There is one chance in 15 000 that the baby will have the disease but, if detected early, a modified diet can drastically reduce the problems of intellectual disability, poor coordination and lack of muscular control that would occur as a result of the build-up of the amino acid phenylalanine in the body. High levels of phenylalanine are toxic, especially to nerve cells. The genetic condition is a result of a mutation in the gene that normally produces an enzyme that breaks down phenylalanine. If the disease is detected during pregnancy, women are counselled about the increased risks to their own and their baby's health.

8 The table below summarises statistics related to the incidences and recurrent risks of having a Down syndrome child and its relationship to the maternal age (the age of the mother). Study the table and then answer the questions that follow.

Table 15.1 Incidences and recurrent risks of having a Down syndrome child related to maternal age

Maternal age	Incidence of Down syndrome	Recurrent risk—the risk of having a second affected child
16–26	1 in 1300	1 in 100
27–34	1 in 700	1 in 100
35–39	1 in 350	1 in 100
40–44	1 in 100	1 in 50
45–47	1 in 30	1 in 20
48	1 in 12	1 in 12

a What steps could you take to ensure the data is accurate and reliable? (2 marks)

b How might you investigate a claim that the increase is the result of the effect of mutagens as humans age? (2 marks)

9 Outline the treatment and management for a non-infectious disease that you have investigated. (6 marks)

10 Using an example evaluate the methods used in an epidemiological study and the benefits of engaging in the study. (8 marks)

ANSWERS

KEY QUESTIONS

Key questions ⊃ p. 201

1 Epidemiology is the systematic and ongoing collection, analysis and interpretation of data about disease. Epidemiology is used to identify correlations between disease and risk factors, leading to an understanding of the causes of diseases and the populations most at risk.

2 The incidence and prevalence of diseases and causes of death are commonly used as indicators to measure a population's health over time. Patterns of incidence and prevalence of non-infectious diseases form the basis for determining the resources needed for health care, education and research, and help evaluate public health interventions.

3 This is a sample answer. When we analysed patterns we looked for trends such as changes over time; patterns in specific populations such as geographical remoteness; socio-economic status; rural, urban or Aboriginal communities; incidences and prevalence in males compared to females in a population; and patterns in relation to age distributions.

Key questions ⊃ p. 203

4 This is a sample answer for lung cancer, a disease caused by environmental exposure. Treatment depends on the type of lung cancer and how advanced the disease is. Treatments involve surgery to remove the tumour and/or radiotherapy and chemotherapy. Surgery is rarely used to treat small cell lung cancer. If surgery or radiotherapy is not suitable, thermal ablation may be used. In this treatment needles are inserted into the cancer to destroy the cells by heating them.

5 Future directions in diabetes management and treatment include the processes outlined below:

- A bionic (artificial) pancreas monitors blood sugar every five minutes and automatically controls blood glucose levels through two pumps which deliver insulin and glucagon.
- Stem cell therapy: researchers have spent years developing methods to turn stem cells into insulin-producing beta cells but so far only progenitor cells have been produced. These are the precursor cells of mature beta cells. Clinical trials have shown that progenitor cells transplanted under the skin of diabetics can produce insulin.
- Improving gut health: a drug called metformin is used by people with diabetes to reduce the amount of glucose made by the liver. Researchers have observed that it also changes the gut bacteria. Faecal transplants in mice using faeces from metformin-treated individuals have shown improved glucose tolerance, suggesting that people may be able to manage diabetes by modifying their diet to improve the gut bacteria.
- Insulin is genetically engineered by placing human insulin into bacteria and then growing the bacteria in large vats.

Key question ⊃ p. 205

6 Sample answer 1: Epidemiological studies to determine the prevalence, cause and clinical significance of Down syndrome (DS) have been conducted for more than 100 years. Studies began in the mid-1800s when several physicians described groups of patients who had intellectual disability and short stature along with specific facial characteristics, including oblique eye fissures, epicanthal folds, a flat nasal bridge and a protruding tongue. J Langdon Down, for whom DS was named, contributed significantly to the epidemiology of this syndrome by emphasising that this set of clinical findings constituted a distinct entity and that affected individuals could be distinguished from other groups with intellectual disabilities. Once DS, also known as trisomy 21, was established as a separate disorder it became possible for epidemiological studies to look for causes.

The first cause to be linked was increased parental age at the birth of the infant with DS. Studies then showed that advanced maternal age, not paternal age or birth order, was the key determining factor for DS. Researchers also suggested that very young mothers have an increased chance of an infant with DS. In the late 1950s, with advances in understanding chromosomes and with the development of karyotyping, the cause of DS was identified as the presence of an extra chromosome 21. It is now estimated that more than 90% of individuals with DS have an extra chromosome 21 as a result of meiotic non-disjunction or the abnormal segregation of chromosomes during gamete formation. DS is now one of the most intensively studied human disorders. The most important risk factor associated with DS is advanced maternal age. The impact of this factor is significant given the trend for women to delay childbirth (Source: S Sherman, EG Allen, LH Bean and SB Freeman, 'Epidemiology of Down Syndrome', *Mental Retardation and Developmental Disabilities Research Reviews*, 2007).

Sample answer 2: Analysis of cancer in twin pairs is a traditional epidemiological method for studying relationships between causes of cancer, genes and environment. A study of data obtained from 44 000 same-sex twin pairs was used to assess cancer risks for co-twins of twins with cancer. There were almost 10 000 pairs in which one of the members had cancer. The analysis of environmental and inherited contributions was based on correlations between identical twins who share the genome completely and non-identical twins, the difference being the assumption that only 50% of the genes are common. The assumption was that the environment is affecting identical and non-identical twins similarly. Statistically significant heritability estimates were detected for colorectum (35% chance), breast (27% chance) and prostate (42% chance) cancers (Source: 'Genetic epidemiology as a tool for studying gene–environment interactions' from www.ncbi.nlm.nih.gov/books/NBK220897/). Note: You reviewed how twin studies are used in genetic studies in Chapter 3.

Key question ⊃ p. 206

7 Epidemiological studies provide rates as measures to characterise a disease, giving valuable information about why and how diseases occur in different populations. This knowledge is used for further scientific research and medical treatments, health management and prevention. For example, cancer epidemiology is the collection, analysis and dissemination of information about the occurrence, distribution and causes of cancer and is used to direct cancer control efforts. The large-scale epidemiological studies expand understanding of the factors that contribute to cancer and protect against the disease.

8 An example of an epidemiological study is the link between vitamin D deficiency during pregnancy and the associated adverse health outcomes in mothers, the developing baby and children later in life. Studies have found that women in some countries within Europe can have severe vitamin D deficiency. This compromises the healthy growth of the baby's skeleton, leading to long-term skeletal problems, low birth weights and poorly developed immune systems, causing problems with immune responses later in life. The benefits of this research are that women at high risk can be targeted, screened and treated with vitamin D supplements.

HSC EXAM-TYPE QUESTIONS

Objective-response questions

1 **D**. Care has to be taken when drawing conclusions because data is in percentages of male deaths and female deaths, not totals. **D** is correct because about 1% of males and 1% of females die from infectious diseases so non-infectious disease deaths far outweigh infectious disease deaths. **A** and **C** are incorrect because information about specific diseases is not included. **B** contains a conclusion that cannot be validly made from the data.

2 **C**. Incidence rates refer to the number of new cases. **A** and **B** are incorrect because prevalence is the number of cases or proportion of cases. **D** is wrong because mortality refers to death rates.

3 **B**. An epidemiological study is the study of the patterns and causes of diseases in populations. It involves statistical analysis and interpretation of large quantities of data. **A** and **D** are incorrect because it is more than a nutritional or disease study: it involves investigating links between diet and disease. **C** is incorrect as the example does not describe a clinical trial.

4 **A**. Stem cell research is an example of future directions and haemophilia is a non-infectious disease. **B** and **C** are incorrect because HIV and malaria are examples of infectious diseases. **D** is incorrect because supplementing diets is not future research.

5 **C**. This option reflects the trends shown in the data provided. **A** is incorrect because the incidence rate of melanoma rose between 1982 and 2002. **B** is wrong because for cancer of the cervix, the incidence rate fell between 1982 and 2001, and then remained steady. **D** is incorrect as the age-standardised rate for breast cancer in 2012 was about 118 new cases per 100 000 females.

Short-answer questions

6 **EM** You are asked to analyse the data and write ONE conclusion. A number of sample answers to this question are provided below.

For coronary heart disease in Australia (✓) Indigenous Australians are more likely to have the disease, be hospitalised for the disease and die from the disease as compared with non-Indigenous Australians. (✓)

You could also have written the following conclusions:

For coronary heart disease in Australia (✓) people in combined remote and very remote areas are more likely to have the disease, be hospitalised for the disease and die from the disease as compared with those living in major cities. (✓) OR For coronary heart disease in Australia (✓) people living in the lowest socio-economic areas are more likely to have the disease, be hospitalised for the disease and die from the disease as compared with those living in the highest socio-economic areas. (✓)

7 **EM** This question asks you to interpret information to construct and complete a summary table. Marks are given for the table heading (✓); the table itself (✓) (✓); and for summarising the information into the correct column. (✓)(✓)

Table A15.1 Summary of the occurrence, cause, the body's response, symptoms and the treatment/management of the disease phenylketonuria

Occurrence	one in 15 000 people in Australia
Cause	genetic condition is a result of a mutation in the gene that normally produces an enzyme that breaks down phenylalanine
Effects	the body fails to break down phenylalanine and therefore it builds up in the body; phenylalanine is toxic to nerve cells
Symptoms	intellectual disability, poor coordination and lack of muscular control if condition is not addressed
Treatment/ Management	a modified diet that is very low in phenylalanine; early detection; genetic counselling

8 **EM** For part **a**, you need to outline steps to confirm the validity and reliability of the information. For part **b**, planning an investigation requires following scientific methods and controlling variables such as those outlined on page 203.

a Check the source to ensure the author has scientific credibility. Check the date to ensure the research is current. (✓) Check who published the information to make sure they are an authoritative source and compare your source with other sources. (✓)

b Compare the incidences of other genetic conditions that may result from mutations linked with the age of the mother. (✓) Identify a range of mutations and use epidemiological criteria to look for correlations. (✓)

9 **EM** This question requires you to organise your answer to include both treatment and management. It is a recall question.

This is a sample answer for mesothelioma, a disease caused by environmental exposure. Pleural mesothelioma is caused by exposure to asbestos. (✓) Currently there is no cure for mesothelioma. Treatment includes active treatment such as chemotherapy, radiotherapy or surgery, or a combination of these treatments that can help some people to achieve a longer period of control over the disease. (✓) Surgery either removes the tumour and the pleura, the affected lung, the diaphragm and a portion of the lining of the heart or removes the pleura and as much disease as possible (✓) in order to manage the symptoms.

When mesothelioma is diagnosed at an advanced stage the main goal of treatment is to manage the symptoms and keep them under control for as long as possible. (✓) Treatments can include surgery to improve breathlessness caused by fluid build-up. Video-assisted thoracoscopic surgery involves removing some of the lining of the chest wall and lung. During the procedure sterile talcum powder is injected into the pleural cavity, causing inflammation that closes the pleural cavity and helps prevent fluid from building up again. (✓) For people not suitable for open surgery, a pleural drain is inserted which allows fluid to be drained one to two times a week. (✓)

10 **EM** This is a higher-order question that asks you to recall an epidemiological study that you have investigated, evaluate the methods used, and outline some benefits. Think about how you will organise your answer and make sure you answer both parts of the question.

The study: The Framingham heart study is a long-term ongoing cardiovascular study which began in 1948 in Framingham, Massachusetts, USA, involving 5209 adult subjects. The study is still going. (✓) The Framingham study first centred attention on the concept of risk factors associated with cardiovascular disease and most recently reported substantial 30-year risk data showing the accumulation of risk over time. (✓)

Evaluation: The methodology is robust because the study has been going for a long time and is now in its third generation of participants (✓) and because it involved a large number of people. (✓) It is also site-specific, which means data can be compared over time for the same place. (✓)

Benefits: Prior to the study nothing was known about hypertension or arteriosclerotic cardiovascular disease. Much of what is now known is based on this longitudinal study; for example, prior to this study it was thought that clogging of the arteries and high cholesterol were natural parts of aging. (✓) The study showed the importance of a healthy diet, regular exercise and confirmed the role of smoking as a high-risk factor. The term *risk factor* itself came from the study. (✓) The Framingham heart study has also shown that risk factors for incidence start very early and accumulate with behavioural, social and economic factors over an individual's lifetime. (✓)

CHAPTER 16 PREVENTION

MODULE 8 NON-INFECTIOUS DISEASE AND DISORDERS

INQUIRY QUESTION:

How can non-infectious diseases be prevented?

Data from epidemiological studies reinforce the importance of both treatment and **prevention** of non-infectious diseases. Better diagnosis and treatment can extend and improve the lives of those individuals who have an established disease but successful prevention of the disease itself and avoiding associated risk factors are required to reduce the incidence of non-infectious disease. In this chapter you will review the effectiveness of current disease prevention methods and strategies.

1 The effectiveness of current disease prevention methods and strategies

» Students use secondary sources to evaluate the effectiveness of current disease prevention methods and develop strategies for the prevention of a non-infectious disease, including but not limited to educational programs and campaigns, and genetic engineering.

Current disease prevention methods

- It is now widely accepted that health is much more than the presence or absence of disease. Health reflects genetic, lifestyle and environmental factors along with cultural influences, socio-economic conditions, the availability and quality of healthcare programs and services, and effective prevention methods.
- In evaluating the effectiveness of current disease prevention methods you would have determined the value of strategies by examining the extent to which the prevention method(s) contributed to a decline in the incidence and prevalence of disease and made judgements against a number of criteria, including the level of prevention.
- The Australian Health Policy Collaboration (see www.apo.org.au and www.phaa.net.au) defines levels of prevention according to a 'stages of disease' continuum. The four levels of prevention are outlined below:
  - Primordial intervention prevents the emergence of social and environmental conditions that can lead to the cause of disease. It includes population-based interventions to prevent the development of risk factors; for example, interventions to decrease the number of children who take up smoking, such as regulation of age eligibility for purchasing cigarettes.
  - Primary intervention aims to limit the incidence of diseases through eliminating or reducing specific risk factors and promoting factors that are protective of health; for example, urban design that provides safe environments for walking, cycling and other physical activities to encourage active lifestyles.
  - Secondary prevention involves reducing the progression of chronic diseases through early detection (usually by screening at an asymptomatic stage) and early intervention; for example, dietary management and the promotion of physical activity for people at risk of diabetes to reduce or delay the onset of that disease.
  - Tertiary prevention involves improving function, preventing complications and minimising the impact of established disease; for example, rehabilitation for people who have established cardiovascular disease and have suffered a heart attack (Source: S Willcox, 'Chronic diseases in Australia: The case for changing course', Australian Health Policy Collaboration Issues paper no. 2014-02, 2014).
- Prevention can be targeted at individuals, at communities—including geographically based communities or specific population groups such as Indigenous people or parents with young children—and at the total population. Secondary and tertiary prevention activities are usually targeted at ndividuals. These are mainly at high-risk individuals who have multiple risk factors. Primordial and primary prevention activities usually aim to influence the total population.
- A **population-based prevention approach** has greater reach and coverage than prevention activities targeted at individuals and provides scope for multi-sector action involving many groups. These include governments, industry and manufacturers, non-government organisations, health professionals, the workplace, community and sporting clubs, schools and health services.

prevention: actions undertaken to reduce or eliminate the onset, causes, complications or recurrence of disease
population-based prevention approach: an approach focused on populations rather than on individuals and on prevention and protection rather than on treatment

- The methods and strategies for prevention are often integrated and include medication and other technologically based strategies, early detection (e.g. screening) and public health interventions, including targeted interventions for specific communities or populations. Other methods and strategies include public policy, regulations and legislation along with education and awareness campaigns, behavioural or lifestyle risk-reduction approaches and, importantly, a **life-course approach** to prevention.
- Most health systems respond to chronic diseases through attempting to identify individuals at high risk and encouraging them to reduce their risk factors, usually late in life, but there is widespread recognition that the development of chronic diseases is based on modifiable risks to which people are exposed much earlier in life. For example, there is a large body of evidence linking under-nutrition in early life to increased chronic disease risk later in life. Low birth weight and rapid weight gain after infancy are linked to increased risk of coronary heart disease and diabetes in adulthood. Figure 16.1 below provides one example of the risk factors for coronary heart disease that people may be exposed to across their lifetime.

KEY QUESTIONS

1 **List two possible challenges encountered when evaluating the effectiveness of current disease prevention methods.**

2 **Briefly outline an example of a population-based health prevention strategy.**

3 **What does a life-course approach to disease prevention involve?**

Answers ➲ p. 220

The effectiveness of disease prevention at a global level

- In Chapters 14 and 15 you reviewed the facts that non-infectious diseases now cause more deaths than infectious diseases and they reduce quality of life and the ability to function. The most common chronic non-infectious (non-communicable) diseases, such as cardiovascular diseases, cancer, diabetes and **chronic obstructive pulmonary disease (COPD)**, share key modifiable risk factors and are therefore largely preventable. Given that non-infectious diseases are an important public health challenge in all countries there has been a worldwide shift to focus on their prevention.
- At a global level integrated chronic disease prevention and control is recognised as the most cost-effective approach. The aim of the integrated approach is to strengthen prevention and control of chronic non-communicable diseases by tackling the major risk factors. An integrated approach responds not only to the need for intervention with the aim of reducing premature mortality and morbidity of chronic non-communicable diseases (NCDs) but also the need to integrate primary, secondary and tertiary prevention, health promotion and related programs across sectors and different areas of research. See www.who.int/chp/about/integrated_cd/en/ for more information.

life-course approach: an approach that recognises many chronic diseases are the result of risks that accumulate throughout an individual's lifetime and that these risks can and must be reduced and prevented at all stages of life

chronic obstructive pulmonary disease (COPD): a long-term lung disease where damage to the lungs, usually because of both emphysema and chronic bronchitis, obstructs oxygen intake and causes increasing shortness of breath; the greatest cause by far is cigarette smoking

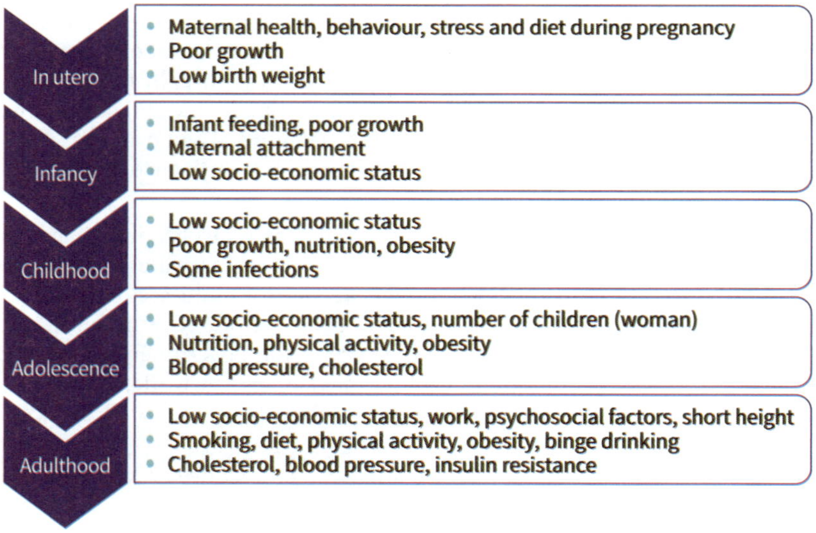

Figure 16.1 Risk factors for coronary heart disease vary at different stages of life
Source: derived from *Australia's Health 2014*

- At the first United Nations high-level meeting on non-infectious disease in 2011 world leaders committed to reducing premature deaths from non-infectious disease by one-third by 2030 and promoting mental health and well being through a global action plan. The World Health Organization (WHO) defined ten national progress indicators to enable all 194 countries to track their effectiveness in moving towards the agreed overall target: for all people, especially the poorest, to live in a world free of the avoidable burden of non-communicable disease. Australia has signed up to the WHO Global Action Plan for the Prevention and Control of Non-communicable Diseases 2013–20.
- Progress on achieving the WHO targets is measured by surveys. The targets include setting national NCD targets for 2025 and strengthening health systems to address NCDs through people-centred primary health care and universal health coverage.
- WHO strategies for reducing risk factors include:
  - reducing affordability of tobacco and alcohol by increasing excise taxes and prices on products
  - limiting exposure to second-hand tobacco smoke in all indoor workplaces, public places and public transport
  - implementing plain packaging and/or large graphic health warnings on all tobacco packages
  - enacting and enforcing comprehensive bans on tobacco advertising, promotion and sponsorship
  - implementing effective mass media campaigns that educate the public about the harms linked to certain behaviours
  - implementing at least one national public awareness and motivational campaign for physical activity, including mass media campaigns for physical activity behavioural change
  - adopting national policies to reduce salt consumption
  - adopting national policies that limit saturated fatty acids and eliminate industrially produced trans fatty acids in the food supply
  - implementing WHO's set of recommendations on marketing of foods and non-alcoholic beverages to children (Source: Non-communicable Diseases Progress Monitor, World Health Organization, 2017). See also the World Health Organization's non-communicable diseases country profiles at www.who.int/nmh/countries/en/.

The effectiveness of disease prevention in Australia

- In Australia there is a solid core of evidence about cost-effective preventive interventions. The evidence is largely provided by the Assessing Cost-Effectiveness (ACE) Prevention study that was conducted over five years on more than 150 preventive health interventions, both population-wide and individual. The study provides evidence on the interventions that have the largest population health impact (through reducing the burden of disease), a ranking of interventions on their cost-effectiveness (including whether they result in cost savings to the health system), and evidence on how to combine and sequence preventive interventions for important health priority areas. The study found that many preventive interventions have very strong cost-effectiveness. These include tax increases on tobacco, alcohol and unhealthy foods as well as mandatory salt limits on processed foods. Together these four interventions would result in fewer years lived with a disability for the Australian population and generate $6 billion of net savings to the health system.
- Another prevention technique is the use of screening programs. Australia has national screening programs for breast, cervical and bowel cancers. The programs target particular populations and age groups where evidence shows screening is most effective at reducing cancer-related morbidity and mortality. Data collected by the Australian Institute of Health and Welfare is used to measure the effectiveness of screening, as outlined in the points below:
  - More than half of all breast cancers detected through BreastScreen Australia are small; small breast cancers are associated with more treatment options and improved survival.
  - For every 1000 women screened for cervical cancer about eight have a high-grade abnormality detected. Detection provides an opportunity for treatment before possible progression to cervical cancer.
  - Bowel cancer is screened using a faecal occult blood test (FOBT). From 1 July 2013 to 30 June 2014, 1.4 million FOBT invitation kits were issued to eligible Australians. Of these nearly 510 000 returned a completed FOBT kit for analysis: a participation rate of 36%. Of the people who returned a valid screening test 7.5% returned a FOBT with a positive screening result and 68% of those had a follow-up **colonoscopy**. Of those participants who underwent a colonoscopy one in 25 was diagnosed with a confirmed or suspected bowel cancer and one in 11 were diagnosed with an advanced adenoma (pre-cancerous tumour).
- In Chapters 14 and 15 you reviewed the link between cervical cancer and the human papillomavirus (HPV). A national HPV vaccination program was introduced for females in 2007 and extended to males in 2013. Australia was the first country to introduce the HPV immunisation program. Of children turning 15 in 2014, 73% of girls were fully immunised and 60% of boys were fully immunised. A new screening program was launched in December 2017 and an improved version of the vaccine for all 12 and 13 year olds was made available in 2018. In 2018 researchers reported that HPV infection rates have fallen to 1% in women aged up to 24 and suggested that the disease could be eliminated within the next few decades.

colonoscopy: a procedure where the inside of the large bowel (colon) is viewed using a long flexible tube inserted through the anus

- One of the biggest prevention success stories in Australia has been the reduction in tobacco smoking rates. Tobacco smoking is the leading risk factor contributing to death and disease in Australia. This includes the risks associated with past tobacco use, current use and exposure to second-hand smoke. Australia has been successful in reducing smoking prevalence over many years through the use of multiple, multi-sector strategies that are sustained and delivered on many fronts. These include:
  - increases in excise duty and taxes
  - advertising bans in print media and at point-of-sale
  - sponsorship bans
  - bans on smoking indoors (smoke-free dining, bans in pubs, clubs, gaming venues and on aircraft) and in many outdoor public spaces
  - the Tackling Indigenous Smoking Initiative
  - price increases and graphic health warnings on packs
  - restrictions on sales to minors
  - public education and media campaigns.
- As a result fewer people, both proportionally and absolutely, are smoking daily and more people have never smoked compared with 20 years ago. In addition fewer people are being exposed to tobacco smoking. This reduction has been particularly marked for younger people with the fall in daily smoking rates over the past decade occurring predominantly among people aged 18 to 49. For more information see page 173 of *Australia's Health Report 2016*.
- For more information see www.health.gov.au/tobacco and www.tobaccoinaustralia.org.au, a website which tracks tobacco control campaigns and their effectiveness.

→ KEY QUESTION

4 **Outline data that provides evidence of the effectiveness of one strategy used in Australia to prevent a named non-infectious disease.**

Answer ➲ p. 220

The effectiveness of educational programs and campaigns

- In Chapter 12 you reviewed public health campaigns for infectious diseases that aimed to communicate information about symptoms, treatment and public health support and also attempted to persuade the public to engage in or stop specific behaviours. Public health campaigns for non-infectious disease prevention include educational programs, mass media campaigns and government sponsorship of activities in sport, arts and popular culture related to the promotion of good health.
- There is a growing body of evidence in Australia about the prevention methods and strategies that work. These evidence-based interventions include taxation, media campaigns and regulation of unhealthy products. Targeted interventions are also cost-effective by reducing the chronic disease burden and hence the pressure on the health system. A report on targeted chronic disease prevention interventions in children and young people at www.preventioncentre.org.au provides evidence of the effectiveness of school-based interventions, classroom-based physical activity, nutrition education and education programs in multiple settings.
- Examples of education programs, media campaigns and targeted interventions are provided in the following table. The effectiveness of one of these campaigns is provided below.
- The 'Slip, Slop, Slap' campaign started in the 1980s with the aim of reducing sun exposure and protecting the population against sunburn and therefore the increased risk of skin cancer. Since the campaign began the incidence of the two most common forms of skin cancer in Australia (basal-cell carcinoma and squamous-cell carcinoma) has declined. However, the incidence of melanoma skin cancer and vitamin D deficiency in Australia has increased.

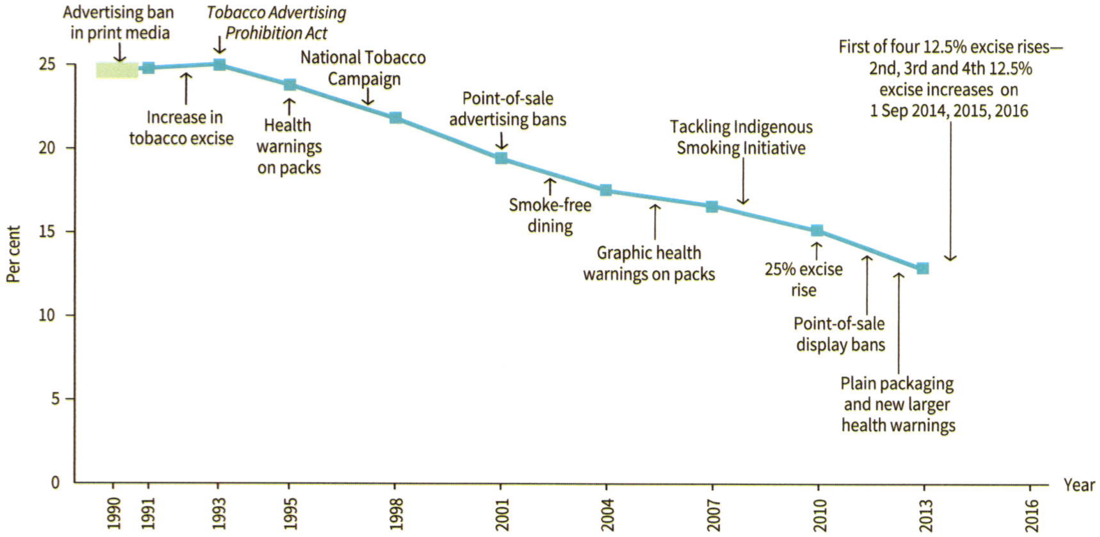

Figure 16.2 Daily smokers aged 14 and over and key tobacco control measures, Australia, 1990–2016
Source: *Australia's Health Report 2016* at www.aihw.gov.au/australias-health/

Table 16.1 Examples of education programs, media campaigns and targeted interventions in Australia

Education program and/or campaign	Target	Website(s)
'Crunch and Sip' fruit and veg month	school-based programs targeting diet and nutrition to prevent diseases such as overweight, obesity and type 2 diabetes in children and young people	www.crunchandsip.com.au also has a healthy lunch box website at www.healthykids.nsw.gov.au
'Jump Rope for Heart'	school-based program aimed at keeping kids healthy and raising funds to fight heart disease	www.heartfoundation.org.au
'Swap It, Don't Stop It' campaign	community-wide campaign to encourage people to make small lifestyle changes for big improvements in health	www.nutritionaustralia.org/act/swap-it-don't-stop-it
'Slip, Slop, Slap' and now extended to 'Slip, Slop, Slap, Seek, Slide'	community-wide sun protection education program and campaigns to encourage people to slip on a shirt, slop on 30+ sunscreen, slap on a hat, seek shade and slide on sunglasses	www.cancer.org.au/sun-protection www.cancer.org.au/preventing-cancer/sun-protection/
'Sun Smart'	community-based, statewide media campaigns to change broader social norms about sun protection includes targeting schools, early childhood centres, parents, workplaces and sports clubs	www.sunsmart.com.au is an augmented reality app that helps you see the strength of UV radiation around you and, by taking a selfie, you can see what could happen to your skin if it was damaged by UV www.generationsunsmart.com.au/posters
'Quit Smoking' and 'I Can Quit' national tobacco campaigns	anti-smoking education and mass media campaigns that target specific groups such as young people and Aboriginal communities	www.quit.org.au and www.cancerinstitute.org.au/quit-smoking (includes 'I Can Quit' website and the Aboriginal Quit Smoking Quitline)
'Don't turn a night out into a nightmare' campaign	national anti–binge drinking campaign targeting teenagers and young adults	www.drinkingnightmare.gov.au www.alcohol.gov.au
'Relay for Life', Cancer Council pink ribbon and 'Australia's Biggest Morning Tea'	education about cancer and support for research into this disease	www.cancercouncil.com.au

- Researchers have studied the possible link between chemicals in sunscreen and melanoma cancer and some epidemiological studies have concluded that the increase in melanoma cancer could not be associated with sunscreens. Doctors now recommend small amounts of exposure to sun to maintain adequate vitamin D levels.

KEY QUESTION

5 Provide an example of the effectiveness of an education program and campaign used in Australia to prevent chronic disease.

Answer p. 220

The effectiveness of genetic engineering

- In Chapters 1, 7and 8 you reviewed genetic technologies and techniques along with biotechnology. Developments in genetic engineering, **molecular genetics**, genetic testing, advanced statistical methods and **high-throughput technologies**, along with progress in studying gene–environment interactions, provide researchers with methods and strategies that potentially help prevent non-infectious diseases. For example, **proteomics** provides potential biomarkers to profile disease risk more accurately while **pharmacogenetics** help researchers develop more effective medications and a better understanding of why certain drugs appear to be more effective in some people.

molecular genetics: the study of the structure and function of genes at the molecular level
high-throughput technologies: the automation of experiments that allows for large-scale high-speed repetition and accuracy; they are important because many genome studies involve large numbers of genes
proteomics: the large-scale study of proteins
pharmacogenetics: the study of genetic differences in drug metabolic pathways

Molecular genetics

- ➔ The association of suspected carcinogenic exposure and cancer risk in populations can be studied with classic epidemiologic techniques (see Chapter 15). However, these techniques are not applicable to the assessment of risk in individuals. Molecular genetics integrates molecular biology, laboratory models, biochemistry and epidemiology to infer individual cancer risk and tries to explain why some people develop cancer and others do not.
- ➔ Molecular genetics is effective in the prevention of non-infectious diseases as outlined in the examples below:
  - By studying large populations of individuals with and without disease researchers can provide data as to which single nucleotide polymorphism (SNP) variations are associated with a disease. The research identifies functional polymorphisms that influence an individual's risk of cancers and focuses on gene products involved in activation and detoxification of carcinogens and DNA repair.
  - Molecular genetics provides evidence such as differences in molecules and genetic pathways in pre-cancerous and malignant cells. By understanding altered genetic pathways researchers will be able to develop novel anti-cancer drugs and understand more about pathways for cancers, such as melanoma initiation and progression.

Genetic testing

- ➔ Genetic testing is effective in the prevention of non-infectious diseases as outlined in the examples below:
  - Mice can be genetically tested to research the melanoma-preventing effectiveness of sunscreens.
  - Genetic testing is used to test for the mutated genes currently known to be associated with the predisposition to certain cancers. For example, in hereditary breast cancer the most common mutations are in genes BRCA1 and BRCA2 but other genes may be identified in the future (see Chapter 15). As a result of these studies women identified at high risk are now given the options of risk-reducing hormone therapy medicines or protective surgery. Protective surgery involves removing one or both healthy breasts (and ovaries), which may reduce the risk by as much as 97%. Removing the ovaries before menopause may reduce the risk by about 50%. Removing ovaries reduces the risk of breast cancer because ovaries are the main source of oestrogen.

Studying gene–environment interactions

- ➔ In Chapter 3 you reviewed how genes and the environment affect phenotypic expression. In Chapter 12 you reviewed the fact that non-infectious diseases can be genetic; that is, the mutation is passed from parent(s) to offspring. A gene mutation can also be acquired through a person's lifetime through exposure to environmental factors.
- ➔ Several environmental factors continue to surface as contributing to incidences of various tumours. The interaction between gene alterations and environmental factors may provide strategies for cancer prevention; for example, more than 80% of genetic epidemiological studies that have been published reveal a reduction in cancer risk with an increase in fruit and vegetable consumption, particularly vegetables containing phytonutrients.

Genetic engineering

- ➔ In Chapter 8 you reviewed genetic engineering as the manipulation of an organism's DNA using genetic techniques (see page 107). You also reviewed techniques and applications used in recombinant DNA technology.
- ➔ The following examples are provided to show how genetic engineering is effective in the prevention of non-infectious diseases. There are also a number of other examples of how genetic engineering has been used in the prevention of diseases in previous chapters.
  - Mice can be engineered to simulate a human genetic disease and this can then be used to develop novel prevention methods.
  - Genetic engineering of primates such as marmosets and macaques has been used in disease prevention; for example, in China researchers have cloned monkeys to test new drugs for a range of diseases before their clinical use. In Japan primates have been genetically engineered to have Parkinson's disease, Alzheimer's disease and motor neurone disease. Researchers investigate the neural circuits that are damaged at each stage as the monkeys' disease develops in order to find ways to predict the onset of symptoms and develop drugs to slow down the progress of the disease.
  - Researchers have identified five diseases that could someday be eliminated with gene-editing technology (see CRISPR-Cas9 on pages 91 and 162). These diseases are HIV, some types of inherited retinitis pigmentosa (a cause of progressive blindness), sickle-cell anaemia, familial hypercholesterolemia (extremely high levels of cholesterol that can cause heart attacks at a young age) and haemophilia.
  - Organisms such as bacteria and viruses can be modified to produce substances such as insulin that can prevent diseases.
  - Genetically modifying genes in human embryos can potentially eliminate the risk of genetic diseases.

➔ KEY QUESTION

6 **Use an example to briefly explain how genetic engineering helps prevent disease.**

Answer ➲ p. 220

CHAPTER SYLLABUS CHECKLIST

Are you able to answer these questions from the syllabus for this chapter? Tick each question as you go through the checklist if you are able to answer it. If you cannot answer a question, turn to the relevant page in the study guide to find the answer. For NESA key word meanings, go to www.educationstandards.nsw.edu.au and search 'key words'.

	FOR A COMPLETE UNDERSTANDING OF THIS TOPIC:	PAGE NO.	✓
1	Can I use secondary sources to evaluate the effectiveness of current disease prevention methods and develop strategies for the prevention of a non-infectious disease?	211–214	
2	Can I use secondary sources to evaluate the effectiveness of current disease prevention methods and develop strategies for the prevention of a non-infectious disease, including educational programs and campaigns?	214–215	
3	Can I use secondary sources to evaluate the effectiveness of current disease prevention methods and develop strategies for the prevention of a non-infectious disease, including genetic engineering?	215–216	

HSC EXAM-TYPE QUESTIONS

Objective-response questions (1 mark each)

1 Read the following to answer this question.

Tanning salons are now banned in most of Australia, a country with the highest incidence of melanoma in the world. (*Medscape medical news*, 20 January 2015)

Which one of the following statements is correct in relation to this statement?

A The banning of tanning salons is an example of changes to legislation to prevent a nutritional disease.
B The banning of tanning salons is an example of an education campaign to prevent a nutritional disease.
C The banning of tanning salons is an example of an education campaign to prevent a disease caused by environmental exposure.
D The banning of tanning salons is an example of changes to legislation to prevent a disease caused by environmental exposure.

2 Preventive health requires early intervention to protect the health of vulnerable children. Which of the following could be a strategy to prevent childhood obesity?

A Ban advertising of unhealthy food and beverages targeted at young children.
B Increase taxes on tobacco products.
C Provide free soft drinks at school.
D Ban advertising of unhealthy food and beverages targeted at adults.

3 Several epidemiological studies have provided analyses of the role that dietary changes may have played in the reduction of coronary heart disease mortality and incidence in a specific population. The research reviewed dietary trends in the decades prior to the decline in coronary heart disease and found that a series of changes in eating patterns that occurred ten to 20 years before the decline could have contributed to the decline. The phrase 'changes in eating patterns' is an example of:

A a risk factor treatment strategy for coronary heart disease.
B a prevention strategy to minimise the risk of coronary heart disease.
C a nutritional disease treatment strategy.
D a prevention and incidence strategy.

4 In China scientists are using CRISPR-Cas9 tools to modify genes associated with Beta thalassaemia, a hereditary blood disease that can be fatal. This is an example of:

A a genetic technology for treatment of cancer.
B genetic engineering to help prevent a genetic disorder.
C gene sequencing for a blood disorder.
D unethical practice.

5 What type of prevention strategy is the 'Sun Smart Slip! Slop! Slap! Seek! Slide!' strategy?

A an intervention strategy
B an education program and legislation campaign
C an education program and media campaign
D a management strategy

Short-answer questions

6 Read the following text and then answer the questions which follow.

Premature death from non-communicable diseases (NCDs) continues to be one of the major development challenges in the 21st century. NCDs kill 15 million women and men between the ages of 30 and 70 each year and leave no country untouched. This burden is rising disproportionately among low-income and lower-middle-income countries, where almost half of premature NCD deaths occur. Within countries these deaths disproportionally affect the poorest and those furthest behind. The NCD epidemic is driven by poverty, globalization of marketing and trade of health-harming products, rapid urbanisation and population growth.

Source: *Non-communicable Diseases Progress Monitor 2017*, World Health Organization

a What are non-communicable diseases? (1 mark)
b Give two examples of products which harm health. (2 marks)
c Describe two interventions for the prevention and control of NCDs. (2 marks)

7 Epidemiological studies have reported links between having a diet rich in folate and lowering the risk of pancreatic cancer and possibly bowel cancer. Based on this reported finding:

a Outline one education program or campaign that could be developed to help prevent pancreatic and bowel cancers. (2 marks)
b Outline one example of genetic engineering research that could be done to prevent pancreatic and bowel cancers. (2 marks)

8 A comprehensive approach to tobacco control has resulted in a major reduction in adult smoking rates in Australia from 35% in 1980 to 13% in 2013. This reduction has contributed to declining death rates from coronary heart disease and stroke over the past 30 years.

Summarise some of the strategies used in the approach to tobacco control and outline similar strategies that could be used for controlling alcohol use. Construct a table to answer the question. (5 marks)

9 Obesity and being overweight are epidemics of the 21st century and some of the biggest challenges confronting Australia's health system.

Imagine you are part of a national working group whose job is to develop a set of integrated prevention strategies and methods with the aim to halt and reverse the rise in overweight and obesity in Australia. List five options you would recommend for development. (5 marks)

10 The causes of chronic diseases are multi-factorial and extend beyond individual behaviours to include many external factors, such as the social and environmental characteristics of the communities in which people live. A population-based approach has important implications for actions required to prevent chronic disease. It moves the focus beyond the behaviour of individuals (whether they smoke and/or drink) to identify how to reduce exposure to risk factors at different stages of life and in different settings.

Construct a table to outline prevention of chronic diseases at three different stages of life: before birth, early childhood and adolescence. In your answer include strategies that go beyond the individual and could address social and environmental factors. (8 marks)

ANSWERS

KEY QUESTIONS

Key questions ➲ p. 212

1 Evaluating the effectiveness of prevention methods can be difficult as it is not easy to directly attribute outcomes to a specific initiative if it involves multiple strategies. Adding to this challenge is the fact that the impact of an intervention will often be delayed, meaning evidence to support the effectiveness of prevention activities may not be available for many years.

2 Population-based health prevention strategies targeting diet and nutrition in Australia include the development of Australian Dietary Guidelines; mandatory fortification of certain food, such as folate fortification of bread; legislative instruments such as the Australia New Zealand Food Standards Code and mandatory food labelling; the National Heart Foundation 'tick' endorsement program and other food labelling and the health star rating systems.

3 Taking a life-course approach to preventing chronic diseases involves reducing risks at each stage of life, including early childhood (and in utero), adolescence and early adulthood. This is well before the onset of many chronic diseases. A population-based approach to prevention that reduces risk factors at each stage of life needs to be balanced with preventive strategies for high-risk populations.

Key question ➲ p. 214

4 A national human papilloma virus (HPV) vaccination program was introduced in 2007 to prevent cervical cancer. Since the program began there has been a fall in HPV-related infection in females, high-grade cervical abnormalities (for example, precancerous lesions of the cervix) in young women, and genital warts among females and males. Data is collected to measure the effectiveness of the strategy; for example, four years after the beginning of the program there was a substantial decrease in vaccine-targeted HPV genotypes among women aged 18 to 24: from 29% in 2004–07 to 6.7% in 2010–11.

Key questions ➲ p. 215

5 An example of an education program and campaign is the national tobacco campaign. Effectiveness studies show that in the first six months of operation the campaign achieved a reduction of 1.4% in smoking prevalence and avoided 32 000 cases of chronic obstructive pulmonary disease, 11 000 cases of acute myocardial infarction, 10 000 cases of lung cancer and 2500 cases of stroke.

Key question ➲ p. 216

6 Genetic engineering is used to modify disease-producing viruses to become anti-cancer drugs; for example, the herpes virus has been modified to try to prevent the reoccurrence of melanoma.

HSC EXAM-TYPE QUESTIONS

Objective-response questions

1 **D**. The banning of tanning salons requires changes to legislation or law. Tanning salons increase exposure to UV radiation and therefore banning them can help prevent skin cancer, a disease caused by environmental exposure. **A** and **B** are incorrect because the ban is not associated with preventing a nutritional disease. **C** is incorrect because the strategy is not an example of an education program.

2 **A**. Obesity is a disease associated with unhealthy diets as well as a lack of physical exercise. Advertising of unhealthy food and beverages normalises unhealthy behaviours for children so targeting advertising and, in particular, bans on advertising of unhealthy food and beverages is a cost-effective way to prevent obesity. **B** is wrong because increasing taxes on tobacco does not prevent obesity. **C** is incorrect because soft drinks are one of the products associated with obesity so they should not be given to children for free. **D** could help but the bans need to be targeted to products advertised to children, not just adults.

3 **B**. Changing eating patterns is a prevention strategy and according to the epidemiological study healthier diets can minimise the risk of coronary heart disease. **A** and **C** are incorrect because these are not treatment strategies. **D** is wrong because even though it is a prevention strategy, incidence refers to the number of new cases of a disease.

4 **B**. Using CRISPR-Cas9 tools is an example of genetic engineering and Beta thalassaemia is a genetic disorder. **A** is incorrect because it is not an example of treatment or a cancer. **C** is wrong because it is not an example of gene sequencing. There is not enough information to determine if it is an ethical or unethical practice so **D** is incorrect.

5 **C**. The 'Slip, Slop, Slap, Seek, Slide' campaign is an example of an education program and media campaign. **A** and **D** are incorrect because intervention and management do not explain the strategy thoroughly enough. **B** is incorrect because the campaign is not an example of a legislation strategy.

Short-answer questions

6 **EM** This is a straight recall question which also requires you to comprehend the paragraph on NCDs.

- a Non-communicable diseases are non-infectious diseases. (✓)
- b Examples of health harming products are tobacco (✓) and sugar. (✓) You could also have alcohol, salt or unhealthy oils such as palm oil.
- c Examples of interventions include increasing excise taxes and prices on products (✓) and mass media campaigns that educate the public about the harms of smoking and excessive alcohol use. (✓)

7 **EM** In your studies you used secondary sources to develop strategies for the prevention of a non-infectious disease. You need to develop different strategies (one education program and options for genetic engineering research) for the particular diseases in the question: pancreatic and bowel cancers.

- a A population-based education program, emphasising the benefits of having folate in the diet, could outline which foods contain folates and emphasise the links between folate in the diet and reduced pancreatic and bowel cancer risk. (✓) The program could include posters, recipes and online information. (✓)
- b Genetic engineering could be used to investigate genetic mechanisms to show how nutrients such as folate influence biochemical pathways in the development of pancreatic and bowel cancer (✓) or to genetically modify plants to contain more folate. (✓)

8 **EM** This question asks you to recall some prevention strategies and then use them to develop prevention strategies to control alcohol use. The table you construct will need a heading, for which one mark is awarded. (✓) Additional marks are given for having at least two tobacco use prevention strategies (✓) (✓) that are then used for alcohol use prevention. (✓) (✓) There are more examples given in the sample answer than are needed.

Table A16.1 Prevention strategies for tobacco use used to develop controls for alcohol use

Tobacco control prevention strategies	Alcohol control prevention strategies
increases in excise duty and taxes on tobacco	increases in excise duty and taxes on alcohol
increasing the price of tobacco	increasing the price of alcohol
banning tobacco companies from sponsorship of sports clubs and sporting events	banning alcohol companies from sponsorship of sports clubs and sporting events and cultural events
campaigns for targeting specific groups such as the Tackling Indigenous Smoking Initiative	campaigns for targeting specific groups such as teenagers and those at risk of binge drinking
restrictions on sales of tobacco to minors	increase in minimum legal drinking age to 21
graphic health warnings on cigarette packs and at points-of-sale	graphic health warnings on bottles and at points-of-sale
national campaign on the effects of smoking on families	national campaign on drink driving

9 **EM** In your studies you used secondary sources to develop strategies for the prevention of a non-infectious disease. You need to develop five strategies to halt or reverse obesity and overweight. Because it is a national strategy your options will need to be population-based as well as targeting individual behaviours. The question assumes that you know the factors that increase the risk of overweight and obesity such as diets high in sugar and fats and a lack of physical activity.

Options could include introducing a tax on sugar, (✓) increasing the price of sugary drinks, (✓) restricting advertising of junk foods on billboards, television and social media, (✓) providing national funding for education programs that focus on weight loss campaigns for children and young people (✓) and developing a national 'walk or ride your bike to school and work' campaign. (✓)

10 **EM** This is a higher-order question that has a number of parts to it: you need to construct a table (✓) and include a heading. (✓) You need to look at different strategies (✓) (✓) (✓) for the three different stages of life (✓) (✓) (✓), making sure that some are social and some are environmental factors. The question assumes that you understand the term *chronic disease*; that is, diseases that tend to be long-lasting and persistent in their symptoms or development. There are more examples given in the sample answer (Table A16.2) than are needed.

Table A16.2 Prevention of chronic diseases at different stages of life

Before birth	Early childhood	Adolescence
improving nutrition of the mother through community-based education and support programs	reducing exposure to tobacco smoke in public places and the home	promoting healthy weight and physical activity through sporting, environmental and cultural activities
targeting individuals to reduce the prevalence of smoking during pregnancy	addressing overweight and obesity through school and community-based programs	social media and education campaigns to reduce the risk of the harmful use of drugs and alcohol
improving access to services for pregnant women, particularly those in remote communities	improving nutrition	social media and education campaigns to reduce the uptake of smoking
	increasing physical activity	providing safe and supportive environments for young people at risk of anxiety and depression
	encouraging sun-safe behaviour	

MODULE 8
NON-INFECTIOUS DISEASE AND DISORDERS

CHAPTER 17 TECHNOLOGIES AND DISORDERS

INQUIRY QUESTION:

How can technologies be used to assist people who experience disorders?

In this chapter you review the **physiology** of disorders along with engineered solutions to assist people with a range of disorders. Solutions often require a multidisciplinary approach and the practical applications of science, technology, engineering and maths.

1 Causes of disorders in the structure and function of organs

» Students explain a range of causes of disorders by investigating the structures and functions of the relevant organs; for example, hearing loss, visual disorders and loss of kidney function.

- In order to design and develop technologies to assist people who experience disorders, medical scientists, engineers and mathematical modellers must understand the structures and functions of the relevant organs and the causes and effects of the associated disorders. In this section you review the structures and functions of the ears, eyes and kidneys and the causes of some of the disorders in these organs.

The structures and functions of the ears

- Our ears function to provide hearing and balance. Our ears and brain detect and interpret changes in the **frequency**, tone and volume of sounds in the immediate environment.
- Humans cannot hear sounds of every frequency; for humans, the range is from 20 to 20 000 Hertz or cycles per second, with the greatest sensitivity around 1000 Hertz.
- Sounds with short wavelengths have high frequencies (high pitch). Sounds with long wavelengths have low frequencies (lower pitch). Sounds also vary in amplitude or loudness.

physiology: the study of organs and their functions and parts
frequency: the number of cycles per second (Hertz or Hz) of vibrations that make up sound

- The loudness of the sound is determined by the energy carried by the wave. The larger the amplitude the louder the sound (see Figure 17.1).

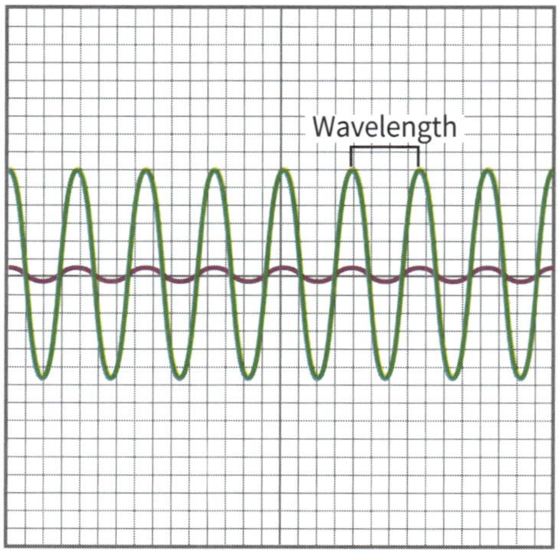

Figure 17.1 Wavelength and frequency of sound waves

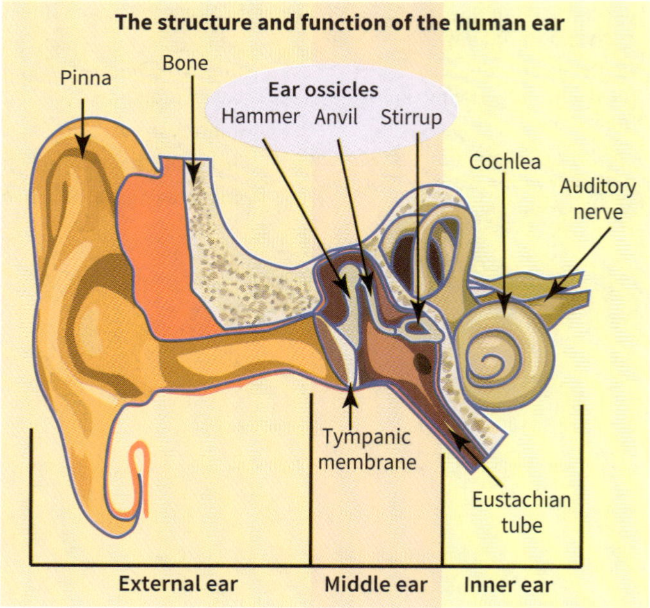

Figure 17.2 Diagram of the human ear

- Remember that sound waves need to travel through a medium, which they do by vibrating the molecules in the medium.

Table 17.1 Structure and function of the parts of the human ear

Structure	Function
pinna (the flap)	collects sound waves from a wide area and funnels the sound into the external ear passage
tympanic membrane (eardrum)	stretched across the end of the auditory canal; separates the outer ear from the middle ear sound waves cause the tympanic membrane to vibrate; these vibrations are then conveyed from the tympanic membrane to the oval window by three tiny bones called the ear ossicles
ear ossicles	three intricately formed bones—the hammer, anvil and stirrup—transmit the sound waves to the inner ear (sound vibrations travel well through bone)
oval window and round window	the oval window and round window are two thin membranes; sound reaches the inner ear at the oval window and pressure is then transmitted in the fluid in the tympanic canal this pressure causes the round window, at the end of the tympanic canal, to bulge outwards; the oval window helps to amplify the pressure of the sound vibrations
cochlea (snail-like spiral coiled tube in the inner ear)	the cochlea contains the receptors for sound and the vestibular apparatus that is associated with a sense of balance; as a result of the round window bulging outwards, fluid in the cochlear tubes vibrates
organ of Corti	contains the auditory receptor cells
auditory nerve	transmits the sound vibrations to the brain

- Sound waves travel through the external ear (ear canal) and cause the eardrum to vibrate. These vibrations are then transmitted by the ear ossicles (**hammer, anvil and stirrup**) of the middle ear to the inner ear. Sound reaches the inner ear at the oval window. Vibrations are transmitted through the oval window to the spiral **cochlea** which contains fluid and nerve endings. Vibrations in the inner ear, displacement of cochlear fluid and movement of hair cells at the organ of Corti produce electrochemical signals. The auditory nerve transmits these neural impulses to the auditory area of the cerebrum in the brain where they are interpreted.

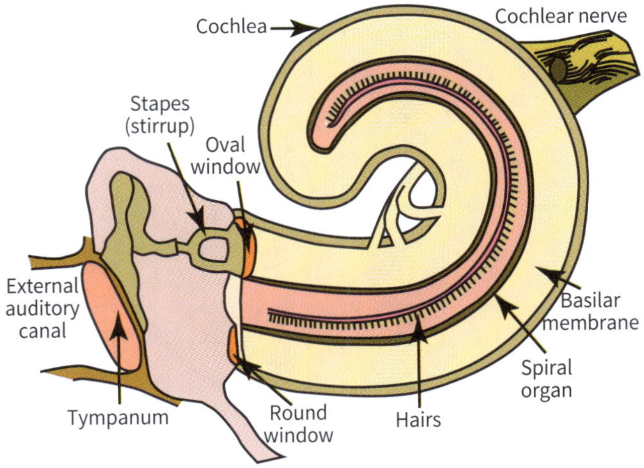

Figure 17.3 Simplified diagram of the cochlea showing hairs or receptor cells in the organ of Corti

hammer, anvil and stirrup: tiny bones in the middle ear; the Latin for these terms (in order) is *malleus*, *incus* and *stapes*
cochlea: the actual organ of hearing; the cochlea receives sound waves and passes them on to the brain

Causes of disorders in the structures and functions of the ears: hearing loss

- Good hearing requires good conduction of the sound waves through the outer ear, proper mechanical working of the ear ossicles, healthy nerve cells and fluids in the inner ear and functional nerve pathways for the signals to the brain. If anything goes wrong with these structures and their functions a person will have some kind of hearing loss.
- There are three types of hearing loss and the type depends on which structures and functions are affected (see Table 17.2).
- An individual's hearing loss varies depending on the pitch (frequency) of the sounds they cannot hear and can be different in each ear. Hearing loss is measured in **decibels** (dB). On average, the quietest sound people can hear is from 0 to 20 dB. More than 90 dB can lead to chronic hearing damage if people are exposed to it every day. A rock concert is about 120 dB.
- The quietest sound heard by people with mild hearing loss is between 20 and 40 dB. For moderate hearing loss the quietest sound is 41 to 60 dB while for severe hearing loss it is 61 to 80 dB. With profound hearing loss it is more than 81 dB.

decibel: a unit that measures the intensity of sound or sound pressure levels

KEY QUESTIONS

1. **Relate the structure and function of the outer, middle, inner ear and brain to the hearing of sound.**
2. **For each part of the ear—outer, middle and inner—explain the cause and effect of one type of hearing loss. Use a table to answer this question.**

Answers ⊃ p. 234

Table 17.2 Types of hearing loss and some common causes

Type of hearing loss	Causes
conductive hearing loss due to problems with the ear canal, eardrum, middle ear or ear ossicles	blocked ear canal caused by excessive earwax which decreases the volume of the sound being transmitted infection in the ear canal perforated eardrum caused by loud noise (e.g. an explosion), poking the inner ear, injury from a severe blow to the head or rapid changes in air pressure such as in scuba diving; the hearing loss is usually temporary while the eardrum heals itself benign tumour fluid in the middle ear from colds middle ear infection (*Otitis media*); this is common in young children, particularly when the Eustachian tube does not fully form damage to ear ossicles, usually caused by serious infections or head injuries; sound waves cannot pass through the eardrum to the inner ear; occasionally babies are born with misshapen ear ossicles abnormal bone growth in the outer or middle ear: otosclerosis is an inherited disorder in which abnormal bone grows onto the stirrup (or stape); a person with otosclerosis will have severe hearing loss if the bone stops moving all together; otosclerosis may sometimes cause **sensorineural hearing loss**
sensorineural hearing loss due to problems with the inner ear	damage to the hairs or nerve cells in the cochlea caused by aging or cumulative exposure to excessive noise or certain **oxotoxic** drugs when the hairs are damaged electrical signals are not transmitted efficiently severe cochlea damage means that large portions of the hair cells of the organ of Corti are damaged or missing; sounds are not able to be sent to the brain and once damaged the hairs do not regenerate head trauma an infection from a virus such as measles, mumps or meningitis malformation of the inner ear **autoimmune** inner ear disease Menier's disease with symptoms that include ringing in the ears (tinnitus), vertigo and loss of balance a tumour adjacent to the auditory nerve can be inherited
mixed hearing loss	a combination of conductive and sensorineural hearing loss to the outer, middle and inner ear or auditory nerve; therefore a possible combination of causes

The structures and functions of the eye

- The eye is responsible for both vision and balance. The structures of the eye relate to its function of admitting light, **refracting** and focusing light to form an image, and converting that image into nerve impulses that are then conveyed to the brain for interpretation (see Figure 17.4). Remember that unlike sound waves, light does not need a medium to travel through. Light can pass through objects, be reflected (bounce back) or refracted.

sensorineural hearing loss: hearing loss caused by a reduction in the function of the organ of Corti or auditory nerve; also known as nerve-related hearing loss
oxotoxic: drugs toxic to the ear, especially the cochlea and auditory nerve
autoimmune: disorders where the immune system makes antibodies that attack the body's own tissues and organs, slowly destroying them
refracting: bending light; light bends when it changes speed as it moves from one medium of a different optical density to another medium

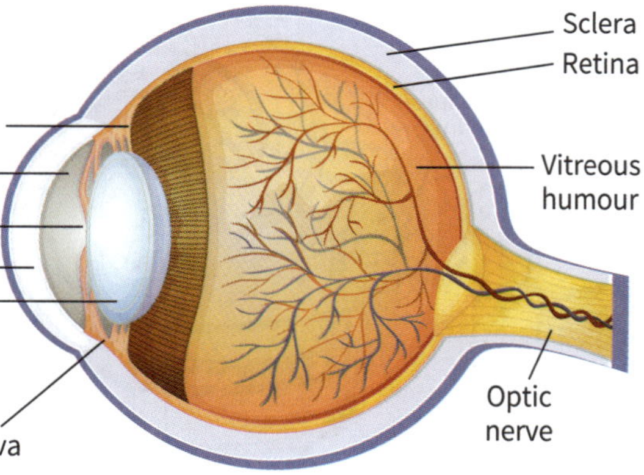

Figure 17.4 Diagram of the human eye

- The conjunctiva protects the eye. The cornea refracts light rays entering the eye.
- The function of the lens is also to refract light, focusing the rays onto the light sensitive (photoreceptor) cells of the retina.

- The shape of the lens changes as the ciliary muscles relax or contract. Lenses refract light differently. A thicker lens has a shorter **focal length**. The natural shape of the lens is round. This shape refracts light more and is used to focus on near objects. For this to occur the ciliary muscle must contract. This is why looking at close objects, such as reading for a long time, is tiring to the eyes.
- To focus on distant objects, the lens is flatter. This requires ciliary muscles to be relaxed, creating a bigger diameter, and less refraction occurs.
- The retina is a complex structure of photoreceptor cells (rods and cones) on the back of the eye. Photoreceptors allow us to see shape, movement and colour. Retinal nerve cells convert the light into nerve impulses.
- Rods contain a photosensitive pigment called **rhodopsin**. Rhodopsin is highly sensitive to light and rods are therefore specialised for night vision.
- There are three types of cones, each containing pigments sensitive to blue, green or red light. Cones are therefore specialised for colour vision.
- The fluid-filled anterior chamber and gel-filled posterior chamber help keep the eyeball in shape.
- With normal vision eyes can focus on both near and distant objects. When all structures are functioning well light passes through clearly and is refracted as required so that images are correctly focused on the retina.

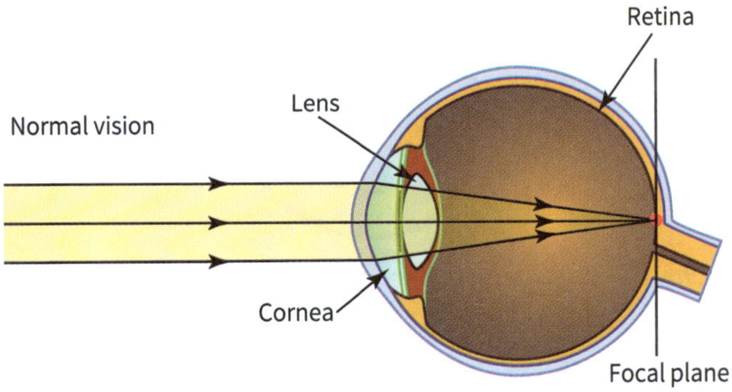

Figure 17.5 Normal vision

focal length: the point at which light rays converge; the measure of the refractive power of a lens
rhodopsin: the pigment in rods that is responsible for the detection of light; used in night vision

Causes of disorders in the structures and functions of the eyes: visual disorders

- Good eyesight requires good transmission of light, correct refraction of light, a precise focal length for focusing the light and functional nerve pathways that take the signal to the brain. When things go wrong with these structures and processes a person has visual disorders or blindness.
- **Hyperopia** is called long-sightedness. When the eye is working well it can focus and produce clear images of distant objects. However, if the cornea is flatter (less curved) or the eyeball is shorter or the lens is positioned further back in the eye than normal then the image is produced (at an imaginary spot) behind the retina. Near objects are not in focus and therefore the person has a visual disorder.

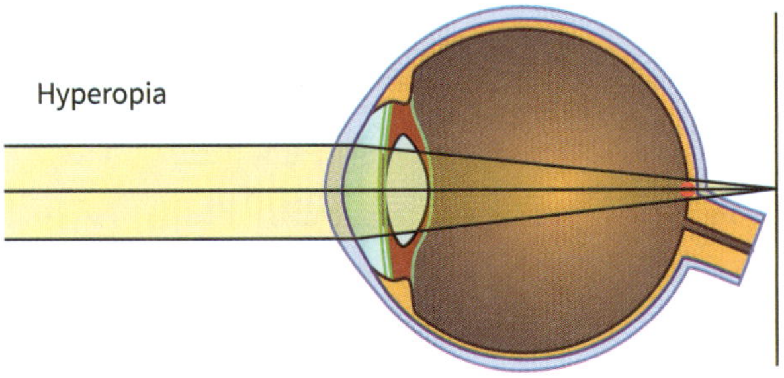

Figure 17.6 Hyperopia

- **Myopia** is the opposite of hyperopia. Myopia is called near- or short-sightedness. When the eye is working well it can focus and produce clear images of objects close up. If the cornea is more curved, the eyeball longer or the lens positioned further forward than normal the image produced is in front of the retina. Distant objects are not in focus and therefore the person has a visual disorder.

hyperopia: light is focused behind the retina instead of directly on it; inability to see near objects clearly; the prefix *hyper* comes from Greek meaning 'over'
myopia: light is focused in front of the retina instead of directly on it; distant objects become blurred

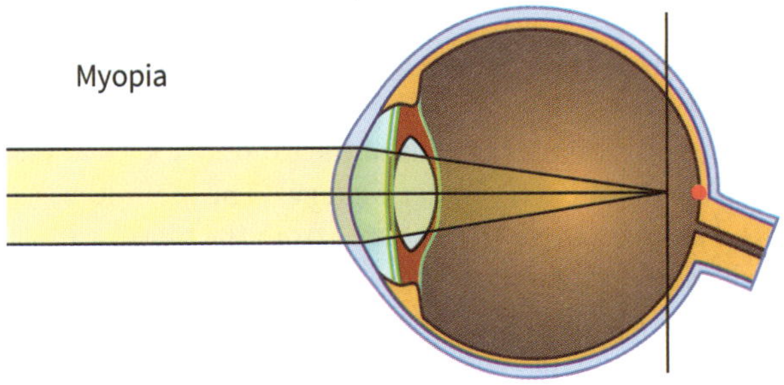

Figure 17.7 Myopia

- Hyperopia and myopia are caused by genetic and environmental factors. Both tend to run in families although close-up work such as reading and eye strain can increase the chances of myopia.
- Pretiopia is a specific disorder of hyperopia where the lens gets harder and less flexible with age. When this happens the lens does not return to its normal shape and this makes focusing on near objects more difficult.
- Other visual disorders and their causes are summarised in the table below.

Table 17.3 Other visual disorders and their causes

Visual disorder	Causes
astigmatism	a refractive error in which the eye does not focus the light evenly on the retina
cataracts	a degenerative condition in which the lens of the eye clouds over, obstructing the passage of light to cause vision loss and potentially blindness; it is the result of deterioration of proteins in the lens
macular degeneration	a progressive deterioration of the macula of the retina (the central inner-lining); it is often related to old age and results in a loss of central vision
diabetic retinopathy	a complication of diabetes caused by prolonged high blood glucose levels; over time high blood glucose levels weaken and damage blood vessels within the retina, starving the retina of oxygen and causing haemorrhage or swelling diabetic retinopathy occurs in 15% of Australians with type 1 diabetes
glaucoma	caused by damage to the optic nerve due to increased pressure within the eyes
night blindness	caused by short-sightedness, cataracts or vitamin A deficiency
childhood blindness	caused by a number of factors such as abnormalities in the eye itself and **congenital** blindness as a result of damage to the brain due to lack of oxygen at birth
colour blindness	a genetic disorder caused by the lack of one or more of the colour-sensitive pigments in the cones the most common type of colour blindness is red–green in which the cones most receptive to red light and green light are missing

➔ KEY QUESTIONS

3 **Myopia and hyperopia are caused by errors in the refraction of light entering the eye. What is refraction and how does it help us see? Draw a simple diagram to explain your answer.**

4 **Draw a table to identify how refraction and/or light transmission can be affected by three different visual disorders.**

Answers ➲ p. 234

The structures and functions of the kidneys

- ➔ Kidneys are complex organs of the excretory system (see pages 40–41 of *Excel Year 11 Biology*). The main functions of the kidneys are to filter wastes from the blood and produce urine without the loss of too much water. The main wastes are the nitrogenous products urea and ammonia. These wastes are produced when proteins are metabolised. They are toxic and must therefore be removed from the body.
- ➔ Kidneys also help in the homeostasis of water, glucose and inorganic salts (see Chapter 13).
- ➔ The composition, quality and quantity of urine produced depends on factors such as diet, physical exercise, metabolism and water intake.
- ➔ The renal artery brings blood carrying wastes and other substances to the kidneys. Urine is formed in the cortex and medulla, collected in the pelvis and delivered to the bladder through the ureter.

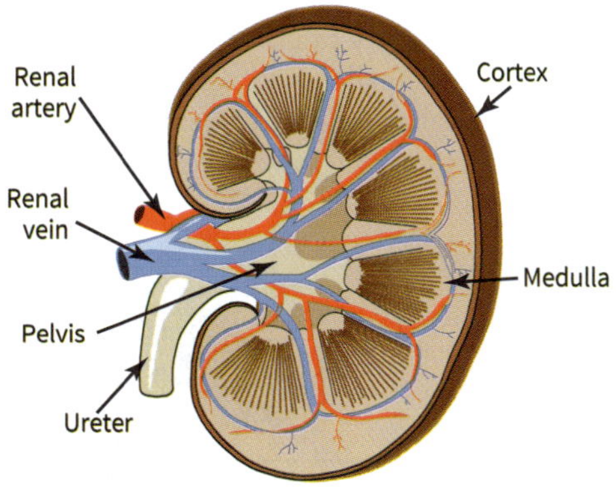

Figure 17.8 Structure of a human kidney

- ➔ The renal vein returns filtered blood to the body.
- ➔ In the human kidney water reabsorption is a passive process of diffusion and osmosis (see pages 17–18 of *Excel Year 11 Biology*). Reabsorption of sodium salts, glucose and amino acids are active processes that require energy.
- ➔ The functional unit of the kidney is the **nephron** (see Figure 17.9).
- ➔ Nephrons are arranged in the outer cortex and central medulla. Each nephron is made up of:
  - a filtering unit
  - areas where both passive and active reabsorption take place
  - a collecting area for urine.

congenital: a disease or disorder present from birth (as compared to a genetic disease or disorder that is inherited)
nephron: the basic filtering unit of the kidney; there are approximately one million nephrons in each kidney

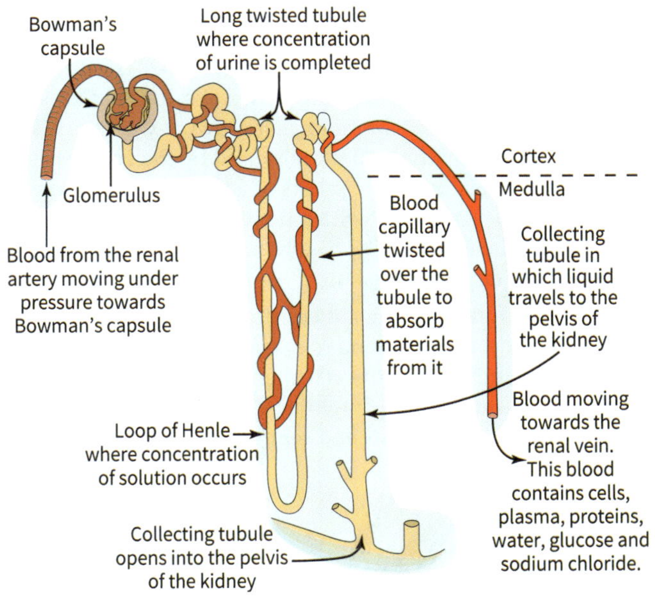

Figure 17.9 Nephron

(Note: The diameters of tubules are not drawn to scale.)

Table 17.4 Parts of a nephron and their function

Parts of a nephron	Function
Bowman's capsule	a cuplike sac that encloses a glomerulus; it functions to carry out the first step in filtration of blood from the urine
glomerulus (glomeruli is plural)	a ball of capillaries within the Bowman's capsule that acts as an 'ultra' filter
long twisted tubules and Loop of Henle	the twisted tubules occur in the cortex; the Loop of Henle is the u-shaped part of the nephron and it extends from the cortex down into the medulla and back; the tubules play a role in the reabsorption of substances
collecting tubule	long straight tubule which opens into the pelvis of the kidney; the wall of the collecting tubule may be permeable or impermeable to water (see below)

- If the wall of the collecting tube is permeable to water, water is reabsorbed into the surrounding tissues. Concentrated urine is excreted and water is conserved. If the wall of the collecting tubule is impermeable to water, no water is reabsorbed and dilute urine is excreted. The permeability of the wall is controlled by the antidiuretic hormone (ADH) (see Chapter 13).
- The adrenal cortex within the adrenal glands produces two hormones: cortisol and aldosterone. Cortisol controls many metabolic processes, including blood pressure. Aldosterone functions to maintain the balance of water and salts in the body. The hormone stimulates the nephrons to decrease reabsorption of potassium and increase reabsorption of sodium into the blood.

Causes of disorders in the structures and functions of the kidneys: loss of kidney function

- A healthy functioning kidney needs structures and processes for filtration, active and passive reabsorption of substances, and diffusion and osmosis. When things go wrong with these structures and processes a person has loss of kidney function.
- Disorders of the kidneys include kidney disease and kidney failure. Kidney disease occurs when the nephrons are damaged. Kidney disease is often called a 'silent disease' as up to 90% of kidney function can be lost before symptoms appear. As a result many people are unaware that they have the condition. When a person has evidence of kidney damage and/or reduced kidney function lasting at least three months this is called chronic kidney disease. Chronic kidney disease is caused by risk factors such as smoking, high blood pressure, overweight and obesity, and diabetes. The most severe form of chronic kidney disease is end-stage kidney disease where people usually require kidney replacement therapy—a kidney transplant or dialysis—to survive.
- Acute kidney injury occurs when an abrupt loss of kidney function causes the body to accumulate waste products and become unable to maintain homeostasis. It is caused by injury, trauma or severe inflammation that restricts blood supply to the tissues.
- **Nephritic disease** or nephritis is an inflammation of the kidney due to infections and toxins but most commonly caused by autoimmune disorders. Pyelonephritis is caused by a bacterial infection of the urinary tract that reaches the renal pelvis of the kidney and causes inflammation.
- **Nephrosis** or nephrotic syndrome is a collection of degenerative disorders of the kidney that result from a problem with the functioning of the glomeruli. When the glomeruli are damaged proteins from the blood leak into the urine. Nephrosis is caused by a primary kidney disease that results in scar tissue forming on the glomerulus or from secondary causes due to diseases such as diabetes. As the kidney function fails extra salts and fluids build up in the body, causing swelling (edema) and high blood pressure.
- Kidney stones are clumps of dissolved minerals that collect on the inner lining of the kidneys. They are formed from the build-up of crystals from substances such as calcium-oxalate and uric acid. The main cause of kidney stones is lack of water in the body.

nephritic disease: swelling and inflammation of the kidneys; also called nephritis; the suffix *itis* means 'inflammation'

nephrosis: a kidney disorder where the body excretes too much protein in the urine; the suffix *osis* means 'abnormal' or 'diseased'

- **Addison's disease** is caused by the destruction or shrinking of the adrenal cortex and therefore the lowering of production of hormones important for maintaining homeostasis. Balance cannot be maintained if the body cannot secrete aldosterone, water and salt. When this homeostasis is upset the volume of blood falls dangerously low and there is a drop in blood pressure. Most cases of Addison's disease are caused by autoimmune disorders. Other causes include tuberculosis and cancer.

KEY QUESTIONS

5 **Relate the structure to the function of one part of the kidney.**

6 **Describe one disorder of the kidneys and its cause.**

Answers ⊃ p. 234

2 Investigating technologies used to assist with the effects of disorders

» Students investigate technologies that are used to assist with the effects of a disorder, including but not limited to hearing loss (cochlear implants, bone conduction implants, hearing aids); visual disorders (spectacles, laser surgery) and loss of kidney function (dialysis).

Technologies that assist with the effects of hearing loss

- Technologies that are used to assist with the effects of hearing loss include **cochlear implants**, **bone conduction implants**, **hearing aids** and assisted hearing devices. With the development of digital and wireless technologies more and improved devices are becoming available.
- The technology used to assist people depends on the type and extent of hearing loss.
- Bone conduction hearing aids are external devices designed for people with conductive or unilateral hearing loss (hearing loss in one ear only) who get no benefit from wearing conventional hearing aids. Bone conduction hearing aids create vibrations that move across the skull to the inner ear where they are sensed by the cochlea as sound. The technology consists of a hearing aid worn behind the ear coupled with a bone conductor vibrator fitted to a headband. The headband holds the vibrating bone conductor to the skull.
- A conventional hearing aid has three basic parts: microphone, amplifier and speaker. The microphone picks up the sound waves, converts them to electrical signals and sends them to the amplifier. The amplifier increases the power of the signals and sends them to the speaker.

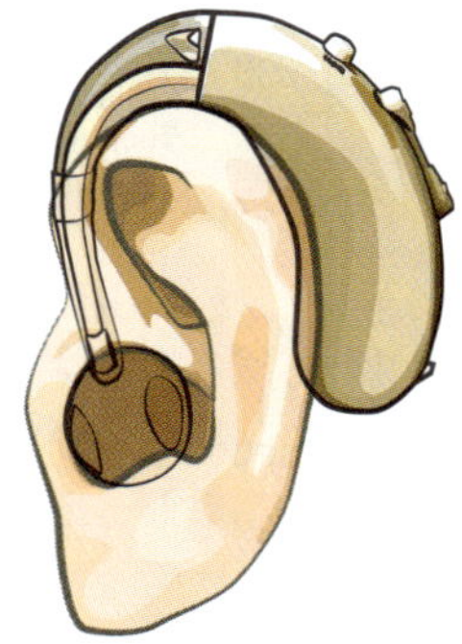

Figure 17.10 Simple model of a hearing aid

Addison's disease: a disorder in which the adrenal glands do not produce enough hormones
cochlear implant: an electronic, surgically implanted device for treating bilateral hearing loss that directly stimulates the cochlea
bone conduction implant: a surgically implanted device that indirectly stimulates the cochlea through vibrating the bones of the skull near the ear
hearing aids: devices worn externally (behind the ear or in the ear canal) that amplify sounds for people with mild to severe hearing loss

Table 17.5 Technologies used to assist people with the effects of hearing loss

Type of hearing loss	Technologies to assist the disorder			
conductive		bone conduction hearing devices and hearing aids		
sensorineural	hearing aids assistive hearing devices	hearing aids and hybrid cochlear implants	hearing aids and cochlear implants	cochlear implants and auditory brain stem implants
mixed hearing loss		middle-ear implants and hearing aids	middle-ear implants and hearing aids	cochlear implants
	Extent of hearing loss ⟷			
	mild	moderate	severe	profound

Source: www.hearnet.org.au

- There are different styles (types) of hearing aid designs that can be used for the different types of hearing loss. Most devices that are located behind the ear are used for people with moderate to severe hearing loss. Hearing aids that are fitted in the ear or in the ear canal are for people with moderate to mild hearing loss. Due to technological advances hearing aids have become more sophisticated; for example, hearing aids can now work with other devices and technologies such as iPhones.
- Assistive hearing devices help amplify sounds, especially when there is a lot of background noise. Some are designed for use in large areas such as classrooms, theatres and airports. Others are designed for personal use in small settings and for one-on-one conversations. All can be used with or without hearing aids or a cochlear implant. For example, personal amplifiers are about the size of a mobile phone. The amplifier increases sound levels and reduces background noise for a listener. Some have directional microphones that can be angled toward a speaker or another source of sound. The amplified sound can be picked up by a receiver that the listener is wearing, either as a headset or as ear buds.
- Middle-ear implants are a more recent technology and are used as an alternative to conventional hearing aids or bone conduction hearing devices. They are made up of an external processor that transmits sound and a surgically implanted internal receiver that is attached to one of the bones of the middle ear or near to the membrane window of the cochlea. The implant works by directly moving the bones of the middle ear or by vibrating the cochlear window.
- Bone conduction implants are used for conductive or unilateral hearing loss. A sound processor worn externally captures sounds and turns them into vibrations. The vibrations are sent to the implant which then transmits the sound through bone vibrations to the inner ear. People using bone conduction implants have good speech discernment once the sound gets to the cochlea.
- A cochlear implant, also called a bionic ear, is used to help people who have severe to profound hearing loss and/or damage to the inner ear or auditory nerve. Cochlear implants work by having a sound processor worn behind the ear which has a microphone that picks up sounds in the person's environment. The sound processor then converts the sound into an electrical impulse that passes through the skin to a receiver surgically inserted and attached to the skull. The receiver relays the electrical impulse to an electrode array that has been surgically inserted into the cochlea to produce the sensation of sound at specific frequencies (pitches). For more information see www.hearing.com.au/cochlear-implants.
- Improvements in cochlear implant technologies are improving people's ability to hear conversations in noisy environments, enjoy music and use the implant processors while swimming. In addition to having remote controllers they are also becoming increasingly compatible with iPods, iPhones and iPads.

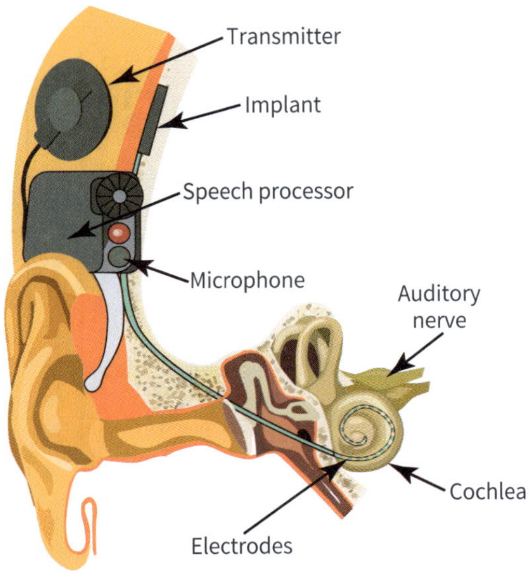

Figure 17.11 Simplified diagram of a cochlear implant

- More recent hybrid technologies combine a hearing aid and a cochlear implant in the one device (see www.cochlear.com). The hearing aid component amplifies low-frequency sounds and sends them via the normal hearing pathway. A cochlear implant processor converts high-frequency sounds to digital information that is sent to the cochlear implant under the skin. The implant sends the electrical signals down an electrode to the cochlea, stimulating the nerve cells. The nerve response is sent to the brain where it is combined with the response from the amplified sound from the hearing aid component.
- An auditory brain stem implant is a surgically implanted electronic device that provides a sense of sound for a person who is profoundly deaf. It uses similar technology to the cochlear implant but instead of stimulating the cochlea it directly stimulates the brain stem. Brain surgery is required to implant the device.

→ KEY QUESTION

7 **Briefly describe two technologies that you have studied that assist with the effects of hearing loss.**

Answer ➲ p. 234

Technologies that assist with the effects of visual disorders

- Whether visual disorders are treatable depends on the cause. Technologies that assist with the effects of visual disorders include spectacles, laser surgery and cataract surgery.
- Prescription spectacles and contact lenses are artificial lenses designed to correct the refractive errors of myopia and hyperopia as shown in the diagrams following.

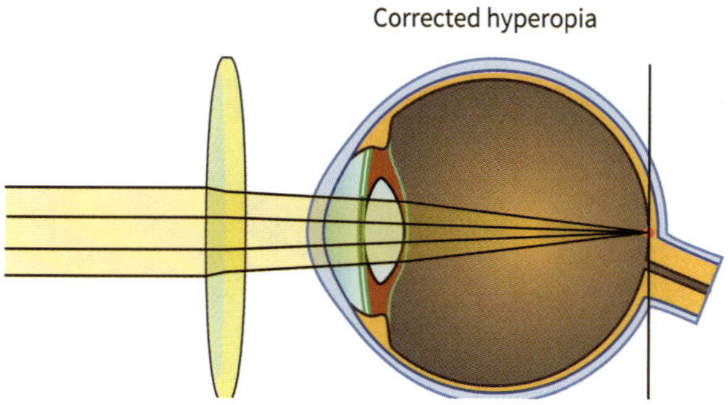

Figure 17.12 Hyperopia corrected with a concave lens

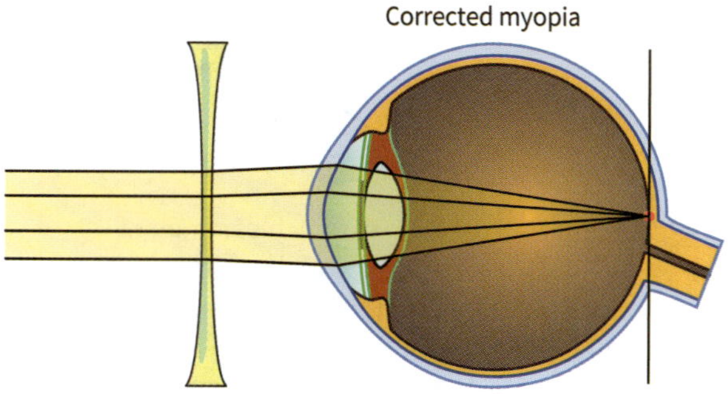

Figure 17.13 Myopia corrected with a convex lens

➔ Laser surgery is a form of vision correction that involves reshaping the cornea to correct refractive error. The technology uses cool temperature laser. LASIK surgery (laser in situ keratomileusis) is the most common form of laser surgery. It involves two steps as outlined below:
- A thin flap is created in the cornea to allow it to be reshaped; the laser then removes the exact amount of tissue needed to reshape the cornea.
- The corneal flap is then replaced.

➔ Cataract surgery involves the removal of the opaque lens and its replacemen with a clear artificial replacement lens called an intraocular lens. For more information see www.visioneyeinstitute.com.au.

➔ KEY QUESTION

8 Briefly outline how technology assists with the effects of three different visual disorders you have studied.

Answer ➲ p. 234

Technologies that assist with the effects of loss of kidney function

➔ Technologies that are used to assist with the effects of loss of kidney function include **dialysis** and kidney transplant.

Dialysis

➔ There are different types of dialysis as outlined below:
- **Haemodialysis** is done at least three times a week and lasts for four to five hours. It can also be done overnight. During haemodialysis needles are used to access the blood. One needle takes the blood out for treatment and the other needle returns the blood. The blood is taken through a special filter called a dialyser, which cleans it. It is then returned clean to the body. The extra water and waste products produced in dialysis travel from the blood through the dialyser and are disposed of. Haemodialysis can take place in the home or at a dialysis clinic.
- **Peritoneal dialysis** is done in the home. There are two types of peritoneal dialysis: the daytime bags, which are also called continuous **ambulatory** peritoneal dialysis, and automated peritoneal dialysis, which is done overnight. Both use a tube which is put into the abdomen. The tube carries special peritoneal dialysis fluid that is used to clean the blood. The fluid must be changed regularly. In between fluid changes the person can carry out normal activities. Most people choose the overnight machine and have the dialysis while they sleep. For more information see kidney.org.au/your-kidneys/support/kidney-disease/treatment/dialysis-453.

➔ Dialysis machines are monitored and controlled by computer technology.

dialysis: an artificial method of removing waste substances from the blood and regulating levels of circulating chemicals; these functions are usually performed by the kidneys
haemodialysis: *haemo* comes from the Greek word *haima* meaning 'blood'; therefore haemodialysis involves dialysis with direct access to the blood
peritoneal dialysis: peritoneum is the tissue that lines the inside of the abdomen; therefore peritoneal dialysis uses the lining of the abdomen to filter blood
ambulatory: walking around

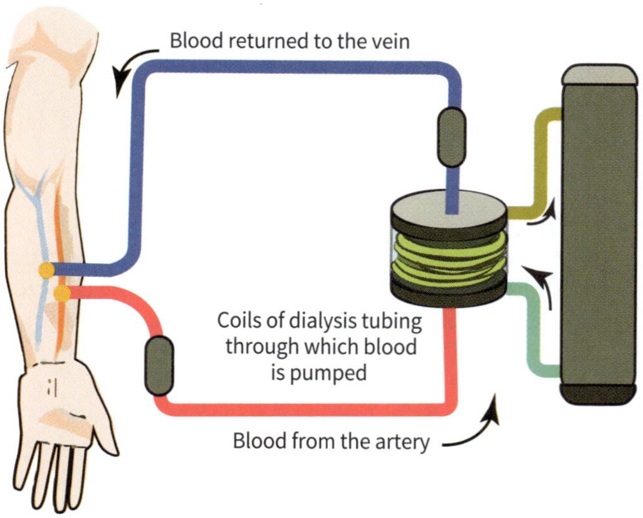

Figure 17.14 Simple representation of renal dialysis

- A kidney transplant involves surgery where a healthy kidney is taken from one person and surgically placed into someone with end-stage kidney disease. The kidney can come from a live or deceased donor.

KEY QUESTION

9 **Name a technology that is used to assist with the effects of loss of kidney function.**

Answer p. 234

3 Evaluating the effectiveness of technologies to manage and assist with the effects of disorders

» Students investigate the effectiveness of a technology that is used to manage and assist with the effects of a disorder.

- When you investigated the effectiveness of technologies you would have determined the value of the technology and considered factors such as the extent to which it manages and assists with the effects of the disorder, the cost and affordability of the technology, potential side effects, risks associated with intervention such as surgery and the longevity or life of the technology; for example, before it needs replacement.

The effectiveness of technologies used to manage and assist with hearing loss

- Hearing loss management not only involves getting the right technology for the type and extent of loss and for an individual's lifestyle, it also involves working with medical specialists to help each person's brain make sense of the new information being sent by the device.
- The advantage of hearing aids is that they are relatively cheap and easy to install. They are life-changing, particularly for children with hearing loss. Hearing aids provide the wearer with better access to sounds and an improved ability to understand speech. With advances in technologies hearing aids have become much smaller and digitised. Digital hearing aids can be programmed for specific listening situations and linked to other technologies.
- Some limitations of hearing aids are outlined in the points below:
  - They do not assist people who have damage to the inner ear or auditory nerve.
  - Because they must provide enough amplification without making sounds painfully loud some models do not suit some individuals and may need frequent adjustments for different intensities of sounds.
  - They usually cannot assist people with ear mould allergies, skin problems, narrow, collapsed or closed ear canals or malformed ears. The technology now used for these people is middle-ear implants but these rely on the person having a working cochlear and auditory nerve.
- Cochlear implants are life-changing for profoundly deaf people. The disadvantages are the initial and ongoing costs for readjustments, the need for an operation to install the implant, the need to wear an external sound processor, and risks associated with surgery such as postoperative side effects that can include droopy face and numbness to the tongue.

The effectiveness of technologies used to manage and assist with visual disorders

- Spectacles are affordable and easy to wear. Ready-made spectacles can now be purchased cheaply and can be effective for the wearer but do not correct for astigmatism. Prescription spectacles are made specifically for each individual's eye disorder and can correct for astigmatism. High-refraction plastic lenses are now used to reduce the weight of thick lenses. Sunglasses are designed to protect the eyes from harmful ultraviolet radiation.
- The cost of laser surgery could be prohibitive. The surgery has side effects that include dry eye and temporary visual disturbances. For LASIK surgery to be successful, patients need normally shaped corneas with adequate thickness. Even if the surgery is successful the vision disorder may not be completely corrected and the person may still need to wear spectacles. Patients with thin, irregular-shaped corneas are not suitable. Laser surgery is not recommended for people with autoimmune disorders or diabetes as both of these conditions may have damaged the cornea.
- Cataract surgery is an example of a technology that is highly effective. It is one of the most frequently performed surgeries in the world and takes less than 30 minutes. Once the opaque lens is replaced with an artificial lens it is not possible to develop another cataract.

The effectiveness of technologies used to manage and assist with loss of kidney function

- The process of dialysis does not match the complexity of the natural kidney and has limits as a long-term substitute. Dialysis must be performed for the rest of a person's life or until they receive a kidney transplant. If the transplant is unsuccessful dialysis can be restarted. Before the advent of dialysis end-stage kidney failure meant death.
- Dialysis treatments must be done frequently and treatments are time consuming. Dialysis treatments can be done in the home provided special plumbing is installed, the machines and all supplies are provided, and the patient or carers are properly trained.
- Continuous ambulatory peritoneal dialysis, in which the person wears a bag connected to a tube inserted in the abdomen, allows the patient to be mobile during treatment. The treatment must be done every day and the bag must be changed four times daily.
- Peritoneal dialysis requires a simple operation to insert a catheter tube into the abdomen. The most common complication of this type of dialysis is infection of the abdominal lining.

- Dialysis does not cure or treat the underlying cause of the kidney failure.
- Dialysis is one the most energy and resource consuming treatments in today's healthcare system. Simple measures including reuse of waste water, recycling of plastics and improving energy efficiency have been introduced across the world to minimise the environmental and economic costs of dialysis. The environmental impacts are outlined in the points below:
  - Haemodialysis consumes large volumes of water.
  - Dialysis produces a large volume of waste: a single haemodialysis session produces about 2.5 kg of solid clinical waste of which approximately 38% is plastic.
  - A great deal of the waste produced by dialysis services is incinerated because it is deemed to be 'infectious' waste. This not only costs a significant amount but leaves a heavy carbon footprint.
  - Dialysis requires a large amount of electricity.

→ KEY QUESTION

10 Evaluate the effectiveness of one technology you have studied.

Answer ⊃ p. 234

CHAPTER SYLLABUS CHECKLIST

Are you able to answer these questions from the syllabus for this chapter? Tick each question as you go through the checklist if you are able to answer it. If you cannot answer a question, turn to the relevant page in the study guide to find the answer. For NESA key word meanings, go to www.educationstandards.nsw.edu.au and search 'key words'.

	FOR A COMPLETE UNDERSTANDING OF THIS TOPIC:	PAGE NO.	✓
1	Can I explain a range of causes of disorders by investigating the structures and functions of the relevant organs; for example, hearing loss, visual disorders and loss of kidney function?	222–228	
2	Can I investigate technologies that are used to assist with the effects of hearing loss?	228–229	
3	Can I investigate technologies that are used to assist with the effects of visual disorders?	229–230	
4	Can I investigate technologies that are used to assist with the effects of loss of kidney function?	230–231	
5	Can I evaluate the effectiveness of a technology that is used to manage and assist with the effects of hearing loss?	231	
6	Can I evaluate the effectiveness of a technology that is used to manage and assist with the effects of visual disorders?	231	
7	Can I evaluate the effectiveness of a technology that is used to manage and assist with the effects of loss of kidney function?	231–232	

HSC EXAM-TYPE QUESTIONS

Objective-response questions (1 mark each)

1 **Why must nitrogenous wastes be removed from the body and what is the organ responsible for doing this?**
- A The molecules are too big to be reabsorbed and the kidney is the organ responsible.
- B The molecules are too big to be reabsorbed and the skin is the organ responsible.
- C Nitrogenous wastes are toxic and the kidney is the organ responsible.
- D Nitrogenous wastes are toxic and the liver is the organ responsible.

2 **What term is used to refer to a visual disorder caused by a refractive error where the distance between the lens and the retina is too great so the image will be focused in front of the retina?**
- A hyperopia or long-sightedness
- B myopia or short-sightedness
- C hyperopia or short-sightedness
- D myopia or long-sightedness

3 **Which of the below is an example of a technology used to assist people with the effects of a hearing disorder?**
- A cataract
- B cochlear implant
- C intraocular surgery
- D bypass surgery

4 **What technology does the simplified diagram below represent?**

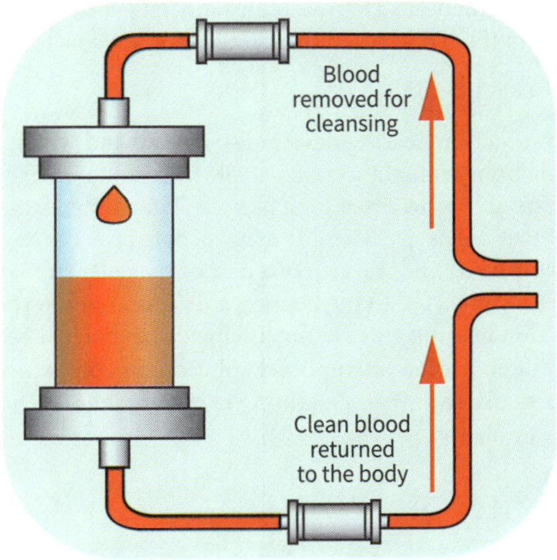

Figure 17.15 Simplified diagram

- A nephritis
- B peritoneal dialysis
- C transplant
- D haemodialysis

5 **Artificial bone is sometimes used in surgery to replace damaged stapes. What disorder would be assisted by this technology?**
- A managing hearing loss in the middle ear
- B managing hearing loss in the ear canal
- C managing a disorder due to nerve cell damage
- D managing bone conduction in the skull

Short-answer questions

6 **A telecoil is a coil of wire that is installed inside many hearing aids and cochlear implants to act as a miniature wireless receiver. It was originally designed to make sounds clearer to a listener over the telephone. Beginning with the receiver:**
- a **use a simple flow diagram to show how a hearing aid works.** (2 marks)
- b **use a simple flow diagram to show how a cochlear implant works.** (2 marks)

7 **Briefly outline two disorders of the kidneys that impair their function.** (4 marks)

8 **From your study of technologies used to assist people who experience disorders describe one technology that you have studied and list two advantages and two limitations.** (5 marks)

9 **Using the ear or the eye as an example, briefly outline how transmission of messages occur and name one disorder that affects the transfer of sound energy in the ear or light energy in the eye.** (6 marks)

10 **Compare and contrast the functions of a healthy kidney with the process of dialysis.** (5 marks)

ANSWERS

KEY QUESTIONS

Key questions ⊃ p. 223

1 Outer ear: sounds are funnelled through the tube-structured ear canal into the middle ear. Middle ear: the middle ear is separated from the outer ear by the structure called the eardrum. The eardrum vibrates in response to sound waves. The vibrations are amplified by tiny bones called the ear ossicles. Inner ear: amplified sound waves are picked up by the cochlea. Tiny hairlike structures or nerve receptor cells pick up the vibrations and pass the message (as an electrical impulse) to the cochlear nerves. Brain: the impulse is sent by cochlear nerves to the auditory nerves in the brain stem. The electrical impulses are interpreted by the brain and the sound is heard.

2

Table A17.1 An example of the cause and effect of hearing loss in the outer, middle and inner ear

Outer ear	Middle ear	Inner ear
impacted wax in the ear canal limits the conduction of sound waves the volume of the sound is reduced	damage to the ear ossicles (tiny bones) can be caused by serious infections or head injury the effect is reduction in the ability of the middle ear to transfer sounds to the inner ear	if the organ of Corti is damaged the nerve cells are damaged and sounds are not able to be sent efficiently to the brain

Key questions ⊃ p. 226

3 Refraction is the bending of light rays when they travel from one medium to another. In the human eye most refraction is carried out by the cornea and the lens. Ciliary muscles allow the lens to change shape. This changes the focal length of the lens and therefore the amount of refraction of the light. In this process light from either near or distant objects is focused and falls on the retina so that a clear image is formed.

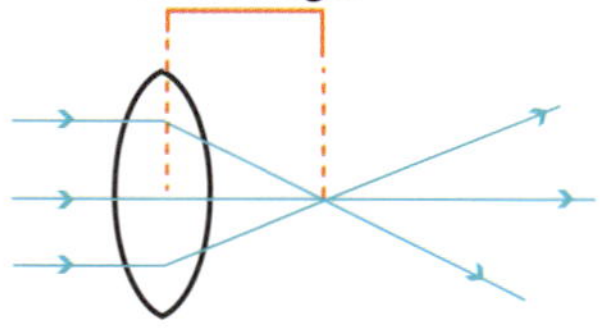

Figure A17.1 Diagram showing light being refracted through a lens and focused at a point

4 Table A17.2 How refraction and/or light transmission can be affected by visual disorders

The visual disorder	How refraction and/or light transmission can be affected by visual disorders
cataracts	the lens of the eye clouds over and becomes opaque, meaning that transmission of light is limited which will eventually cause vision loss and, potentially, blindness
hyperopia	refraction does not occur correctly: light is focused behind the retina instead of directly on it so the person has problems seeing near objects clearly
myopia	refraction does not occur correctly: light is focused in front of the retina instead of directly on it so the person has problems seeing distant objects

Key questions ⊃ p. 228

5 The kidneys filter and remove waste from the blood. The nephrons are the functional units inside the kidney. Each nephron filters a small amount of blood. The glomerulus in the nephron lets fluid and wastes pass through but prevents blood cells and large molecules such as proteins from passing through. The filtered fluid passes through tubules which reabsorb salts such as sodium back into the bloodstream and remove wastes from the blood.

6 Nephrosis is a collection of degenerative disorders that result from a problem with the functioning of the glomeruli. When the glomeruli are damaged proteins from the blood leak into the urine.

Key question ⊃ p. 229

7 Hearing aids amplify the sound and are worn externally. Cochlear implants do not amplify sounds: they replace the hearing function in the inner ear. Incoming signals are processed into electrical signals that go directly to the auditory nerves by bypassing the dead or damaged part of the inner ear.

The three main components of an implant are:
- an external sound processor and coil worn behind the ear
- the cochlear implant receiver-stimulator implanted just under the skin behind the ear
- a headset.

Key question ⊃ p. 230

8 Myopia is corrected with spectacles fitted with a concave lens. This diverges the light so that the image falls correctly on the retina. Hyperopia is corrected with spectacles fitted with a convex lens. This converges the light rays so that a clear image forms. To treat cataracts the opaque lens is replaced with an intraocular lens in a relatively simple operation.

Key question ⊃ p. 231

9 When a person's kidneys stop working they require either a kidney transplant or dialysis. Dialysis is an artificial process that removes waste from the blood when the kidneys are not able to do this.

Key question ⊃ p. 232

10 Bone conduction hearing devices use a headband which holds the vibrating bone conductor to the skull. This can make them uncomfortable to wear for long periods. Due to the discomfort and the fact that the device is visible many people prefer to have the bone conductor fitted into the arm of a pair of specially strengthened spectacles connected to the hearing aid's sound processor. Bone conduction hearing aids are not as efficient as normal hearing aids and are often used as a temporary solution for conductive hearing loss, particularly for people waiting to have a bone-anchored hearing implant.

HSC EXAM-TYPE QUESTIONS

Objective-response questions

1 **C**. The kidney is the organ responsible for removing toxic nitrogenous wastes. **A** and **B** are incorrect because nitrogenous wastes should not be reabsorbed into the blood. **D** is incorrect because the liver is not the organ responsible.

2 **B**. Myopia or short-sightedness is a disorder in which the image produced is in front of the retina. **A** is incorrect because hyperopia or long-sightedness is a disorder in which the image is produced (at an imaginary spot) behind the retina. **C** is incorrect because hyperopia is long- not short-sightedness. **D** is wrong because myopia is short-sightedness.

3 **B**. A cochlear implant is a technology used to assist people with a hearing disorder. **A** is incorrect because a cataract is a disorder of the eye, not a technology. **C** is wrong because intraocular surgery uses a technology to assist in a visual disorder. **D** is incorrect because bypass surgery is used for heart disorders.

4 **D**. The diagram is a simple representation of haemodialysis in which blood is taken from the body and put through a special filter called a dialyser, which cleans it and then returns the blood to the body. **A** is incorrect because nephritis is a disorder of the kidney and is not represented by the diagram. **B** is wrong because peritoneal dialysis does not involve taking blood directly but instead uses the lining of the abdomen to filter blood. **C** is incorrect because the diagram does not represent a transplant.

5 **A**. The stapes or stirrup is one of the tiny bones in the middle ear and if damaged causes hearing loss. **B** is wrong because stapes are not in the ear canal or outer ear. **C** is wrong because it would not help with nerve damage. **D** is wrong because the question does not refer to bones in the skull.

Short-answer questions

6 **EM** This question asks you to recall your understanding of how two technologies—hearing aids and cochlear implants—work and represent your understanding as flow diagrams. Make sure you begin each flow diagram with the receiver.

a Receiver picks up the sound waves → converts the sound to electrical signals → sends sounds to an amplifier to increase the signal → (✓) signal is sent to the speaker so that the amplified sound can travel through the ear canal (✓)

b Receiver picks up the sound waves → converts the sound into an electrical impulse → signal passes through the skin to the implant → (✓) signal is relayed to the electrode array that has been surgically inserted into the cochlea (✓)

7 **EM** This is a recall question. Make sure that you name two disorders and how they impair the function of the kidney.

Nephritis is inflammation of the kidney (✓) due to infections and toxins but most commonly caused by autoimmune disorders. Severe inflammation restricts blood supply to the kidneys and compromises the filtering function of the nephrons. (✓)

Nephrosis results from a problem with the functioning of the glomeruli. (✓) When the glomeruli are damaged proteins from the blood leak into the urine. As the kidney function fails extra salts and fluids build up in the body, causing swelling and high blood pressure. (✓)

8 **EM** This question relates to the part of your study where you evaluated the effectiveness of technologies used to manage and assist in the effects of disorders. Make sure you name the technology as well as what it does and provide two advantages and two limitations. This is a good example where dot points would help to organise your answer.

The technology is laser surgery to correct refractive errors in vision disorders. (✓)

Advantages: it has been proven to work in most cases and vision is corrected almost immediately (✓) with very little pain and recovery is quick. (✓)

Limitations: may not be suitable for people with a high degree of short-sightedness or with thin, irregular-shaped corneas. (✓) It is a technically complex surgery with risks of side effects that include dry eye and temporary visual disturbances. There is also the risk that even though people have paid for the surgery the problem may not be completely corrected and the person may still need to wear spectacles some of the time. (✓)

9 **EM** You are asked to choose either the ear or the eye to answer this question. Make sure you answer both parts of the question for the organ that you choose. The sample answer relates to the ears.

Sound waves travel through the ear by causing different parts of the ear to vibrate; (✓) that is, sound travels through the external ear and causes the eardrum to vibrate. These vibrations are transmitted by the ear ossicles to the inner ear. (✓) Vibrations are transmitted through the oval window to the spiral cochlea which contains fluid and nerve endings. Vibrations in the inner ear produce electrochemical signals. The auditory nerve transmits these neural impulses to the brain, where they are interpreted. (✓)

Conductive hearing loss (✓) is due to problems with the ear canal, eardrum, middle ear or ear ossicles, such as a blocked ear canal caused by excessive earwax that limits the sound from being conducted through the canal, decreasing the volume of the sound being transmitted. (✓) A perforated eardrum (✓) limits the vibrations of the eardrum's function in transmitting sounds from the outer ear to the middle ear. (✓)

10 **EM** This higher-order question asks you to compare and contrast. When you compare think about how the two are similar and, when you contrast, about how they are different.

In dialysis a selectively permeable membrane prevents large molecules such as proteins from leaving the blood. (✓) The glomerulus in the nephron of a healthy kidney also acts as a filter, preventing red blood cells and large molecules such as proteins from leaving the blood. (✓) In a healthy kidney substances needed by the body are reabsorbed back into the blood (✓) whereas in dialysis these substances are selectively added back into the blood. (✓) Dialysis is less efficient than normal healthy kidneys. (✓) Dialysis only involves passive transport whereas kidneys rely on active and passive reabsorption. (✓)

SAMPLE HSC EXAMINATION 1

Try to complete these papers as if they are the real thing. These are the instructions you need to follow in the HSC Exam:

General instructions

- Reading time: 5 minutes
- Working time: 3 hours
- Write using black pen.
- Draw diagrams using a pencil.
- For questions in Section II, show all relevant working in questions involving calculations.
- NESA approved calculators may be used.

Total marks: 100

Section I: 20 marks

Section II: 80 marks

- Attempt all questions.

Section I: 20 marks

Attempt Questions 1–20.
Allow about 35 minutes for this section.

1 **What is the ultimate source of natural genetic variation?**
A crossing over
B meiosis
C recombinant DNA
D mutation

2 **During the 1854 cholera epidemic in London an English doctor, John Snow, methodically investigated the rate of cholera deaths in two neighbourhoods. The first was a site where water was drawn from a pump downstream from an area Snow suspected was contaminated with cholera. The second was a site where a pump supplied water upstream and away from the contaminated site. Five hundred deaths occurred in the neighbourhood of the downstream pump in ten days. No cases of cholera occurred in the neighbourhood upstream of the pump. Once the pump handle at the downstream site was removed, the water supply from that pump stopped and the number of cholera cases went down. Snow proposed that cholera was not spread by miasma but by contaminated water.**

This case study is an example of:
A an early epidemiological study.
B an early study of non-infectious disease.
C a study of stream ecosystems.
D a study of microbiology.

3 **Which of the following processes are all involved in the continuity of life?**
A meiosis, DNA replication, binary fission, fertilisation
B homeostasis, osmoregulation, mitosis, implantation
C ovulation, budding, spore production, respiration
D excretion, circulation, hormone production, implantation

4 **An infectious disease can be thought of as:**
A an immune response.
B a non-communicable disease.
C a disruption to homeostatic mechanisms.
D a serious outbreak threat.

5 **DNA is found:**
A in the cell membrane.
B bound to proteins in chromosomes in the nucleus of eukaryotes.
C as unbound circular DNA in the cytosol of eukaryotes.
D in the mitochondria and chloroplasts of prokaryotic cells.

The following table provides WHO UNICEF data on the prevalence of a number of infectious diseases in the Southeast Asia region in the years 2012 to 2016.
The following graph shows polio cases worldwide in 1988 and 2014.

Analyse the data in the table and the graph to answer Questions 6 and 7.

Disease	2016	2015	2014	2013	2012
diptheria	4016	2504	7666	4080	3953
measles	27 530	48 888	42 899	24 564	31 582
mumps	31 739	42 937	38 327	36 352	47 086
pertussis	43 141	29 813	54 953	37 602	45 847
polio	0	2	0	0	0
rubella	10 361	6515	9690	10 434	6877
encephalitis (Japanese)	3500	3320	2831	1356	282

Source: www.who.int/immunization/monitoring_surveillance/data/gs_gloprofile.pdf

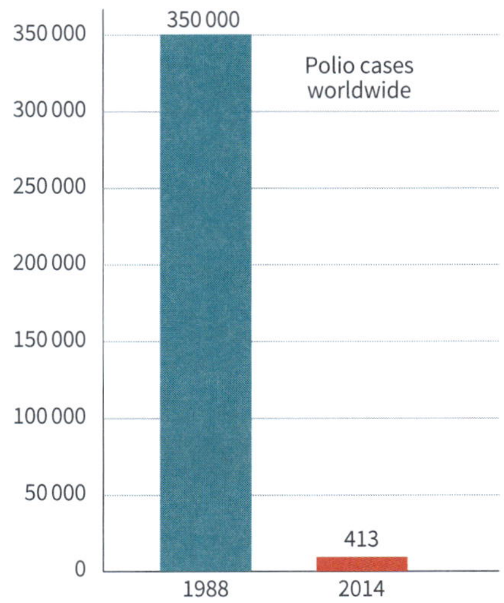

Source: https://www.unicef.org.au/blog/news-and-insights/april-2016/this-infographic-proves-how-effective-vaccines-are

6 **The prevalence of which disease shows a continually increasing trend from 2012 to 2016?**

A encephalitis **B** rubella
C measles **D** diptheria

7 **What is the most likely explanation for the prevalence of data for polio as compared to other diseases?**

A The disease was never prevalent in the Southeast Asia region.
B The disease was possibly prevalent until 2012 and then disappeared.
C The disease was prevalent but has been eliminated through monitoring control.
D The disease was prevalent but has been almost eliminated through immunisation.

8 **Which of the following is a modern technique used in biotechnology?**

A mutation breeding
B the use of yeast to make bread
C the use of plasmids as vectors
D the use of a biological control to reduce prickly pear

9 **What is the main cause of changes in allele frequency in the gene pool of a population?**

A natural selection **B** asexual reproduction
C gene expression **D** random segregation

10 **Which of the following is an example of a chemical change that occurs in the cells and tissues of animals in response to the presence of pathogens?**

A Pathogens cause an increase in the diameter of blood vessels.
B Pathogens cause an increase in the permeability of blood vessels.
C Pathogens cause tissues at the site of an infection to become hot.
D Pathogens cause the production of cytokines to set up the inflammatory response.

11 **Use the following table, which shows the inheritance of blood types in humans, to answer the question which follows.**

Phenotype	Genotype
O	OO
A	AA, AO
B	BB, BO
AB	AB

A father has an O blood type and the mother has an A blood type. What are the possible blood types of their offspring?

A O, A, B **B** A, B, AB
C A, O **D** AB, A

12 **The following graph shows percentage rates of overweight and obesity for Australian boys and girls measured in the 2011 to 2012 *Australian Health Survey*. Use the information to answer the question which follows.**

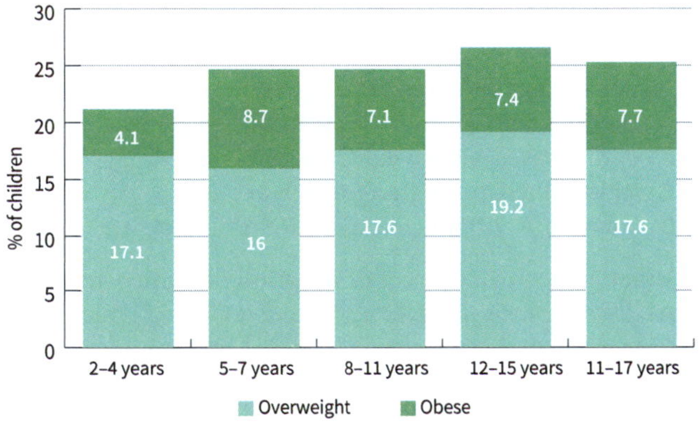

Source: *Australian Health Survey, 2011–12*

Which of the following statements provides a correct analysis of the data shown?

A About one in four Australian children were overweight or obese, with the highest percentage in the 16–17 years age group.
B About one in four Australian children were overweight or obese, with the highest percentage in the 12–15 years age group.
C Rates of overweight are the same for all age groups.
D Rates of obesity are highest for the 16–17 years age group.

13 **Which of the following is an example of a technology used to assist people with a visual disorder?**

A laser surgery **B** cochlear implant
C haemodialysis **D** nephrosis

14 **What is an advantage of external fertilisation over internal fertilisation?**

A increased chance of fertilisation
B better suited to terrestrial environments
C facilitates widespread dispersal of zygotes
D fewer gametes produced

15 **Which of the below is the correct sequence in polypeptide synthesis?**
- A transcription, tRNA moves from the nucleus to ribosome in cytoplasm, translation
- B transcription, mRNA moves from nucleus to ribosome in cytoplasm, translation
- C translation, mRNA moves from the nucleus to ribosome in cytoplasm, transcription
- D transcription, mRNA moves from ribosome in cytoplasm to the nucleus, translation

16 **The 'Slip, Slop, Slap' campaign started in the 1980s in Australia with the aim of preventing skin cancer. Since then the incidence of the two most common forms of skin cancer has declined. This provides an example of:**
- A an increased understanding of the genetics of skin cancer.
- B the effectiveness of an educational campaign for a genetic disease.
- C the effectiveness of an educational campaign for a disease caused by environmental exposure.
- D the effectiveness of targeting specific populations.

17 **Which of the following are all mutagens?**
- A gamma rays, oncoviruses, alkylating agents
- B radio waves, visible light, microwaves
- C chlorophyll, progesterone, polypeptides
- D vitamin A, ribonucleic acid, DNA polymerase

18 **What is the most correct description of homeostasis?**
- A Homeostasis involves a stimulus model in which change in the external or internal environment is detected and appropriate responses occur through neural pathways.
- B Homeostasis involves a stimulus-response model in which change in the external or internal environment is detected and appropriate responses occur through hormones.
- C Homeostasis involves a stimulus-response model in which change in the external or internal environment is detected and appropriate responses occur via negative feedback.
- D Homeostasis involves a nerve pathway model in which change in the external or internal environment is detected and appropriate responses occur via negative feedback.

19 **The image below represents which type of pathogen?**

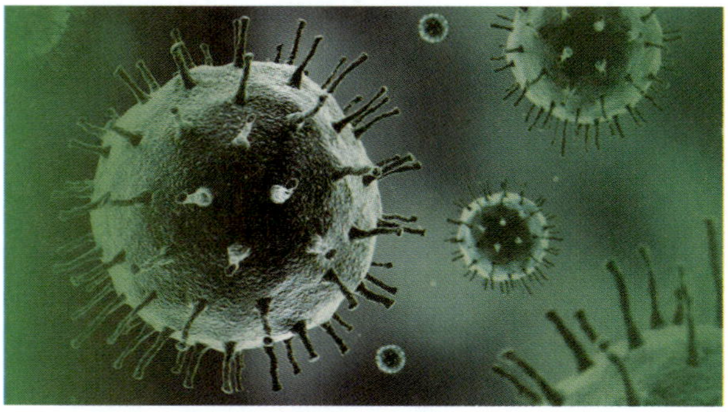

- A bacteria
- B prion
- C fungi
- D virus

20 **Which of the following is a use of biotechnology in agriculture that could increase biodiversity?**
- A genetic drift
- B gene flow
- C gene cloning
- D genetic erosion

Section II: 80 marks

Attempt questions 21–32.
Allow about 2 hours and 25 minutes for this section.
Instructions:
- In the HSC Exam you will answer the questions in the spaces provided. These spaces provide guidance for the expected length of response.
- Show all relevant working in questions involving calculations.

21 **Briefly outline how population genetics has helped in understanding human evolution.** (3 marks)

22 **Construct a simple model to summarise plant and animal responses to infection.** (4 marks)

23 **Different pathogens have different modes of transmission. For four different types of pathogens that you classified in your study provide an example of the pathogen and the infectious disease that it causes and list each pathogen's mode of transmission. Construct a table to organise the answer.** (4 marks)

24 **Use the graph and your own analysis to explain the role of hormones in the progression of pregnancy and birth in mammals.** (8 marks)

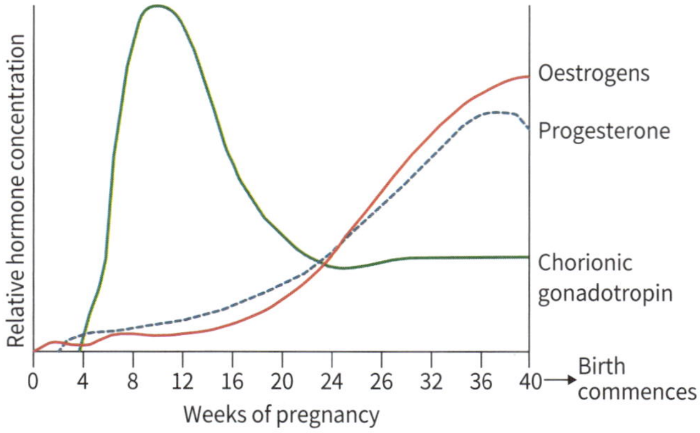

Source: By OpenStax College [CC BY 3.0 (http://creativecommons.org/licenses/by/3.0)], via Wikimedia Commons

25 Construct a diagram and explain how the structure of DNA supports the process of DNA replication yet at the same time has allowed for point mutations to occur. (8 marks)

26 During your study you were asked to investigate the structure and function of proteins in living organisms. One approach to this task is to analyse diseases caused by faulty proteins. The following table provides some data about diseases associated with faulty proteins. Study the table and pedigree diagram below and then answer the questions which follow.

Symptoms	Cause/Defective protein
Cystic fibrosis	
fluids such as sweat, digestive juices and mucus are very thick often results in lung infections, digestive problems and a significantly reduced life expectancy	the protein cystic fibrosis transmembrane conductance regulator (CFTR) is defective, most commonly as a result of a loss of the amino acid phenylalanine at the 508th position on the protein the disease is inherited as autosomal recessive
Motor neurone disease	
muscle weakness, initially in hands or feet, followed by loss of speech, difficulty swallowing, muscle twitching, cramps, fatigue, weight loss, eventual paralysis and death about 10% of sufferers have inherited the condition but most cases arise without a known cause	nerves activating muscles degenerate Dr Justin Yerbury (a researcher and sufferer of the disease) believes that failure of proteins to be recycled within neurones results in protein accumulation to toxic levels; these proteins spread to adjacent neurones, progressing the disease

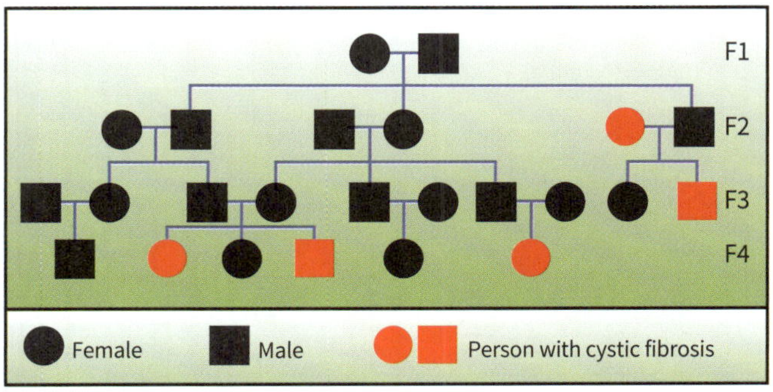

a Use the information from the table and your own research to explain the relationship between the structure and function of proteins. (3 marks)

b Describe the genetic change that would result in the inheritance of cystic fibrosis if neither parent in the first generation carried the allele for cystic fibrosis. (2 marks)

c Use the pedigree diagram to construct a Punnett square diagram to show the gametes and offspring of the F3 generation cross between the two cousins that produced three F4 offspring. (4 marks)

27 The graph below provides information about the homeostasis of blood glucose levels after a high carbohydrate meal. The black line is glucose, the blue line is insulin and the red line is glucagon. Study the graph and then answer the question that follows.

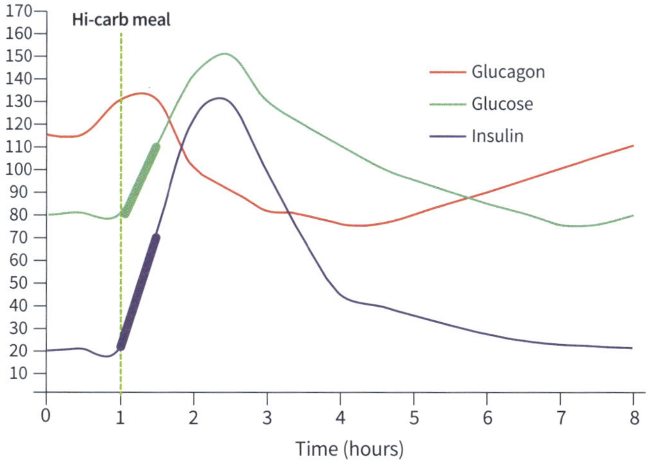

Interpret the graph to explain what happens to glucose levels, insulin levels and glucagon levels after the high carbohydrate meal. (6 marks)

28 UNICEF estimates that about 3% of all babies in the world are born with genetic disorders.

a What is a genetic disorder? Provide an example of the cause and effects of a genetic disorder that you have studied. (3 marks)

b Provide one example of how genetic engineering could prevent genetic diseases and briefly outline an ethical concern associated with this genetic technique. (3 marks)

29 The following points provide information about some procedures that can be employed to prevent the spread of the infectious disease malaria. Malaria is widespread in the tropics and half of the world's population is potentially exposed to the disease. In 2015 there were approximately 212 million malaria cases and an estimated 429 000 malaria deaths.

- Pesticides are used in swampy areas to reduce the breeding of mosquitoes.
- Surface minnows released in waterways eat the mosquito larvae and reduce the mosquito population.
- People entering a malaria-infected area can take a course of chloroquine that interferes with the development of the parasite. A virulent form of the malaria protozoan is resistant to chloroquine.
- Education programs encourage the use of insect repellents, protective clothing and mosquito nets over bedding.
- A vaccine called PlasProtecT consists of inactivated human malaria parasites which have been prevented from growing and causing a malaria infection. The vaccine is currently in early human trials.

- Artemisinin is a drug that has been used in Chinese medicine for 1500 years and is extracted from a wormwood shrub called *Artemisia annua*. It is grown by 100 000 small farmers in Asia and Africa and also extracted from wild plants in China. In combination with other drugs it is effective against multidrug resistant strains of the malaria protozoan.
- A strain of yeast has been genetically modified to produce a precursor to artemisinin in order to ensure reliability and reduce the price of the drug. A 50 000-litre chemical reactor could provide sufficient quantities of the drug for the world though it remains more expensive than the naturally grown and extracted artemisinin.
- CRISPR-Cas9 gene editing and gene drive technology could be combined to produce mosquitoes unable to carry the protozoan that would, when released, out-compete and eliminate the wild-type mosquito within a few generations.

a Use the information above to explain how social, economic and cultural contexts may influence biotechnology. (6 marks)

b Construct a table to evaluate each procedure described. (8 marks

30 The following predictions are about future developments in biotechnology. Read these and then answer the question that follows.

Personalised medicine becomes normal. Doctors use understanding of a person's genetic profile to predict which medicine will work best for them.

Gene therapy will be used to correct the impact of defective genes, reducing the problems of inherited disease.

Compare these statements with your own research into the future directions of biotechnology and the potential benefits for society of research using gene technologies. (6 marks)

31 The diagram below comes from the Australian Academy of Science website and shows a model of how the human immune system responds to exposure to a pathogen. Interpret the model to answer the questions that follow.

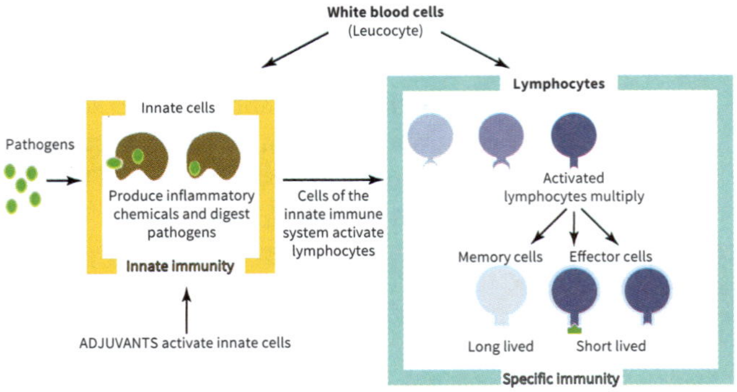

Source: www.science.org.au/learning/general-audience/science-booklets/science-immunisation/1-what-immunisation

a In the diagram what does the image  represent? (2 marks)

b Briefly explain the difference between innate immunity and specific immunity. (2 marks)

c Name two types of lymphocytes and their function. (2 marks)

d Explain the part of the diagram that states: ADJUVANTS activate innate cells. (2 marks)

32 In your course work you investigated practical applications of STEM (science, technology, engineering and maths). Describe one practical application of STEM that is used in the treatment of a named non-infectious disease or disorder and briefly outline some advances in the technology that have improved the effectiveness of the treatment. (4 marks)

SAMPLE HSC EXAMINATION 2

Try to complete these papers as if they are the real thing. These are the instructions you need to follow in the HSC Exam:

General instructions

- Reading time: 5 minutes
- Working time: 3 hours
- Write using black pen.
- Draw diagrams using a pencil.
- For questions in Section II, show all relevant working in questions involving calculations.
- NESA approved calculators may be used.

Total marks: 100

Section I: 20 marks

Section II: 80 marks

- Attempt all questions.

Section I: 20 marks

Attempt Questions 1–20.
Allow about 35 minutes for this section.

1 **What is the term used when pathogens spread directly from host to host?**

A vehicle transmission
B vector-borne transmission
C direct transmission
D fomite transmission

2 **Which of the below is an example of the application of scientific knowledge to assist the reproduction of plants?**

A cloning by embryo splitting
B sex reversal in sequential hermaphrodites
C propagation by tissue culture
D artificial insemination

3 **Which of the following chemical changes occurs in the cells and tissues of plants in response to the presence of pathogens?**

A reinforcement of the cell wall
B production of resistance peptides
C cytoplasmic streaming
D stomatal closure

4 **What is a key factor in the difference between binary fission in bacteria with that in protists?**

A Bacteria are unicellular and protists are multicellular.
B Bacteria reproduce sexually whereas protists reproduce asexually.
C Protists are much smaller than bacteria.
D Bacteria are prokaryotic and protists are eukaryotic.

5 **Which of the following is an adaptation of a pathogen that facilitates its entry into a host?**

A The outer surface of many microorganisms is composed of molecules that do not occur in their host.
B A pathogen causes an inflammation response.
C Some bacteria coat themselves with a thick polysaccharide capsule that is not recognised by phagocytes.
D Some macrophages have granules that contain anti-microbial enzymes.

6 **What is the effect of genetic drift on the gene pool of a population?**

A Alleles that support adaptation to the environment become more frequent.
B Alleles that cause disease are likely to be eliminated.
C Allele frequency changes because of random events.
D Allele frequency increases because of immigrants.

7 **Binge drinking among teenagers and young adults is a community-wide problem. Binge drinking can increase the risk of chronic diseases later in life. You are asked to develop strategies for the prevention of binge drinking. What strategy could be most effective?**

A raising funds to help fight diseases related to alcohol use at a young age
B education about supporting families with chronic diseases related to alcohol use at a young age
C a social marketing education campaign primarily targeting parents of teenagers
D a social marketing education campaign primarily targeting teenagers and young adults

8 **What is a likely potential benefit for society of research into genetic technologies?**

A increased life expectancy and better health
B reduced use of animals in biological testing
C reduced numbers of deleterious mutations occurring
D an expansion in worldwide biodiversity

9 **Which of the following is an example of a chromosomal mutation?**

- **A** frameshift mutation
- **B** aneuploidy
- **C** nucleotide substitution
- **D** nonsense mutation

The figure below represents the formation of insulin, which controls blood glucose levels. This figure may assist with your answers to Questions 10, 11 and 12.

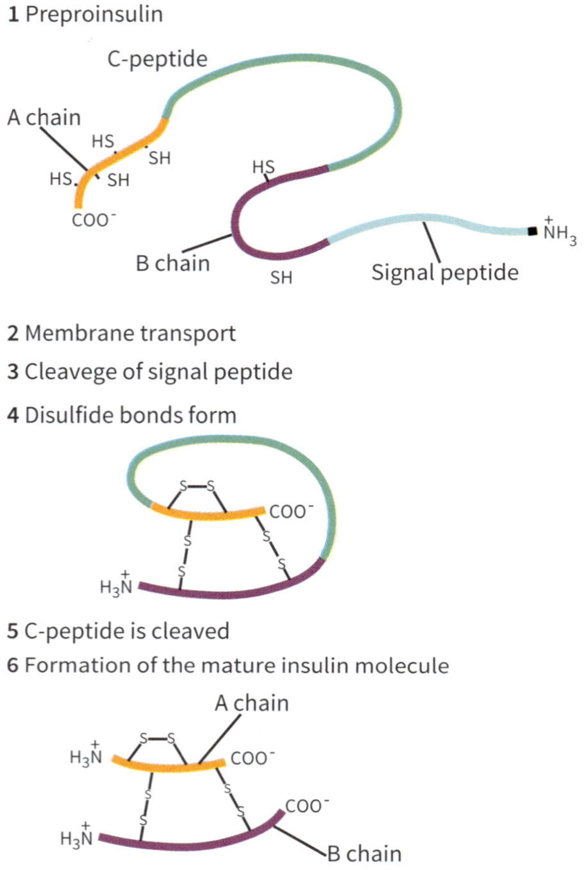

10 **What is the process that immediately precedes Step 1 in the figure above?**

- **A** polypeptide synthesis
- **B** DNA replication
- **C** crossing over
- **D** chromosome translocation

11 **Some people describe the hormone insulin as a 'key' that unlocks the cell to let glucose enter through the cell membrane. What is the best conclusion about the role of insulin?**

- **A** Insulin is an enzyme that catalyses chemical reactions.
- **B** Insulin is a protein whose structure is important for its function.
- **C** Insulin is part of the cell membrane that attracts glucose.
- **D** Insulin reacts with glucose to release cellular energy.

12 **What is the name of a non-infectious disease in which the body becomes resistant to the normal effects of insulin or gradually loses the ability to produce enough insulin?**

- **A** gestational diabetes
- **B** type 2 diabetes
- **C** chronic disease
- **D** pre-diabetes

13 **One way to assess the effectiveness of cloning is to:**

- **A** count the number of offspring.
- **B** screen for antibiotic resistance.
- **C** compare the number of attempts made with the number of genetically identical products.
- **D** determine the number of endangered species that have been saved.

14 **What are the processes that cause variation during cell replication?**

- **A** DNA replication and mutation
- **B** crossing over and random segregation
- **C** cytokinesis and chromatid replication
- **D** oncoviruses and electromagnetic radiation

15 **The following graph provides epidemiological data of dementia in the Australian population. Dementia describes a syndrome characterised by the gradual impairment of brain function.**

What is the graph an example of?

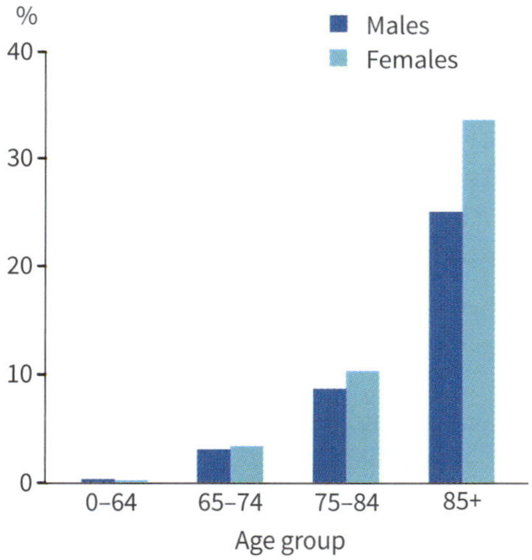

Source: Australian Institute of Health and Welfare 2016

- **A** prevalence data
- **B** mortality data
- **C** incidence data
- **D** population aging data

16 What do the terms *antigen* and *antibody* mean?

A Antigen refers to the invading microbe and the antibody is the pathogen.

B Antigen refers to the invading pathogen and the antibody is the protein that neutralises the pathogen.

C Antigen is the white blood cells that make antibodies that neutralise the pathogen.

D Antigens are toxins and antibodies help the toxins to work.

17 What does the figure below represent?

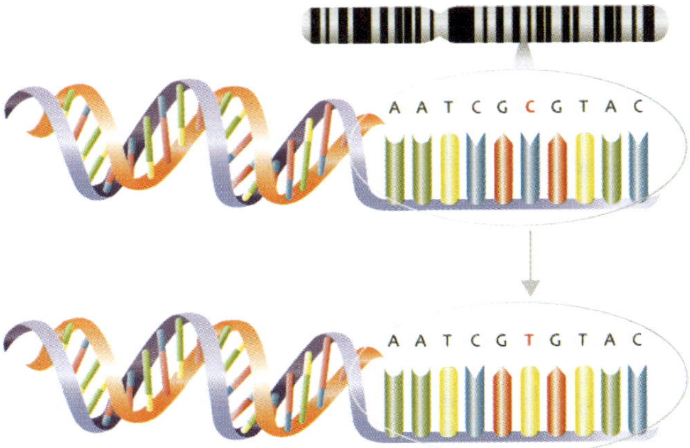

Source: image by NHS National Genetics and Genomics Education Centre (Flickr) [CC BY 2.0 (http://creativecommons.org/licenses/by/2.0)], via Wikimedia Commons

A chromosomal mutation and a translocation

B nucleotide base deletion and a nonsense mutation

C nucleotide base insertion and a frameshift mutation

D nucleotide base substitution and a single nucleotide polymorphism

18 Which process is essential for the continuity of a species?

A cell replication

B fertilisation

C implantation

D binary fission

19 In your study you investigated the economic costs and costs to productivity of diseases in populations. The following data shows Australian expenditure on patients admitted to hospital in 2004–05, 2008–09 and 2012–13. Study the data and answer the question which follows.

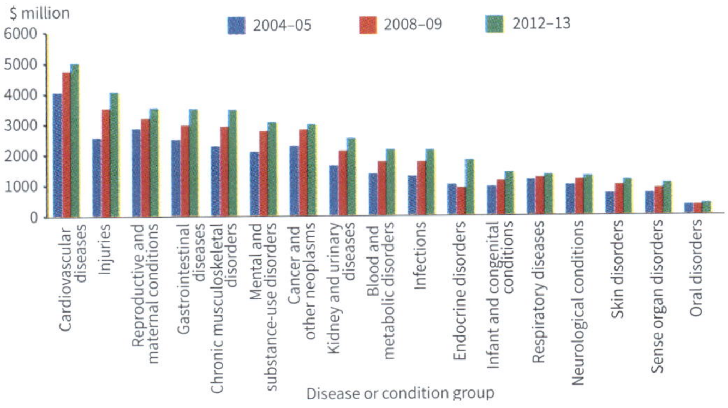

Source: AIHW disease expenditure database

What disease or condition group was responsible for a total expenditure of around $13.8 million in the three time periods?

A cancer

B cardiovascular disease

C muscular skeletal disorders

D respiratory disease

20 Which of the following diseases could be most easily prevented and kept out of Australia by quarantine?

A sickle-cell anaemia

B smallpox

C lung cancer

D foot-and-mouth disease

Section II: 80 marks

Attempt questions 21–34.

Allow about 2 hours and 25 minutes for this section.

Instructions:

- In the HSC Exam you will answer the questions in the spaces provided. These spaces provide guidance for the expected length of response.
- Show all relevant working in questions involving calculations.

21 Oils stored in the oil glands of leaves, such as eucalyptus oil, are examples of a preformed chemical defence of eucalyptus plants. Eucalyptus oil is also a traditional bush medicine used by Aboriginal people and is now used in a wide range of cleaning products and pharmaceuticals such as throat lozenges.

a Explain the relationship between eucalyptus oil as a preformed defence in eucalypts and its use by Aboriginal people as a bush medicine. (2 marks)

b Briefly outline the contemporary application of Aboriginal protocols in the development of medicines in Australia. (2 marks)

22 Analyse the features of fertilisation and implantation in terms of how they contribute to continuity of species. (4 marks)

23 In 2008 about two million new worldwide cases of cancer were attributed to infectious agents such as certain viruses, bacteria and parasites. This equates to approximately 16% of all cancers worldwide. The proportion is higher in developing countries, where 23% of all cancers were attributed to infections, and lower in developed countries, where 7.4% of cancers were attributed to infectious spread. Study the table on the following page which details the number of new cancer types in 2008 that were attributable to infectious agents and then answer the questions that follow.

Cancer type	Number of new cancer cases	Number of new cancer cases attributed to infection in less developed regions	Number attributed to infection in more developed regions
stomach	870 000	470 000	180 000
liver	750 000	510 000	69 000
cervix/uterus	530 000	450 000	77 000
vagina	27 000	4100	7500

a What term is used to refer to numbers of new cases of a disease? (1 mark)

b Suggest why the proportion of new cases was higher in less developed regions. Justify your answer using information from the data. (2 marks)

c From your understanding of infectious and non-infectious diseases describe two strategies used to prevent cancers attributable to infection. (2 marks)

24 Study the figure below and answers the questions which follow.

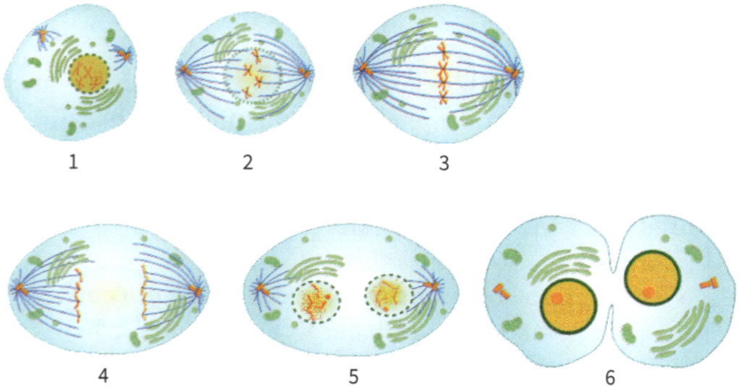

Source: images by LadyofHats (own work, public domain), via Wikimedia Commons

a Name the type of cell replication occurring in the figure above and describe a key process that you modelled in your studies for each of the numbered steps. (3 marks)

b Name the other main type of cell division that you modelled and contrast the above processes and products with its processes and products. (2 marks)

25 Maintenance of homeostasis requires more than a simple reflex. The model below shows the more complex feedback mechanism required to maintain homeostasis.

stimulus → receptor → afferent pathway → control centre → efferent pathway → effector → response → negative feedback

a Use the model to describe the negative feedback loop that maintains homeostasis for either temperature or glucose. (4 marks)

b Computer models of human thermoregulation responses have been developed for use in the design of clothing such as military suits. Simulating and modelling the thermoregulatory system also enables scientists to study the effects of extreme environments on the human body. From your understanding of trends and patterns in structural and physiological adaptations in endotherms that assist in thermoregulation describe some adaptations that would be considered in the design, simulations and modelling? (4 marks)

26 Construct diagrams to compare the structure and role of DNA with protein structure and function. (8 marks)

27 The World Health Organization has prepared a list of diseases that pose a serious threat of a global outbreak. The list includes the following diseases: Ebola (a fruit bat virus), Zika (a monkey virus), SARS (a bat virus), Lassa fever (carried by rats) and Nipah (a bat virus).

a What is the term for an outbreak of disease? (1 mark)

b Briefly explain how these diseases have become a global threat. (2 marks)

c Describe the key factors in the transmission of a disease during an outbreak. (2 marks)

d Outline three current strategies that are used to predict and control outbreaks of disease. (3 marks)

28 The figure below demonstrates how gene drives (Cas9) and any linked alleles (Cargo) can 'take over' genetic control by cutting and repairing to insert the linked allele. Blue represents the mosquito with the wild allele and red represents the mosquito with the altered allele; that is, the gene drive and altered allele in the right-hand diagram. Study the figure and then answer the question that follows.

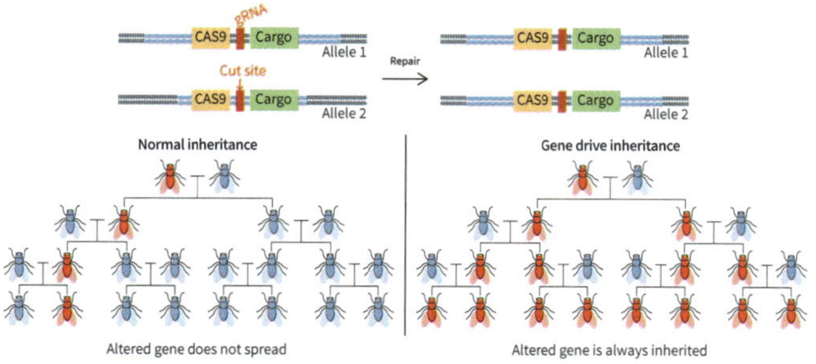

Source: image by Mariuswalter (own work), [CC BY-SA 4.0 (https://creativecommons.org/licenses/by-sa/4.0)], via Wikimedia Commons

Compare the mechanism of gene drives with the normal inheritance pattern demonstrated in the figure. (6 marks)

29 In your study you conducted a practical investigation relating to microbial testing of water or food samples. Below is a summary of one technique used to investigate microbes in food samples. Read the procedure and then answer parts a and b.

Spread plate procedure
In the spread plate procedure sterile, melted nutrient agar is poured into a number of Petri dishes and allowed to cool. Small amounts of different food samples are placed on the nutrient agar. The lid is secured onto each Petri dish and the dishes incubated at a certain temperature for a specific period of time. Microbes may appear on the surface of the agar. Microbiologists can pick off samples of the microbes with a sterile inoculating needle or loop and use these samples to inoculate fresh nutrient agar plates.

a Outline experimental procedures that you would use to ensure that the investigation is valid. (2 marks)

b Outline the safety precautions that you would follow when carrying out this procedure. (2 marks)

30 Compare the processes, technologies and outcomes of DNA sequencing with DNA profiling. (6 marks)

31 A researcher is trying to develop new varieties of flowers by mutation breeding. Evaluate an example of each of the three different types of mutagens that might be considered to promote the mutant varieties. (6 marks)

32 Supporters of the use of growth hormones in the livestock industry list the benefits of increased productivity, reduced production costs and improved food affordability. Recombinant bovine growth hormone (rBGH) is a synthetically produced hormone that has been shown to increase milk yield between 10–15%. Australia, New Zealand and the European Union have banned the use of rBGH based on evidence that it increases the risk of health issues in cows and because of concerns regarding milk contamination. Countries such as the United States allow it.

Use this information and your own investigations to answer the questions below.

a Outline the impact of social and economic contexts on biotechnology. (3 marks)

b Discuss the ethical uses of biotechnology. (2 marks)

33 In the early 1900s in Colorado a dentist named Frederick McKay started to investigate brown stains that were prolific on the teeth of locals. His investigation revealed that the individuals affected were resistant to tooth decay. He eventually traced the cause to high levels of fluoride that occurred in the water supply. Thirty years later the first pilot study into fluoridation was conducted in Grand Rapids. Fluoridation involves adding fluoride to drinking water. Children born after the fluoridation commenced showed a 60% decrease in the number of dental cavities and an improvement in oral health.

In your studies you analysed the role of epidemiology in understanding the incidence and prevalence of nutritional diseases and those caused by environmental exposure. Compare the results of one of your analyses with the case study described above. (5 marks)

34 Evaluate the effectiveness of a current technology that is used to assist with the effects of one of the following disorders: hearing loss, visual disorders or loss of kidney function. Describe future directions for further research and technological development in relation to the chosen disorder. (6 marks)

ANSWERS

SAMPLE HSC EXAMINATION 1

Section I

1 **D**. Mutation results in the production of new alleles (the source of natural genetic variation). **A** and **B** are wrong because crossing over and meiosis produce new combinations of alleles but not new alleles. **C** is incorrect because recombinant DNA technology results from human intervention.

2 **A**. Snow's methods of gathering data, looking at relationships and hypothesising about causes of disease are part of epidemiological studies. **B** is wrong because cholera is an infectious disease. **C** is incorrect because it is not an ecosystem study. **D** is wrong because historically the definite links between microbes and bacterial infection had not been made in 1854.

3 **A**. All the processes in this option are involved in either sexual or asexual reproduction. **B** is wrong because homeostasis and osmoregulation are involved in metabolism, not reproduction. **C** is incorrect as respiration is essential for the release of energy in cells, not reproduction specifically. **D** is wrong because excretion and circulation are not directly involved in reproduction.

4 **C**. Infectious diseases cause a disruption to the body's homeostasis. **A** is wrong because the body responds to infectious disease through an immune response. **B** is wrong because non-communicable diseases are non-infectious diseases. **D** is incorrect because not all infectious diseases lead to serious outbreaks.

5 **B**. This option describes how DNA occurs in chromosomes. **A** is wrong because the cell membrane is a phospholipid bilayer embedded with proteins and cholesterol. **C** is incorrect because DNA is unbound in circular plasmids in prokaryotic cells, not eukaryotic cells. **D** is wrong because DNA occurs in mitochondria and chloroplasts in eukaryotic cells, not prokaryotic cells.

6 **A**. Encephalitis increases continually from 2012 to 2016. **B**, **C** and **D** are incorrect because the data shows that that the pattern of prevalence of these diseases fluctuates between 2012 and 2016.

7 **D**. Polio has been almost eliminated through global immunisation programs as shown on the graph. **B** is not as accurate an answer as **D** (refer to the graph). The disease was prevalent in Southeast Asia so **A** is wrong. **C** is incorrect because monitoring is not going to almost eliminate diseases such as polio.

8 **C**. Modern biotechnology can involve the use of bacterial enzymes, plasmids as vectors and techniques including gel electrophoresis, bacterial transformations and the Polymerase Chain Reaction (PCR). **A**, **B** and **D** are wrong because these current techniques and applications of biotechnology have been used for either decades or centuries.

9 **A**. When selection pressures in the environment confer a selective advantage on a specific phenotype to enhance its survival and reproduction the frequency of an allele in a gene pool can change. This has been occurring throughout the course of evolution. **B** is wrong because asexual reproduction produces offspring genetically identical to parents. **C** is wrong because gene expression involves processes that convert the genotype into a phenotype of an organism and does not change alleles. **D** is incorrect because random segregation of homologous chromosomes does not confer advantage or disadvantage on any particular alleles or chromosomes.

10 **D**. Production of cytokines is a chemical change. The inflammatory response occurs quickly in animals in response to the presence of pathogens and produces both physical and chemical changes in the cells and tissues. **A**, **B** and **C** are examples of physical changes that occur in the cells and tissues of animals in response to the presence of pathogens.

11 **C**. Because O is recessive the father must have genotype OO. The mother could be AO or AA. The possible offspring would be OO or OA. **A**, **B** and **D** are wrong because neither parent carries the B allele so B or AB blood types are not possible.

12 **B**. The percentage for all age groups is about 25% (around one in four). The highest percentage is in the 12–15 years age group. **A** is partly correct but the highest percentage group is wrong. **C** is wrong because the rates are not exactly the same for all age groups. **D** is incorrect because the highest percentage is not in the 16–17 years age group.

13 **A**. Laser surgery is one type of technology used to correct visual disorders. **B** is wrong because cochlear implants are a technology used to assist with the effects of hearing loss. **C** is incorrect because haemodialysis is a technology used to assist with loss of kidney function. **D** is wrong because nephrosis is a disorder of the kidneys.

14 **C**. This option is correct as external fertilisation usually occurs in water and water currents can disperse gametes and zygotes. **A** is incorrect because there is greater loss of gametes so less chance of fertilisation. **B** is wrong because it is better suited to aquatic environments. **D** is wrong as generally far more gametes are produced and released.

15 **B**. This option correctly identifies that transcription occurs before translation and that transcription results in mRNA that travels from the nucleus to the ribosome. **A** is wrong because transcription does not produce tRNA. **C** is wrong because transcription occurs first. **D** is incorrect because mRNA moves *from* the nucleus not *to* the nucleus.

16 **C**. 'Slip, Slop, Slap' is an educational campaign that has been effective in reducing environmental exposure to the sun and therefore preventing skin cancer. **A** is wrong because, even though increased understanding of genetics is important, 'Slip, Slop, Slap' does not provide an example of this. **B** is incorrect because the campaign is aimed at preventing a disease caused by environmental exposure. **D** is wrong because the campaign targeted the entire Australian population.

17 **A**. This option correctly identifies physical, natural and chemical mutagens. **B** is wrong as it lists forms of electromagnetic radiation that are not mutagens. **C** and **D** are incorrect because they list biochemicals that are not mutagens.

18 **C**. Homeostasis involves a stimulus-response model in which change in the external or internal environment is detected and appropriate responses occur via negative feedback. **A** is wrong because homeostasis is stimulus and response, and involves more than neural pathways. **B** is wrong because homeostasis involves more than hormones. **D** is wrong because homeostasis involves more than a nerve pathway model.

19 **D**. The diagram is of a virus so options **A**, **B** and **C** are incorrect.

20 **B**. Gene flow is the increase in biodiversity of a population because of immigrants. This is not necessarily a good thing if it is, for example, a result of invasion of a natural ecosystem by a crop. **A** is incorrect because genetic drift relates to changes in allele frequency in a population due to random events. **C** is wrong because gene cloning creates copies of genes that are identical. **D** is wrong because genetic erosion is the decline of genetic diversity.

Section II

21 **EM** The question asks for a brief outline so you could consider using dot points to link population genetics to a better understanding of human evolution.

Population genetics investigates factors that influence the change of frequency of alleles over time and across locations and populations. Factors in human evolution include mutation, natural selection, genetic drift, gene flow and, particularly, migration. (✓)

Studies of DNA of ancestral groups such as Denisovans and Neanderthals have often relied on analysis and comparison of mitochondrial DNA which tends to be relatively well preserved and can be used as a biological clock because it is inherited through the maternal line. Fossils and archaeological remains in the Hohlenstein-Stadel cave in Germany could not be carbon dated but mitochondrial DNA put the fossilised bone to be about 124 000 years old. (✓)

Sudden reductions in population due to catastrophes such as the Toba super-volcano in Indonesia 70 000 years ago are believed to have reduced human diversity. (✓)

22 **EM** To answer this question you need to know that all plants and animals have an innate immune response to infections and vertebrates also have an adaptive immune response. Your model could be a flow diagram. Marks are given for the construction of the model (✓) (✓) and for correct information comparing plant and animal immune responses. (✓) (✓) A sample model is given below.

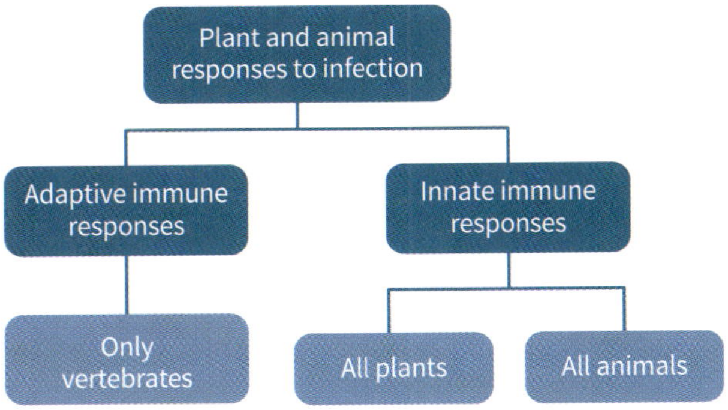

23 **EM** This is a simple recall question that asks you to organise your responses in a table. One mark (✓) is given for each correct answer for the four different types of pathogens. Note that this is a sample answer and you may have chosen other pathogens such as prions or fungi. Make sure you answer all parts of the question.

Type of pathogen	Example of disease it causes	Mode of transmission
bacteria	*Vibrio cholerae* causes cholera	indirect transmission through contaminated water
virus	flavivirus causes yellow fever	vector transmission by mosquitoes
protozoan	*Plasmodium falciparum* causes malaria	vector transmission by mosquitoes
macroparasite	tapeworms cause infection of the digestive tract	indirect transmission through undercooked food or water contaminated with tapeworm eggs and larvae

24 **EM** To answer this question you should first analyse the graph and the trends shown and outline the role of the three hormones in maintaining pregnancy. The drop in the level of progesterone is a clue to the explanation of the role of falling progesterone in triggering birth. The high mark value of this question indicates you need to explain thoroughly.

Progesterone is called the pregnancy hormone. It helps to facilitate the implantation of the embryo by maintaining the uterus lining. Other roles include the suppression of muscle contractions of the uterus. (✓) Falling levels of progesterone are part of the process that initiates birth. (✓)

A number of different forms of oestrogen are produced during pregnancy and they complement the role of progesterone. They prepare the body for birth by strengthening the smooth muscle of the uterus and initiating the development of mammary glands ready for milk production. (✓)

Oestrogens support progesterone in inhibiting the pituitary gland from its role in initiating ovulation and also by suppressing immune rejection. (✓)

Chorionic gonadotrophin is produced in high levels early in pregnancy and its levels fall in later stages. It helps to maintain the corpus luteum and hence levels of progesterone and oestrogens in the early stages of pregnancy. (✓) Chorionic gonadotrophin induces the production of the hormone relaxin that helps with expansion of maternal joints to allow room for the growing foetus. (✓)

In addition to some of the oestrogens oxytocin is the hormone that orchestrates the contractions that result in birth. (✓) Relaxin helps with the expansion of the pelvis during birth but also facilitates the dilation of the cervix. (✓)

25 **EM** You need to identify the key components and structure of DNA, explain how replication occurs in relation to the structure and construct a diagram that summarises the key steps. You receive one mark for each of the two elements of the diagram. After you construct the diagram describe the changes that occur in point mutations in relation to the structure.

DNA is a polymer made up of nucleotide monomers. Each nucleotide contains a deoxyribose sugar joined to a phosphate and also joined to one of four nitrogenous bases: adenine and guanine are purine bases and thymine and cytosine are pyrimidine bases. (✓) The entire macromolecule consists of two chains of these nucleotides running in opposite directions and which are held together by hydrogen bonds between the nitrogen bases. (✓) Adenine attracts thymine and cytosine attracts guanine. The hydrogen bonds are not nearly as strong as the covalent bonds that hold each strand.

During DNA replication the sugar-phosphate backbones holding the nucleotides are peeled open when enzymes break the weak hydrogen bonds between the bases. (✓) The two single strands then become templates and, with the help of enzymes, the slotting in of matching nucleotides occurs so that two replicas of the original DNA are formed. (✓) Each molecule will reform into the double helix shown in the diagram that accompanies the question.

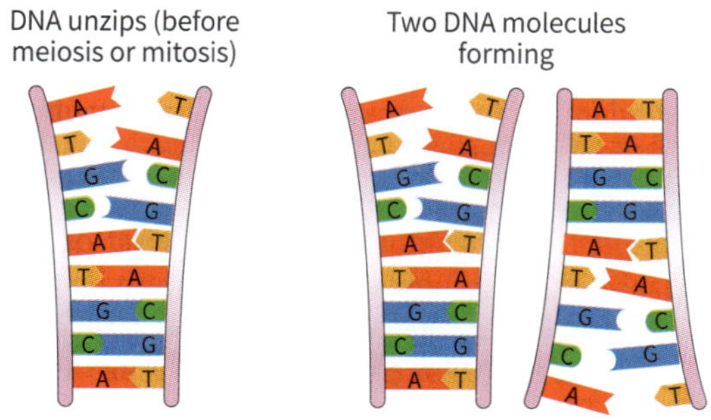

A point mutation is a mistake in one of the bases in a nucleotide of DNA. This can occur because of the action of a mutagen or if a mistake occurs during replication and the normal correction processes do not work. (✓) If the change is a substitution of one base for another there is only one change in the DNA molecule. If one additional nucleotide is inserted or a nucleotide deleted the whole of the following sequence of DNA will be altered. (✓)

26 a EM You need to make clear connections between structure and function for the two listed diseases and other examples you have researched.

Cystic fibrosis is caused by a defective transport protein that impacts the functioning of membranes in a range of body organs. Change in the primary structure of the protein impacts on mucus secretions; that is, on protein functioning throughout the body. (✓)

Motor neurone disease is not well understood but the suggestion that protein levels can become toxic because of a failure to break down the proteins emphasises their importance in cell functioning. (✓)

Diseases such as prion diseases also demonstrate that the structure of proteins is critical to their proper functioning as defective proteins destroy the brain's structure and functioning. (✓)

b EM You need to identify and provide information about the cause of genetic change.

The genetic change is a point mutation (✓) and in the case of cystic fibrosis a single nucleotide polymorphism results in a substituted amino acid in the polypeptide. (✓)

c EM You need to demonstrate your ability to interpret a pedigree diagram to construct a Punnett square and clearly indicate the probability of the outcome.

Let C represent the allele for normal transmembrane protein and c the defective form that causes cystic fibrosis. Both parents must be heterozygous for cystic fibrosis. (✓)

Male gametes→ Female gametes↓	C	c
C	CC	cC (✓)
c	Cc	cc (✓)

Offspring have a one in four probability of having cystic fibrosis and a one in two chance of being a heterozygous carrier. (✓)

27 EM To answer this question you need to understand the role of glucagon and insulin in maintaining homeostasis for glucose. Remember that the pancreas produces two hormones—insulin and glucagon—that regulate the amount of glucose in the blood. Insulin lowers blood glucose (blood sugar) levels by facilitating glucose transport from the blood into cells. Glucagon raises blood glucose levels by metabolising glycogen into glucose. When the system is functioning properly there is always some insulin and some glucagon being produced to create a balance between glucose release into the blood and glucose uptake into cells.

Very soon after a high carbohydrate meal glucose levels in the blood and insulin levels are higher. (✓) This rise is because the rise in glucose is detected by beta cells in the pancreas. These cells produce more insulin. (✓) When the glucose levels fall so too do the insulin levels. (✓)

Before the meal glucagon levels are high because no glucose is being absorbed by the small intestine. (✓) At the 1.5-hour time point on the graph, when the glucose levels reach the high end of the normal glucose range, glucagon levels start to fall (✓) and continue to do so until glucose levels return back to the normal range of 70 to 110 mg/dl. At this point the glucagon concentration levels off and then slowly begins to rise again. (✓)

28 a EM This question is straight recall but you need to make sure you answer all parts.

A genetic disorder is an inherited disorder or disease caused by genetic mutations passed down from parent(s) to offspring (✓); for example, Huntington's disease is disease caused by a faulty gene that makes an abnormal, toxic version of the Huntington protein. (✓) The toxic protein collects in the brain, causing damage to brain and nerve cells. (✓)

b EM This part of the question brings together your understanding of genetic techniques that can be used to prevent disease as well as your study of ethical issues. Make sure you answer both parts of the question; that is, choose an example of a genetic technique and provide one ethical concern for that technique.

Researchers are studying how genetic engineering can alter genes in human embryos to potentially eliminate risks of genetic diseases; that is, prevent genetic disorders from being passed from one generation to the next by genetically modifying the DNA of the embryo. (✓)

Ethical concerns are that the technology will be developed to manipulate human traits and alter the genetic sequencing of human embryos based on preference. This is sometimes called manipulating 'designer babies' and goes against the nature of heredity. (✓) Those opposed to the technology being used this way say that scientists should use the technology only for treating medical problems and not for choosing traits in embryos. (✓)

29 a EM Demonstrate your understanding of social, economic and cultural contexts from the examples given for various approaches to managing malaria.

Social contexts include the many small farmers in Asia and collectors in China who rely on the natural source of the anti-malarial drug artemisinin (✓) and whose livelihoods are threatened by competition by the use of yeast and gene technology. (✓)

Economic contexts include the high and/or prohibitive costs of production of the anti-malarial drug artemisinin and the research and development costs of using gene drives and CRISPR-Cas 9 to develop mosquitoes that are unable to be vectors of the malarial parasite. (✓) Having patents or intellectual property rights that control the use of gene technologies could lead to situations in which the needed drugs were too expensive for populations in Asia and Africa. (✓)

Cultural contexts include belief systems that impact on the willingness to adopt medical interventions or the use of environmentally damaging pesticides. Without pesticides there would need to be increased emphasis on simple preventative measures such as mosquito nets or using minnows as biological controls. (✓) Some belief systems are opposed to the use of vaccines to prevent infectious disease. (✓)

b EM The term *evaluate* means to make judgements based on criteria. In this question you need to evaluate the benefits or advantages and possible negative consequences and/or limitations of each procedure; that is, does it effectively prevent the spread of the disease malaria? In answering this part of the question it would be helpful to organise your answer in a table. One mark is given for the evaluation of each procedure.

	Evaluation	
Procedure described	**Benefits/Advantages**	**Consequences/Limitations**
pesticides	more available and affordable than other control methods	can have negative impacts on human health and the environment; resistance of mosquitoes to pesticides is a limitation
surface minnows	a cost-effective procedure	may have negative consequences on biodiversity
chloroquine	generally effective against some *Plasmodium* species and is safe for children and pregnant women and cheaper than some other treatments	protozoan resistance to the drug means that the virulent forms will survive to reproduce, which could result in an increase in the population of virulent form of protozoans
education programs	cost-effective forms of prevention and can have a greater reach and coverage than some other methods as well as targeting those communities at highest risk	availability of netting, insect repellent and protective clothing may be a problem in some regions, particularly rural and remote areas
vaccines	have been shown to be highly effective for the prevention, control and even eradication of some infectious diseases that were once common globally	the high cost of vaccines and community mistrust could be limiting factors, particularly in developing countries
artemisinin from natural sources	naturally occurring and traditionally used useful for supplementing treatment, especially where there are drug-resistant *Plasmodium* may be cost effective	affordability is an issue as supplies of the drug may be limited increasing production of artemisinin could be a solution but may impact negatively on farming areas and communities drug companies will use the plant for commercial gain without the informed consent of communities and without any benefits flowing back to communities that have traditionally used the plant
genetic engineering to produce the precursor of artemisinin	genetic engineering increases production of artemisinin and could make the treatment more affordable genetic engineering could also increase the amount of anti-malarial chemicals that the plant can produce	the drug is limited by its cost and availability
CRISPR-Cas9 gene editing and gene drive technology	engineering the genes of vectors such as mosquitoes can control the spread of diseases	the use of genetic engineering to control a disease vector raises ethical questions; for example, modified genes could escape the species boundary of the mosquito and cause problems in other species, leading to unforeseen ecological consequences

30 EM These predictions focus on the potential medical applications of biotechnology. As a comparison, possible directions in agricultural and industrial applications could provide contrasting ideas. Sample answers are also given for these two areas.

Medical applications of gene technologies, such as production of human insulin and other hormones and peptides, have been accepted more widely than applications in food production so it is reasonable to project that the more personalised approach to medicine described would ensue. (✓)

There may be more pressure on acceptance of GM food as the population grows and current agricultural practices result in environmental degradation. (✓) Crops could be grown that are more nutritional and in a much wider range of environmental conditions to address human food demand. (✓) Food could be produced in bioreactors so that natural environments can be conserved. (✓)

Industrial applications could see the further development of renewable energy resources and reduce greenhouse gases to better manage global warming. (✓) Waste could become a dated concept with all products and/or by-products being used in different ways. (✓)

31 EM The question tests your knowledge of how the human immune system responds to exposure to a pathogen. It requires recall of the innate and adaptive (specific) immune systems as well as vaccines.

a The image shows phagocytosis of a pathogen (✓) by a white blood cell called a macrophage. (✓)

b The innate immune system in the human body is the general defense system and is not specialised for specific pathogens. (✓) The specific immunity refers to the adaptive immune system, which acts specifically against certain antigens. (✓)

c B cells produce antibodies (✓) and T cells release certain chemicals or act as killer cells. (✓)

d Adjuvants are ingredients in vaccines. (✓) Adjuvants amplify immune responses by targeting the defence cells and chemicals of the innate defence system. (✓)

32 EM Make sure you name the practical application as well as the non-infectious disease or disorder it is used to treat. You could consider using dot points for your answer.

A cochlear implant or bionic ear is used to help people who have severe to profound hearing loss and/or damage to the inner ear or auditory nerve. (✓)

Cochlear implants work by wearing a sound processor behind the ear which has a microphone that picks up sounds in the person's environment. (✓) The sound processor then converts the sound into an electrical impulse that passes through the skin to a receiver surgically inserted and attached to the skull. The receiver relays the electrical impulse to an electrode array that has been surgically inserted into the cochlea to produce the sensation of sound at specific frequencies (pitches). (✓)

Improvements in cochlear implant technologies include applications that are compatible with iPods, iPhones and iPads. Hybrid technologies are being developed that combine a hearing aid and a cochlear implant in the one device. (✓)

SAMPLE HSC EXAMINATION II

Section I

1 C. Direct transmission is direct contact from one host to another. **A**, **B** and **D** are wrong because these are all indirect forms of transmission.

2 C. Tissue culture uses understanding of plant mitosis and sterile conditions to create large numbers of plant clones. **A**, **B** and **D** all relate to techniques that are used for animal reproduction.

3 B. One of the chemical responses of plants to the presence of pathogens is the production of specific receptors called resistance peptides that kill pathogens or stop them spreading. **A**, **C** and **D** are physical plant responses and therefore incorrect.

4 D. Bacteria are prokaryotic and lack a nucleus whereas protists are eukaryotic. Bacterial fission is different from mitosis. Mitosis is the basis of binary fission in protists. **A**, **B** and **C** are all incorrect statements.

5 C. If the bacterium is not recognised by phagocytes it is able to get past the body's first line of defence and enter into the bloodstream. **A** is incorrect because when a pathogen is composed of molecules that are different to the host's it will be recognised as non-self and an immune response will be triggered. **B** is incorrect because an inflammation response is designed to either kill the pathogen or prevent it from spreading. **D** is wrong because macrophages are part of the body's immune defence which prevent the entry of pathogens into the host.

6 C. Random drift causes changes in allele frequency because of events such as random mating. **A** and **B** relate to changes in frequency resulting from natural selection. **D** is wrong because gene flow is a result of immigrants bringing different alleles into a population.

7 D. The primary target for prevention needs to be teenagers and young adults. An education campaign through social media has been shown to be effective. **A** and **B** are wrong because they are not strategies for prevention. **C** is partly correct but the primary target needs to be the young people rather than the parents.

8 A. This option correctly identifies that better health and increased life expectancy are benefits for society that could reasonably be projected into the future. **B** and **D** are incorrect because they do not list benefits for society. **C** is wrong because genetic technologies will not directly eliminate mutagens though they may make screening and treating better.

9 B. Aneuploidy is an additional or missing chromosome, usually resulting from non-disjunction in cell replication. **A**, **C** and **D** are all examples of point mutations.

10 A. The diagram shows the transition from the polypeptide to the protein (insulin). **B** and **C** are incorrect because they are processes that occur in cell replication. **D** is wrong because the diagram represents a molecule, not a chromosome.

11 B. The structure of proteins is vital for their functioning. **A** is wrong because proteins have many roles in addition to being enzymes. **C** and **D** are wrong as they are not functions of hormonal proteins.

12 B. Type 2 diabetes is a disease in which the body becomes resistant to the normal effects of insulin or gradually loses the ability to produce enough insulin. **A** is wrong because gestational diabetes is a form that is first diagnosed during pregnancy. **C** is wrong because chronic disease is a general term for diseases that tend to be long lasting and persistent in their symptoms or development. **D** is incorrect because pre-diabetes refers to impaired glucose tolerance. This is a condition in which blood glucose levels are higher than normal but less than required for a diagnosis of diabetes.

13 C. Gene and whole organism cloning are gauged by the percentage success rate. **A** is wrong as it only refers to whole organism cloning. **B** is incorrect as antibiotic resistance is often linked to a cloned gene. Successful clones can be screened but the process does not work for whole organism cloning. **D** is wrong because it refers to only one possible application of whole organism cloning.

14 B. Sources of variation in cell replication are crossing over and random segregation (and mutation). **A** is wrong because it includes DNA replication and **C** is wrong because both processes result in replication. **D** is incorrect because it lists mutagens.

15 A. Prevalence is the number of cases or the proportion of cases in a population in a given time period. **B** is wrong because mortality refers to numbers of deaths. **C** is incorrect because incidence refers to the number of new cases in a specific time period. **D** is wrong because even though the number of cases of dementia is increasing—as a result of the number of older people as a proportion of the population increasing—this is not about population aging.

16 B. An antigen is any substance, microbe or pathogen that is usually foreign to an organism's own body and thus triggers an immune response. An antibody is a protein capable of interacting with a specific antigen. **A** is only partly correct and therefore not the best option: antigen does refer to the invading pathogen but the antibody is not the pathogen. **C** is wrong because antigens are not white blood cells and antibodies are mainly made by plasma cells or B cells. **D** is incorrect because even though antigens can refer to toxins produced by pathogens, antibodies attach to and neutralise toxins.

17 D. A single nucleotide base substitution in at least 1% of population is an SNP. **A** is wrong as the change is not large enough to be a chromosomal mutation. **B** and **C** are incorrect as the figure involves substitution as opposed to the insertion or deletion of a nucleotide.

18 A. Without cell replication multicellular organisms could not grow and unicellular organisms could not reproduce. **B** and **C** are wrong because they relate to sexual reproduction, which is not universal across all life forms. **D** is incorrect because it refers only to reproduction in bacteria and some fungi.

19 B. According to the graph, expenditure on cardiovascular disease totalled around 13.8 million dollars. All other diseases were less than that in terms of expenditure so options **A**, **C** and **D** are incorrect.

20 **D.** Foot-and-mouth disease is a disease in cattle that has been successfully kept out of Australia by quarantine. **A** is wrong because sickle-cell anaemia is a genetic disease. **B** is incorrect because smallpox has been controlled by vaccination. **C** is wrong because lung cancer is caused primarily by smoking and quarantine is ineffective in stopping smoking.

Section II

21 **a** **EM** The question asks you to relate two different understandings: plant responses to infections and Aboriginal bush medicines.

Eucalyptus oil contains antibacterial properties and is used by Eucalypts as a preformed chemical defence against the entry of pathogens. (✓) This makes it an effective bush medicine that has been traditionally used by Indigenous Australians. (✓)

b **EM** This is a recall question.

Aboriginal protocols are used in the development of medicines to ensure recognition and protection of Indigenous cultural and intellectual property (✓) and as guides to ethical practices and the equitable sharing of benefits that come from using bush medicines in the pharmaceutical industry. (✓)

22 **EM** You need to describe the key events in fertilisation and implantation and methodically examine them for their role in producing new generations of offspring.

Fertilisation is a key event in sexual reproduction because without the union of the male and female gametes offspring would not eventuate. If fertilisation failed across a species then the long-term continuity of that species would be at risk. (✓) Because it is a source of genetic diversity fertilisation is a key process contributing to the continuity of species. (✓)

Implantation is the process by which the embryo embeds in the lining of the uterus of placental and marsupial mammals to allow development of the foetus, making it essential for continuity of the species in these groups. (✓)

Species that reproduce asexually such as fungi (budding, spores), bacteria (binary fission) and protists (binary fission, budding) and plants reproducing vegetatively do not rely on either fertilisation or implantation for their asexual reproduction. The continuity of a species relies on genetic diversity so asexual reproduction may not provide enough diversity to support natural selection in a changing environment. (✓)

23 **a** **EM** This is a straight recall question.

The answer is incidence of disease. (✓)

b **EM** When answering this question make sure you justify or support your answer using the data provided.

The data shows that in developing countries the higher proportion was due to infections. (✓) This suggests that infections are more likely where there is poor sanitation and contaminated water, and where there are lower standards of health care and understanding of infectious agents. (✓)

c **EM** There are a number of ways to answer this question. Make sure the strategy is about prevention rather than treatment. Naming two strategies will result in two marks.

Examples include vaccination (e.g. vaccination against human papillomavirus (HPV) which causes cervical cancer), education campaigns to prevent transmission of the infectious agent, (✓) improved sanitation and clean water supplies to reduce the risks of infection, and community-wide screening programs to detect early signs of the cancer. (✓)

24 **a** **EM** You need to interpret the steps in the model of mitosis, identify key events represented and give information about the changes that are occurring.

1 DNA has replicated, nuclear membrane disappears, chromatids condense, joined by centromeres, centrioles form
2 centrioles move apart, joined by spindle fibres (✓)
3 chromatids joined to spindle at centromeres and are lined up along the 'Equator'
4 centrioles move apart and the chromatids separate, going to opposite poles (✓)
5 centrioles and spindles break down and the nuclear membrane starts to form
6 cytokinesis completes the mitosis cell division to divide cytoplasm between two cells (✓)

b **EM** This question requires that you identify meiosis as the other type of cell division and highlight how it differs from mitosis.

Mitosis, as described in part **a** above, is cell replication that results in two identical cells from the one parent cell. (✓) Meiosis is the cell division which results in the reduction of chromosome numbers by half. Genetic variation results from meiosis because of random segregation and crossing over. Meiosis results in four unique haploid cells from the one diploid parent cell. (✓)

25 **a** **EM** This is a sample answer. Half a mark is given for correctly completing each part of the model.

Stimulus ↓	increase in body temperature when exercising or when external temperature is hot
Receptors ↓	thermoreceptors in the skin detect change in temperature; receptors in the hypothalamus detect rise in blood temperature
Afferent pathway to control centre ↓	heat loss centre in the hypothalamus is activated (heat promoting centre shuts off)
Control centre	hypothalamus
Efferent pathway from the control centre to the receptors ↓	message sent from the hypothalamus to blood vessels in the skin and sweat glands (for example)
Receptors such as glands, blood vessels and skeletal muscles ↓	blood vessels dilate, capillaries become flushed with blood and sweat glands are activated
Response ↓	heat radiates from the skin's surface so that the body cools down; evaporative heat loss due to perspiration
Negative feedback	blood temperature drops

b EM **This is a higher-order question that asks you to apply your understanding of structural and physiological adaptations for thermoregulation. You need to know that humans are endotherms in order to answer this question correctly.**

Adaptations considered would include structural and physiological adaptations that help control the exchange of heat with the environment. (✓) These include insulation or small surface area to minimise heat loss (✓) along with blood flow (✓); for example, reducing blood flow to extremities to keep the core of the body warm. (✓)

26 EM **To answer this question you need to explain the structure and function of DNA, then the structure and function of proteins, and then outline the relationship between DNA and proteins. Diagrams should show the double helical structure of DNA and contrast that with the variable structure of proteins, such as the secondary (alpha and beta sheets), tertiary and quaternary arrangements.**

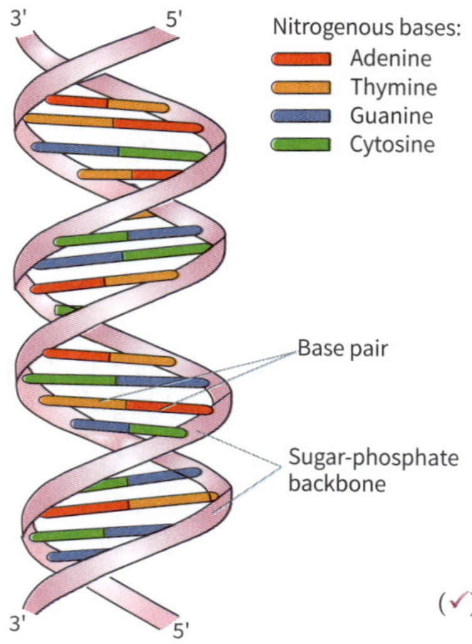

(✓)

DNA is the genetic code and the structure of the molecule has to ensure its replication and allow for the synthesis of polypeptides. (✓) DNA is a polymer made of nucleotide monomers in two strands running in opposite directions. The four bases on the nucleotides are arranged in specific pairs held together by hydrogen bonds. During polypeptide synthesis each triplet of bases is the code for the insertion of a specific amino acid in a sequence that forms a polypeptide. (✓) The double helical shape encloses the base pairs as a form of protection from mutagens in order to ensure stability of the code.

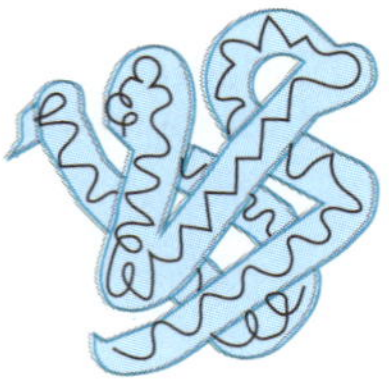

(✓)

Proteins are complex polymers made up of various numbers of polypeptide chains. Each polypeptide chain consists of up to 20 different amino acids (monomers) forming the primary structure. (✓) The amino acids interact with each other, forming beta sheets and alpha helixes at various places along a polypeptide which results in its secondary structure. These may fold together to form the tertiary structure (as shown in the diagram). Finally, a protein may form from at least one or more polypeptide chains interacting together. (✓)

Compared with DNA, protein structures are much more varied. Some may be long and can interact together, such as actin and myosin which bring about muscle contraction. Other proteins have a specific shape to interact with other molecules in order to bring about chemical changes (enzymes). Other proteins store and transport (e.g. haemoglobin) while others control entry into cells, provide mechanical support, control metabolism (hormones) or recognise foreign substances (antibodies).The structures of proteins suit their many purposes. (✓)

27 a EM **This is a straight recall question. You could answer with either epidemic or pandemic.**

An epidemic is the rapid spread of infectious disease to a large number of people in a particular population in a relatively short period of time. (✓) A pandemic is the rapid global spread of infectious disease. (✓)

b EM **Make sure you relate your answer to mobility. Note: The diseases in the question are all zoonotic diseases; that is, diseases that can spread between animals and humans.**

The speed and volume of the mobility of individuals and human population movement is unprecedented in human history. Given the speed of travel it is possible for an infected individual to travel the globe within the incubation period of a disease, thereby causing an exponential increase in contacts. (✓)

Global changes in land and animal use have significantly increased the risk of zoonotic diseases in humans. Increases in human population make it easier for diseases to spread and bring more people in closer contact with other animals. (✓)

c EM **Many factors can contribute to the spread of disease during an epidemic. Sample answers are provided. Consider using dot points for your answer. One mark is given for each correct point. Note: More information is given below than you would need to provide to obtain two marks.**

Transmission of a disease during an epidemic is usually rapid. Key factors that increase exposure to pathogens and cause the spread of disease are:

- a continual source of the disease; for example, a reservoir such as a contaminated water supply
- environmental factors such as changes in climate
- loss of healthcare systems and shortage of medical supplies; for example, as a result of civil war
- natural disasters such as flooding, which can damage sewerage system infrastructure
- an increase in the amount of the pathogen or new strains of the pathogen that can be spread from animals to humans
- introduction of the pathogen into an area where the pathogen has not been before; for example, because of deforestation
- slow responses by health agencies at the start of the epidemic.

d EM **As with part c, consider using dot points for your answer. One mark is given for each correct point.**

- Modelling with computers is used to predict future occurrences of disease along with projecting and simulating how a disease will progress and spread. Modelling can also simulate the effects of possible interventions. (✓)
- Restriction measures include travel restrictions and increased quarantine surveillance. (✓)
- Event-based surveillance systems can systematically scan a multitude of informal sources; for example, news reports, online newspapers, social media and internet-based searches can all assist in ensuring a rapid response to outbreaks. (✓)

28 **EM** **You need to interpret the pedigree diagram showing normal inheritance of a mutation and compare it with that showing the pattern of inheritance with the added gene drive.**

Using the pedigree diagram for normal inheritance it is apparent that the normal or wild type allele is recessive (✓) and the fly with the altered allele was heterozygous dominant meaning that half of its offspring showed the phenotype for the altered allele. (✓) This pattern of inheritance predicts that 12.5% of offspring will have the altered allele in the third generation as demonstrated in the pedigree diagram. (✓)

With the gene drive inheritance shown in the figure all offspring of the parents with altered alleles have the altered allele in their phenotype. (✓) This means that even when the offspring mate with wild-type flies all offspring have the altered allele. This is because the gene drive 'edits' the allele on the other homologous chromosome, converting it to the altered allele. (✓) As the altered allele attached to the gene drive always becomes homozygous it is always passed on to the offspring. In the offspring the gene drive makes the allele homozygous and so on through generations. (✓)

29 **a** **EM** **This is a straight recall question which demonstrates your understanding of scientific method. Validity involves designing your investigations to use independent, dependent and controlled variables to ensure a valid procedure. Below is a sample answer.**

I would ensure controlled variables; that is, the factors that are kept constant during the procedure. This would include having the same amount of agar in each Petri dish, ensuring the same temperature for each Petri dish (✓) and having the same amount of food sample and the same time for incubation. (✓)

b **EM** **This is a straight recall question which demonstrates your understanding of safe work practices and managing the risks associated with inoculating nutrient agar plates and growing microbes. One mark is given for each correct procedure. Your answer could have included any two of the points below.**

Safe work practices and risk management include:

- the use of protective clothing; for example, enclosed shoes, masks, lab coat, gloves and protective eyewear
- sterilising flaming transfer loops, wires or needles immediately before and after transfer of samples
- ensuing the correct labelling of all samples
- thorough cleaning of work surfaces
- washing hands thoroughly before leaving the laboratory
- autoclaving agar plates at the end of the investigation to ensure harmful microbes are destroyed.

30 **EM** **The answer is given below as an explanation but you may find that that the use of dot points or a table will help with the three aspects of comparing DNA sequencing with profiling.**

DNA sequencing enables mapping of species genomes by determining the order of the nucleotide bases on all DNA in an organism (✓) while DNA profiling uses the unique genetic make-up of individuals for analysis of crime scene samples, unidentified remains and confirming family relationships. (✓)

DNA profiling relies on comparing sections of non-coding DNA that are highly variable in length; that is, variable number tandem repeats (VNTRs) and short tandem repeats (STRs). Particular enzymes called nucleases break the DNA from the sample into these fragments. (✓) DNA sequencing also uses enzymes and involves breaking fragments at particular nucleotide bases that are identified in some way or by synthesising fragments from a single DNA strand that acts as a template. (✓)

Both DNA sequencing and DNA profiling have used technologies such as the polymerase chain reaction to increase the quantity of DNA being analysed along with electrophoresis for separation and comparison of the lengths of the DNA fragments. (✓) Both DNA sequencing and DNA profiling have undergone dramatic improvements in the technologies applied and reductions in the time taken to complete the processes. (✓)

31 **EM** **The key to answering this question is to identify the types of mutagens—chemical, electromagnetic and natural—and make judgements as to how they could be used to best promote mutations in plants.**

Gamma rays are a high-energy, high-frequency, short wavelength form of ionising electromagnetic radiation. (✓) Gamma rays can be sourced from radioisotopes such as potassium-40. Gamma rays have been shown to produce mutations in both seeds and pollen, and have successfully been used in the production of mutant rice lines. Gamma rays often result in chromosome aberrations and safety precautions must be used when handling radioisotopes. (✓)

Chemical mutagens used in plant breeding are often the alkylating agents such as ethyl methanesulfonate (EMS). (✓) These agents add alkyl (hydrocarbon chains) to the bases of DNA, causing changes such as base substitutions, but are generally much easier and safer to handle than gamma rays. (✓)

Natural mutagens include certain viruses that are known to induce point mutations (insertions, deletions and substitutions) in DNA, particularly in mammalian cells. (✓) Transposable elements such as transposons, retrotransposons and retroviruses are mutagens that could be classified as natural because they are self-replicating segments of DNA that excise and/or insert themselves within the genome. Transposable elements have an advantage in mutation breeding of plants because the changes they promote are not random; that is, there are places where they are more likely to insert themselves or replicate. (✓)

32 **a** **EM** **You need to present ideas on how the social and economic benefits flowing to certain individuals and groups will impact on the support and use of biotechnology.**

The acceptance and use of biotechnology is influenced by social contexts. Some individuals and groups in society have concerns regarding animal welfare, the use of chemicals in the environment and the methods used in food production. These can restrict the acceptance of some applications of biotechnology. (✓) Some groups may fear the impact of altering natural processes in animals by the use of rBGH or be concerned that food labelling may not accurately describe what they are purchasing and whether the food contains high levels of hormones. (✓)

Farmers who can afford rBGH and benefit from it may support its use whereas others who cannot afford it or have concerns for animal health may oppose it. (✓)

Economic and social contexts interact. If societal pressures reduce sales of a product, thereby making it not economically viable, then the application of the specific biotechnology becomes defunct. (✓)

b **EM** **Arguments about what is ethically right or wrong with the uses of biotechnology need to be developed in your response to this question.**

Ethical uses of biotechnology are about ensuring that benefits to society and the environment outweigh any harm, particularly to societal groups or animals that may not have any choice about the applications of biotechnology. (✓) An ethical use of biotechnology could be the development of biofuels that reverse the impact of global warming. Other uses could include more effective bioremediation of land or the protection or restoration of damaged ecosystems such as coral reefs. (✓)

33 **EM** When you compare the studies and results show how they are similar and how they are different. Remember that prevalence measures the number or proportion of cases in a population at a given time.

The sample study chosen for this answer is the epidemiology of vitamin D deficiency, a nutritional disease caused by insufficient exposure to the sun, often as a result of the use of sunscreen, skin pigmentation and clothing habits. Vitamin D is a fat-soluble vitamin needed for calcium maintenance, bone health and homeostasis. (✓) Deficiency is related to diseases such as osteoporosis, hypertension, diabetes and cancer. Deficiency is measured by levels of the blood serum 25(OH)D. (✓)

Available prevalence data shows that vitamin D deficiency is a global problem, particularly in girls and women from Middle Eastern countries but also in adults in many European, South American and African countries. One billion people worldwide—about 15% of the population—are vitamin D deficient or insufficient. (✓)

Epidemiological studies play a role in reducing the prevalence of non-infectious diseases or disorders. The benefits of epidemiological studies are evident in both the fluoridation and vitamin D deficiency studies. McKay's study led to targeted public health intervention that reduced tooth decay and improved oral health. (✓) Epidemiological studies of vitamin D deficiency allow scientists to identify risk factors, target those populations that need to be screened, and determine the resources needed for health care: in this case diet supplementation to at-risk populations. Studies have shown that vitamin D3 supplements are more effective in raising 25(OH)D levels than vitamin D supplements. (✓)

34 **EM** You only need to choose one disorder. When you evaluate effectiveness consider the value of the technology and other factors such as the extent to which it assists with the effects of the disorder, as well as its cost and affordability. Make sure you answer both parts of the question: you need to describe and evaluate a current technology and describe a possible future research and technology. The sample answer provided relates to visual disorders.

Example of current technology: the visual disorder is cataracts, which are the world's leading cause of blindness. Cataracts are caused when proteins in the lens clump together, making the lens cloudy and blocking and distorting the light entering the eye. (✓)

Surgery can easily restore normal sight. Cataract surgery is an example of a highly effective technology as the treatment is simple. It is one of the most frequently performed surgeries in the world and takes less than 30 minutes. Once the cloudy (opaque) natural lens is replaced with a clear artificial lens it is not possible to develop another cataract. The cost in developing countries is about $15 to $100 (✓) and the major challenge is to get the treatment to those who need it. (✓)

Example of future technology: the visual disorder is retinitis pigmentosa, a disease in which photoreceptor cells in the retina die because of a gene mutation and vision dims from the outside to form a tunnel of light. (✓) Disorders of the retina are harder to treat than disorders of the lens. A bionic eye that uses a microchip containing thousands of photodiodes (to replace the function of the photoreceptor cells) has been developed and is currently in the trial stage. The microchip is implanted among dead photoreceptor cells in the retina's macular region, allowing the optic nerve to pick up electrical signals. (✓) The technology requires high costs for research and development and the surgery is complex. Another drawback is that the implant does not restore full vision and the patient must learn to interpret the new visual data. (✓)

INDEX

H

I

J

K

L

M

N

O

P

Reprinted 2019, 2021, 2024

ISBN 978 1 74125 674 1

Pascal Press
PO Box 250
Glebe NSW 2037
(02) 9198 1748
www.pascalpress.com.au

Publisher: Vivienne Joannou
Project editor: Mark Dixon
Edited by Ian Rohr
Answers checked by Sarah Barrand
Indexed by Puddingburn Publishing Services
Cover and page design by Sonia Woo
Typeset by Grizzly Graphics (Leanne Richters)
Illustrated by Sefko Obucina and Leanne Richters
Printed by Vivar Printing/Green Giant Press

Students
All care has been taken in the preparation of this study guide, but please check with your teacher or the NSW Education Standards Authority about the exact requirements of the course you are studying as these can change from year to year.

The validity and appropriateness of the internet addresses (URLs) in this book were checked at the time of publication. Due to the dynamic nature of the internet, the publisher cannot accept responsibility for the continued validity or content of these web addresses.

CONTENTS (1)

CONTENTS (3)

REPRODUCTION

1 The asexual mechanisms of reproduction that ensure the continuity of a species

- **Asexual reproduction involves** a single parent producing genetically identical offspring (clones).

  Examples include binary fission in bacteria and some protists; budding in yeasts and some protists; spore production in plants and fungi; regeneration of simple animals such as flatworms; parthenogenesis in some animals such as aphids; and bulbs, corms, rhizomes and stolons in plants.
- Asexual reproduction results from mitosis or fission cell divisions.
  - The **advantage** is rapid population increase.
  - The **disadvantage** is a lack of genetic diversity.

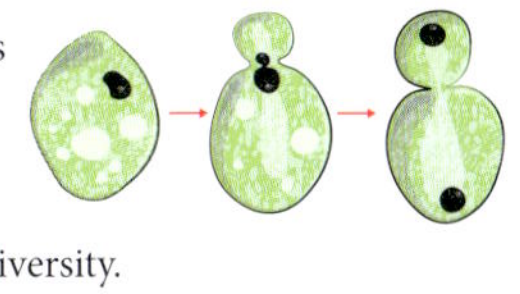

MODULE 5: HEREDITY

See ➲ *Excel* Year 12 Biology p. 4

REPRODUCTION

3 Fertilisation and implantation in mammals

- **Fertilisation** usually occurs in the oviduct of female mammals.
- The **oocyte completes** meiosis to become an ovum after spermatozoon penetrates the cell membrane.
- The **union of a spermatozoon and oocyte** and the resultant unification of the pronuclei results in a zygote (new genetically unique offspring) or fertilisation.
- The **zona pellucid ensures** the entry of spermatozoon of the same species and only one enters the oocyte.
- **Implantation occurs** in the endometrial lining of the uterus of placental mammals.
- The **trophoblast develops** into the placenta in placental mammals to support the nutrition of the foetus.
- **Hormones play a supportive role** in preventing the immune rejection of the foetus.

MODULE 5: HEREDITY

See ➲ *Excel* Year 12 Biology p. 12

REPRODUCTION

5 The impact of scientific knowledge on the manipulation of plant and animal reproduction in agriculture 1

- **Vegetative propagation techniques result** in rapid expansion of genetically identical plant populations.
- **Selective breeding** has improved the agricultural potential of many plants.
- **Seed banks** have been developed to preserve genetic diversity.
- **Polyploidy** can be used to create seedless (sterile) plants but also to create fertile hybrids.
- **Cytoplasmic male sterility** has been used to help ensure crossbreeding is not contaminated.
- **Mutation breeding** (the use of mutagens such as gamma rays) has been used to expand the variation in plants in the hope of producing favourable mutants for plant breeding.
- **Genetic modification** (the insertion of genes from one species into another) has been used to improve the economic value of some plants.

MODULE 5: HEREDITY

See ➲ *Excel* Year 12 Biology p. 16

CELL REPLICATION

7 Modelling cell replication: mitosis

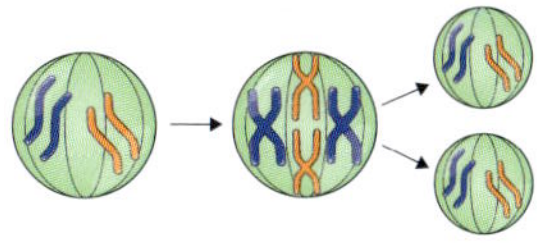

- **DNA replicates** → nuclear membrane breaks down and nucleolus disappears → chromatin condenses into chromosomes that have duplicated into two chromatids held together by centromere → centrosomes separate forming spindle → chromosomes move to middle (equator) → chromatids pulled by spindle go to opposite poles, becoming chromosomes → nuclear membrane reforms → cytokinesis → two cells identical with parent cell.
- **Mitosis is essential** for growth, repair and asexual reproduction.

MODULE 5: HEREDITY

See ➲ *Excel* Year 12 Biology p. 26

CELL REPLICATION

9 Modelling cell replication: nucleotide composition, hydrogen bonding and pairing

- **DNA is a polymer** made up of nucleotide monomers.
- A **nucleotide** contains a deoxyribose sugar joined to a phosphate group and a nitrogenous base.
- The **sugar-phosphate chain forms** an external backbone for the DNA strand and the nitrogenous bases radiate towards the centre of the helical molecule, joined to the sugar in the backbone.

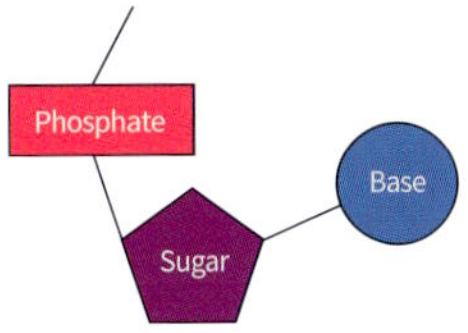

- **Two purine bases**: guanine (G) and adenine (A) and two pyrimidine bases—thymine (T) and cytosine (C)—are the nitrogenous bases.
- **Hydrogen bonds result** in the nitrogenous bases pairing faithfully: A with T and G with C.

MODULE 5: HEREDITY

See ➲ *Excel* Year 12 Biology p. 30

CELL REPLICATION

11 Modelling cell replication: DNA replication

- **Cell replication** cannot occur without DNA replication.
- **Two DNA strands** unzip.
- **Primers** slot into place.
- The **leading strand** has nucleotides lock in and bonds form to create a continuous copy of DNA.
- The **lagging strand** has short segments of nucleotides sectioned in, the gaps are filled and bonds formed.

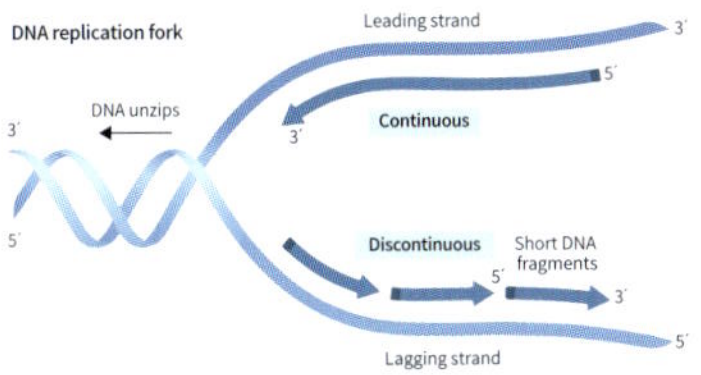

- **Enzymes control the process**, check for errors and correct any that are detected.

MODULE 5: HEREDITY

See ➲ *Excel* Year 12 Biology p. 33

DNA AND POLYPEPTIDE SYNTHESIS

13 Comparing and modelling DNA in prokaryotes and eukaryotes

- **Similarities** include:
  - **DNA combined** with proteins
  - **same role** (to make polypeptides)
  - have **extrachromosomal DNA** in either plasmids (prokaryotes), mitochondria or chloroplasts (eukaryotes)
  - **DNA replicates** before cell division.
- **Differences in eukaryotes** include:
  - large sections of '**non-coding' DNA**
  - **chromatin formed** from DNA closely linked with the histone proteins
  - densely packed **in nucleus as linear chromosomes** in mitosis
  - **multiple copies of each gene** (usually two) as chromosomes occur in pairs.
- **Differences in prokaryotes** include:
  - **circular chromosome** in nucleoid region
  - very little **'non-coding' DNA**
  - **one copy** of each gene.

MODULE 5: HEREDITY

See ➲ *Excel* Year 12 Biology p. 38

REPRODUCTION

The sexual mechanisms of reproduction that ensure the continuity of a species

2

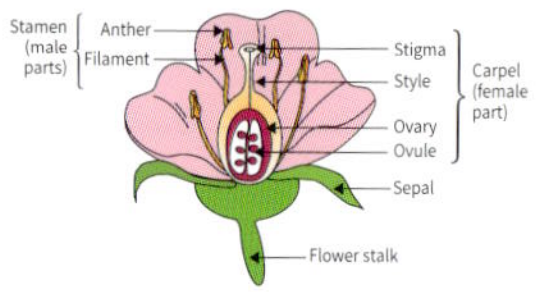

- **Sexual reproduction involves** parents of different genders producing haploid gametes that unite in fertilisation to form the diploid zygote.

  Examples include seeds from flowers, cones in advanced land plants and zygote production in animals.
- **Internal fertilisation results** within the female from the release of semen from the penis into the female reproductive tract.
- **External fertilisation results** from males and females releasing gametes into the outside environment (usually water).
- **Stamens in flowers** and **male cones release pollen** that is transferred to the female stigma.

See ➲ *Excel* Year 12 Biology p. 6

CONTENTS (4)

CONTENTS (2)

CELL REPLICATION

Modelling cell replication: meiosis

8

- **DNA replicates** → nuclear membrane and nucleolus disappear → chromatin condenses into chromosomes (two chromatids held together by centromere) → centrosomes separate forming spindle → chromosomes move to middle → homologous pairs come together and crossing over occurs → homologous pairs of chromosomes pulled apart by spindle go to opposite poles and cell becomes haploid → nuclear membrane reforms → cytokinesis → nuclear membrane breaks down → chromatin condenses into two chromatids held together by centromeres → centrosomes separate forming spindle → chromosomes move to middle → chromatids pulled by spindle go to opposite poles, becoming chromosomes → nuclear membrane reforms → cytokinesis → four haploid cells that are genetically unique.
- **Meiosis is essential** for sexual reproduction, gamete production and genetic diversity.

See ➲ *Excel* Year 12 Biology p. 28

REPRODUCTION

The impact of scientific knowledge on the manipulation of plant and animal reproduction in agriculture 2

6

- **Selective breeding has resulted** in many new strains of useful livestock.

  Examples of techniques in animals include artificial insemination (AI), ovulation induction, in-vitro fertilisation (IVF), multiple ovulation and embryo transfer (MOET), juvenile in-vitro fertilisation and embryo transfer (JIVET), and somatic cell nuclear transfer cloning (SCNT).
- **Genetic modification** has been used in aquaculture to improve the productivity of salmon as a food source.
- **Sex reversal** of sequentially hermaphroditic fish has been used in aquaculture to improve productivity.
- **Sterile triploid fish** have been bred to ensure that genetically modified fish do not interbreed with wild fish.

See ➲ *Excel* Year 12 Biology p. 19

REPRODUCTION

Hormonal control of pregnancy and birth in mammals

4

- **Progesterone** is the pregnancy hormone that prepares the uterus so that implantation and subsequent development of the foetus is sustained.
- **Oestrogens support changes** during pregnancy, such as the preparation of breasts for lactation and the uterus for contractions in childbirth.
- **Birth is initiated** because of stress to the foetus, the release of the corticotrophin-releasing hormone and the dominance of oestrogens over progesterone.

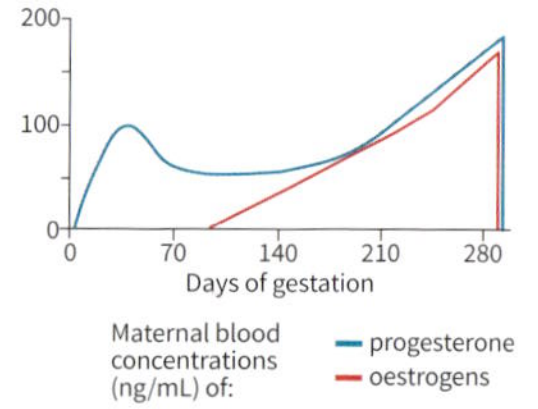

- **Oxytocin** is the main hormone of birth and causes contractions of the uterus.
- **Relaxin** is a hormone that plays a role in pregnancy and birth by relaxing the pelvis and joints to allow passage of the foetus.

See ➲ *Excel* Year 12 Biology p. 14

DNA AND POLYPEPTIDE SYNTHESIS

Process of polypeptide synthesis: transcription

14

- **Transcription carries** the genetic code in the form of mRNA from the nucleus to the ribosome.
- **DNA is edited** so the 'coding' DNA becomes the mRNA and the 'non-coding' DNA is **removed**.

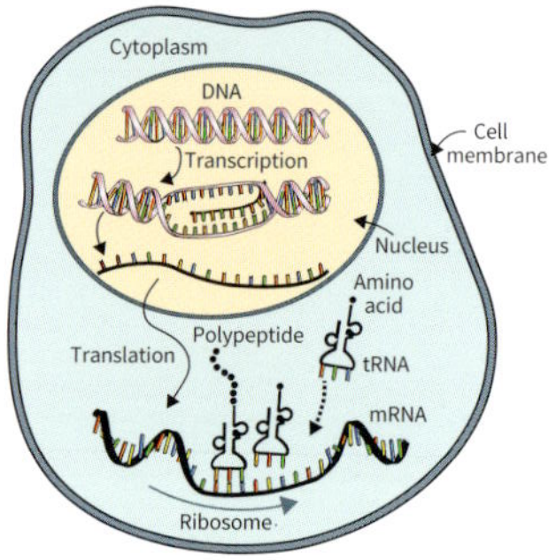

See ➲ *Excel* Year 12 Biology p. 39

CELL REPLICATION

Assessing the effect of cell replication on the continuity of species

12

- **Asexual reproduction** through mitosis and binary fission creates large numbers of similar offspring with predictable characteristics.
- **Continuity** depends on the large numbers and a stable environment.
- **Sexual reproduction** through meiosis and fertilisation creates diverse offspring.
- **Variation helps continuity** of a species in a changing environment.
- **Organisms undergoing sexual reproduction** have to achieve sexual maturity.
- **Sexual maturity** requires cell replication, expansion and differentiation.

See ➲ *Excel* Year 12 Biology p. 34

CELL REPLICATION

Modelling cell replication: the Watson and Crick DNA model

10

- A **double helix is formed** from two strands of DNA nucleotides that spiral in opposite directions.
- The **nitrogenous base pairing ensures** that the correct nucleotides slot into place during replication.
- **Models of DNA** must show that the pairing bases—A with T, G with C—fit each other but no other base.

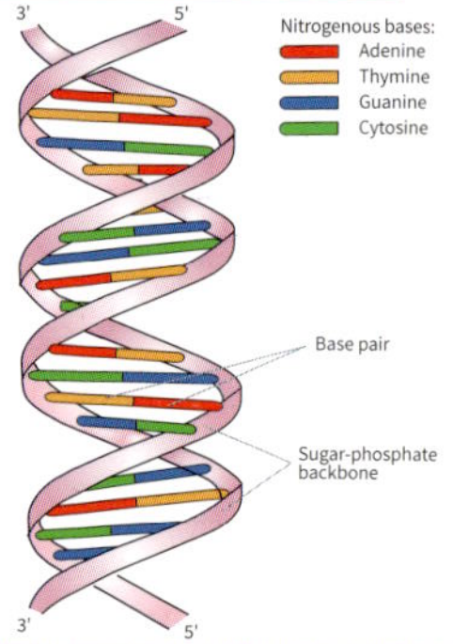

See ➲ *Excel* Year 12 Biology p. 32

DNA AND POLYPEPTIDE SYNTHESIS

15 Process of polypeptide synthesis: translation

- → **tRNA and the ribosome provide** the structures that allow the translation of mRNA in the cytoplasm to form the amino acid sequence of the polypeptide chain.

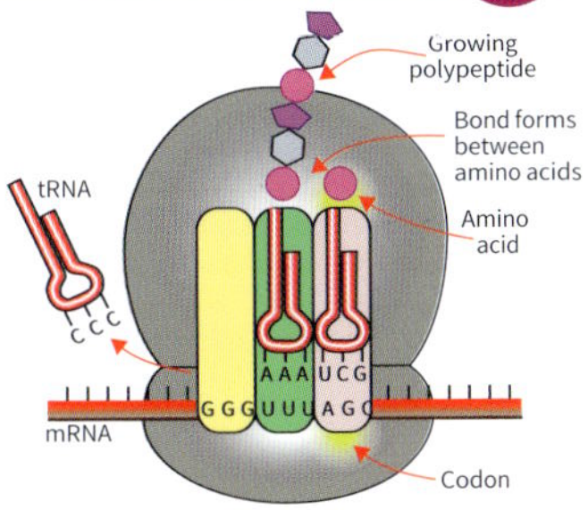

 See Excel Year 12 Biology p. 42

DNA AND POLYPEPTIDE SYNTHESIS

17 How genes and the environment affect phenotype expression

- → **Genes impact** on phenotype through polypeptide synthesis.
- → **Gene regulation** means some genes are active and others inactive. The same DNA results in highly differentiated cells.
- → **Epigenetic** studies reveal the role of gene regulation on changes in phenotype.
- → **Environmental factors** such as diet impact on gene regulation through 'temporary' changes to DNA, such as methylation and histones.
- → **Environment and lifestyle** may also impact through the length of telomeres, which directly impact on cell aging and some diseases such as cancer.

 See Excel Year 12 Biology p. 43

GENETIC VARIATION

19 Variation resulting from random segregation and crossing over

- → In the early stages of meiosis **random segregation** of each member of the homologous pairs of chromosomes results in haploid cells and produces highly varied gametes.
- → **Crossing over restructures** some chromatids and may produce new combinations of alleles along a chromosome.
- → **Both processes result** in new combinations of alleles in gametes.

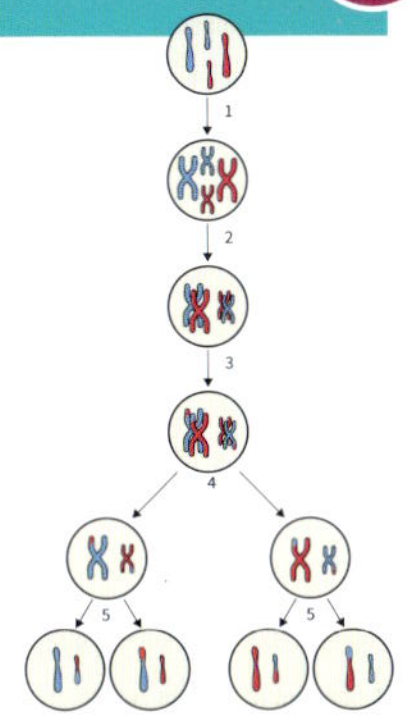

 See Excel Year 12 Biology p. 52

GENETIC VARIATION

21 Autosomal inheritance and sex linkage

- → **Autosomal inheritance** is the inheritance of genes carried on normal (autosomal) chromosomes that impacts males and females equally.
- → **Alleles can interact in a dominance/recessive relationship** in which the dominant allele is expressed in the hybrid (heterozygous form) and the recessive phenotype is only expressed in the homozygous genotype.
- → **Sex chromosomes**, such as the X and Y chromosomes in humans, determine the sex of the offspring but the X chromosome also carries some genes which are expressed more frequently in males that only have one X.

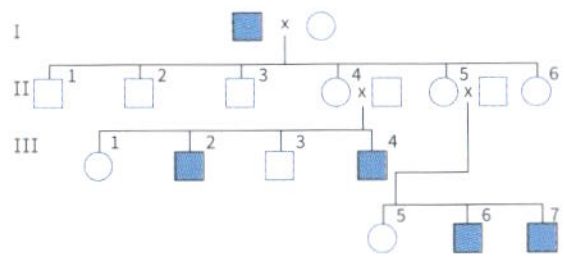

- → The **pattern of sex-linked inheritance** is different in males to females.

 See Excel Year 12 Biology p. 55

GENETIC VARIATION

23 Frequency data of characteristics in populations

- → **Genetic diversity** is reflected in the frequency of characteristics across a population.
- → **Frequencies of characteristics** are collected, recorded and presented to reveal data on phenotypes in local populations.
  Observations could include characteristics such as eye colour, hair colour, tongue-rolling, or attached or detached ear lobes.
- → **Data is limited** because phenotypes do not always indicate the genotype.
- → **Sampling may be insufficient** to reflect the whole population gene frequency.
- → **Allele frequency** is a measure of the genetic diversity of a gene pool.

 See Excel Year 12 Biology p. 60

INHERITANCE PATTERNS IN A POPULATION

25 Technologies used to determine the inheritance patterns in a population 1

- → **DNA sequencing uses technologies** such as:
  - the polymerase chain reaction (PCR)
  - sequencing by synthesis (SBS)
  - sequencing by oligonucleotide ligation and detection (SOLiD)
  - labelling; for example, use of fluorescent dyes
  - separating by electrophoresis
  - base detection; for example, via the use of dyes or by light emission.

  Examples of use include the detection of single nucleotide polymorphisms (SNPs) as indicators of genetic diversity and disease risk and in the Human Genome Project.

 See Excel Year 12 Biology p. 68

INHERITANCE PATTERNS IN A POPULATION

27 Data analysis from a large-scale collaborative project for conservation management

- → **Population genetics investigates** factors that influence the change of frequency of alleles over time and across locations and populations.
- → **Allele frequency changes** with mutations, natural selection, gene drift and gene flow.
- → **Single nucleotide polymorphisms (SNPs)** are used as an indicator of levels of genetic diversity and to compare the genetic diversity of species.
- → **Sample size** is important for obtaining reliable data for studying population genetics.
- → **Genetic diversity** is important for the survival of endangered species and small or isolated populations.
- → **Conservation management** needs to consider genetic diversity as well as habitat requirements.
- → **Tasmanian devils** are at risk from a contagious cancer that is thought to be linked to the low diversity of genes associated with its immune system.

 See Excel Year 12 Biology p. 71

MUTATION

29 How a range of mutagens work

- → **High-energy electromagnetic radiation** includes gamma rays, X-rays and some ultraviolet rays.
- → **Electromagnetic radiation passes** through tissues to break DNA strands and chromosomes or alter nucleotide bases by changing atoms.
- → **Chemical mutagens** substitute bases and modify the backbone of DNA and often prevent it from replicating. Some chemical mutagens break chromosomes.
- → **Naturally occurring mutagens** include viruses, chemicals in plants (alkaloids) and moulds (aflatoxin) that change DNA structure.
- → **Reactive oxygen species (ROS) interact** with other mutagens to break DNA.

 See Excel Year 12 Biology p. 80

MUTATION

31 Somatic versus germ-line mutations, mutations in 'coding' versus 'non-coding' DNA segments

- → **Somatic (body cell) mutations can result** in diseases such as cancer but are not passed on to the next generation.
- → **Germ-line mutations can be passed on** to offspring and thereby alter allele frequency in the gene pool. Many chromosome mutations in the germ-line result in sterility.
- → **'Coding' DNA segments result** in the production of polypeptides so mutations may result from the failure to produce a functional polypeptide.
- → **'Non-coding' DNA segments** may have no known function or regulate gene expression. If regulatory, mutations may result from the failure to produce a functional polypeptide.

 See Excel Year 12 Biology p. 84

GENETIC VARIATION

20 Variation resulting from mutation and fertilisation

➔ **Mutations** are the ultimate source of genetic variation.

➔ **Punnett square diagrams show** the possible genotype of gametes and possible combinations on fertilisation.

➔ Each **fertilisation event is independent** of any other.

Male gametes→ Female gametes↓	T	t
T	TT	Tt
t	Tt	tt

See ➲ *Excel* Year 12 Biology p. 54

DNA AND POLYPEPTIDE SYNTHESIS

18 The structure and function of proteins in living things

➔ The **role of proteins as catalysts** (enzymes) means they control reactions (metabolism) in cells.

➔ **Proteins control cell composition** through their role in the cell membrane.

➔ **Transport proteins** in body fluids such as haemoglobin in blood help transport substances around the body.

➔ **Proteins strengthen cellular structures and enable movement** by providing support, by interacting and by changing shape.

➔ Many **hormones are proteins that control processes** such as reproduction, growth and the storage and release of glucose.

➔ **Proteins provide for storage**, such as oxygen stored in ferritin.

➔ **Immune response results** from the ability of some proteins to recognise other molecules.

See ➲ *Excel* Year 12 Biology p. 44

DNA AND POLYPEPTIDE SYNTHESIS

16 Function and importance of polypeptide synthesis

➔ **Provides the main component of proteins** which are essential to the structure and functioning of cells, tissues and organisms.

➔ **Provides a mechanism** for the action of inheritance.

➔ **Antibodies are proteins formed** when polypeptides are made in response to the presence of antigens.

➔ The figure shows that the **immune response is dependent on polypeptide synthesis** to protect the body from infection.

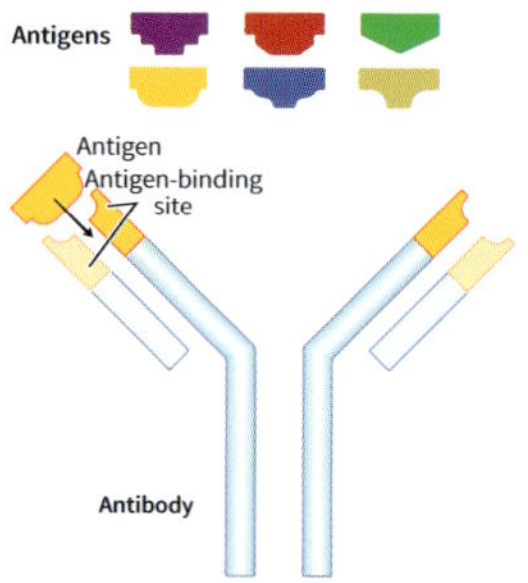

See ➲ *Excel* Year 12 Biology p. 42

INHERITANCE PATTERNS IN A POPULATION

26 Technologies used to determine the inheritance patterns in a population 2

➔ **DNA profiling uses technologies** such as:
- the PCR
- separation by electrophoresis.

Examples of use include determining the probability of an identity (such as matching crime scene evidence to individuals) or a familial relationship by comparing highly variable parts of the DNA of individuals.

➔ These **variable parts** include:
- **restriction fragment length polymorphism (RFLP) analysis relies** on varying lengths of RFLPs because they contain variable numbers of tandem repeats or mini-satellites.
- **short tandem repeats (STRs)** are loci that are targeted and STRs are amplified using PCR and compared in an alternative method of profiling.

➔ **DNA samples can be contaminated** and care must be taken to avoid this.

See ➲ *Excel* Year 12 Biology p. 70

GENETIC VARIATION

24 Analysing single nucleotide polymorphisms

➔ A **single nucleotide polymorphism (SNP)** is an alternative form of an allele resulting from a single base substitution (mutation) in the DNA which occurs in at least 1% of the population.

➔ **HbS** is a SNP of normal haemoglobin (HbA).

➔ The **sickle-cell trait** is the result of HbS in the heterozygous form.

➔ **Sickle-cell anaemia** is the result in the homozygous form.

➔ **90% of human variation in DNA** occurs as SNPs are found in 'non-coding' as well as 'coding' sections of DNA.

See ➲ *Excel* Year 12 Biology p. 61

GENETIC VARIATION

22 Co-dominance, incomplete dominance and multiple alleles

➔ **Alleles can interact in a co-dominant manner** in which both alleles are expressed (e.g. roan cattle).

➔ In **incomplete dominance** the heterozygous form has an intermediate or a blended phenotype, different from either parent (shown in diagram).

➔ **Multiple alleles** is the inheritance of a characteristic in which there are more than two form of alleles so there is a great variety of possible phenotypes and genotypes (e.g. human blood types may be A or B (co-dominant) or O (recessive to both A and B).

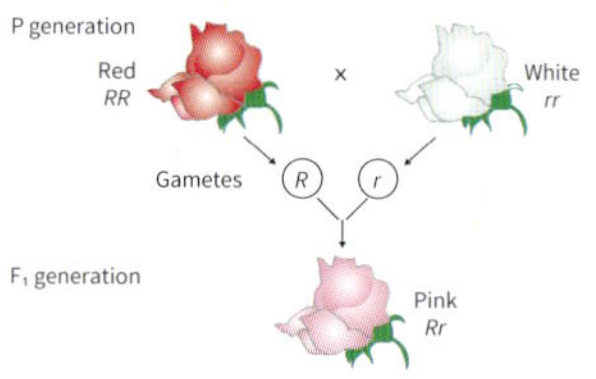

See ➲ *Excel* Year 12 Biology p. 58

MUTATION

32 Genetic variation relating to fertilisation, meiosis and mutation and the effect of mutation, gene flow and genetic drift on the gene pool of populations

➔ **Fertilisation and meiosis result** in recombinations of existing alleles.

➔ **Mutations produce new alleles** and have the potential to change the gene pool and influence evolution.

➔ **Mutation increases variation** in the gene pool.

➔ **Gene flow** from immigrants can increase genetic diversity.

➔ **Genetic drift** is a random change that mainly impacts very small populations.

See ➲ *Excel* Year 12 Biology p. 85

MUTATION

30 Different types of mutation: causes, processes and effects

➔ **Point mutations** involve a change in one nucleotide base. They include:
- **Substitution of a nucleotide** may result in a different amino acid in a polypeptide chain.
- **Insertion or deletion of a nucleotide base** results in a frameshift that has major harmful impacts on polypeptide synthesis.

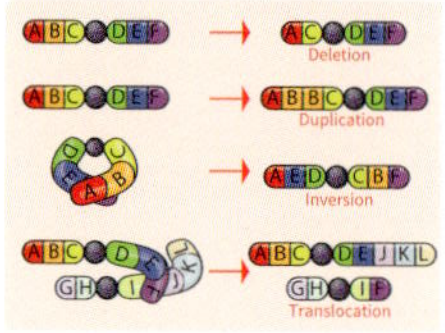

➔ **Chromosomal mutations result** in whole sections of chromosomes being deleted, inserted, inverted or translocated to other chromosomes. These rearrangements are usually very harmful to organisms.

➔ **Non-disjunction of homologous chromosomes results** in a change in chromosome number (aneuploidy) that is usually very harmful or fatal.

See ➲ *Excel* Year 12 Biology p. 82

INHERITANCE PATTERNS IN A POPULATION

28 Data analysis of population genetics for understanding inheritance of a disease or disorder and human evolution

➔ Many diseases have both **genetic and environmental causes** so population genetics data can help determine the genetic factors that contribute.

➔ **Genome-wide association (GWA) studies** have helped to collect and analyse data on the genetic components of disease.

➔ The **frequency of alleles for sickle-cell anaemia** resulting from a SNP that changes haemoglobin is higher in areas where there is exposure to malaria as it provides some resistance to the protozoan pathogen.

➔ In **human evolution mtDNA** is used as a molecular clock to determine the date of the last common ancestor.

➔ **DNA from mitochondria and the nucleus** are used to compare modern humans with recent ancestors such as Neanderthals and Denisovans.

See ➲ *Excel* Year 12 Biology p. 73

BIOTECHNOLOGY

Past uses and applications of biotechnology — 33

- **Agriculture and aquaculture has resulted** in selective breeding of crops and animals over the last 10 000 years, improving the quantity and quality of food.
- **Fermentation has resulted** in improved preservation and an increased range of foods such as beer, wine, cheese, yoghurt and soy products, as well as organic acids such as lactic acid.
- **Medical applications of biotechnology** such as MAbs, antibiotics, vaccines and some gene technology have saved lives.
- **Biotechnology has improved** human health and increased the population size.

 See *Excel* Year 12 Biology p. 90

BIOTECHNOLOGY

Future uses and applications of biotechnology — 35

- **Widespread use in research** and for medical applications.
- **Genetic techniques that do not create transgenic organisms** could be increasingly used.
- **Personalised medicine** will result from the selection of the most effective medicine for each individual.
- **Better biofuels and waste recycling/reduction** could be developed.
- A wider range of more **environmentally friendly bioplastics** could be developed.
- **Human stem cells** may replace the need for animal models in research.
- **Bioluminescence** may be developed as an alternative light source.

 See *Excel* Year 12 Biology p. 94

BIOTECHNOLOGY

Future directions of the use of biotechnology — 37

- **Many factors influence** possible future directions. Four options include:
  - **Proliferation of present trends**; for example, increased use of gene technologies such as transgenic species until their use becomes accepted for agriculture, industry and medicine.
  - **Social and cultural resistance continues** so new techniques use understanding of genetics without producing transgenic crops and animals.
  - **Bioreactors become** responsible for food production because of the breakdown of agricultural and aquaculture ecosystems.
  - A 'brave new world' of **bio-computers** and **synthetic biology** along with **bioluminescence** for household lighting.

 See *Excel* Year 12 Biology p. 96

GENETIC TECHNOLOGIES

Uses and advantages of current genetic technologies — 39

- **Current genetic technologies** include:
  - **mutation breeding** (improved productivity of many grain crops)
  - **marker-assisted breeding** (improved pond yields of black tiger prawns by over 50%)
  - **genetic engineering** (using gene editing tools and deleting genes to introduce new characteristics into species)
  - **CRISPR-Cas9** (a precise and more reliable gene editing tool which improves the efficiency of genetic engineering)
  - **gene drives** increase the impact of an introduced gene; could potentially stop the transmission of the malaria protozoan by mosquitoes
  - **gene silencing or RNAi** (research and disease control)
  - **chromosome engineering** (improved crops such as soybeans and maize)
  - **microbial vectors** to introduce new genes into plants for improved productivity or production of new substances such as vaccines.

 See *Excel* Year 12 Biology p. 103

GENETIC TECHNOLOGIES

The effectiveness of cloning — 41

- **Gene cloning** makes copies of genes for use in genetic engineering and research or for therapeutic cloning.
- **Genes for antibiotic resistance** are often combined with the gene being cloned so screening with antibiotics ensures that only the cloned genes are obtained.
- **Whole organism cloning creates** copies of plants and animals that are genetically identical to the parents and thus have predictable characteristics.
- Whole organism cloning **generally has a high failure rate** so is used for brood stock, not food production.
- **Therapeutic cloning uses stem cells** to make corrections to organisms that are compatible with the organism.
- **Barriers to its effectiveness still remain** though it has the potential for treating conditions such as Parkinson's disease.

 See *Excel* Year 12 Biology p. 106

GENETIC TECHNOLOGIES

Applications of recombinant DNA technology — 43

- **Rapidly and predictably produces organisms** with known characteristics.
- Applied to **agriculture, medicine and industry**.

  Examples include insect resistance (Bt cotton, potatoes, maize and rice); herbicide resistance (canola, soybeans, maize, cotton); longer shelf life (FlavrSavr tomato, non-browning apples); increased nutrition (Golden Rice); disease resistance (virus-resistant papaya, squash, potatoes); stress tolerance (DroughtGard maize, drought-resistant sugar cane); appearance (blue carnations); modified oils for biofuels and healthier foods; bioremediation (plants can remove mercury); pharmaceuticals (insulin, safer vaccines, antibodies, blood-clotting agents); biosensors and better medical diagnostic tests; use of research animals for studying diseases such as cancer (knockout mice).

 See *Excel* Year 12 Biology p. 108

GENETIC TECHNOLOGIES

Benefits of using genetic technologies in industry — 45

- **Benefits in industry** include:
  - **Bioreactors** can produce mass quantities of various organic compounds.
  - **Highly-sensitive analytical procedures** can detect substances such as pollutants.
  - **Renewable and better fuels** can be produced along with products to clean up existing pollution.
  - **Natural resources can be recovered** with minimum production of pollutants.
  - **Biodegradable plastics and bioplastics** reduce demands on the petroleum industry and promote recycling.
  - There can be **more efficient and safe extraction** of metals from ores.

 See *Excel* Year 12 Biology p. 109

GENETIC TECHNOLOGIES

Influences of social contexts on a range of biotechnologies — 47

- The **social context relates** to which groups in society would benefit or suffer detriment from biotechnologies.
- **Perceptions about the benefits and risks** are a part of the social context.
- **Education is a major factor** in social understanding and acceptance.
- **Societal concerns** often revolve around the use and impact of technology in terms of widening the gap between rich and poor within and across societies. This may impact on the acceptance of biotechnologies.
- **Farmers and gene technology companies** may benefit but consumers may be taking a risk (e.g. allergic reactions) from the foods they consume.
- **Perceptions of better health for consumers** via more nutritious foods promotes acceptance but there are concerns as to whether these will be available to poorer people.

 See *Excel* Year 12 Biology p. 110

CAUSES OF INFECTIOUS DISEASE

How are diseases transmitted? — 49

- A **disease** is any condition that impairs normal living processes.
- **Infectious diseases** differ from other diseases, such as genetic and lifestyle diseases, in that they are caused by invasion by a pathogen and can be transmitted from one host to another.
- A **pathogen** is any organism living in or on another organism that is capable of causing disease.
- **Transmission must occur** for an infection to spread. For an infectious disease to be transmitted it must:
  - **escape from its host** or from where it normally lives and multiples
  - **survive transmission** from one host to another; some pathogens can only be transmitted by direct contact because they cannot survive outside the host
  - **enter the new host** or **escape from the new host**.

 See *Excel* Year 12 Biology p. 117

BIOTECHNOLOGY

Potential changes in Earth's biodiversity due to genetic techniques

38

- **Genetic erosion** is loss of biodiversity because of the increased use of clones and monoculture cultivation.
- **Genetic pollution introduces** new genes into species and changes the characteristics of species and ecosystems.
- **Biodiversity can be sustained** through:
  - improved **waste management**
  - use of **bioremediation**
  - **metabarcoding to help** better manage endangered species
  - **using light biosensors** for rapid detection of pollution
  - **better climate control improving** the outlook for natural communities.
- The **use of gene drives and RNAi** could eliminate some infectious diseases, thereby helping to conserve some species at risk.
- Some **environments could be protected and restored** through more humane and effective ways of eliminating feral species.

See ➲ *Excel* Year 12 Biology p. 97

BIOTECHNOLOGY

Social implications and ethical uses of plant and animal examples of biotechnology

36

- **Social implications** include concerns over:
  - **equity of access** to new technologies and their applications
  - **displacement** of traditional lifestyles
  - **loss of consumer autonomy**; for example, poorly labelled food
  - **population changes** that impact on the planet and quality of life
  - **potential abuse** (e.g. bioterrorism).
- **Ethical concerns** include advances in understanding diseases at the expense of animals; for example, the use of knockout mice in labs and RNAi to produce virus-free pig hearts for xenotransplantation.
- **Environmental uses of biotechnology** might include the restoration of ecosystems such as coral reefs, better waste management and addressing global warming by engineering plants with increased rates of photosynthesis or for biofuels.
- **Alternate protein-rich foods** might be produced in bioreactors.

See ➲ *Excel* Year 12 Biology p. 95

BIOTECHNOLOGY

Present uses and applications of biotechnology

34

- **Expanded agriculture and aquaculture dramatically increases** environmental impacts, improves human health, and increases human lifespan and population size.
- The increased use of a wide range of genetic technologies has impacted on agriculture, aquaculture, medicine and industry.

  Examples include ELISA, rAbs, RNAi gene silencing, marker-assisted breeding and CRISPR-Cas9 for more precise gene modification.
- **Increasing widespread use of gene technologies** has produced more productive plants and animals (see card 43).
- **Expanding medical uses** for diagnosis, production of pharmaceuticals and research (see card 43).
- **Industrial uses of biotechnology** include production of biofuels, bioplastics and biodegradable plastics, bioremediation, expanded waste recycling and developing new metal extraction processes.

See ➲ *Excel* Year 12 Biology p. 93

GENETIC TECHNOLOGIES

Benefits of using genetic technologies in agriculture and medicine

44

- **Benefits in agriculture** include:
  - more **rapid, precise and cost-effective breeding** of organisms
  - more **nutritious foods** can be produced
  - **pesticides** can be used less or more effectively or can reduce the need for soil cultivation.
- **Benefits in medicine** include:
  - **safer vaccines and pharmaceutical treatments** such as insulin, human growth hormones, follistim (for treating infertility), human albumin, monoclonal antibodies and blood clotting and thinning factors.
  - **better analytical techniques**, including ELISA and western blot technologies that can be used for research and diagnosis
  - **better understanding** of disease causes and the role of certain genes
  - more **effective cancer treatments** with fewer side effects.

See ➲ *Excel* Year 12 Biology p. 109

GENETIC TECHNOLOGIES

Techniques used in recombinant DNA technology

42

- **Desirable genes are cloned** (shown on right) before hosts are transformed and crossbred to ensure the new organisms are homozygous.
- **New techniques** include use of CRISPR-Cas 9 to precisely locate and edit parts of the genome and gene drives to increase the frequency of desired alleles.

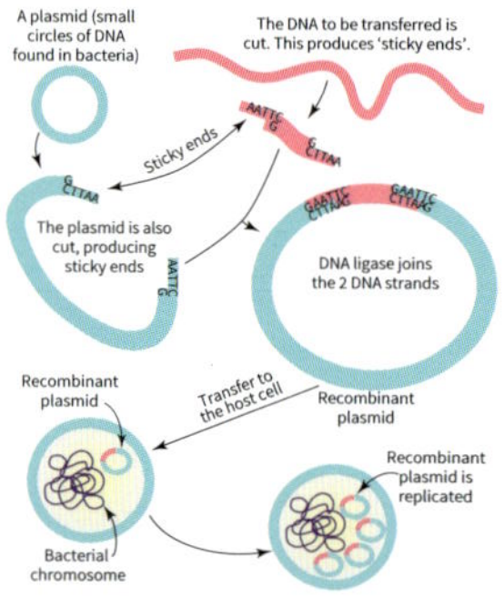

See ➲ *Excel* Year 12 Biology p. 107

GENETIC TECHNOLOGIES

Processes and outcomes of reproductive technologies

40

- **Artificial pollination involves** the collection of pollen from desirable plants, removing stamens or otherwise sterilising flowers that have the desirable female gametes and transferring pollen on to stigma.
- **Artificial insemination (AI) involves** the collection of semen from desirable male animals and its insertion into desirable female breeding stock.
- **AI maximises the breeding potential** of male and female animals and can be used to shorten generation intervals in some livestock.
- **In-vitro fertilisation (IVF) involves** the collection of semen and oocytes and their artificial combination. The resultant embryos are then introduced into prepared females, possibly after genetic screening.
- **Reproductive technologies increase** the chances of having healthy offspring.

See ➲ *Excel* Year 12 Biology p. 105

CAUSES OF INFECTIOUS DISEASE

A variety of infectious diseases caused by pathogens

50

- **Pathogens include microscopic pathogens**; for example, bacteria, viruses, fungi and protozoans.
- **Bacteria** are prokaryotic, unicellular organisms about 1000 nanometres in size. An example of a disease-causing bacteria is *Bordatella pertussis* which causes whooping cough.
- **Viruses** are non-cellular pathogens requiring living host cells to replicate. Their size range is 20 to 40 nanometres. An example is rhinovirus, which causes colds.
- **Fungi** are eukaryotic organisms. An example of a disease-causing fungi is *Phytophthora cinnamomi*, which causes root rot in plants.
- **Protozoa** are eukaryotic, unicellular organisms. An example is entamoeba, which causes amoebic dysentery.
- **Prions** are non-cellular pathogens. They are a defective form of a protein molecule found mostly in the brain.
- **Pathogens also include macroscopic pathogens**; for example, parasites such as arthropods, flatworms and roundworms.

See ➲ *Excel* Year 12 Biology p. 118

GENETIC TECHNOLOGIES

Influences of cultural and economic contexts on a range of biotechnologies

48

- The **economic context** relates to the potential of biotechnologies to create wealth for some groups or companies and the impact that competition and additional costs versus returns may have on producers and users.
- **Intellectual rights and patents** may control who benefits.
- The **cultural context** encompasses the values and beliefs that are associated with the practices of biotechnology.
- **Beliefs that genetically modified foods are unnatural** or that GM is 'playing God' may impede acceptance.
- **Growing trends towards veganism** reflect cultural shifts towards concern for animal welfare.
- The production of **plant- or microbe-based protein foods** that are meat or dairy substitutes along with fibres that are alternatives to wool may increase acceptance of biotechnologies.

See ➲ *Excel* Year 12 Biology p. 110

GENETIC TECHNOLOGIES

Effect on biodiversity of using biotechnology in agriculture

46

- **Biodiversity has been reduced** through the use of traditional biotechnology in agriculture and aquaculture, especially through habitat destruction and the growing of monocultures.
- The **use of new genetic technologies** in modern biotechnology, including gene technologies, initially increases biodiversity. This is because **new genetic combinations** not possible in nature are introduced.
- **Less environmental degradation may result** from the use of some applications that require less cultivation, produce less pollution, result in less harvesting of wild populations or restore damaged environments.
- **Non-target species may be impacted** by applications such as insect-resistant crops.
- **Gene pollution may reduce biodiversity** in the long term; for example, the spread of genes resistant to herbicides.

See ➲ *Excel* Year 12 Biology p. 110

CAUSES OF INFECTIOUS DISEASE

Modes of transmission of infectious diseases

51

- **Different modes of transmission** are used by different pathogens.
- **Direct contact**, **indirect contact** and **vector transmission** are the main modes of transmission.
- **Direct transmission includes** direct contact between people through touch, saliva, mucus, blood and other bodily fluids.
- **Indirect transmission includes** transmission from environmental reservoirs, such as soil and water, or from contaminated objects, such as towels or contaminated medical supplies.
- **Vector transmission** can be mechanical or biological.
- **Biological vectors** include insects and other arthropods.
  Example: malaria is transmitted (spread) from person to person by the Anopheles species of the female mosquito carrying the protozoan *Plasmodium falciparum*.

See Excel Year 12 Biology p. 124

CAUSES OF INFECTIOUS DISEASE

Microbial testing of food and water samples

53

- In **practical investigations you would have managed the risks** associated with inoculating nutrient agar plates and growing microbes, autoclaving, and the safe disposal of products.
- Employed **safe work practices** as per below.
  Examples include wearing protective clothing and eyewear; taking care with flaming transfer loops, wires and needles; correctly labelling all samples; thoroughly cleaning work surfaces; and washing hands thoroughly before leaving the laboratory.
- **Valid procedures** would have been ensured by designing your investigation to use variables and experimental controls.
- **Variables** in an investigation are factors that can be changed, maintained or measured, such as temperature, concentration or pH of substrate, and time.
- **Experimental control**, also referred to as the **controlled variable**, is the variable that is kept constant during an investigation.

See Excel Year 12 Biology p. 123

CAUSES OF INFECTIOUS DISEASE

The causes and effects of diseases on agricultural production

55

- **Australia is free of the world's worst plant and animal diseases**, such as foot-and-mouth disease and Avian Influenza H5N1.
- Animal pests and diseases are a **major threat** to Australia's livestock and poultry industries.
- **Outbreaks can be devastating** to farmers and communities and have a negative impact on Australia's export industries.
- As an example, **sheep lice can cost producers** in NSW around $120 million per year in lost production and treatment costs.
  Examples of the impact of sheep lice include the following: lousy sheep produce about 10% less wool; their fleeces have broken wool fibres, making them more felted or matted and less valuable; sheep are stressed through the discomfort of irritated skin and this causes rubbing and affects feed intake, which damages their hides and makes them more susceptible to flystrike.

See Excel Year 12 Biology p. 128

RESPONSES TO PATHOGENS

How does a plant or animal respond to infection?

57

- **To cause infection** pathogens must enter a host and reproduce or replicate.
- Plants and animals have **evolved innate defences** to prevent the entry of pathogens.
- **Highly specialised responses** to pathogens occur when they enter tissues and cells.
- **Innate immune responses** are general, non-specific defence and response mechanisms. These mechanisms are primed and ready to fight an infection at all times.
- **All plants and animals** have innate immune systems.
- To **mobilise a response** the immune system must be able to distinguish self from non-self.
- **Non-self includes** foreign particles and pathogens. This means that the host must be able to first recognise and then protect the body against non-self particles and pathogens.

See Excel Year 12 Biology p. 137

RESPONSES TO PATHOGENS

The response of a named Australian plant to a named pathogen

59

- Some of the **preformed defences and induced responses** of Eucalyptus to the fungal pathogen *Phytopthera cinnamomi* are shown below.

Passive defences (preformed)		Active defences or induced responses
Physical barriers	Chemical barriers	
thick cuticle and bark; waxy leaves and dry leaf surfaces vertically hanging leaves prevent the formation of moisture which also provides some protection from water moulds	stored oils function as preformed chemical defences, not only against defoliating animals but also as antifungal and antibacterial agents or for priming defences in both the host and neighbouring plants	the formation of barrier zones in new tissue produced by the vascular cambium; these zones protect the healthy sapwood by separating it from the adjacent infected or damaged tissue; in some Eucalypts these zones contain protective wound-sealing gums periderm separates damaged tissue from healthy tissue and prevents spread of the disease(s)

See Excel Year 12 Biology p. 139

RESPONSES TO PATHOGENS

The physical or chemical changes that occur in the cells and tissues of animals in the presence of pathogens 1

61

- **Macrophages immediately recognise** pathogens which cross protective barriers and begin replicating.
- Macrophages then **release cytokines** which along with other chemicals (e.g. histamine) set up the inflammatory response.
- Macrophages and **neutrophils produce toxins** to help kill ingested pathogens.

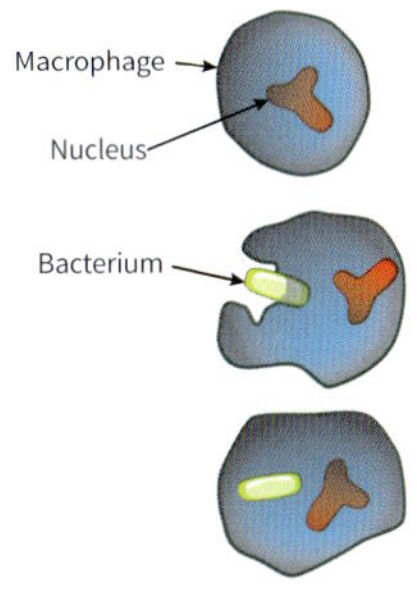

See Excel Year 12 Biology p. 141

IMMUNITY

The innate and adaptive immune systems in the human body

63

- The **human immune system responds** to exposure to a pathogen:
  - through the **innate immune system** (a fast-acting, non-specific and broadly effective general defense)
  - through the **adaptive immune system** (specialised and specific).

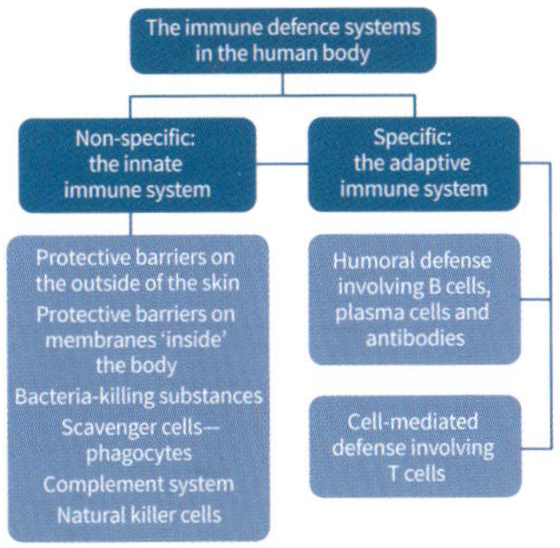

See Excel Year 12 Biology p. 147

IMMUNITY

The adaptive immune system in the human body

65

- The **adaptive immune response** takes longer to respond but has advantages over the innate system, including that:
  - it **targets the pathogen** more accurately
  - it **acts specifically** against certain antigens.
- The adaptive immune system **can 'remember' the antigen (the pathogen)** because it produces memory cells.
- After **first (primary) contact with the antigen** it takes several days for the adaptive immune system to respond.
- After a **second infection with the same pathogen** the adaptive immune system recognises it and immediately releases the antibodies needed to fight the pathogen. The defense response is quicker and symptoms are usually weaker or absent.
- The adaptive immune system is made up of:
  - **antibodies** (as soluble proteins in the blood)
  - **B cells**
  - **T cells.**

See Excel Year 12 Biology p. 150

IMMUNITY

B cells (B lymphocytes) and T cells (T lymphocytes)

67

- **White blood cells differentiate** into different types of cells, including lymphocytes.
- **B cells (B for bone marrow)** are lymphocytes that continue their development in the bone marrow.
- **T cells (T for thymus)** are lymphocytes that migrate from the bone marrow and continue their development in the thymus.
- **Plasma cells** are specialist cells produced by the presence of an antigen which stimulates the B cells to produce more B cells and plasma cells. Plasma cells make large amounts of antibody.
- **Memory cells** are also produced by B cells.
- **T cells destroy antigens** directly. They have structures on their surfaces that pathogens can bind to.
- A **pathogen that exactly fits a T cell stimulates** this T cell to multiply quickly and develop into specialised T cells.

See Excel Year 12 Biology p. 151

CAUSES OF INFECTIOUS DISEASE

The adaptations of different pathogens that facilitate their entry into and transmission between hosts

56

- **Pathogens have adaptations** to exploit their hosts' biology, behaviours and ecology and this allows them to live in and move between hosts. Examples include bacteria, viruses and protozoa that have evolved the ability to survive in arthropod vectors, such as insects and ticks, and use these vector to gain entry into the host when the vector bites the host or sucks its blood.
- Pathogens also **use the life cycle of their vectors** to spread from one host to another.
- **Bacterial and viral infections cause** mucus build-up and, therefore, coughing and sneezing. Coughing and sneezing help to clear the pathogens from the respiratory tract but the sneezing also helps to transmit the disease.

See ➲ *Excel* Year 12 Biology p. 129

CAUSES OF INFECTIOUS DISEASE

The work of Robert Koch and Louis Pasteur

54

- **Pasteur discovered** that microbes such as bacteria and moulds can cause contamination and disease.
- **Koch showed** that bacteria were the cause of a disease called anthrax in horses, sheep and cows, and of tuberculosis in humans.
- **Koch's postulates** are rules of procedure for showing that a particular microorganism is the cause of a particular disease.
- The **four steps in Koch's postulates** are:
  1. **Establish the presence** of the specific microorganism in all organisms with the disease.
  2. **Grow the microorganism** in a pure culture and inject it into a healthy organism.
  3. **Isolate the microorganism and culture it** once that organism has developed symptoms.
  4. **Compare this culture** with the original culture. If they are the same then this shows that the microorganism caused the disease.

See ➲ *Excel* Year 12 Biology p. 126

CAUSES OF INFECTIOUS DISEASE

Transmission of a disease during an epidemic

52

- An **epidemic is the rapid spread of infectious disease** to a large number of people in a particular population in a relatively short period of time. The transmission rate exceeds the extent of the disease normally found in the population.
- A **pandemic** is an epidemic that spreads through an entire nation and other countries.
- **Many factors increase exposure to pathogens** and cause the rapid spread of disease.
- **Two key factors** are:
  - the **presence of a continual source of the disease**, such as contaminated drinking water
  - the **ease of spread of the pathogen**; pathogens can spread more easily and rapidly when humans are in close proximity with each other—through overcrowding—and with other animals such as farm animals.

See ➲ *Excel* Year 12 Biology p. 125

RESPONSES TO PATHOGENS

The physical or chemical changes that occur in the cells and tissues of animals in the presence of pathogens 2

62

- The **inflammatory response results** in the processes outlined below:
  - There is an **increase in the diameter of blood vessels**, increasing blood flow and bringing more white blood cells and defence compounds to the infected site.
  - The cells lining the blood vessels **secrete adhesion chemicals** that help white blood cells bind to the site of infection.
  - **Increased permeability of blood vessels allows** for the white blood cells and defence proteins to move out of the bloodstream and accumulate in the infected tissues.
  - **Fever and an increase in temperature of the tissues and body temperature** help the immune system fight infection.
  - **Secretion of proteins that bring about blood clotting** help prevent the pathogen from entering the bloodstream and spreading to other parts of the body.

See ➲ *Excel* Year 12 Biology p. 147

RESPONSES TO PATHOGENS

The responses of animals to pathogens

60

- **Preformed defences help prevent** pathogens from entering the host.
- **Immune responses are rapidly activated** when a pathogen crosses protective barriers and begins to replicate.
- These **fast-acting mechanisms** pre-exist in all animals.

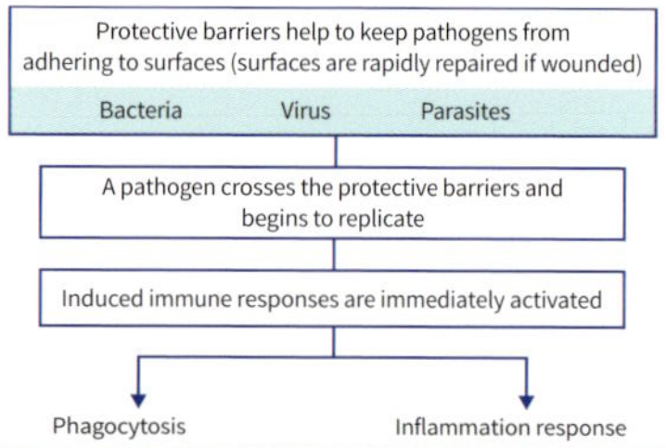

See ➲ *Excel* Year 12 Biology p. 140

RESPONSES TO PATHOGENS

The responses of plants to pathogens

58

- The **passive and active responses** of plants to pathogens are shown in this diagram.

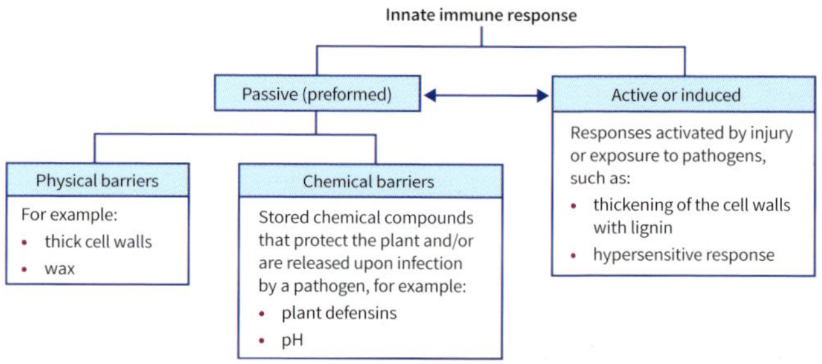

See ➲ *Excel* Year 12 Biology p. 139

IMMUNITY

Humoral and cell-mediated immune responses

68

- The **adaptive immune system has several parts** that react in different ways, depending on the place in the body where the pathogen is and the type of foreign substance or cell.
- If the **infection is outside the cell** in the tissues, antibodies are made. This is called the humoral immune response.
- If the **infection is in the cell**, a cell-mediated immune response is needed.
- The **humoral immune response** involves B cells that recognise antigens and produce the plasma cells that make antibodies. The antibodies are released and circulate through the body to 'attack' antigens.
- The **cell-mediated immune response** involves mostly T cells. Cell-mediated immunity does not involve antibodies.
- It involves the **activation of T lymphocytes and the release of various cytokines** in response to an antigen.

See ➲ *Excel* Year 12 Biology p. 152

IMMUNITY

Antibodies

66

- When **an antigen enters the body** the adaptive immune system produces antibodies against it. Each antibody has a unique binding site shape which locks onto the specific shape of the antigen. The antibodies destroy the antigen, which is then engulfed and digested by macrophages.
- Antibodies **work in three ways**:
  - they **neutralise the pathogen** by either attaching directly to the surface through the antigen–antibody complex or by attaching to the toxins produced by bacteria
  - they **activate other defense cells** such as phagocytes
  - they **activate the complement system**.
- There are **different forms of antibodies** in the human body; for example, Immunoglobulin G (IgG) is an antibody which is mainly produced towards the end of a new infection.

See ➲ *Excel* Year 12 Biology p. 150

IMMUNITY

The innate immune system in the human body

64

- The **innate immune system** is made up of the following:
  - the **skin and all mucous membranes** form external and internal barriers
  - **defense cells** are formed from white blood cells such as phagocytes; macrophages are the first phagocytes to encounter pathogens in the tissues and they are reinforced by neutrophils
  - **macrophages and neutrophils** recognise pathogens by means of cell-surface receptors that distinguish self from non-self
  - various substances circulate in the blood and in body fluids, such as cytokines and antimicrobial peptides called **defensins**
  - the **complement system** involves a system of soluble proteins that activate one another in a chain reaction
  - **natural killer (NK) cells** specialise in identifying cells infected by a virus or that have become cancerous.

See ➲ *Excel* Year 12 Biology p. 147

IMMUNITY

69 How the immune system responds after primary exposure to a pathogen, including innate and acquired immunity

- **Innate immunity** is naturally present and does not involve response to an antigen or confer long-term immunity.
- **Acquired immunity** is achieved through the adaptive immune system and antibodies, B cells and T cells. It can confer long-term immunity, which is why some diseases can only be contracted once in a lifetime.

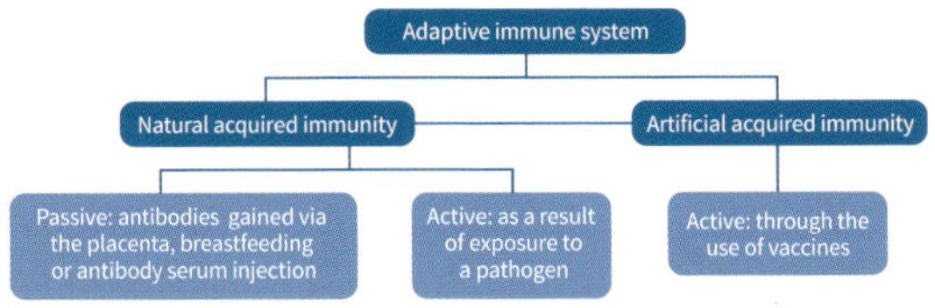

 See Excel Year 12 Biology p. 153

PREVENTION, TREATMENT AND CONTROL

71 The range of interrelated factors involved in limiting local, regional and global spread of an infectious disease

- To **limit the spread of an infectious disease** it is important to understand a number of factors, such as:
  - the main **mode of transmission**
  - how the **pathogen reproduces or replicates**
  - **the type of setting** where spread is most likely to occur
  - **individual and population susceptibility** to the disease
  - **transmission at different** stages of infection
  - **seasonality** of infection.
- Factors involved in **limiting the spread at the local level** include personal hygiene and public health responses.
- Factors involved in **limiting the spread at the regional level** include surveillance and area quarantining.
- Factors involved in **limiting the spread at the global level** include global surveillance, trade restrictions and travel warnings.

 See Excel Year 12 Biology p. 159

PREVENTION, TREATMENT AND CONTROL

73 Procedures employed to prevent the spread of disease 2

- **Passive immunity** is provided when a person is given antibodies.
- **Active immunity** results when exposure to a disease triggers the immune system to produce antibodies. It is long lasting and sometimes lifelong.
- **Artificial active acquired immunity** comes about through vaccination.
- **Vaccination is a procedure** that uses vaccines to prevent diseases in populations via immunisation programs.
- **Vaccines are effective in stopping infections**, which reduces the need for antibiotics and therefore helps limit the spread of antibiotic resistance.
- **Antigens and adjuvants** are the two key ingredients in vaccines.
- Antigens **cause the immune system to produce antibodies** and/or T cells against a specific pathogen or its toxin.
- Adjuvants **amplify immune responses** by targeting the defence cells and chemicals of the innate defence system.

 See Excel Year 12 Biology p. 160

PREVENTION, TREATMENT AND CONTROL

75 Environmental management and quarantine methods used to control an epidemic or pandemic

- **Environmental management and quarantine methods** aim to control the spread of a disease by reducing exposure to the pathogen through regulation of the environment and modification of behaviours.
- In **evaluating methods to control an epidemic or pandemic** you would have made a judgement to determine the effectiveness of the management and control based on some sort of criteria or measures.
- The **rate and extent of spread and severity** of disease determines the urgency and scope of the response. A fast-spreading epidemic or pandemic with high morbidity and mortality requires more extreme measures.
- As a **first priority** the disease must be identified, contained and controlled at its source as quickly as possible.

 See Excel Year 12 Biology p. 164

PREVENTION, TREATMENT AND CONTROL

77 Data relating to the incidence and prevalence of infectious disease in populations 2

- **Malaria and dengue fever** are the most common mosquito-borne diseases in the Southeast Asia region.
- **Dengue** is a **viral infection**.
- **Malaria** is a **disease caused by protozoa**.
- **In Southeast Asia approximately 70% of the population of 1.8 billion people** is at some risk of malaria, with 26% at high risk; 460 million people inhabit areas with a reported incidence of >1 case per 1000 population per year.
- The **majority of confirmed cases** in the region are due to *Plasmodium falciparum* (a unicellular parasite) although the proportion varies greatly among countries.
- **Scientists are concerned** that the morbidity and mortality from these diseases will rise with the rise in global temperatures, drug resistance and international travel.

 See Excel Year 12 Biology p. 165

PREVENTION, TREATMENT AND CONTROL

79 The contemporary application of Aboriginal protocols in the development of particular medicines and biological materials in Australia

- **Aboriginal people have developed a close and unique connection** with the lands and environments in which they live.
- They have **established distinct systems of knowledge, innovation and practices** relating to the uses and management of biological diversity.
- Much of this **traditional knowledge forms an important contribution to research and development**, particularly in areas such as pharmaceuticals and agricultural and cosmetic products.
- **Research and development of traditional medicines and biological materials** requires partnerships between Aboriginal communities and industries or professions and involves the use of Aboriginal protocols.
- **Protocols** are ways of behaving, communicating and showing respect.
- The **lack of recognition of Indigenous cultural and intellectual property rights** is a concern.

 See Excel Year 12 Biology p. 169

HOMEOSTASIS

81 Constructing and interpreting negative feedback loops 1

- **Homeostasis can be explained using a stimulus-response model** in which a change in the external or internal environmental conditions is detected and appropriate responses occur.
- These responses occur via a **negative feedback mechanism**, which involves:
  - a **stimulus**, which is a change in the environment
  - a **receptor** that detects the change in the environment
  - a **control centre** that controls the response
  - an **effector** that receives the message and carries out the response
  - a **response** that mitigates the original stimulus, hence the negative feedback.
- When you **constructed negative feedback loops** you would have put the parts of the stimulus-response pathway together.
- **Flow and circular diagrams** would most likely have been used.

 See Excel Year 12 Biology p. 177

HOMEOSTASIS

83 Mechanisms used by organisms to maintain their internal environment within tolerance limits 1

- **Endotherms generate** their own body heat internally and therefore maintain a constant body temperature.
- **Body temperatures are controlled** largely by altering metabolic processes and by adaptive mechanisms that control the rate of heat exchange with the environment.
- **Thermoregulation** has the **following general trends and patterns:**
  - **Changing behaviour**; for example, curling up into a ball to keep warm.
  - **Increasing or decreasing metabolic heat production**; for example, muscle contraction through behaviours such as shivering and rubbing hands increases muscle activity and boosts heat production.
  - **Controlling the exchange of heat with the environment** includes structural and physiological circulatory mechanisms such as altering blood flow patterns; for example, vasoconstriction of blood vessels that supply the skin reduces blood flow and helps retain heat.

 See Excel Year 12 Biology p. 179

HOMEOSTASIS

85 Mechanisms used by organisms to maintain their internal environment within tolerance limits 3

- **Mechanisms in plants that allow water balance to be maintained** are complex.
- **Different plants have different structures** for water uptake or minimising water loss and different responses in times of water stress; for example, thin leaves with a waxy cuticle, sunken stomata or stomata on the underside of the leaf.
- **Plants often grow sparsely to minimise competition** for limited water and have shallow extensive root systems to enable water uptake as soon as it rains.
- **Other plants have deep root systems** to access water from deeper soil layers.
- **Other mechanisms** include:
  - **decreasing photosynthesis** in response to water stress (occurs through a number of mechanisms such as closing stomata)
  - **dropping leaves** in response to water stress through the production of abscisic acid.

 See Excel Year 12 Biology p. 181

PREVENTION, TREATMENT AND CONTROL

74 The use of pharmaceuticals as treatment strategies for the control of infectious disease

- **Pharmaceuticals are manufactured compounds** used as medicinal drugs.
- **Antivirals treat a narrow range of viruses** and include drugs for the treatment of influenza, HIV, herpes, hepatitis B and hepatitis C.
- **Antibiotics target** bacterial pathogens. Antibiotics do not work against viral infections such as the cold or influenza.
- **Antibiotic- and antiviral-resistant pathogens** are key challenges to preventing and treating some infectious diseases.
- **Substantial reductions in infectious disease morbidity and mortality** have been achieved through the use of pharmaceuticals, access to better health care and preventive practices such as improved sanitation.
- **Morbidity refers** to the rate of disease in a population.
- **Mortality refers** to the number of deaths in a population.

See ➲ *Excel* Year 12 Biology p. 163

PREVENTION, TREATMENT AND CONTROL

72 Procedures employed to prevent the spread of disease 1

- **Hygiene practices** can reduce the risk of objects, materials or food becoming contaminated and infectious diseases being contracted.
- **Quarantine rules limit or control human mobility** through travel restrictions and impede the introduction of disease through surveillance, border security and screening.
- **Public health campaigns** include health promotion information, immunisation campaigns and disease prevention programs.
- **The use of pesticides** can prevent and control disease vectors, such as mosquitoes, and control insect infestations and plant diseases in agriculture. Resistance of insects to pesticides is now one of the greatest problems impacting on the global control of vector-borne diseases.
- **Genetic engineering** makes it possible to engineer the genes of vectors such as mosquitoes in order to control the spread of diseases such as malaria, dengue and Zika virus.

See ➲ *Excel* Year 12 Biology p. 160

PREVENTION, TREATMENT AND CONTROL

70 How can the spread of infectious diseases be controlled?

- **Prevention of infectious disease** refers to a range of interrelated measures which reduce the risks of, and ultimately stop, infections by pathogens in the first place.

  Examples of prevention include vaccination, good hygiene practices, health campaigns and genetic engineering.
- **Treatment of infectious disease** depends on a range of factors, including the type of pathogen, its virulence, the incubation period, the severity of the disease and the treatment drugs available.

  Examples of treatments include antibiotics and antivirals.
- **Control of infectious disease** refers to procedures employed to contain and stop the spread of disease.

  Examples of control include surveillance and early detection, quarantine methods and vector control.

See ➲ *Excel* Year 12 Biology p. 158

HOMEOSTASIS

80 How an organism's internal environment is maintained in response to a changing external environment

- **Organisms** must be **able to maintain system structure and function** in the face of changes in their external and internal environments in order to survive.
- **Homeostasis involves** the mechanisms that control an organism's responses to environmental change.
- **Changes in temperature and water availability**, along with diseases, presents significant challenges for organisms and requires coordinated system responses.

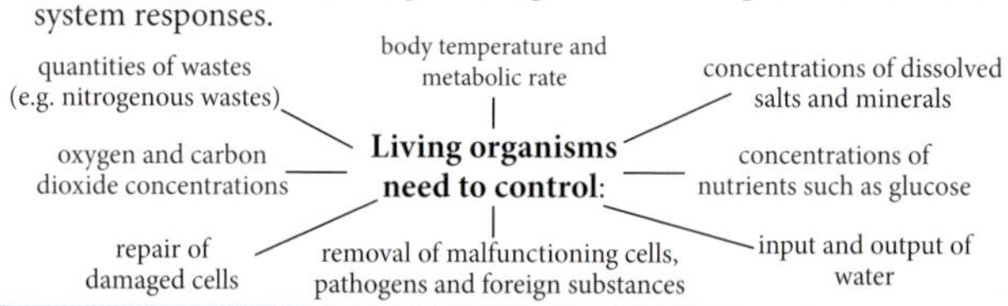

See ➲ *Excel* Year 12 Biology p. 177

PREVENTION, TREATMENT AND CONTROL

78 Historical, culturally diverse and current strategies to predict and control the spread of disease

- In **evaluating strategies to predict and control the spread of disease** you would have made judgements about models and systems and how well they help predict outbreaks of disease.
- **Historically** there was very little understanding of causes of diseases and therefore minimal knowledge of how to prevent, treat or control them.
- **Culturally diverse** strategies need to be understood when trying to predict and control the spread of disease in communities.

  Examples of **current strategies** include modelling to predict future occurrences of disease; projections and simulations of how a disease will progress and spread; identifying hotspots of emerging (novel) infectious diseases to provide early warnings; web-based surveillance systems; and technologies that provide enormous and fast processing capacity and data storage of disease-related data.

See ➲ *Excel* Year 12 Biology p. 167

PREVENTION, TREATMENT AND CONTROL

76 Data relating to the incidence and prevalence of infectious disease in populations 1

- **Incidence is the number** of new cases of a disease at a given time
- **Prevalence measures** the number or proportion of cases of a disease at a given time.
- **Mobility or movement of individuals** influences the incidence and prevalence of disease by changing the frequency of contacts between infected and susceptible individuals. More susceptible individuals include those with no immunity to the disease or the non-immunised.
- The **proportion of the population immunised** significantly reduces the incidence and prevalence of infectious diseases.
- The **speed and volume of mobility of individuals and human populations** is unprecedented in human history.
- **Analysis of data** can help in the prediction of potential epidemics as well as in the evaluation of strategies to control the spread of disease.

See ➲ *Excel* Year 12 Biology p. 164

CAUSES AND EFFECTS

86 The causes and effects of non-infectious diseases in humans 1

- **Non-infectious diseases** are diseases and disorders caused by something other than a pathogenic organism.
- **Non-communicable diseases (NCDs)** is another name for these diseases and disorders.
- NCDs have now **exceeded infectious diseases** as the leading cause of death and disability worldwide.
- **Causes of NCDs** include genetic disorders, environmental exposure, poor nutrition and cancer.
- A **combination of factors** can be the cause of some NCDs.
- **Genetic disorders** are inherited diseases caused by genetic mutations passed down from parent(s) to offspring.

  Examples include sickle-cell anaemia (caused by a gene mutation) and Down syndrome (caused by an additional chromosome 21).

See ➲ *Excel* Year 12 Biology p. 187

HOMEOSTASIS

84 Mechanisms used by organisms to maintain their internal environment within tolerance limits 2

- **Internal coordination systems** that allow homeostasis to be maintained are hormones and neural pathways.
- **Hormones are produced** by the endocrine system. Hormones alter the metabolism of target cells, tissues or organs by increasing or decreasing their activity.
- Hormones are **produced in endocrine glands** and released into the bloodstream, where they reach target cells via the circulatory system.
- **Neural pathways** consist of the central nervous system (the brain and spinal cord) and peripheral nerves.
- The **neurone or nerve cell** is the basic unit of the nervous system. Nerve cells transmit signals by electrochemical changes in their membranes.
- The **nervous system provides** rapid, short-term coordination of internal organ systems.
- The **hormonal system provides** slower, longer-lasting response coordination.

See ➲ *Excel* Year 12 Biology p. 180

HOMEOSTASIS

82 Constructing and interpreting negative feedback loops 2

- Negative feedback loops that **maintain homeostasis for temperature**: thermoreceptors detect temperature changes. External temperature changes are detected by receptors in the skin. Internal changes are detected by receptors in the hypothalamus. These inputs are then sent via the nervous system to the appropriate effectors such as muscles and glands. The subsequent response brings about changes depending on whether it is hot or cold.
- Negative feedback loops that **maintain homeostasis for glucose**: the pancreas produces two hormones, insulin and glucagon, and these regulate the amount of glucose in the blood. Insulin lowers blood glucose (blood sugar) levels by facilitating glucose transport from the blood into cells. Glucagon raises blood glucose levels by metabolising glycogen into glucose.

See ➲ *Excel* Year 12 Biology p. 177

CAUSES AND EFFECTS

87 The causes and effects of non-infectious diseases in humans 2

→ **Diseases caused by environmental exposure** include those resulting from exposure to a range of potential mutagens in the environment, such as hydrocarbons, pesticides, certain foods, pollution and cigarette smoke.
Examples include pleural mesothelioma (a cancer caused mainly by exposure to asbestos) and skin cancers (e.g. malignant melanoma caused by excessive exposure to ultraviolet radiation).

→ **Nutritional diseases are caused by**:
- **nutritional deficiency** of an important compound such as a protein, vitamin or mineral and/or undernutrition; for example, night blindness caused by vitamin A deficiency.
- **excesses in diet and other behavioural risk factors**, such as unhealthy eating and lack of physical exercise; for example, diabetes and obesity.
- **eating disorders**; for example, bulimia and anorexia nervosa.

MODULE 8: NON-INFECTIOUS DISEASE AND DISORDERS See *Excel* Year 12 Biology p. 188

CAUSES AND EFFECTS

89 Data that shows the incidence, prevalence and mortality rates of non-infectious disease

→ An **example of data for lung cancer** in Australia is provided in the points below:
- **In 2014 there were 8251 deaths** from lung cancer in Australia. In 2017 it was estimated that this would increase to 9021 deaths.
- **In 2014, 11 556 new cases** of lung cancer were diagnosed in Australia. This accounted for close to 9% of all cancers diagnosed.
- **In 2015 there were 8 466 deaths** caused by lung cancer.
- Data estimates showed that in **2017, 12 434 new cases** would be diagnosed.

→ The **incidence rate** is expected to increase with age. It was estimated that in 2017 the risk of an individual being diagnosed with lung cancer before their 85th birthday would be one in 17.

MODULE 8: NON-INFECTIOUS DISEASE AND DISORDERS See *Excel* Year 12 Biology p. 192

EPIDEMIOLOGY

91 The treatment/management and possible future directions for further research of non-infectious diseases

→ The **treatment and management of a disease caused by environmental exposure** is outlined below.
Example: the treatment for melanoma depends on the stage of the disease and the severity of symptoms. Surgery is the main initial treatment. Additional treatments can include chemotherapy, radiotherapy, targeted therapy and immunotherapy. Future directions for further research include development of a monoclonal antibody targeting the CDK4 antigen in order to treat advanced melanoma.

→ The **treatment and management of a nutritional disease** is outlined below.
Example: type 2 diabetes is a largely preventable disease often caused by modifiable risk factors. It can be managed initially through changes in lifestyle, including a healthy diet to manage blood glucose levels and body weight, and regular exercise to help insulin work more effectively. Future directions for further research include improving gut health.

MODULE 8: NON-INFECTIOUS DISEASE AND DISORDERS See *Excel* Year 12 Biology p. 201

EPIDEMIOLOGY

93 The benefits of engaging in an epidemiological study

→ **Benefits include** being able to identify risk factors and to determine the resources needed for health care, education and research in order to develop targeted public health interventions that are cost and resource effective.

→ An **example of a study and its benefits** is provided below.
Example: cervical cancer is the fourth most common worldwide cancer in women. In the last two decades epidemiological studies have shown that cervical cancer is a rare outcome of a common sexually transmitted infection caused by some types of human papillomavirus (HPV). Epidemiological studies have clearly and consistently shown the key determinants of infection in women. Findings from the studies have resulted in the use of pap smears and, more recently, screening and vaccination strategies for the prevention of cervical cancer.

MODULE 8: NON-INFECTIOUS DISEASE AND DISORDERS See *Excel* Year 12 Biology p. 205

PREVENTION

95 The effectiveness of current disease prevention methods and strategies 2

→ In Australia there is a solid core of evidence about **cost-effective preventive interventions**. The evidence is largely provided by the Assessing Cost-Effectiveness (ACE) Prevention study, which was conducted over five years on more than 150 preventive health interventions, both population wide and individual. The study found that many preventive interventions have very strong cost-effectiveness.
Examples include tax increases on tobacco, alcohol and unhealthy foods as well as mandatory salt limits on processed foods.

→ The **effectiveness of preventive strategies** is measured by data collected by the Australian Institute of Health and Welfare.
Examples include screening for bowel, cervical and breast cancer and the national human papillomavirus (HPV) vaccination program to protect against cervical cancer.

MODULE 8: NON-INFECTIOUS DISEASE AND DISORDERS See *Excel* Year 12 Biology p. 212

PREVENTION

97 The effectiveness of current disease prevention methods and strategies 4

→ **Developments in genetic technologies**, including genetic engineering, molecular genetics, genetic testing, advanced statistical methods, high-throughput technologies and progress in studying gene-environment interactions provide researchers with methods and strategies that potentially help prevent non-infectious diseases.

→ **Molecular genetics provides** evidence that molecules and genetic pathways differ in pre-cancerous and malignant cells.

→ **Genetic testing is used** to test for the mutated genes currently known to be associated with the predisposition to certain cancers; for example, in hereditary breast cancer the most common mutations are in genes BRCA1 and BRCA2.

→ **Genetic engineering helps** prevent disease; for example, the engineering of mice to simulate a human genetic disease can then be used to develop novel prevention methods.

MODULE 8: NON-INFECTIOUS DISEASE AND DISORDERS See *Excel* Year 12 Biology p. 215

TECHNOLOGIES AND DISORDERS

99 Causes of disorders in the structure and function of organs: eyes

→ Our eyes **provide sight and balance**.

→ Some **common disorders** are outlined in the table below.

Visual disorder	Causes
myopia or near-sightedness (short-sighted)	a refractive error caused when the distance between the lens and the retina is too great
hyperopia or far-sightedness (long-sighted)	a refractive error caused when the distance between the lens and the retina is too short
cataracts	a degenerative condition in which the lens of the eye clouds over
macular degeneration	a progressive deterioration of the macula of the retina (the central inner lining of the eye)
glaucoma	caused by damage to the optic nerve due to increased pressure within the eyes

MODULE 8: NON-INFECTIOUS DISEASE AND DISORDERS See *Excel* Year 12 Biology p. 224

TECHNOLOGIES AND DISORDERS

101 Technologies that are used to assist with the effects of disorders: hearing loss

→ **Hearing aids** consist of a microphone that picks up sound waves, converts them to electrical signals and sends them to the amplifier which increases the power of the signals and sends them to a speaker.

→ **Bone conduction implants** are used for conductive hearing loss. A sound processor captures sounds and turns them into vibrations. The vibrations are sent to the implant which transmits the sound through bone vibrations to the inner ear.

→ **Cochlear implants** or bionic ears are used to help people who have severe to profound hearing loss. A sound processor worn behind the ear picks up sounds and converts these into an electrical impulse which is sent to a receiver attached to the skull. The receiver relays the electrical impulse to an electrode array surgically inserted into the cochlea.

MODULE 8: NON-INFECTIOUS DISEASE AND DISORDERS See *Excel* Year 12 Biology p. 228

TECHNOLOGIES AND DISORDERS

103 Technologies that are used to assist with the effects of disorders: loss of kidney function

→ **Haemodialysis** is done at least three times a week and lasts for four to five hours. During haemodialysis, needles are used to access the blood. One needle takes the blood out for treatment and the other needle returns the blood. The blood is taken through a special filter called a dialyser, which cleans it before it is returned to the body.

→ **Peritoneal dialysis** is done in the home and uses a tube that is put into the stomach. Special peritoneal dialysis fluid is used to clean the blood. The fluid must be changed regularly.

→ A **kidney transplant involves surgery** where a healthy kidney is taken from one person and surgically placed into someone with end-stage kidney disease.

MODULE 8: NON-INFECTIOUS DISEASE AND DISORDERS See *Excel* Year 12 Biology p. 230

EPIDEMIOLOGY

Methods used in an epidemiological study

92

- To **evaluate methods used in an epidemiological study** you make judgements using criteria such as the value and scale of the study, the length of time involved, the specificity of the population and the site, and an assessment of the evidence and any limitations in the study.

  Example: the linking of cause and effect may require many studies involving large populations over long periods of time along with rigorous scientific method employed in order to eliminate variables.
- **Scientific methodology checks the validity and reliability** of the study by using criteria such as the questions below:
  - Was it **established** that a problem exists?
  - Was as much **data collected** as possible?
  - Are there **patterns and trends**?
  - Is there a **relationship** between the potential disease-causing agent and the disease?

See ➲ *Excel* Year 12 Biology p. 203

EPIDEMIOLOGY

Patterns of non-infectious diseases in populations

90

- **Epidemiology** is the systematic and ongoing collection, analysis and interpretation of data about epidemics. Epidemiology identifies correlations between disease and risk factors to understand the causes of diseases and the populations most at risk.
- **Analysis of patterns** of non-infectious diseases in populations means identifying trends in data.

  Examples include changes, over time, of patterns in specific populations and age distributions.
- This graph shows the **WHO's profile of Australia's mortality and risk factors** related to non-communicable diseases in 2014.

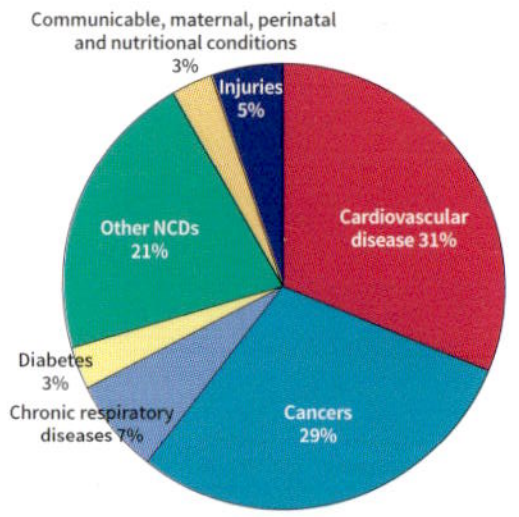

See ➲ *Excel* Year 12 Biology p. 199

CAUSES AND EFFECTS

The causes and effects of non-infectious diseases in humans 3

88

- **Cancer** is a disease of the body's cells.
- **Cells normally grow and multiply in a controlled way** but if something causes a mutation to occur in the cells' DNA (or mitochondria) this control can be lost.
- **Cancer or tumour is the term used** to describe the collections of these cells growing and potentially spreading within the body.

  Examples of cancers include breast cancer and lung cancer. Breast cancer is the abnormal growth of the cells lining the breast lobules or ducts. The main risk factor for breast cancer is a strong family history of the disease. Lung cancer is a malignant tumour starting in the tissue of one or both lungs; tobacco smoking causes the majority of lung cancers.

See ➲ *Excel* Year 12 Biology p. 190

TECHNOLOGIES AND DISORDERS

Causes of disorders in the structure and function of organs: ears

98

- Our ears **provide hearing and balance**.
- **Disorders in the structure and function of the ear** results in hearing loss or deafness.
- **Conductive hearing loss** is due to problems with the ear canal, eardrum, middle ear or ear ossicles and is caused by middle-ear infections and damage to ear ossicles.
- **Sensorineural hearing loss** is due to problems of the inner ear, such as damage to the hairs or nerve cells in the cochlea caused by aging, or cumulative exposure to excessive noise or certain drugs. When the cells are damaged electrical signals are not transmitted efficiently.
- **Mixed hearing loss** is a combination of conductive and sensorineural hearing loss to the outer, middle and inner ear or auditory nerve.

See ➲ *Excel* Year 12 Biology p. 222

PREVENTION

The effectiveness of current disease prevention methods and strategies 3

96

- The World Health Organization (WHO) has a **global action plan** for preventing and controlling non-infectious disease.
- A **growing body of evidence** exists in Australia about the prevention methods and strategies that work.

  Examples include evidence-based interventions, including taxation; media campaigns; regulation of unhealthy products; school-based interventions; classroom-based physical activity; education programs in multiple settings; nutrition education; and targeted interventions that reduce the chronic disease burden and hence the pressure on the health system.

  Examples of education programs, media campaigns and targeted interventions in Australia include 'Slip, Slop, Slap, Seek, Slide' (community-wide sun protection education) campaigns and 'Quit smoking' and 'I Can Quit' national tobacco campaigns (anti-smoking education and mass media campaigns that target specific groups such as the young and Aboriginal communities).

See ➲ *Excel* Year 12 Biology p. 214

PREVENTION

The effectiveness of current disease prevention methods and strategies 1

94

- **Prevention of non-infectious diseases** refers to actions to reduce or eliminate the onset, causes, complications or recurrence of disease.
- A **population-based approach** focuses on populations rather than on individuals and on prevention and protection rather than on treatment.
- A **life-course approach** to prevention recognises that many chronic diseases are the result of risks that accumulate throughout an individual's lifetime and that these risks can and must be reduced and prevented at all stages of life.
- In **evaluating the effectiveness of current disease prevention methods** you would have determined the value of strategies by examining the extent to which the prevention method(s) contributed to a decline in the incidence and prevalence of disease and made judgements against a number of criteria, including the level of prevention.

See ➲ *Excel* Year 12 Biology p. 211

TECHNOLOGIES AND DISORDERS

The effectiveness of technologies used to manage and assist with the effects of disorders

104

- **Investigating effectiveness** means determining the value of the technology and factors such as the extent to which it manages and assists with the effects of the disorder, the cost and affordability, potential side effects, risks associated with intervention (e.g. surgery) and the life of the technology.
- **Hearing aids** are relatively cheap and easy to install. With advances in technologies they have become much smaller and digitised.
- **Cataract surgery** is highly effective and takes less than 30 minutes to perform. Once the opaque lens is replaced with an artificial lens it is not possible to develop another cataract.
- Because the **process of dialysis** does not match the complexity of the natural kidney it must be performed for the rest of a person's life or until they receive a kidney transplant.

See ➲ *Excel* Year 12 Biology p. 231

TECHNOLOGIES AND DISORDERS

Technologies that are used to assist with the effects of disorders: visual disorders

102

- **Spectacles and contact lenses** are artificial lenses designed to correct the refractive errors of myopia and hyperopia.
- **Laser surgery** is a form of vision correction that involves reshaping the cornea to correct refractive error. The technology uses a cool-temperature laser and LASIK surgery is the most common form.
- **Cataract surgery** involves the removal of the opaque lens and its replacement with a clear artificial lens (called an intraocular lens).

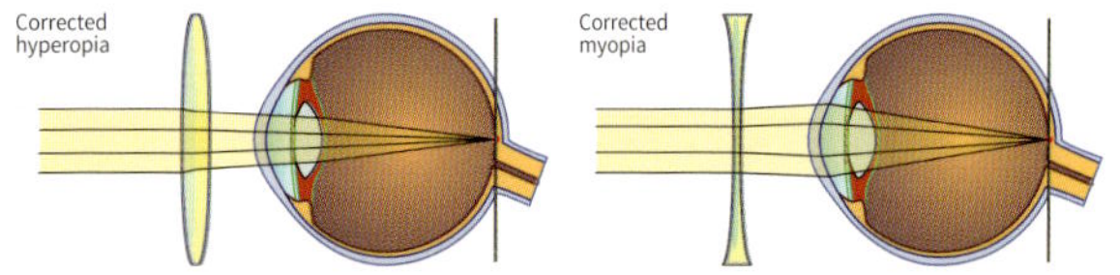

See ➲ *Excel* Year 12 Biology p. 229

TECHNOLOGIES AND DISORDERS

Causes of disorders in the structure and function of organs: kidneys

100

- **Kidneys are complex organs** of the excretory system.
- The **main functions of kidneys** are to filter blood of wastes and produce urine without the loss of too much water. Kidneys also help in the homeostasis of water, glucose and inorganic salts.
- The **nephron** is the functional unit of the kidney.

  Examples of disorders of the kidneys include chronic kidney disease caused by risk factors such as smoking, high blood pressure, obesity and diabetes; acute kidney injury caused by injury, trauma or severe inflammation that restricts blood supply to the tissues; and nephritic disease or inflammation of the kidney.
- **Nephrosis results** from a problem with the functioning of the glomeruli. When the glomeruli are damaged, proteins from the blood leak into the urine.

See ➲ *Excel* Year 12 Biology p. 226